Birkhäuser

Lecture Notes in Geosystems Mathematics and Computing

More information about this series at https://link.springer.com/bookseries/15481

Fernando Sansò • Daniele Sampietro

Analysis of the Gravity Field

Direct and Inverse Problems

Birkhäuser

Fernando Sansò
Civil & Environmental Engineering
Polytechnic University of Milan
Milan, Italy

Daniele Sampietro
Geomatics Research & Development srl
Lomazzo, Italy

ISSN 2730-5996 ISSN 2512-3211 (electronic)
Lecture Notes in Geosystems Mathematics and Computing
ISBN 978-3-030-74355-0 ISBN 978-3-030-74353-6 (eBook)
https://doi.org/10.1007/978-3-030-74353-6

Mathematics Subject Classification: 86A22, 86A20, 60G60

This book is published under the imprint Birkhäuser, www.birkhauser-science.com by the registered company Springer Nature Switzerland AG
The registered company address is: Gewerbestrasse 11, 6330 Cham, Switzerland

Introduction

This book grows out of an experience of the authors in teaching a course of "Gravity Interpretation" at Politecnico di Milano in the years 2010–2014, as well as from practical works done in this area with oil companies together with a Politecnico di Milano spin-off.

The matter has a long history in the development of modern sciences, going back to Isaac Newton, and accompanying the evolution of geodesy, planetary geophysics, and then applied geophysics, as well as of the important mathematical branch of inverse and improperly posed problems (among others, standard textbooks on the topic are [11, 28, 32, 35, 46, 103, 116, 139]).

Here we made a choice: although it is interesting to study the distribution of masses inside the Earth at a global level (the global and regional inversion is studied for instance in [31, 140]), in this book, the inverse methods are focused on a local-level interpretation, which is a typical theme in the search for natural resources, namely, in exploration geophysics.

This has the important consequence that a fundamental analysis tool can be used, namely Fourier integral/series, as opposed to the spherical (or ellipsoidal) harmonic analysis.

For this second aspect, we send the interested reader to the geodesy text. "Determination of the Geoid: theory and methods" [133], where the spherical harmonic analysis is explained in due detail; here we just borrow the general concept of a global gravity model necessary to reduce the observations to a data set much more sensitive to local mass anomalies than to global features.

Another important choice characterizing the book comes from our firm belief that in gravity modeling and interpretation, a continuous rather than discrete approach should give comparable results, so that computable discrete models could always be seen as an approximation of continuous models. Hence, both aspects are typically included in the presentation, and rules to pass from a continuous to a discrete model are here and there given explicitly. When writing a book like this, one of the difficult issues is to decide what is the appropriate level of presentation of the matter. The same argument in fact could be seen from the mathematical standpoint as a nice example (one of the oldest) to which apply the theory of deterministic regularization or of random fields and stochastic optimization, or, from the other side, as a very practical issue to be handled by an operator with a lot of experience, guessing the internal mass distribution, starting from a basic scenario provided by geologists, adjusted by small corrections based on a trial-and-error approach.

Our choice has been to proceed assuming the reader to be familiar with basic mathematical notions (calculus and maybe a first course on advanced mathematics) and some material from statistics and probability theory, as described in the prerequisites presented in Appendix A, and then to build a theory where most of the mathematical tools are developed autonomously with proofs that give control on the matter.

This explains the decision to include into the book a chapter with a rigorous exposition of Fourier transform, which is so important in our field and often treated in geophysical books mostly in terms of practical rules. The same holds true for Bayesian estimation and its application to the optimization theory, namely the so-called Monte Carlo Markov Chain methods, which are presented with important proofs, rigorous though tailored to our gravity inversion problem.

When the analysis, like the Tikhonov-Morozov theory, or the probability theory for generalized random fields, became harder, we decided to shift the matter to Appendices B and C, so that along the text readers can concentrate on solution algorithms and then separately decide whether they want to have a deeper understanding of the mathematical background.

The logical structure of the book is as follows:

- In Part I, beyond giving the basic definition of what the gravity field is and describing its properties in terms of potential theory, we show how to compute gravity from mass distributions, exploiting the effects of simple geometrical forms and simplified density models, for example, constant densities. This part is closed by a chapter on Fourier transform and its important applications in forward gravity modeling.
- In Part II, we focus on data and their preprocessing/processing. In particular, after recalling the rigorous definition of gravity of the Earth and the relation between the gravity potential and the observable quantities, we describe the first preprocessing of gravity observations, aiming at eliminating instrumental drifts and unwanted time-dependent signals of short period, primarily tidal effects. Then the elimination of the attraction of local, topographic masses, necessary to discern the gravity anomalies coming from a deeper mass distribution, is presented with a careful explanation on how to perform it with a simultaneous use of global gravity models.

  Finally, a useful "gravity anomalies predictor" is described, allowing to interpolate the gravity field at points of a grid, on the horizontal plane, which is the form in which the subsequent analysis for the inversion can be more easily performed.
- In Part III, we first show how to interpret a gravity anomaly field in terms of elementary geometrical figures supporting a constant density distribution. In this respect, part of the literature on bounds on minimize/maximize depth, including

the theory of Parker's ideal body, is reviewed. Then the large indeterminacy of the inverse gravimetric problem is properly characterized in mathematical terms, as well as its intrinsic instability related to the downward continuation of the gravity potential. Uniqueness theorems for simplified models, like the two-layer model with constant density, are presented too. Finally, the general inversion problem is attacked, applying both a deterministic approach and a stochastic optimization approach. In this respect, let us emphasize that the matter discussed in the book does not exhaust all possible approaches to gravity inversion; nevertheless, the Tikhonov deterministic approach is the most widely applied, and the stochastic MCMC method is, in our opinion, the most innovative and capable of future developments. So, we have preferred to restrict the material discussed, but giving self-contained view on its mathematical foundation.

- In Appendix A, some reminders of linear algebra, statistics, and probability theory are reported.
- In Appendix B, the theory of generalized random fields is recalled, a chapter of probability theory that is essential for building models simultaneously valid in continuous and discrete contexts.
- In Appendix C, the functional background of Tikhonov's regularization theory is presented for both nonlinear and linear problems.

Most of the chapters are provided with examples that serve a double scope: to clarify the material of the section by synthetic but very simple cases, or to show a realistic application to give the reader a feeling of achievable results.

Moreover, most of the chapters are endowed with exercises having a double purpose too: to let the readers realize whether they have properly understood a subject, or to shift some more difficult or lengthy proofs of propositions only stated in the main text.

A final warning concerns the choice of the frame in which we describe the interpretation process. Contrary to the tradition of applied geophysics, our Cartesian triad has always the z-axis pointing up, instead of down. This is not only because the background of the authors is geodesy, but also because we don't like to invert the verse of z when describing the gravity field across the Earth surface.

Contents

Symbols

G	Universal gravitational constant, $G \cong 6.67430 \cdot 10^{-11}\ \mathrm{m^3s^{-2}kg^{-1}}$
$B \backslash A$	The relative complement of A with respect to a set B, or the difference of sets B and A, is the set of elements in B but not in A
∇	Gradient operator
$\nabla.$	Divergence operator
$E\{x\}$	Average of x
RHS	Right hand side
LHS	Left hand side
Δ	Laplacian
$\mathring{A}$	Interior of A
a	Equatorial radius of the reference ellipsoid (6378137 Km for WGS84)
$\bar{A}$	Closure of A
∂A	Boundary of A
$\mathrm{A^c}$	Complement of A
$A \div B$	$(A \backslash B) \cup (B \backslash A)$
σ	Point on the unit sphere of spherical coordinates $\sigma = (\lambda, \varphi)$ or $\sigma = (\lambda, \theta)$, with φ spherical latitude, $\theta = \frac{\pi}{2} - \varphi$
$d\sigma$	Area element on the unit sphere
d_2x	Area element in R^2
d_3x	Volume element in R^3
e^2	Squared first eccentricity of the reference ellipsoid $e^2 = 6.69438002290.10^3$
$\boldsymbol{e}_{QP}$	$\boldsymbol{e}_{QP} = \boldsymbol{x}_P - \boldsymbol{x}_Q$
e_{QP}	$e_{QP} = \lvert \boldsymbol{x}_P - \boldsymbol{x}_Q \rvert$
ω	Angular velocity of the Earth $\omega = 7.292115.10^{-5}$ radss^{-1}
$\boldsymbol{r}_p, \boldsymbol{x}_p$	Position vector of the point P
$\mathscr{R}(A)$	Range of the operator A
$\mathscr{N}(A)$	Null space of the operator A
$\overline{\lim}$	Upper limit
$\underline{\lim}$	Lower limit
Inf	Infimum

Sup	Supremum
$\|f\|_L$	Norm of f in the L Banach space
$\langle x, y \rangle$	Scalar product between x and y in some real Hilbert space
x^*	Conjugate of the complex number x
$S^{\perp}$	Orthogonal complement of S
$\mathscr{D}(A)$	Domain of A
Δ_σ	Laplace-Beltrami operator
J_0	Bessel function of order 0
J_1	Bessel function of order 1

Part I
Forward Modelling of the Gravity Field

Chapter 1
Gravitation

1.1 Outline of the Chapter

This chapter introduces the basic concepts of gravitational forces, such as Newton's law, the concept of gravitational potential and the ordinary elements of vector calculus used to describe the first properties of the gravitational field.

This is a standard introductory material common to most books concerned with the gravity field, attempting to provide a self contained presentation of the material. It is for this reason that the chapter is strongly reminiscent of the first sections of a book of one of the authors [133].

The last section, though, introduces examples tuned to the needs of later reading.

1.2 The Newton Law of Gravitation

In the year 1686 I. Newton, in his *Philosophiae Naturalis Principia Mathematica*, formulated one of the fundamental laws of physics, astronomy and geodesy, namely his celebrated law of gravitational attraction:

- any two point masses M_P, M_Q, in an inertial system, attract each other with a force proportional to the value of the masses and inversely proportional to the square of the distance between them:

$$F = G\frac{M_P M_Q}{\ell_{PQ}^2}; \tag{1.1}$$

  the proportionality constant G is the universal gravitational constant, with value

$$G \cong 6.67430 \cdot 10^{-11}\mathrm{m}^3\mathrm{s}^{-2}\mathrm{kg}^{-1},$$

  determined with an accuracy of $\pm 0.15 \cdot 10^{-14}\mathrm{m}^3\mathrm{s}^{-2}\mathrm{kg}^{-1}$;

F. Sansò, D. Sampietro, *Analysis of the Gravity Field*, Lecture Notes in Geosystems Mathematics and Computing, https://doi.org/10.1007/978-3-030-74353-6_1

- the direction of the gravitational force exerted by M_Q on M_P is along the straight line passing through P and Q and is pointing from P to Q, so that (1.1) can be written in the vector form

$$\boldsymbol{F}_{QP} = -GM_P M_Q \frac{\boldsymbol{\ell}_{QP}}{\ell_{QP}^3}. \tag{1.2}$$

This law has quite a general validity in classical physics and certainly covers the aim of describing the gravity field of the Earth, in relation to its sources.

The proportionality of $\boldsymbol{F}_{QP}$ to the mass M_P, i.e., the mass that feels the force, suggests to decompose (1.2) as

$$\boldsymbol{F}_{QP} = M_P \boldsymbol{g}(P, Q), \tag{1.3}$$

where

$$\boldsymbol{g}(P, Q) = -GM_Q \frac{\boldsymbol{\ell}_{QP}}{\ell_{QP}^3}, \tag{1.4}$$

can be considered as a field acting on the proof mass M_P, i.e., something generated by the mass M_Q, independently of the probe, namely M_P, transforming it into a force.

The field $\boldsymbol{g}(P, Q)$ is the gravitational field generated by the mass M_Q placed at point Q in the 3D space and computed at the point P. Indeed, $\boldsymbol{g}$ has the dimension of an acceleration.

If we assume that we have more masses M_{Qi} ($i = 1, 2 \ldots N$) and we probe their total gravitational force on a mass M_P, since forces can be just added vectorially, the same will be true for the respective gravitational fields and we can find the total field at P as

$$\boldsymbol{g}(P) = -G \sum_{i=1}^{N} M_{Qi} \frac{\boldsymbol{\ell}_{Q_i P}}{\ell_{Q_i P}^3}. \tag{1.5}$$

The above generalizes to the case of many infinitesimal mass elements distributed along a line L, a surface S, or a bulky body B, namely

$$\boldsymbol{g}(P) = -G \int_L \frac{\boldsymbol{\ell}_{QP}}{\ell_{QP}^3} dM(Q), \tag{1.6}$$

$$\boldsymbol{g}(P) = -G \int_S \frac{\boldsymbol{\ell}_{QP}}{\ell_{QP}^3} dM(Q), \tag{1.7}$$

$$\boldsymbol{g}(P) = -G \int_B \frac{\boldsymbol{\ell}_{QP}}{\ell_{QP}^3} dM(Q). \tag{1.8}$$

Usually the mass distributions on L, S and B will be represented respectively by linear, surface and body densities respectively, namely

$$dM(Q) = \lambda(Q)dL_Q \ (dL_Q = \text{line element of } L) \tag{1.9}$$

in (1.6),

$$dM(Q) = \alpha(Q)dS_Q \ (dS_Q = \text{surface element of } S) \tag{1.10}$$

in (1.7), and

$$dM(Q) = \rho(Q)dB_Q \ (dB_Q = \text{volume element of } B). \tag{1.11}$$

in (1.8).

Let us further observe that in the above formulas, the vector $\boldsymbol{\ell}_{QP}$ has Cartesian components $x_P - x_Q, y_P - y_Q, z_P - z_Q$ and ℓ_{QP} is its modulus, i.e.

$$|\boldsymbol{\ell}_{QP}| = \ell_{QP} = \sqrt{(x_P - x_Q)^2 + (y_Q - y_Q)^2 + (z_p - z_Q)^2}. \tag{1.12}$$

Remark 1.1 Since ℓ_{QP} vanishes for $Q = P$, we see that (1.6), (1.7), and (1.8) are singular integral operators that can loose their meaning when P is on L, S or B, respectively. This is in fact the case when we have an, even smooth, linear distribution $\lambda(Q)$. When we are dealing with a surface distribution, $\boldsymbol{g}$ can still be defined on S by the so-called Cauchy principal value of integrals, i.e., by excluding a small circular region on S of radius ε around P and then letting $\varepsilon \to 0$.

The body integral (1.8), on the contrary, has only a weak singularity and $\boldsymbol{g}(P)$ can be defined in B, provided that B has a regular geometry (e.g., B is a simply connected domain with a smooth boundary surface S) and $\rho(Q)$ is also well-behaved (e.g., it is measurable and bounded). In fact, if we use a spherical coordinate system centered at P to write (1.8) and we put

$$\begin{cases} Q = P + r\boldsymbol{e}_\sigma \\ \boldsymbol{e}_\sigma = -\dfrac{\boldsymbol{\ell}_{QP}}{r} \end{cases} \tag{1.13}$$

we can write

$$\boldsymbol{g}(P) = -g \int_B \frac{\boldsymbol{e}_\sigma}{r^2} \rho(P + r\boldsymbol{e}_\sigma)\, dB. \tag{1.14}$$

Now if we assume for instance that P is inside B, so that there is a sphere B_ε with center P and radius ε included in B, we can split (1.14) as

$$\int_B \bullet dB = \int_{B\setminus B_\varepsilon} \bullet dB + \int_{B_\varepsilon} \bullet dB. \tag{1.15}$$

Considering that the first integral on the right-hand side (RHS) is regular in P, because now P is outside its masses, i.e., $\ell_{QP} > \varepsilon$ when Q is in $B\setminus B_\varepsilon$, we can focus on the second integral and write it as

$$g_\varepsilon(P) = -G \int_\sigma \int_0^\varepsilon \frac{\boldsymbol{e}_\sigma}{r^2} \rho(P + r\boldsymbol{e}_\sigma) r^2 dr d\sigma \tag{1.16}$$

where σ is the unit sphere, $d\sigma$ is its area element, that in spherical coordinates (λ, φ) reads $d\sigma = \cos\varphi\, d\varphi d\lambda$, and $\boldsymbol{e}_\sigma$ is the unit vector pointing to the point on σ of spherical coordinates (λ, φ). Now it is clear that in (1.16) singularities cancel out and the integral has a finite value.

1.3 The Gradient Operator and the Gravitational Potential

The gradient is a vector differential operator that transforms a function $f(P)$ into a vector field $\boldsymbol{F}(P)$.

Definition 1.1 Consider a function $f(P)$, smooth in the sense that it is continuous in P and such that if we move from P to $P + dP$ we get an increase

$$\Delta f = f(P + dP) - f(P) \tag{1.17}$$

that admits the representation[1]

$$\Delta f = \boldsymbol{F} \cdot dP + o(|dP|); \tag{1.18}$$

then by definition we have

$$\boldsymbol{F}(P) = \text{grad } f(P) = \nabla f(P) \tag{1.19}$$

and

$$df(P) = \boldsymbol{F}(P) \cdot dP = \nabla f(P) \cdot dP. \tag{1.20}$$

[1] Note: remember that $o(|dP|)$ means that $\frac{o(|dP|)}{|dP|} \to 0$ when $|dP| \to 0$.

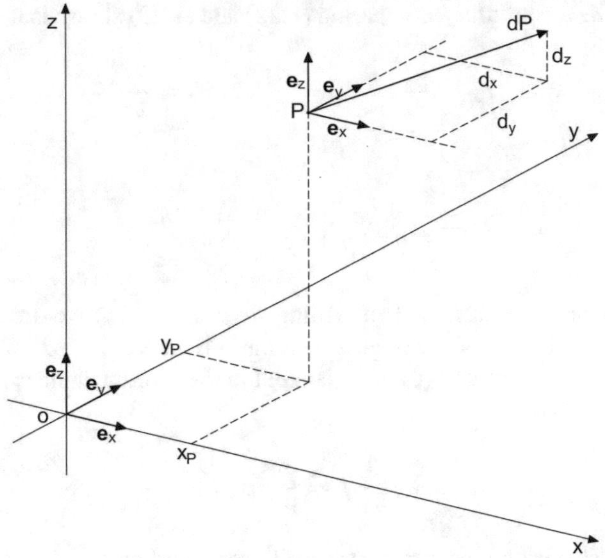

Fig. 1.1 The Cartesian coordinate system

In (1.18) $|dP|$ is the length of the vector $d\boldsymbol{P}$, which points from P to $P + dP$, $\boldsymbol{F} \cdot d\boldsymbol{P}$ is the scalar product

$$\boldsymbol{F} \cdot dP = |\boldsymbol{F}| \cdot |dP| \cos \vartheta \tag{1.21}$$

where ϑ is the angle between $\boldsymbol{F}$ and dP. From the intrinsic definition given above we can derive the form of the operator ∇ in various systems of coordinates.

Example 1.1 (Cartesian Coordinates) By taking a Cartesian coordinate system (see Fig. 1.1), characterized by the three unit vectors $(\boldsymbol{e}_X, \boldsymbol{e}_Y, \boldsymbol{e}_Z)$ we specialize (1.20) by taking subsequently $dP = dx\, \boldsymbol{e}_X$, $dP = dy\, \boldsymbol{e}_Y$, $dP = dz\, \boldsymbol{e}_Z$. We find then

$$\boldsymbol{F} \cdot dx\, \boldsymbol{e}_X = F_X dx, \quad \boldsymbol{F} \cdot dy\, \boldsymbol{e}_Y = F_Y d_Y, \quad \boldsymbol{F} \cdot dz\, \boldsymbol{e}_Z = F_Z d_z,$$

where (F_X, F_Y, F_Z) are the Cartesian components of $\boldsymbol{F}$. Therefore, for a general infinitesimal shift $dP = dx\, \boldsymbol{e}_X + dy\, \boldsymbol{e}_Y + dz\, \boldsymbol{e}_Z$, we get

$$\nabla f \cdot dP = \boldsymbol{F} \cdot d\boldsymbol{P} = F_X dx + F_Y dy + F_Z dz. \tag{1.22}$$

On the other hand, by the theorem of the total differential,

$$df = \frac{\partial f}{\partial x} dx + \frac{\partial f}{\partial y} dy + \frac{\partial f}{\partial z} dz. \tag{1.23}$$

Since *dx, dy, dz* are arbitrary, comparing (1.22) and (1.23) shows that

$$F_x = \frac{\partial f}{\partial x}, \quad F_y = \frac{\partial f}{\partial y}, \quad F_z = \frac{\partial f}{\partial z}, \tag{1.24}$$

or that

$$\nabla = \boldsymbol{e}_X \frac{\partial}{\partial x} + \boldsymbol{e}_Y \frac{\partial}{\partial y} + \boldsymbol{e}_Z \frac{\partial}{\partial z}. \tag{1.25}$$

One important remark is that in writing the form (1.25) we are free to invert the order in the products of the vectors with the derivatives. In fact, in the Cartesian coordinates the unit vectors $\boldsymbol{e}_X, \boldsymbol{e}_Y, \boldsymbol{e}_Z$ tangent to the coordinate lines at every point, are constant, so that

$$\boldsymbol{e}_X \frac{\partial}{\partial x} f = \frac{\partial}{\partial x}(\boldsymbol{e}_X f), \tag{1.26}$$

and so forth. This is no longer true for curvilinear coordinates.

Example 1.2 (Cylindrical Coordinates) In this coordinate system (see Fig. 1.2)

$$dP = dr\, \boldsymbol{e}_r + r d\lambda\, \boldsymbol{e}_\lambda + dz\, \boldsymbol{e}_Z \tag{1.27}$$

so that

$$\nabla f \cdot dP = \boldsymbol{F} \cdot dP = F_r dr + r F_\lambda d\lambda + F_z dz, \tag{1.28}$$

where

$$F_r = \boldsymbol{F} \cdot \boldsymbol{e}_r, \; F_\lambda = \boldsymbol{F} \cdot \boldsymbol{e}_\lambda, \; F_z = \boldsymbol{F} \cdot \boldsymbol{e}_Z.$$

On the other hand,

$$f(P) = f(r, \lambda, z)$$

so that

$$df = \frac{\partial f}{\partial r} dr + \frac{\partial f}{\partial \lambda} d\lambda + \frac{\partial f}{\partial z} dz. \tag{1.29}$$

Comparing (1.28) and (1.29) we see that

$$F_r = \frac{\partial f}{\partial r}, \quad F_\lambda = \frac{1}{r}\frac{\partial f}{\partial \lambda}, \quad F_z = \frac{\partial f}{\partial z}, \tag{1.30}$$

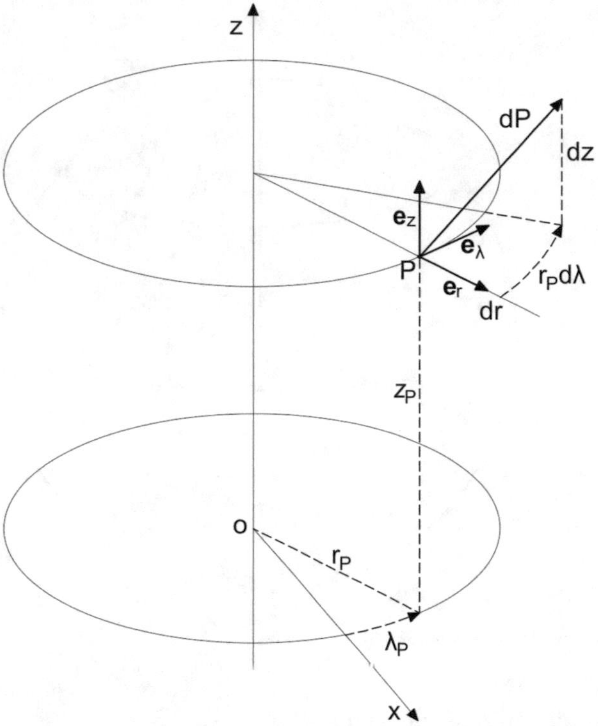

Fig. 1.2 The cylindrical coordinate system

which yields the expression of the gradient in cylindrical coordinates

$$\nabla = \boldsymbol{e}_r \frac{\partial}{\partial r} + \boldsymbol{e}_\lambda \frac{1}{r} \frac{\partial}{\partial \lambda} + \boldsymbol{e}_Z \frac{\partial}{\partial z}. \tag{1.31}$$

We note that now $\boldsymbol{e}_r$ and $\boldsymbol{e}_\lambda$ are no longer constant in space, so that they cannot be commuted with the respective partial derivatives.

Example 1.3 (Spherical Coordinates) In spherical coordinates (see Fig. 1.3) we have

$$dP = dr\boldsymbol{e}_r + r \cos \varphi \, d\lambda \boldsymbol{e}_\lambda + rd\, \varphi \boldsymbol{e}_\varphi \tag{1.32}$$

Let us note that the coordinate r in spherical coordinates is different from that of cylindrical coordinates, although we use here the same letter. The former is the distance of P from the origin, the latter from the Z-axis.

From (1.32) we can compute

$$\nabla f \cdot dP = \boldsymbol{F} \cdot dP = F_r dr + r \cos \varphi \, F_\lambda d\lambda + r F_\varphi d\varphi \tag{1.33}$$

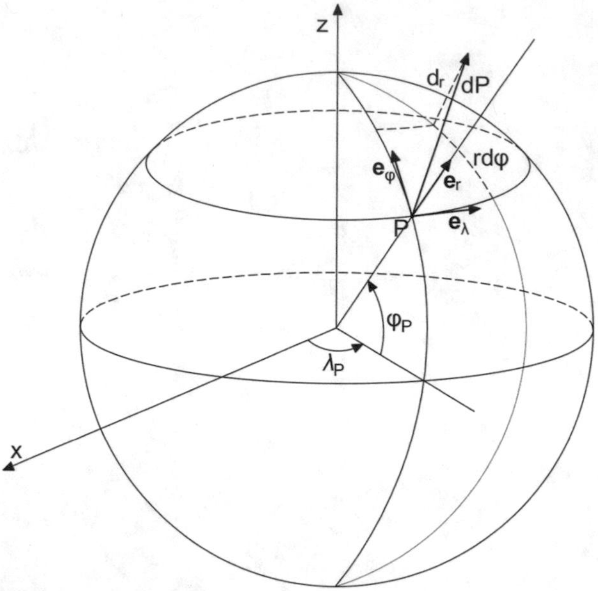

Fig. 1.3 The spherical coordinate system

that compared with the total differential of f gives

$$F_r = \frac{\partial f}{\partial r}, \quad F_\lambda = \frac{1}{r\cos\varphi}\frac{\partial f}{\partial \lambda}, \quad F_\varphi = \frac{1}{r}\frac{\partial f}{\partial \varphi},$$

or

$$\nabla = \boldsymbol{e}_r \frac{\partial}{\partial r} + \frac{1}{r\cos\varphi}\boldsymbol{e}_\lambda \frac{\partial}{\partial \lambda} + \frac{1}{r}\frac{\partial}{\partial \varphi} \tag{1.34}$$

Now we need an important statement in order to introduce a new function, namely, the gravitational potential.

Proposition 1.1 *We have*

$$-\frac{\boldsymbol{\ell}_{QP}}{\ell_{QP}^3} = \nabla_P \frac{1}{\ell_{QP}} \tag{1.35}$$

where ∇_P is the gradient evaluated at the point P.

Proof It is sufficient to prove (1.35) in a specific coordinate system, for instance in Cartesian coordinates.

In fact, recalling (1.12), we have

$$\frac{\partial}{\partial x_p}\frac{1}{\ell_{QP}} = -\frac{x_P - x_Q}{\left[(x_P - x_Q)^2 + (y_P - y_Q)^2 + (z_P - z_Q)^2\right]^{3/2}} = -\frac{\boldsymbol{\ell}_{QP}}{\ell_{QP}^3}\cdot \boldsymbol{e}_X. \tag{1.36}$$

The same holds for the other two components. □

Proposition 1.1 leads to the important conclusion that for the field $\boldsymbol{g}$ generated by one point mass M_Q one can write

$$\boldsymbol{g}(P) = -GM_Q\frac{\boldsymbol{\ell}_{QP}}{\ell_{QP}^3} = \nabla_P\frac{GM_Q}{\ell_{QP}} = \nabla_P u(P). \tag{1.37}$$

i.e., $\boldsymbol{g}$ is the gradient of the function u, which is called the potential of the field $\boldsymbol{g}$:

$$\boldsymbol{g}(P) = \nabla u(P). \tag{1.38}$$

The same is easily generalized to the gravitational field generated by line, surface, or body mass distributions.

To explain, take the case of an extended body and notice that

$$\begin{aligned} \boldsymbol{g}(P) &= -G\int_B \rho(Q)\frac{\boldsymbol{\ell}_{QP}}{\ell_{QP}^3}dB_Q = \\ &= G\int_B \rho(Q)\nabla_P\frac{1}{\ell_{QP}}dB_Q = \\ &= \nabla_P\left(G\int_B\frac{\rho(Q)}{\ell_{QP}}dB_Q\right) \end{aligned} \tag{1.39}$$

i.e.,

$$\boldsymbol{g}(P) = \nabla u(P), \tag{1.40}$$

with

$$u(P) = G\int_B\frac{\rho(Q)}{\ell_{QP}}dB_Q. \tag{1.41}$$

Let us observe explicitly that the result (1.40) and (1.41) comes from the fact that ∇_p and $\int \cdot dB_Q$ are operations that act on functions of the coordinates of different points and, as such, they commute one with one another, under regularity conditions sufficiently general for our purposes.

So we reached the following conclusion.

Proposition 1.2 *The Newtonian gravitational field is a potential field, i.e., it is the gradient of a potential function.*

Remark 1.2 From the physical point of view a potential field enjoys some important properties. In fact, it is clear that if we integrate $\boldsymbol{g}(P)\cdot dP = du$ along a line joining two points, P_1 and P_2, we get

$$\int_L g(P)\cdot dP = \int_L du = u(P_2) - u(P_1), \tag{1.42}$$

i.e., the integral, which represents the work done by the field in moving a unit proof mass $M_P = 1$ from P_1 to P_2, is independent of the path followed. This opens the way to the definition of potential energy and to energy conservation theorems typical of classical mechanics.

From our particular purposes, it is enough to observe that the knowledge of the gravitational potential u of a certain mass distribution is equivalent to that of its gravitational field, since the latter can always be computed by taking the ∇ of the former.

1.4 The Divergence Operator: Gauss Theorem

Let us take a smooth vector field $\boldsymbol{F}(P)$, e.g., in the sense that, in Cartesian coordinates, the components of $\boldsymbol{F}$ are differentiable functions of the coordinates of P. It is then expedient to define the divergence of $\boldsymbol{F}$ as a differential operator that generates a new scalar function from $\boldsymbol{F}(P)$.

Definition 1.2 The divergence of the field $\boldsymbol{F}(P)$ is defined as

$$\operatorname{div}\boldsymbol{F}(P) = \nabla\cdot\boldsymbol{F}(P). \tag{1.43}$$

We note that, using the RHS of (1.43), we can compute the divergence of $\boldsymbol{F}$ in various coordinate systems

Example 1.4 (Cartesian Coordinates) One has

$$\operatorname{div}\boldsymbol{F} = (\boldsymbol{e}_X\frac{\partial}{\partial x} + \boldsymbol{e}_Y\frac{\partial}{\partial y} + \boldsymbol{e}_Z\frac{\partial}{\partial z})\cdot(F_X\boldsymbol{e}_X + F_Y\boldsymbol{e}_Y + F_Z\boldsymbol{e}_Z). \tag{1.44}$$

But recalling that $\boldsymbol{e}_X$, $\boldsymbol{e}_Y$, $\boldsymbol{e}_Z$ are constant vectors that constitute a unit orthogonal triad, one gets

$$\operatorname{div}\boldsymbol{F} = \nabla\cdot\boldsymbol{F} = \frac{\partial}{\partial x}F_X + \frac{\partial}{\partial y}F_Y + \frac{\partial}{\partial z}F_Z. \tag{1.45}$$

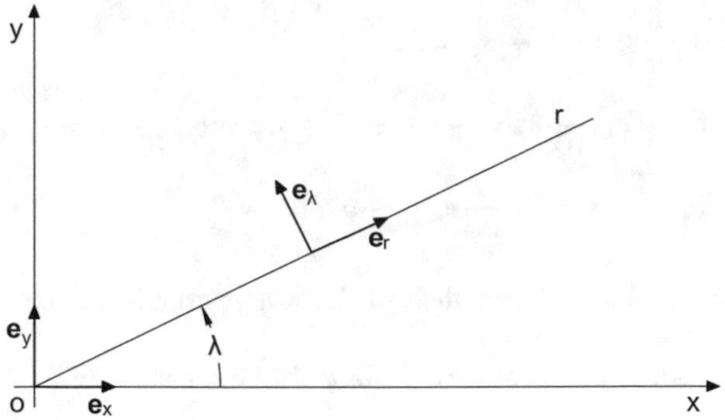

Fig. 1.4 Relation between $\boldsymbol{e}_r$, $\boldsymbol{e}_\lambda$ and $\boldsymbol{e}_x$, $\boldsymbol{e}_y$

Example 1.5 (Cylindrical Coordinates) To compute $\nabla \cdot \boldsymbol{F}$ in cylindrical coordinates, it is convenient to preliminarily establish the following relations:

$$\frac{\partial}{\partial \lambda}\boldsymbol{e}_r = \boldsymbol{e}_\lambda, \ \frac{\partial}{\partial r}\boldsymbol{e}_r = 0, \tag{1.46}$$

$$\frac{\partial}{\partial \lambda}\boldsymbol{e}_\lambda = -\boldsymbol{e}_r, \ \frac{\partial}{\partial r}\boldsymbol{e}_\lambda = 0, \tag{1.47}$$

Furthermore, neither $\boldsymbol{e}_r$ nor $\boldsymbol{e}_\lambda$ depend on z and $\boldsymbol{e}_Z$ is a constant vector at any point P. The relations (1.46), (1.47) are easy to prove, either geometrically or by observing that (see Fig. 1.4)

$$\boldsymbol{e}_r = \cos\lambda\, \boldsymbol{e}_X + \sin\lambda\, \boldsymbol{e}_Y$$

$$\boldsymbol{e}_\lambda = -\sin\lambda\, \boldsymbol{e}_X + \cos\lambda\, \boldsymbol{e}_Y$$

and then taking the required derivatives. Now we are able to compute

$$\begin{aligned} \nabla \cdot \boldsymbol{F} &= \left(\boldsymbol{e}_r \frac{\partial}{\partial r} + \boldsymbol{e}_\lambda \frac{1}{r}\frac{\partial}{\partial \lambda} + \boldsymbol{e}_Z \frac{\partial}{\partial z}\right) \cdot (\boldsymbol{e}_r F_r + \boldsymbol{e}_\lambda F_\lambda + \boldsymbol{e}_Z F_z) = \\ &= \frac{\partial F_r}{\partial r} + \frac{1}{r}F_r + \frac{1}{r}\frac{\partial}{\partial \lambda}F_\lambda + \frac{\partial F_z}{\partial z}. \end{aligned} \tag{1.48}$$

Example 1.6 (Spherical Coordinates) To compute the divergence in spherical coordinates, one has to use the differential table

$$\begin{cases} \dfrac{\partial}{\partial r}\boldsymbol{e}_r = 0, \dfrac{\partial}{\partial r}\boldsymbol{e}_\lambda = 0, \dfrac{\partial}{\partial r}\boldsymbol{e}_\varphi = 0, \\ \dfrac{\partial}{\partial \lambda}\boldsymbol{e}_r = \cos\varphi\,\boldsymbol{e}_\lambda, \dfrac{\partial}{\partial \lambda}\boldsymbol{e}_\lambda = -\cos\varphi\,\boldsymbol{e}_r + \sin\varphi\,\boldsymbol{e}_\varphi, \dfrac{\partial}{\partial \lambda}\boldsymbol{e}_\varphi = -\sin\varphi\,\boldsymbol{e}_\lambda, \\ \dfrac{\partial}{\partial \varphi}\boldsymbol{e}_r = \boldsymbol{e}_\varphi, \dfrac{\partial}{\partial \varphi}\boldsymbol{e}_\lambda = 0, \dfrac{\partial}{\partial \varphi}\boldsymbol{e}_\varphi = -\boldsymbol{e}_r. \end{cases}$$

This in turn can be obtained from the well-known geometric relations

$$\begin{aligned} \boldsymbol{e}_r &= \cos\varphi\cos\lambda\,\boldsymbol{e}_X + \cos\varphi\sin\lambda\,\boldsymbol{e}_Y + \sin\varphi\,\boldsymbol{e}_Z \\ \boldsymbol{e}_\varphi &= -\sin\varphi\cos\lambda\,\boldsymbol{e}_X - \sin\varphi\sin\lambda\,\boldsymbol{e}_Y + \cos\varphi\,\boldsymbol{e}_Z \\ \boldsymbol{e}_\lambda &= -\sin\lambda\boldsymbol{e}_X + \cos\lambda\,\boldsymbol{e}_Y \end{aligned}$$

So we find, after some careful computation,

$$\operatorname{div}\boldsymbol{F} = \nabla\cdot\boldsymbol{F} = \frac{\partial F_r}{\partial r} + \frac{1}{r\cos\varphi}\frac{\partial F_\lambda}{\partial\lambda} + \frac{1}{r}\frac{\partial F_\varphi}{\partial\varphi} + \frac{2}{r}F_r - \frac{\tan\varphi}{r}F_\varphi. \tag{1.49}$$

We are ready now to present one of the fundamental theorems of the entire vector calculus, namely, Gauss' theorem.

Theorem 1.1 (Gauss) *Consider a volume V in $\mathbb{R}^3$ with a smooth boundary endowed with a continuous normal* **n** *and assume that a field* $\boldsymbol{F}$ *is defined on V such that $\nabla\cdot\boldsymbol{F}$ is at least an integrable function there; then*

$$\int_V \nabla\cdot\boldsymbol{F}(P)dV_P = \int_S \boldsymbol{F}(P)\cdot\boldsymbol{n}_P\,dS_P \tag{1.50}$$

Here $\boldsymbol{n}_P$ must be understood as the outward normal to S at P.

Proof Although the theorem is valid under even more general conditions, in particular for bodies with edges and corners on the boundary, like for instance a cube, we shall give a proof under milder conditions, assuming that S is "normal" to the (x, y) plane, in the sense that every parallel to the Z-axis pierces S at most two points P_1 and P_2 (see Fig. 1.5), such that the outward normal $\boldsymbol{n}_2$ to S in P_2 (the upper surface) looks up, while the outward normal $\boldsymbol{n}_1$ to S in P_1 (the lower surface) looks down. We note that in this case the boundary S is split into two surfaces: the upper, S_+, and the lower, S_-, that are contiguous on S with a common boundary that is projected to the boundary of the projection S_0 of S on the (x, y)-plane.

We note as well that S_- and S_+ could be defined by means of functions $Z_+(x, y)$, $Z_-(x, y)$, as

$$\begin{aligned} &P(x, y, z_-) \in S_-; \quad z = Z_-(x, y), \quad (x, y) \in S_0 \\ &P(x, y, z_+) \in S_+; \quad z = Z_+(x, y), \quad (x, y) \in S_0. \end{aligned}$$

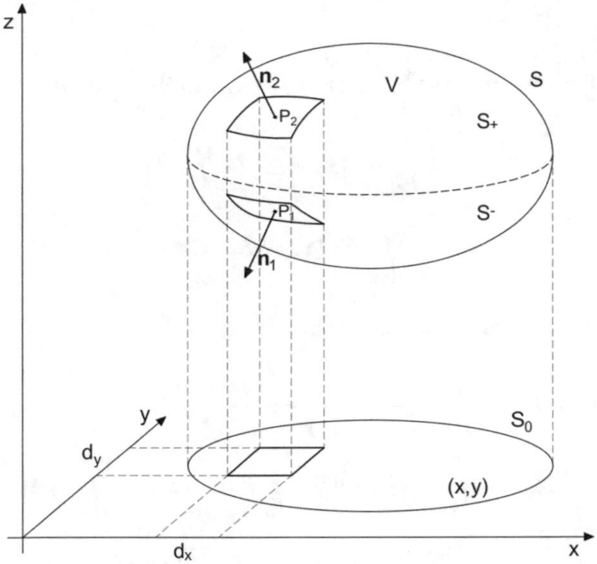

Fig. 1.5 A domain V projecting on S_0 on the (x, y) plane, with an upper surface S_+ and a lower surface S_-

We prove preliminarily (1.50) for a field $\boldsymbol{F}$ that has only one non-zero component, for instance F_Z. It may well happen that the line separating S_- and S_+ is stretched becoming a cylindrical lateral boundary, yet even in this case the subsequent reasoning holds equally well, because the normal to the latter face is then orthogonal to $\boldsymbol{e}_Z$. Namely, we want to prove that

$$\int_V \frac{\partial F_Z(P)}{\partial_z} dV_P = \int_S F_Z(P) \boldsymbol{n}_P \cdot \boldsymbol{e}_Z ds. \tag{1.51}$$

But

$$\begin{aligned}\int_V \frac{\partial F_Z(P)}{\partial z} dV_P &= \int_{S_0} \int_{Z_-(x,y)}^{Z_+(x,y)} \frac{\partial F_Z(P)}{\partial z} dz dx dy = \\ &= \int_{S_0} (F_Z\,[x, y, Z_+(x, y)] - F_Z\,[x, y, Z_-(x, y)]) dx dy.\end{aligned} \tag{1.52}$$

On the other hand, we notice that, looking at Fig. 1.5, we have

$$dx dy = dS_2 \boldsymbol{n}_2 \cdot \boldsymbol{e}_Z = -dS_1 \boldsymbol{n}_1 \cdot \boldsymbol{e}_Z \tag{1.53}$$

so that

$$\int_S F_Z(P)\boldsymbol{n}_P \cdot \boldsymbol{e}_Z dS = \int_{S_+} F_Z(P)\boldsymbol{n}_{P_2} \cdot \boldsymbol{e}_Z dS_2 + \int_{S_-} F_Z \boldsymbol{n}_{P_1} \cdot \boldsymbol{e}_Z dS_1 =$$

$$= \int_{S_0} F_Z\left[x, y, Z_+(x, y)\right] dxdy$$

$$- \int_{S_0} F_Z\left[x, y, Z_-(x, y)\right] dxdy,$$

therefore (1.51) is proved. But then we also have

$$\int_V \nabla \cdot \boldsymbol{F} dV = \int_V \left(\frac{\partial F_X}{\partial x} + \frac{\partial F_Y}{\partial y} + \frac{\partial F_Z}{\partial z}\right) dV =$$

$$= \int_S F_X \boldsymbol{e}_X \cdot \boldsymbol{n} dS + \int_S F_Y \boldsymbol{e}_Y \cdot \boldsymbol{n} dS + \int_S F_Z \boldsymbol{e}_Z \cdot \boldsymbol{n} dS$$

$$= \int (F_X \boldsymbol{e}_X + F_Y \boldsymbol{e}_Y + F_Z \boldsymbol{e}_Z) \cdot \boldsymbol{n} dS = \int \boldsymbol{F} \cdot \boldsymbol{n} dS.$$

□

Let us recall that the RHS of (1.50) is also called the flux of the field $\boldsymbol{F}$ through the surface S. In hydraulics for instance, when $\boldsymbol{F}$ is the velocity field of a water flow, which is ideally an incompressible fluid with constant density, the flux is related to the variation of the water mass in V, which is always zero if there are no sources or sinks in V. In the case of a gravitational field, as we shall see in the next section, the flux of $\boldsymbol{g}$ through S is related to the sources of g in V, namely to the mass contained in such a volume.

1.5 The Laplacian Operator, Poisson's Equation and Green's Identities

By definition, the Laplacian is a second-order differential operator acting on scalar functions that are twice differentiable, producing another scalar function according to the rule

$$\text{Lap}\, f(P) = \nabla \cdot (\nabla f) \equiv \Delta f. \tag{1.54}$$

Since we already found the form of the gradient and divergence in various coordinates, it is only a matter of patience to compute the Laplacian in the same coordinates. We give here only the results.

Example 1.7 In Cartesian coordinates,

$$\Delta f = \frac{\partial^2 f}{\partial x^2} + \frac{\partial^2 f}{\partial y^2} + \frac{\partial^2 f}{\partial z^2} \tag{1.55}$$

Example 1.8 In cylindrical coordinates,

$$\Delta f = \frac{\partial^2 f}{\partial r^2} + \frac{1}{r}\frac{\partial f}{\partial \lambda} + \frac{1}{r^2}\frac{\partial^2 f}{\partial \lambda^2} + \frac{\partial^2 f}{\partial z^2} \tag{1.56}$$

Example 1.9 In spherical coordinates,

$$\Delta f = \frac{\partial^2 f}{\partial r^2} + \frac{2}{r}\frac{\partial f}{\partial \lambda} + \frac{1}{r^2}\left(\frac{\partial^2 f}{\partial \varphi^2} - \tan\varphi \frac{\partial f}{\partial \varphi} + \frac{1}{\cos^2\varphi}\frac{\partial^2 f}{\partial \lambda^2}\right) \tag{1.57}$$

Let us note that the purely angular part of Δ, which can be considered as a kind of Laplacian on the surface of the unit sphere, is also called the Laplace-Beltrami operator and denoted as

$$\Delta_\sigma = \frac{\partial^2}{\partial \varphi} - \tan\varphi \frac{\partial}{\partial \varphi} + \frac{1}{\cos^2\varphi}\frac{\partial^2}{\partial \lambda^2}. \tag{1.58}$$

Remark 1.3 In the context of Newton's theory the Laplace operators is important due to the fact that, since

$$\boldsymbol{g} = \nabla u \tag{1.59}$$

we have also

$$\nabla \cdot \boldsymbol{g} = \nabla \cdot \nabla u = \Delta u. \tag{1.60}$$

In particular, making use of the gravitational potential u and of (1.59), we see that Gauss' theorem takes the form

$$\int_V \Delta u \, dV = \int_S \frac{\partial u}{\partial n} \, dS \tag{1.61}$$

where, by definition,

$$\frac{\partial u}{\partial n} = \boldsymbol{n} \cdot \nabla u. \tag{1.62}$$

To fix ideas, let us stipulate that when we speak of a Newtonian gravitational field $\boldsymbol{g}$ or Newtonian potential u, we think of the field generated by a mass distribution of a bounded body B, with a measurable bounded density $\rho(Q)$, so that

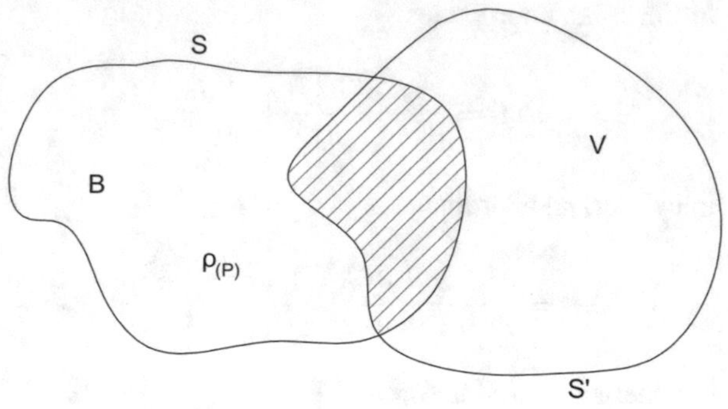

Fig. 1.6 The body B, the set V and their intersection; the integral of $\rho(P)$ on $B \cap V$ is the total mass contained in V

$$\mathbf{g}(P) = -G \int_B \frac{\boldsymbol{\ell}_{QP}}{\ell_{QP}^3} \rho(Q) dB_Q \tag{1.63}$$

and

$$u(P) = G \int_B \frac{1}{\ell_{QP}} \rho(Q) dB_Q. \tag{1.64}$$

Although the theorem we are going to prove holds under very general conditions, we will prove it for a field of the form (1.63). Our aim is to prove the following proposition.

Proposition 1.3 *Let S' be any regular surface, boundary of a set V which is in an arbitrary position in space, outside or inside B, or partly overlapping it (see Fig. 1.6); let M be the mass distributed in B, generating a gravitational field $\mathbf{g}(P)$, as in (1.63), and M_V part of this mass that falls also in V, namely*

$$M_V = \int_{B \cap V} \rho(P) dB_P. \tag{1.65}$$

Then we have

$$\int_{S'} \mathbf{g}(P) \cdot \boldsymbol{n}_P \, dS'_P = -4\pi G M_V. \tag{1.66}$$

Proof Let us define the function

$$\chi_V(P) = \begin{cases} 1, & \text{if } P \in \overset{\circ}{V}, \\ 1/2, & \text{if } P \in S' \\ 0, & \text{if } P \in \Omega \equiv \overline{V}^c, \end{cases} \tag{1.67}$$

where $\overset{\circ}{V}$ is the interior of V, S' is the boundary of V and Ω is the exterior of the closure $\overline{V}$ of V. Assume that the relation

$$\int_{S'} \frac{\boldsymbol{\ell}_{QP}}{\ell_{QP}^3} \cdot \boldsymbol{n}_P \, dS'_P = 4\pi \chi_V(Q) \tag{1.68}$$

is true, as it will soon be proved. Then using (1.63) in the LHS of (1.68), we can write

$$\begin{aligned}\int_{S'} \boldsymbol{g}(P) \cdot \boldsymbol{n}_P dS'_P &= -G \int_{S'} \left(\int_B \frac{\boldsymbol{\ell}_{QP}}{\ell_{QP}^3} \rho(Q) dB_Q \right) \cdot \boldsymbol{n}_P dS'_P = \\ &= -G \int_B \left(\int_{S'} \frac{\boldsymbol{\ell}_{QP} \cdot \boldsymbol{n}_P}{\ell_{QP}^3} dS'_P \right) \rho(Q) dB_Q = \\ &= -4\pi G \int_B \chi_V(Q) \rho(Q) dB_q = \\ &= -4\pi G \int_{B \cap V} \rho(Q) dB_Q = -4\pi G M_V .\end{aligned}$$

□

The next Lemma 1.1 is required to complete the proof of Proposition 1.3.

Lemma 1.1 *The relation* (1.68) *holds true.*

Proof We split the proof into three parts, according to whether $Q \in V$, $Q \in S'$ or, $Q \in \Omega = \overline{V}^c$.

All three cases derive from the following elementary geometric consideration. Let Q be given and let dS'_P be an elementary area on S' at $P \neq Q$, with normal $\boldsymbol{n}_P$ (see Fig. 1.7). We have

$$dS_0 = dS'_P \boldsymbol{n}_P \cdot \boldsymbol{e}_{QP} = dS'_P \boldsymbol{n}_P \cdot \frac{\boldsymbol{\ell}_{QP}}{\ell_{QP}}$$

and

$$d\sigma = \frac{dS_0}{\ell_{QP}^2},$$

so that

$$d\sigma = \frac{\boldsymbol{\ell}_{QP} \cdot \boldsymbol{n}_P}{\ell_{QP}^3} dS'_P . \tag{1.69}$$

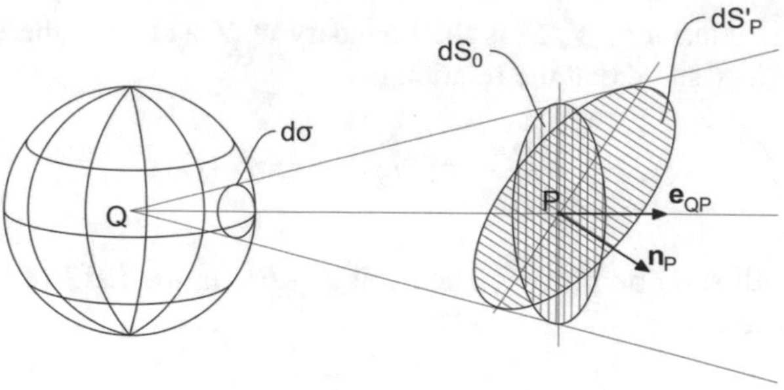

Fig. 1.7 The area dS'_P projected by a cone with vertex Q; the normal $\boldsymbol{n}_P$ to dS'_P; the projection dS_0 of dS'_P on a plane orthogonal to the direction $\boldsymbol{e}_{QP}$

We note that, since dS'_P and $d\sigma$ are positive surface measures, the sign of (1.69) should be turned to minus in the event that $\boldsymbol{n}_P \cdot \boldsymbol{e}_{PQ}$ is negative. Now let us consider an infinitesimal cone issuing from Q and cutting an infinitesimal surface $d\sigma$ on the unit sphere. When this cone cuts also S', we have three cases (see Fig. 1.8):

(a) Q lies outside V; then the cone generally cuts an even number of infinitesimal surfaces on S' (in Fig. 1.8a, dS_1, dS_2, dS_3, dS_4). However, they can be organized in couples such that when one enters V, the outward normal $\boldsymbol{n}_P$ is opposite to $\boldsymbol{e}_{QP}$ and when one leaves V, the outward normal $\boldsymbol{n}_P$ is in the same direction of $\boldsymbol{e}_{QP}$. Since each surface projects onto the same $d\sigma$, according to (1.69), but the two contributions are opposite in sign, we see that they cancel and the same happens for the whole area of points on the unit sphere that are on straight lines hitting S'. Moreover, the whole surface S' is swept by the union of all infinitesimal surfaces dS'. So we have proved

$$\int_{S'} \frac{\boldsymbol{\ell}_{QP} \cdot \boldsymbol{n}_P}{\ell^3_{QP}} dS'_P = 0, \quad Q \in \Omega. \tag{1.70}$$

(b) Q is on the surface S'. In this case (see Fig. 1.8, b) the lines issuing from Q through the unit hemisphere pointing into V meet S' generally at an odd number of points (see in Fig. 1.8 the centers of dS_1, dS_2, dS_3).

Of the infinitesimal areas, couples corresponding to entrance and exit of the line from V, annihilate their contributions, so that only one $d\sigma$ is left. Once this is integrated on the hemisphere one gets its area, i.e., 2π,

$$\int_{S'} \frac{\boldsymbol{\ell}_{QP} \cdot \boldsymbol{n}_P}{\ell^3_{QP}} dS'_P = 2\pi \quad Q \in S'. \tag{1.71}$$

This happens because we have assumed S' to be regular, i.e., it admits tangent plane at Q. Of course would Q be the vertex of a cone on S', formula (1.71) should be suitably changed.

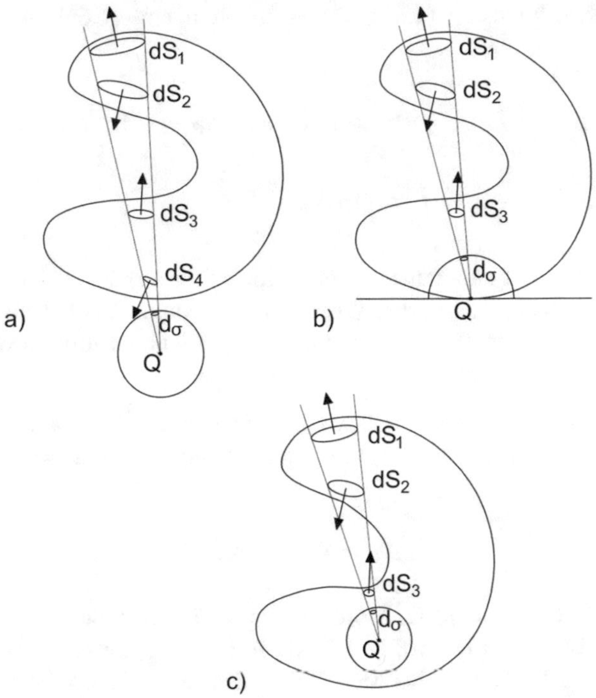

Fig. 1.8 Interactions of an infinitesimal cone with the surface S', when the vertex Q lies: (**a**) outside V, (**b**) on S', (**c**) inside V

It can happen that a line from Q towards the exterior of V, hits S', but then this occurs an even number of times so that the same reasoning as in a) holds and the exterior unit hemisphere never contributes to the integral (1.71).

(c) Finally, when we consider the case that Q lies in V (see Fig. 1.8 c)), we see that a similar reasoning as in (b) holds, with the only difference that now the infinitesimal $d\sigma$ has to run over the whole unit sphere in order for its projections on S' to sweep the whole surface. So, arguing as in (a) and (b) we find

$$\int_{S'} \frac{\boldsymbol{\ell}_{QP} \cdot \boldsymbol{n}_P}{\ell_{QP}^3} dS'_P = 4\pi. \tag{1.72}$$

This completes the proof of the Lemma. □

The relation (1.66) opens the way to results of great importance for us, to which we give the dignity of a Theorem.

Theorem 1.2 (Poisson) *The gravitational potential u generated by a* 3D *mass density $\rho(P)$ satisfies the following equation, called* the Poissons equation

$$\Delta u = -4\pi G\rho(P). \tag{1.73}$$

Proof We have, by using (1.61), Gauss theorem and (1.66), for any V, with boundary S',

$$\int_V \Delta u\, dV = \int_V \nabla \cdot \boldsymbol{g}\, dV = \int_{S'} \boldsymbol{g} \cdot \boldsymbol{n}_P\, dS_P = -4\pi G M_V = \\ = -4\pi G \int_V \rho(Q) dV. \tag{1.74}$$

Since the first and last integrals in (1.74) coincide for any V, we have, by the mean value theorem, that (1.73) is true. When ρ is not continuous but just integrable, a suitable Lebesgue theorem guarantees that (1.73) still holds almost everywhere, i.e., outside a set of zero measure. □

Remark 1.4 We can observe that when u is generated by a finite total mass distributed in a bounded body B, the potential satisfies in $\Omega = \overline{B}^{c}$ the Laplace equation

$$\Delta u = 0, \quad P \in \Omega; \tag{1.75}$$

we say that u is harmonic in Ω. Moreover, since B is bounded, it is contained in a sphere of suitable radius R (see Fig. 1.9). So if we take P outside such a sphere, and we let r_P denote the distance between P and the center of the sphere, we have

$$r_P + R > \ell_{QP} > r_P - R, \quad \forall Q \in B. \tag{1.76}$$

Therefore,

$$\frac{r_P}{r_P + R} < \frac{r_P}{\ell_{QP}} < \frac{r_P}{r_P - R}$$

showing that, for $r_P \to +\infty$,

$$\frac{r_P}{\ell_{QP}} \to 1 \tag{1.77}$$

uniformly and boundedly outside any sphere of radius larger than R.

Therefore we can assert that

$$u(P) = G \int_B \frac{\rho(Q)}{\ell_{QP}} dB_Q = \frac{G}{r_P} \int_B \frac{r_P}{\ell_{QP}} \rho(Q) dB_Q \tag{1.78}$$

has, when $r_p \to \infty$, the asymptotic behaviour:

$$u(P) \sim \frac{G}{r_P} \int_B \rho(Q) dB_Q = \frac{GM}{r_P}; \tag{1.79}$$

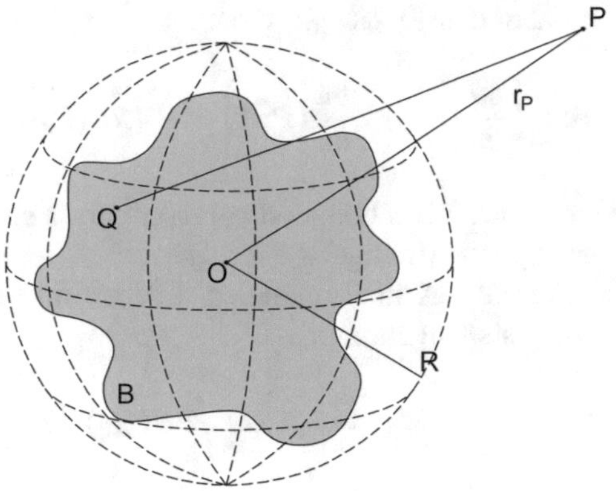

Fig. 1.9 The sphere of radius R containing B

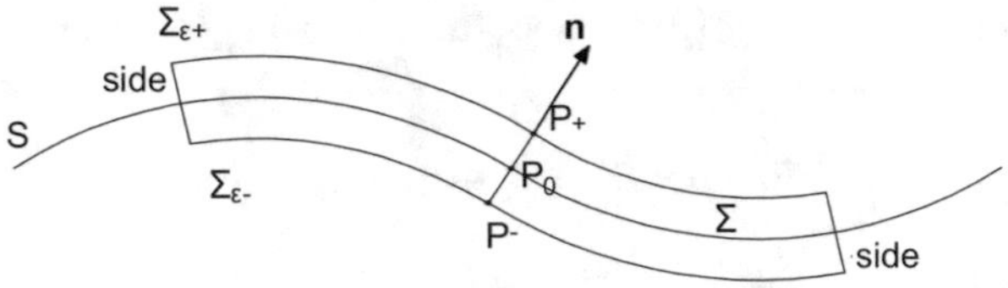

Fig. 1.10 The surface S and one patch Σ; the box $\Sigma' = \Sigma_{\varepsilon+} + \cup \Sigma_{\varepsilon-}$

we say that $u(P)$ is harmonic in Ω and regular at infinity.
A similar reasoning shows that

$$|\boldsymbol{g}(P)| = |\nabla u(P)| \sim \frac{GM}{r_P^2}, \tag{1.80}$$

so that at large distance the field generated by the mass in B is seen as the field generated by M concentrated as a point mass at the origin.

Remark 1.5 Another important consequence of (1.66) is the so-called jump relation of the normal derivative of the single-layer potential. Since we will need it later, in Chap. 8, we provide here a short proof of such a relation.

Let us consider a single mass layer, with density $\alpha(P_0)$ and bounded curvature. Let us further consider a box Σ' constructed from a patch $\Sigma \subset S$, with two surfaces $\Sigma_{\varepsilon+}$, $\Sigma_{\varepsilon-}$ defined by:

$$\Sigma_{\varepsilon\pm} = \{P = P_0 + \varepsilon \boldsymbol{n}(P_0),\, P_0 \in \Sigma\} \tag{1.81}$$

and the side walls that connect the two (see Fig. 1.10). We want to show that if

$$u(P) = G \int_S \frac{\alpha(Q_0)}{\ell_{PQ_0}} dS_{Q_0}, \tag{1.82}$$

the single-layer potential of density $\alpha(Q_0)$, then

$$\lim_{\varepsilon\to 0}\left[\frac{\partial u}{\partial n}(P_+)-\frac{\partial u}{\partial n}(P_-)\right]=-4\pi\alpha(P_0). \tag{1.83}$$

We first of all note that, since S has bounded curvature, for ε small enough the two surfaces $\Sigma_{\varepsilon\pm}$ are unambiguously defined.

Now if we reason exactly as in Proposition 1.3, we first of all come to the conclusion that, for every ε small enough,

$$\int_{\Sigma'}\frac{\partial u}{\partial n'}d\Sigma'=-4\pi GM_{\Sigma'}=-4\pi G\int_{\Sigma}\alpha(Q_0)\,d\Sigma_{Q_0}, \tag{1.84}$$

because the total mass contained in Σ' is exactly the one spread on Σ.

On the other hand, we have

$$\begin{aligned}\int_{\Sigma'}\frac{\partial u}{\partial n'}d\Sigma'=&\int_{\Sigma_{\varepsilon+}}\nabla u\cdot \boldsymbol{n}'(P_+)d\Sigma_+ +\int_{\Sigma_{\varepsilon-}}\nabla u\cdot \boldsymbol{n}'(P_-)d\Sigma_- +\\&+\int_{\text{Side}}\nabla u\cdot \boldsymbol{n}'d\Sigma'=I_1+I_2+I_3.\end{aligned} \tag{1.85}$$

At the same time, observe that

$$\begin{aligned}d\Sigma_\pm&=d\Sigma_{P_0}(1+O(\varepsilon)),\\ \boldsymbol{n}'(P_+)&=\boldsymbol{n}'(P_0),\ \ \boldsymbol{n}'(P_-)=-\boldsymbol{n}'(P_0);\end{aligned} \tag{1.86}$$

in addition one could patiently prove that, for a single-layer potential u with $\alpha\in L^1(S)$ (see [84], ch. II,§14)

$$\frac{\partial u}{\partial n}(P_\pm)=\frac{\partial u}{\partial n}(P_{0\pm})+O(\varepsilon), \tag{1.87}$$

when $P_{0\pm}$ is just the limit of $P_\pm$ from above and from below S, respectively. So putting together (1.86) and (1.87) we see that

$$\lim_{\varepsilon\to 0}I_1=\int_{\Sigma}\frac{\partial u}{\partial n}(P_{0+})\,d\Sigma_{P_0}$$

and

$$\lim_{\varepsilon\to 0}I_2=-\int_{\Sigma}\frac{\partial u}{\partial n}(P_{0-})d\Sigma_{P_0}.$$

Furthermore, one easily sees that

$$\lim_{\varepsilon\to 0}I_3=0. \tag{1.88}$$

All these relations say that

$$\lim_{\varepsilon \to 0} \int_{\Sigma'} \frac{\partial u}{\partial n'} d\Sigma' = \int_{\Sigma} \left(\frac{\partial u}{\partial n}(P_{0+}) - \frac{\partial u}{\partial n}(P_{0-}) \right) d\Sigma_{P_0}. \tag{1.89}$$

Comparing (1.84) and (1.89), given the arbitrariness of Σ in S, we conclude that (1.83) holds almost everywhere on S.

We come now briefly to three identities, well-known in potential theory (see [63], §1.5; [133], §1.6), called *Green's identities.*

First Green's Identity From the differential identity

$$\nabla \cdot (u\nabla v) = \nabla u \cdot \nabla v + u\, \Delta v, \tag{1.90}$$

valid for any two regular, twice continuously differentiable functions in B, we obtain by integrating over B and applying Gauss's theorem (1.50)

$$\int_B \nabla u \cdot \nabla v\, dB + \int_B u \Delta v\, dB = \int_B \nabla \cdot (u\nabla v) dB = \int_S u \frac{\partial v}{\partial n} dS, \tag{1.91}$$

known as *first Green's identity.*

It is remarkable that if v happens to be harmonic in B (i.e., $\Delta v = 0$ in B), (1.91) implies that

$$\int_B \nabla u \cdot \nabla v\, dB = \int_S u \frac{\partial v}{\partial n} dS. \tag{1.92}$$

Now if we assume that also u is harmonic in B, we see that the LHS of (1.92) is symmetric with respect to switching u and v; we must then have the same in the RHS and we conclude that for u, v both harmonic,

$$\int_S u \frac{\partial v}{\partial n} dS = \int_S v \frac{\partial u}{\partial n} dS, \tag{1.93}$$

which can be regarded as a corollary to the first Green's identity.

Second Green's Identity The differential identity

$$\nabla \cdot (u\nabla v - v\nabla u) = u\, \Delta v - v \Delta u \tag{1.94}$$

and the Gauss theorem imply that

$$\int_B (u\Delta v - v\Delta u) dB = \int_B \nabla \cdot (u\nabla v - v\nabla u) dB = \int_S (u\frac{\partial v}{\partial n} - v\frac{\partial u}{\partial n}) dS, \tag{1.95}$$

known as *second Green's identity.*

Let us note that a similar identity can be also applied to Ω, the exterior of S. Only, in this case, since the normal to S as seen from Ω would be opposite to $\boldsymbol{n}$, the outward normal of S, one has to change the sign of the last terms in (1.95), namely

$$\int_{\Omega}(u\,\Delta v - v\Delta u)d\Omega = -\int_S\left(u\frac{\partial v}{\partial n} - v\frac{\partial u}{\partial n}\right)dS; \tag{1.96}$$

indeed, for (1.96) to be valid one has to assume that u, v tend to zero suitably fast at infinity.

Note that if both u and v are harmonic (in fact regular harmonic) one derives that the left-hand side (LHS) is zero and then one finds again (1.96).

Third Green's Identity We establish this without an explicit proof. We only mention that this can be obtained by applying the second Green identity, with $v(Q) = \frac{1}{\ell_{PQ}}$, $P \in B$, to the body B with a spherical cavity around P, B_ε, that tends to zero with ε.

The identity reads

$$P \in B, \quad u(P) = -\frac{1}{4\pi}\int_B \frac{\Delta u}{\ell_{PQ}}dB_Q + \frac{1}{4\pi}\int_S\left(\frac{1}{\ell_{PQ}}\frac{\partial u}{\partial n_Q} - u\frac{\partial}{\partial n_Q}\frac{1}{\ell_{PQ}}\right)dS. \tag{1.97}$$

In this case, too, we have a version for the outer domain Ω, for which the sign of the surface integral has to be reversed, i.e.,

$$P \in \Omega,\ u(P) = -\frac{1}{4\pi}\int_{\Omega}\frac{\Delta u}{\ell_{PQ}}dB_Q - \frac{1}{4\pi}\int_S\left(\frac{1}{\ell_{PQ}}\frac{\partial u}{\partial n_Q} - u\frac{\partial}{\partial n_Q}\frac{1}{\ell_{PQ}}\right)dS_Q. \tag{1.98}$$

Two remarks are in order, concerning the third Green's identity. Namely assume that u is twice continuously differentiable in B and that further $u|_S = \frac{\partial u}{\partial n}\big|_S = 0$, then from (1.97) we see that

$$u(P) = -\frac{1}{4\pi}\int_B \frac{\Delta u}{\ell_{PQ}}dB_Q. \tag{1.99}$$

However, since in the language of distribution theory we have that the distributional Laplacian of a function g, Δg, is given by definition by the identity

$$\int (\Delta f)\varphi\, dB = \int f\,\Delta\varphi\, dB \tag{1.100}$$

which has to hold for any regular φ ($\varphi \in C_0^\infty(B)$), that vanishes outside some closed set strictly contained in the interior of B, (1.99) suggests that $\Delta\frac{1}{\ell_{PQ}}$ is a specific distribution. This is the famous *Dirac's delta*, defined by

$$\int_B \delta(P, Q)\varphi(Q)dB_Q \equiv \varphi(P) \tag{1.101}$$

for any φ as those described above. Therefore, equipped with these notions, we see from (1.99) that

$$\Delta \frac{1}{\ell_{PQ}} = -4\pi\delta(P, Q). \tag{1.102}$$

Comparing with Poisson's equation (1.73), one could say that $\frac{1}{\ell_{PQ}}$ is in fact, as a function of Q, the potential of a point mass, with mass $\frac{1}{G}$, placed at the point P and $\delta(P, Q)$ is in some sense the "density" of this point mass.

1.6 The Gravitational Field of Some Elementary Bodies

In this section we show how it is possible to derive expressions for the gravitational field for three elementary examples, without exercises of integration, but rather by applying some symmetry arguments and Gauss' theorem or Proposition 1.3.

Example 1.10 (The Slab) A slab is a mass distribution of constant positive density ρ, between two parallel planes that for convenience we place in a Cartesian system at height $z = 0$ and at height $z = -H$, respectively (see Fig. 1.11)

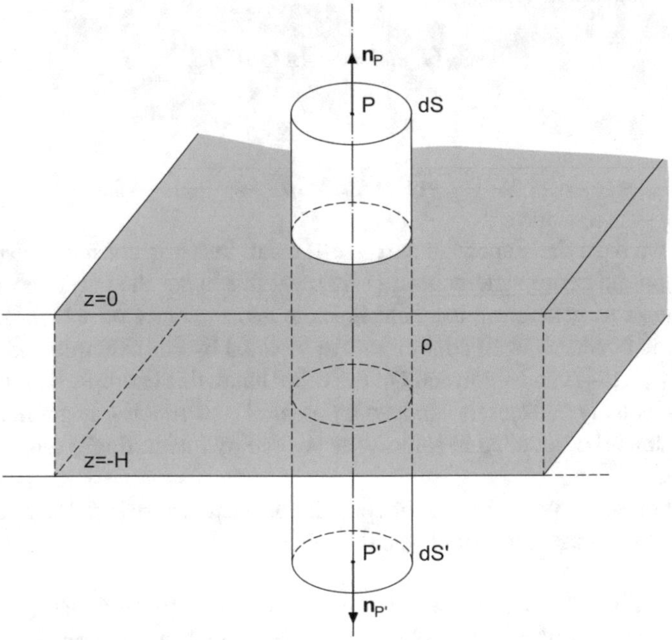

Fig. 1.11 The gravitation field of a slab

For symmetry reasons we see that at any point P in space the vector $\boldsymbol{g}$ has to be parallel to the Z-axis. In fact a rotation around a vertical axis passing through P leaves unchanged the mass distribution and the same has to be for the gravity vector generated by it. So we need to identify only $g(P)$, since then $\boldsymbol{g}(P) = -g(P)\boldsymbol{e}_Z$ in the upper half-space, while $\boldsymbol{g}(P') = g(P')\boldsymbol{e}_Z$ in the lower half-space. Furthermore, if P and P' are placed at the same distance from the slab, since the mass distribution is invariant under a reflection with respect to the middle plane, at height $z = -\frac{H}{2}$, we must also have

$$g(P') = g(P).$$

Now take a small tube of flux as in Fig. 1.11. Since $\boldsymbol{g}$ is directed as $\pm\boldsymbol{e}_Z$, there is no flux through the lateral wall. The flux through the top is $\boldsymbol{g}(P) \cdot \boldsymbol{n}_P dS = -g(P)dS$. Similarly, through the bottom we have the flux $\boldsymbol{g}(P') \cdot \boldsymbol{n}'_P dS' = -g(P')dS' = -g(P)dS$. So the total flux through the cylinder is

$$\Phi = -2g(P)dS. \tag{1.103}$$

On the other hand, the mass cut by the cylinder from the slab is

$$M = \rho dS \cdot H.$$

Hence, by Proposition 1.3,

$$-2g(P)dS = -4\pi G\rho H dS,$$

namely

$$g(P) = 2\pi G\rho \cdot H. \tag{1.104}$$

Therefore we have the somewhat surprising result that g is constant in space. This seems to contradict our statement in (1.80), yet this is not the case, because (1.80) holds if the mass generating the field is finite and the body on which the mass is distributed is bounded. Both conditions are violated by our example and this is the reason why (1.104) can be correct. On the other hand, this example has still another peculiarity, namely one can find indeed a potential u of which $\boldsymbol{g}$ is gradient, yet this potential cannot be bounded at infinity, as we see by integrating g along a parallel to Z.

Despite those features, the slab of constant density, also referred to as a Bouguer slab in literature, has many geophysical applications.

Remark 1.6 Formula (1.104) can also be used to find the gravitation field for a slab where the density ρ is a function of z only. In this case in fact we can simply decompose the slab into horizontal layers of infinitesimal width dz, each of which produces a small attraction

$$dg = 2\pi G\rho(z)dz$$

always directed along $-\boldsymbol{e}_Z$, in the upper half-space. Then the total $\boldsymbol{g}$ in P will be again directed as $-\boldsymbol{e}_Z$ and its modulus will just be given by

$$g(P) = 2\pi G \int_{-H}^{0} \rho(z)dz \quad (1.105)$$

If we write

$$\int_{-H}^{0} \rho(z)dz = H \cdot \overline{\rho}$$

we see that the slab gravitation has the same form as that of a uniform slab with the same width and a mean density $\overline{\rho}$. We emphasize that this is the first time that we meet the general principle that purely vertical variations of density cannot be distinguished from the knowledge of g in the upper half-space. This is the reflection of a much larger indeterminacy of the inversion of gravity, which will be examined in Part III of the book.

Example 1.11 (The Sphere) Consider a homogeneous sphere with a positive mass density ρ. By symmetry under rotations around the center of the sphere we see that $\boldsymbol{g}(P)$ is necessarily directed as $\boldsymbol{e}_r(P)$, i.e., (Fig. 1.12),

$$\mathbf{g}(P) = -g(P)\boldsymbol{e}_r(P); \quad (1.106)$$

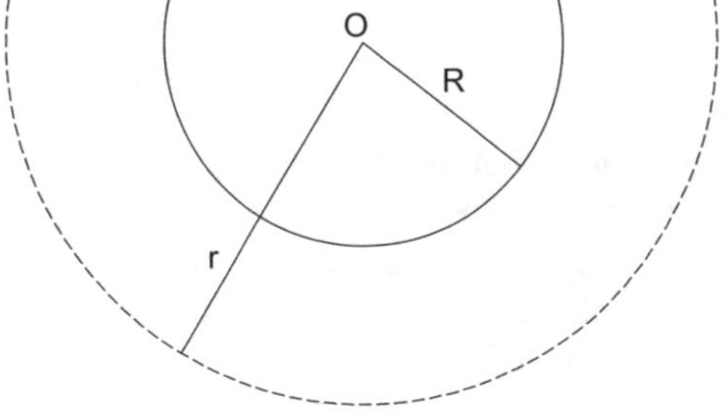

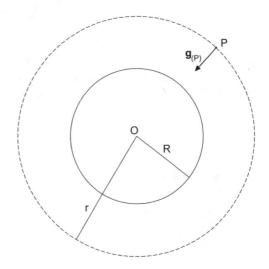

Fig. 1.12 The homogenous sphere and its gravitational field

so we need only to identify $g(P)$. Again by symmetry we know that $g(P)$ can depend only on r_P and therefore it is constant on the sphere of radius r_P. Therefore, the flux of g through such a sphere is

$$\int \boldsymbol{g} \cdot \boldsymbol{n} dS = -g \cdot 4\pi r_P^2 \tag{1.107}$$

On the other hand, the mass contained in the sphere of radius r_P is

$$M(r_P) = \begin{cases} M, & \text{if } r_P > R, \\ \frac{4}{3}\pi r_P^3 \rho, & \text{if } r_P \le R, \end{cases} \tag{1.108}$$

so, by applying Gauss' theorem, we get

$$g(r_P) = \begin{cases} \frac{GM}{r_P^2}, & \text{if } r_P > R, \\ \frac{4}{3}\pi G\rho r_P, & \text{if } r_P \le R. \end{cases} \tag{1.109}$$

Thus, the gravitational field of the sphere, as seen from outside, is just equal to the one of a point mass concentrated at the origin. Furthermore, at the center of the sphere the gravitational attraction goes to zero. This time the gravitational field admits a potential u which we can find by integrating g along a radius from P to infinity (see Fig. 1.13). In fact, knowing that $u(\infty) = 0$, we find

$$u(r_P) = \int_{r_P}^{+\infty} -g(r)dr = \begin{cases} \frac{GM}{r_P}, & \text{if } r_P > R, \\ \frac{GM}{2R}\left(3 - \frac{r_P^2}{R^2}\right), & \text{if } r_P \le R. \end{cases} \tag{1.110}$$

Remark 1.7 As we have noticed, we find that the gravitation of the sphere seen from outside is the same as that of a point of the same mass M, placed at the origin. Now if we consider two spheres of the same density and of radius $R + dR$ and R, respectively, the difference of the attraction of the two when $r_P > R + dR$ is

$$dg = \frac{GdM}{r_P^2} = \frac{4\pi\rho G R^2 dR}{r_P^2}. \tag{1.111}$$

We note that, on the contrary, in the interior of the spherical layer the value of $g(r_P)$, as far as $r_P < R$, is the same for both and therefore their difference is zero.

By the way, formula (1.111) says that again the layer generates the same external field as that of a point of mass dM concentrated at the origin.

It follows that, even if we have a layered sphere with ρ depending on R, $\rho = \rho(R)$ $(0 \le R \le \overline{R})$, the exterior field is always equal to that of a point where all the masses have been concentrated.

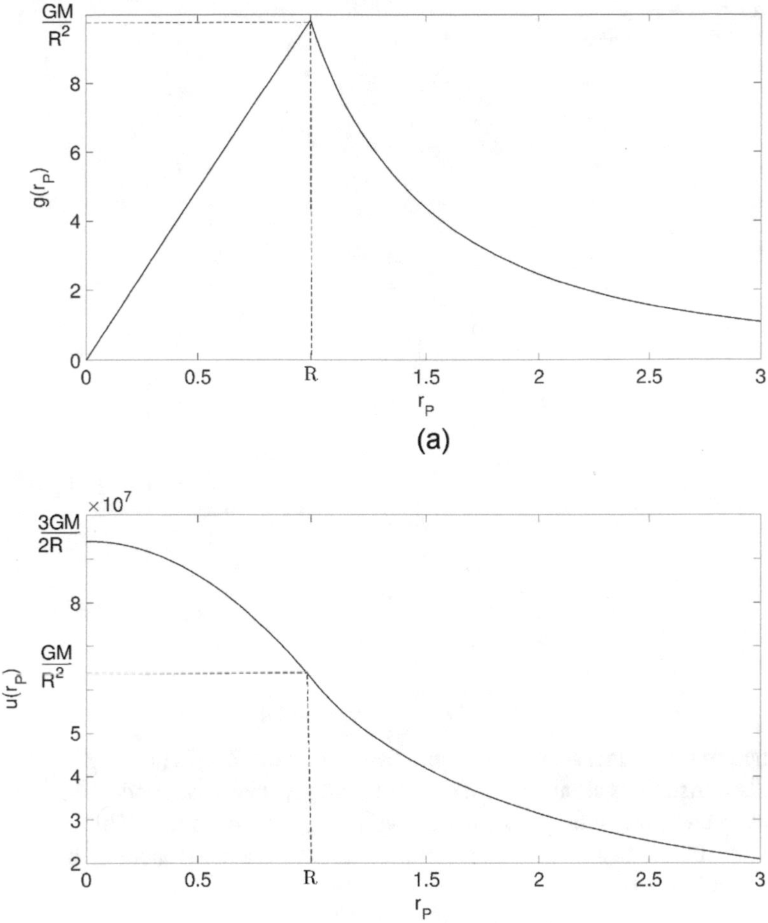

Fig. 1.13 Plots of (**a**) the function $g(r)$, (**b**) the function $u(r)$

Once more, we find that a purely vertical variation of density (this time along the radius, which by the way is the direction of $\boldsymbol{g}$) cannot be distinguished from the knowledge of g on the sphere or outside.

It is interesting to observe that this is a theorem proved by Newton, in a completely different way, to verify that his law could correctly predict the ratio of the accelerations of a mass at the surface of the Earth, supposed to be a spherically layered body, and of the Moon due to its approximate circular motion around the Earth. The success of this computation has been the first direct proof of the validity of Newton's law.

Example 1.12 (The Cylinder) We want to find the gravitational attraction of a homogenous infinite circular cylinder of density ρ. We observe that the mass

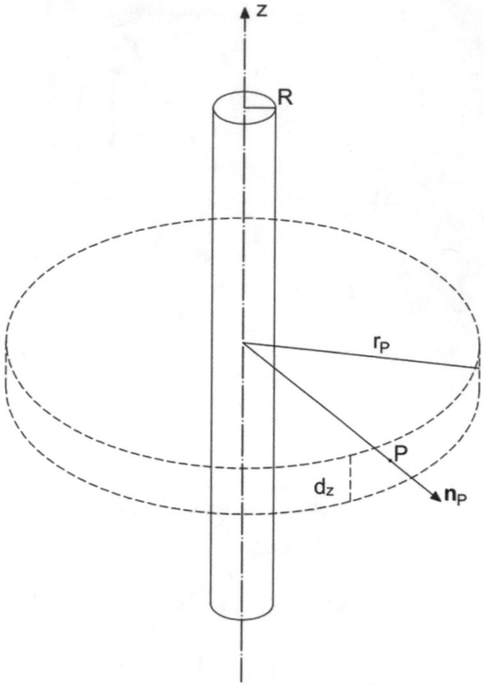

Fig. 1.14 The gravitational attraction of a homogenous cylinder

distribution is now invariant under rotations around the Z-axis (see Fig. 1.14) as well as under translations along the Z-axis. Moreover, it is invariant under reflection with respect to any horizontal plane. It follows that $\boldsymbol{g}$ has to lie in the horizontal plane (i.e., $g_Z = 0$) and that it has to point towards Z, i.e., the component g_λ in cylindrical coordinates has to be zero.

Moreover, clearly g is function of r (the horizontal distance of P from Z) only. Therefore, if we take the flux of $\boldsymbol{g}$ through an infinitesimal cylinder of height dz (see Fig. 1.14), since there is no flux through the top and bottom discs, we have

$$\Phi(\boldsymbol{g}) = -2\pi g r_P dz. \tag{1.112}$$

On the other hand, the mass contained in such a cylinder is

$$M(r_P) = \begin{cases} \pi R^2 dz\rho, & \text{if } r_P > R, \\ \pi r_p^2 dz\rho, & \text{if } r_P \le R. \end{cases} \tag{1.113}$$

Equating $\Phi(\boldsymbol{g})$ and $-4\pi G M(r_P)$, following Proposition 1.3, we get

$$g(r_P) = 2\pi G\rho \begin{cases} \frac{R^2}{r_P}, & \text{if } r_P > R, \\ r_P, & \text{if } r_P \le R. \end{cases} \tag{1.114}$$

Remark 1.8 Also in this case we can observe that the integral, e.g., along a radius r to infinity is not convergent, so that we cannot define a regular potential of (1.39). This is again due to the fact that the total mass of the body is infinite and the body itself is not bounded.

1.7 Exercises

Exercise 1 Consider the function $f(x, y)$ given in the (x, y)-plane by the formula

$$f(x, y) = a_1 + a_2 x + a_3 y + a_4 x^2 + a_5 y^2 + a_6 xy,$$

with $a_1 = 1$, $a_2 = 1$, $a_3 = 2$, $a_4 = 1$, $a_5 = 1$, and $a_6 = 2$.

- Compute analytically the gradient and the second derivatives at the origin;
- use discretization formulas on a grid of 9 points of side 1 at the origin to numerically compute the gradient and compare the result with the analytical one;
- observe that the discretization formulas for the second mixed derivatives at the origin are

$$\left.\frac{\partial^2 f}{\partial x \partial y}(x, y)\right|_{0,0} \approx \frac{f(1,1) - f(1,-1) - f(-1,1) + f(-1,-1)}{4}$$

$$\left.\frac{\partial^2 f}{\partial y \partial x}(x, y)\right|_{0,0} \approx \frac{f(1,1) - f(-1,1) - f(1,-1) + f(-1,-1)}{4}$$

and compute the discrete mixed derivative.

Solution

- The gradient in a Cartesian coordinate system is given by (1.25), therefore in this case it is

$$\nabla f = (a_2 + 2a_4 x + a_6 y)\, \boldsymbol{e}_X + (a_3 + 2a_5 y + a_6 x)\, \boldsymbol{e}_Y,$$

which evaluated at the origin gives $\nabla f|_{0,0} = a_2 \boldsymbol{e}_X + a_3 \boldsymbol{e}_Y = \boldsymbol{e}_X + 2\boldsymbol{e}_Y$.

The second derivatives, which are constant everywhere in the (x, y)-plane, are

$$\frac{\partial^2 f}{\partial x^2}(x, y) = 2a_4 = 2;$$

$$\frac{\partial^2 f}{\partial y^2}(x, y) = 2a_5 = 2;$$

$$\frac{\partial^2 f}{\partial x \partial y}(x, y) = \frac{\partial}{\partial y}(a_2 + 2a_4x + a_6y) = a_6 = 2;$$

$$\frac{\partial^2 f}{\partial y \partial x}(x, y) = \frac{\partial}{\partial x}(a_3 + 2a_5y + a_6x) = a_6 = 2.$$

- The values of the function on a grid of 9 points close to the origin are

$$\begin{bmatrix} f(-1,1) & f(0,1) & f(1,1) \\ f(-1,0) & f(0,0) & f(1,0) \\ f(-1,-1) & f(0,-1) & f(1,-1) \end{bmatrix} = \begin{bmatrix} 2 & 4 & 8 \\ 1 & 1 & 3 \\ 2 & 0 & 0 \end{bmatrix}.$$

In order to compute the gradient at the origin we have to compute the first derivatives at $x = 0$, $y = 0$:

$$\left.\frac{\partial f}{\partial x}(x, y)\right|_{0,0} \approx \frac{f(1,0) - f(-1,0)}{2} = 1$$

$$\left.\frac{\partial f}{\partial y}(x, y)\right|_{0,0} \approx \frac{f(0,1) - f(0,-1)}{2} = 2;$$

therefore, the approximated gradient is $\nabla f|_{0,0} \approx \boldsymbol{e}_X + 2\boldsymbol{e}_Y$.

Note that, in this particular example, the value of the first and second derivatives are computed without any error also from the discretization formulas. This is not true in general. In fact, one can compute the discretization error by means of the Taylor expansion of $f(x + \Delta x, y + \Delta y)$:

$$f(x + \Delta x, y + \Delta y) = f(x, y) + \frac{\partial f(x, y)}{\partial x}\Delta x + \frac{\partial f(x, y)}{\partial y}\Delta y + \\ + \frac{1}{2}\left[\frac{\partial^2 f(x, y)}{\partial x^2}\Delta x^2 + \frac{\partial^2 f(x, y)}{\partial y^2}\Delta y^2 \\ +2\frac{\partial^2 f(x, y)}{\partial x \partial y}\Delta x \Delta y\right].$$

Note that higher order terms are in general equal to 0 in this case since higher order derivatives are zero. We can evaluate the discretization error, for instance of the first derivative with respect to x, as $\frac{f(x+\Delta x, y) - f(x-\Delta x, y)}{2\Delta x}$ by replacing $f(x - \Delta x, y)$ and $f(x + \Delta x, y)$ by their Taylor expansion. Since all the second derivatives are constant we get that the discretization error is zero.

- Applying the discretization formulas we obtain: $\left.\frac{\partial^2 f}{\partial x \partial y}(x, y)\right|_{0,0} = \left.\frac{\partial^2 f}{\partial y \partial x}(x, y)\right|_{0,0}$ $= 2$.

```
% Matlab code for Exercise 1
x = 0; deltax = 1;
y = 0; deltay = 1;
% define the function
a1 = 1; a2 = 1; a3 = 2; a4 = 1; a5 = 1; a6 = 2;
f_xy = inline('1 + 1 * x + 2 * y + 1 * x.^ 2 + 1 * y.^2 +2 .* x .* y','x','y');
% compute analytical first derivatives
dfdx = a2 + 2 * a4 * x + a6 * y
dfdy = a3 + 2 * a5 * y + a6 * x
% compute analytical second derivatives
dfdxdy = a6; dfdydx = a6
% compute discrete first derivative
adfdx = (f_xy( x + deltax,0) -f_xy( x - deltax,0)) / (2 * deltax)
adfdy = (f_xy(0, y + deltay) - f_xy(0, y - deltay)) / (2 * deltay)
% compute discrete second mixed derivative
adf2dxdy = (f_xy( x + deltax, y + deltay) - f_xy( x + deltax, y - deltay) -...
f_xy( x - deltax, y + deltay) + f_xy( x - deltax, y - deltay)) / (4 * deltax *...
deltay)
adf2dxdy = (f_xy( x + deltax, y + deltay) - f_xy( x - deltax, y + deltay) -...
f_xy( x + deltax, y - deltay) + f_xy( x - deltax, y - deltay)) / (4 * deltax *...
deltay)
```

Exercise 2

- Prove that the gradient of a potential $u\ (x, y)$ is a vector always orthogonal to equipotential lines.
- Verify it with the potential

$$u(x, y) = \frac{1}{1 + x^2 + y^2}.$$

  Observe that the gradient is always pointing in the direction of increasing potential.

Solution

- In order to prove that the gradient is always orthogonal to equipotential surfaces we start from (1.17) where we have seen that, by the definition of gradient,

$$\nabla u(P) \cdot dP = u(P + dP) - u(P).$$

  Considering now the potential u (x, y) in the Cartesian plane we have that

$$\nabla u(x, y) \cdot (dx\boldsymbol{e}_X + dy\boldsymbol{e}_Y) = u(x + dx, y + dy) - u(x, y).$$

If now we suppose that the points P and $P + dP$ lie on the same equipotential surfaces, the difference $u(P + dP) - u(P)$ is equal to 0. As a consequence $\nabla u(P) \cdot dP$ is null for every dP tangent to the surface, which means that $\nabla u(P)$ is perpendicular to dP.

- The simplest way to prove that the gradient of $u(x, y) = \frac{1}{1+x^2+y^2}$ is perpendicular to equipotential surfaces is to move from the Cartesian to the polar coordinate system in the plane (x,y), setting

$$\begin{cases} x = r \cos\lambda, \\ y = r \sin\lambda. \end{cases}$$

Then $u(x, y) = \frac{1}{1+x^2+y^2}$ becomes $u(r, \lambda) = \frac{1}{1+r^2}$. Equipotential lines of this simple functions are defined by $r =$ const and infinitesimal vectors lying on an equipotential surface have the form $dP \sim (0, rd\lambda)$. We can now compute the gradient in polar coordinates by recalling equation (1.49):

$$\nabla u(r, \lambda) = \frac{\partial u(r, \lambda)}{\partial r} \boldsymbol{e}_r + \frac{1}{r} \frac{\partial u(r, \lambda)}{\partial \lambda} \boldsymbol{e}_\lambda.$$

This yields

$$\frac{\partial u(r, \lambda)}{\partial r} = \frac{-2r}{(r^2 + 1)^2}$$

and

$$\frac{\partial u(r, \lambda)}{\partial \lambda} = 0.$$

So $\nabla u(r, \lambda) \sim \left(\frac{-2r}{(r^2+1)^2}, 0 \right)$, which is always perpendicular to $dP \sim (0, rd\lambda)$.

Note also that the function $u(r, \lambda) = \frac{1}{1+r^2}$ has its maximum at the origin and is decreasing with r while $\nabla u(r, \lambda)$ is pointing toward $-r$.

Exercise 3 Consider a wire of linear density ρ such that $\rho d\xi = dm$, distributed on the X-axis between $-L$ and L.

Compute its gravitational potential and verify that $u_X(0, y, z) = 0$, $u_Y(x, 0, z) = 0$.

Solution From the definition of the gravitational potential we have:

$$u(x, y, z) = \mu \int_{-L}^{L} \frac{d\xi}{\sqrt{(x - \xi)^2 + y^2 + z^2}},$$

where $\mu = G\rho$. Notice that the following symmetries hold:

$$u(-x, y, z) = u(x, y, z),$$
$$u(x, -y, z) = u(x, y, z),$$
$$u(x, y, -z) = u(x, y, z).$$

Denote $s^2 = y^2 + z^2$ and put $u_X(x, y, z) = \mu F(x, s)$. In order to solve the exercise, we need to compute $F(x, s)$ for $x \geq 0$ and $z \geq 0$:

$$F(x, s) = \int_{-L}^{L} \frac{d\xi}{\sqrt{(x-\xi)^2 + s^2}}, x \geq 0 \quad \text{and} \quad z \geq 0.$$

We substitute now $t = x - \xi$; then we can write:

$$\int \frac{dt}{\sqrt{t^2 + s^2}} = \log\left(\sqrt{t^2 + s^2} + t\right).$$

This can be proven by the change of variable $t = s\tan(u)$. It follows that $dt = s\sec^2(u)du$ and $\sqrt{(s^2 + t^2)} = \sqrt{(s^2\tan^2(u) + s^2)} = s\sec(u)$. Therefore we have to calculate

$$\int \sec(u)\, du.$$

Multiply and divide the integrand by $\tan(u)+\sec(u)$ and denote $p = \tan(u)+\sec(u)$. Since $dp = [\sec^2(u) + \tan(u)\sec(u)]\, du$, we are left with:

$$\int \frac{1}{p} dp = \log(p).$$

Substituting back u, t and ξ one gets the final expression:

$$F(x, s) = \log\left(\sqrt{(L-x)^2 + s^2} + L - x\right) - \log\left(\sqrt{(L+x)^2 + s^2} - L - x\right) =$$
$$= \log \frac{\sqrt{(L-x)^2 + s^2} + L - x}{\sqrt{(L+x)^2 + s^2} - (L+x)}.$$

Chapter 2
The Vertical Gravitational Signal of Homogeneous Bodies, Bounded in the Horizontal Plane

2.1 Outline of the Chapter

The scope of gravity interpretation is to infer a certain distribution of mass anomalies (i.e., inhomogeneities) distributed below the $z = 0$ plane, from gravity measurements on the $z = 0$ plane. Sometimes such anomalies correspond to a geological model in which a certain material, e.g., an iron ore, or a salt dome, with density $\overline{\rho}$, is trapped into a homogeneous background of density ρ_0 (see Fig. 2.1). In such a case a first solution to the problem of finding the anomalous body B is to assume that B has a certain simple shape, for instance a sphere, and then to infer the position and the size of B. This can be done by knowing what is the family of gravitational signals in the Z (vertical) direction generated by this particular shape and comparing it with the observed g_Z, on $z = 0$. We note that, in this classical example, the observed g_Z can be considered as generated by a homogeneous slab of density ρ_0, i.e. with constant magnitude $2\pi G\rho_0 H$, plus the attraction of B, with anomalous density $\rho = \overline{\rho} - \rho_0$.

For this reason we explore the g_Z signal for a number of simple bodies, because such a knowledge will put us in the position to perform a quick look interpretation.

This is why we propose this chapter, which in a sense resembles a collection of integration exercises. The idea is to reduce the dimensions of the integration domain. So, after a first general section showing how to switch from a volume to a surface integral (by first integrating in the Z-direction), we present a number of cases.

If instead we first integrate in a horizontal section of the body and then in the Z variable, we obtain the basis for a technique known in geophysics as *stack of laminas*.

In the last section we introduce the case that the field to be interpreted presents a significant invariance in one direction, which we take as that of the Y-axis.

Accordingly, if we assume that the underlying density is also invariant under translations along the Y-axis, we can derive simplified formulas for the gravitational acceleration g_Z.

F. Sansò, D. Sampietro, *Analysis of the Gravity Field*, Lecture Notes in Geosystems Mathematics and Computing, https://doi.org/10.1007/978-3-030-74353-6_2

This is known as the 2D theory, in which one profile of g_Z data alone (along the X-axis) contains all the information on the underlying geometry.

2.2 Bounded Homogeneous Bodies: Reducing the Vertical Attraction to a Single Layer Potential

Let us consider a situation like the one displayed in Fig. 2.1, where the body B, with density $\overline{\rho}$, is bounded in depth and horizontal extent. If the homogeneous surrounding slab has a known density ρ_0, we can decompose the vertical gravitational signal $g_Z(P)$, produced on the reference plane $z = 0$, into the known contribution of the slab and the differential contribution of B with density

$$\rho = \overline{\rho} - \rho_0, \tag{2.1}$$

so that the problem of deriving $g_Z(P)$ is now reduced to its calculation for the body B with density ρ, embedded in an empty space. A first remark is that, since ρ is constant by hypothesis, in all Newton formulas one can take it out of the integral. Throughout the book we shall put $\mu = G\rho$ whenever it is applicable.

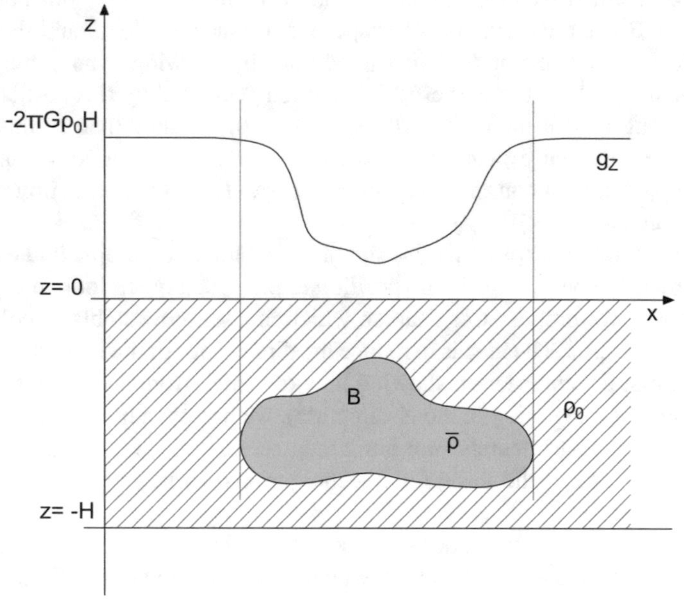

Fig. 2.1 The gravitational signal of a mass anomaly

Noting that the differential identity

$$\nabla \cdot (\boldsymbol{c} f) = \boldsymbol{c} \cdot \nabla f \tag{2.2}$$

holds for any constant vector $\boldsymbol{c}$, we can apply the Gauss theorem to the definition of $g_Z(P)$, obtaining

$$\begin{aligned} g_Z(P) &= \boldsymbol{e}_Z \cdot \boldsymbol{g}(P) = \boldsymbol{e}_Z \cdot \mu \nabla_P \int_B \frac{1}{\ell_{PQ}} dB_Q = \\ &= -\mu \int_B \boldsymbol{e}_Z \cdot \nabla_Q \frac{1}{\ell_{PQ}} dB_Q = -\mu \int_B \nabla \cdot \left(\frac{\boldsymbol{e}_Z}{\ell_{PQ}} \right) dB_Q = \\ &= -\mu \int_S \frac{\boldsymbol{e}_Z \cdot \boldsymbol{n}_Q}{\ell_{PQ}} dS. \end{aligned} \tag{2.3}$$

Thus, we proved the following theorem.

Theorem 2.1 *The vertical attraction $g_Z(P)$ of a bounded homogenous body of density ρ is equivalent to the opposite of the potential generated by a single layer on the boundary S of B, with surface density*

$$\alpha(Q) = \boldsymbol{e}_Z \cdot \boldsymbol{n}_Q. \tag{2.4}$$

Whenever useful in the sequel, we shall adopt the notation

$$U_S(P) = \mu \int_S \frac{\boldsymbol{n}_Q \cdot \boldsymbol{e}_Z}{\ell_{PQ}} dS \tag{2.5}$$

for such a potential.

Remark 2.1 By reasoning as in Sect. 1.5 it is easy to see that we can split S into two parts S_+, S_- such that

$$S_+ \equiv \{Q \in S; \boldsymbol{n}_Q \cdot \boldsymbol{e}_Z > 0\}, \quad S_- \equiv \{Q \in S; \boldsymbol{n}_Q \cdot \boldsymbol{e}_Z < 0\}. \tag{2.6}$$

We note that S could as well contain an area $S_\perp$ where $\boldsymbol{n}_Q \cdot \boldsymbol{e}_Z = 0$, yet such a part of S can never contribute to the integral (2.5).

Furthermore, we observe that (see Fig. 2.2)

$$\boldsymbol{n}_Q \cdot \boldsymbol{e}_Z dS_Q = dS_0 \text{ if } Q \in S_+; \quad \boldsymbol{n}_Q \cdot \boldsymbol{e}_Z dS_Q = -dS_0 \text{ if } Q \in S_-.$$

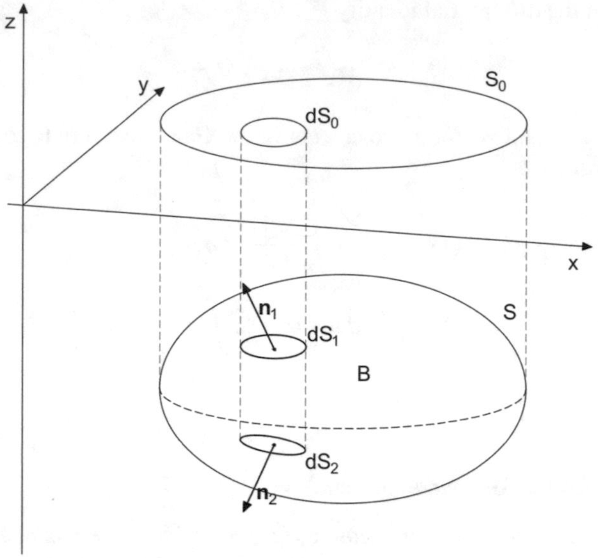

Fig. 2.2 The projection of the integral(2.5) on the (x, y)-plane

Therefore we can write

$$\begin{aligned} U_S(P) &= \mu \int_{S_+} \frac{\boldsymbol{n}_Q \cdot \boldsymbol{e}_Z}{\ell_{PQ}} dS_Q + \mu \int_{S_-} \frac{\boldsymbol{n}_Q \cdot \boldsymbol{e}_Z}{\ell_{PQ}} dS_Q = \\ &= \mu \int_S \left\{ \frac{1}{\ell_{PQ_{0+}}} - \frac{1}{\ell_{PQ_{0-}}} \right\} dS_0 \\ &= U_+(P) - U_-(P), \end{aligned} \tag{2.7}$$

where

$$\begin{cases} \ell_{PQ_{0+}} = [r_{PQ_0}^2 + Z_+^2(Q_0)]^{-1/2}, \\ \ell_{PQ_{0-}} = [r_{PQ_0}^2 + Z_-^2(Q_0)]^{-1/2}, \end{cases} \tag{2.8}$$

with $P \equiv (x, y)$, $Q_0 \equiv (\xi, \eta)$ points on the (x, y)-plane, and

$$r_{PQ_0} = [(x - \xi)^2 + (y - \eta)^2]^{1/2}$$
$$z = Z_\pm(Q_0) = Z_\pm(\xi, \eta) \text{ equations of } S_\pm;$$

let us notice that with this notation we expect $0 > Z_+ > Z_-$.

As formula (2.7) demonstrates, the calculus of the volume integral of $g_Z(P)$ reduces to the calculus of an integral on the (x, y)-plane.

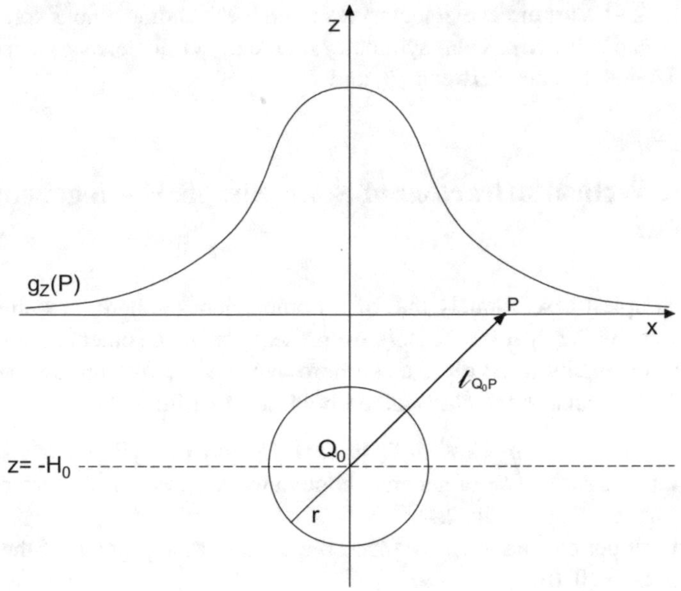

Fig. 2.3 The vertical attraction of a homogeneous sphere

Remark 2.2 One has to remember that the use of (2.7) does not always leads to a true simplification of the calculation of g_Z. This is particularly evident for the case of a homogenous sphere, where the application of (2.7) would need the use of more complicated elliptic integrals, while g_Z can be directly computed from

$$g_Z = \boldsymbol{g} \cdot \boldsymbol{e}_Z,$$

considering that $\boldsymbol{g}$ is known from Example 1.11. Namely, for future use, also looking at Fig. 2.3, we find that

$$\boldsymbol{g}(P) = -\frac{GM}{\ell^3_{Q_0P}} \boldsymbol{\ell}_{Q_0P} \quad (M = \frac{4}{3}\pi R^3 \rho),$$

so that

$$\begin{aligned} g_Z(P) &= -\frac{GM}{\ell^3_{Q_0P}} \boldsymbol{e}_Z \cdot \boldsymbol{\ell}_{Q_0P} = \\ &= -\frac{GM \cdot H_0}{[r^2 + H_0^2]^{3/2}}. \end{aligned} \tag{2.9}$$

Formula (2.9) is the direct elementary solution for the homogenous sphere, which as we can see, displays a circular symmetry on the (x, y)-plane, as g_Z depends only on the horizontal distance between Q_0 and P.

2.3 The Vertical Attraction of Some Simple Homogeneous Bodies

Since the simplest case, namely that of a homogeneous sphere, has already been treated in Remark 2.2 by formula (2.9), we present next some other examples. When proofs are too lengthy to be reproduced here, we send the interested reader to the literature, in particular to the classical textbook of MacMillan [78].

Example 2.1 (The General Vertical Cylinder) Consider a cylinder whose top and bottom is a plane surface S_0 of general shape in the (x, y)-plane and lateral surface parallel to the Z-axis (see Fig. 2.4).

To simplify our calculations, we place the computation point P at the origin of coordinates $O \equiv (0, 0)$.

We can then compute g_Z at any point P by a simple translation of the plane. We can directly exploit the Remark 2.1 to claim that

$$g_Z(P) = U_S(P) = -U_+(P) + U_-(P), \tag{2.10}$$

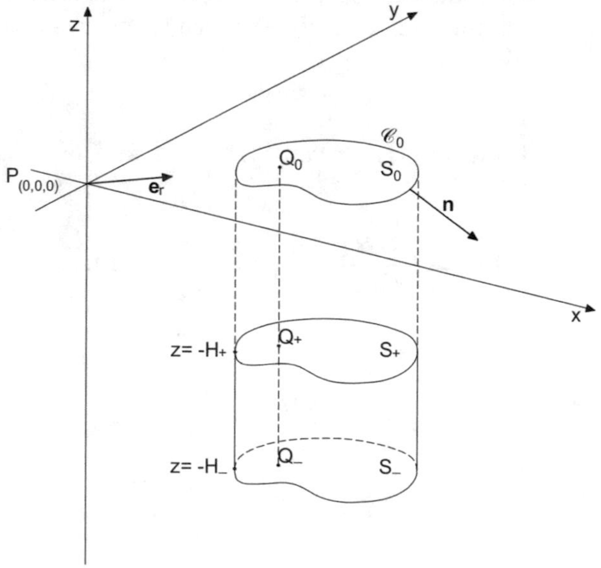

Fig. 2.4 A general vertical cylinder under the (x, y)-plane

where

$$U_\pm(P) = \mu \int_{S_0} \frac{1}{\ell_{PQ\pm}} dS_0, \tag{2.11}$$

$$\ell_{PQ\pm} = \sqrt{r_{PQ_0}^2 + H_\pm^2}. \tag{2.12}$$

Note that we can simply put $r_{PQ_0} = r$ in the integral (2.11) because P coincides with the origin. The relevant integral to be calculated is therefore

$$U(P) = \mu \int_{S_0} \frac{1}{\ell_{PQ}} dS_0, \tag{2.13}$$

with Q in the plane $z = -H$, further specifying then whether we should use $H = H_+$ or $H = H_-$, to compute U_+ or U_-.

The point is that the surface integral (2.13) can be further reduced to a line integral on the contour $\mathscr{C}_0$. This is accomplished through the following proposition.

Proposition 2.1 *Let*

$$\nabla_2 = \boldsymbol{e}_r \frac{\partial}{\partial r} + \frac{1}{r} \boldsymbol{e}_\vartheta \frac{\partial}{\partial \vartheta} \tag{2.14}$$

be the gradient operator in $\mathbb{R}^2$ *(i.e., on the* (x, y)*-plane). Then formally*

$$\nabla_2 \cdot \left(\frac{\ell_{PQ}}{r} \boldsymbol{e}_r \right) = \frac{1}{\ell_{PQ}}, \tag{2.15}$$

where $\boldsymbol{e}_r$ *is the unit radial vector pointing to* Q_0.

Proof One has

$$\nabla_2 \cdot \left(\frac{\ell_{PQ}}{r} \boldsymbol{e}_r \right) = \boldsymbol{e}_r \cdot \nabla_2 \left(\frac{\ell_{PQ}}{r} \right) + \frac{\ell_{PQ}}{r} \nabla_2 \cdot \boldsymbol{e}_r. \tag{2.16}$$

But

$$\boldsymbol{e}_r \cdot \nabla_2 \left(\frac{\ell_{PQ}}{r} \right) = \frac{\partial}{\partial r} \sqrt{\frac{r^2 + H^2}{r}} = -\frac{H^2}{r^2 \ell_{PQ}},$$

$$\nabla_2 \cdot \boldsymbol{e}_r = \frac{\nabla_2 \cdot \boldsymbol{r}}{r} - \boldsymbol{r} \cdot \nabla_2 \frac{1}{r} = \frac{1}{r}.$$

Inserting these expression into (2.16) we get (2.15). □

The relation (2.15) can now be exploited, when S_0 does not include P, by applying the Gauss theorem in $\mathbb{R}^2$, so that, calling $\boldsymbol{n}$ the outward normal to $\mathscr{C}_0$ (see Fig. 2.4) and ds the (positive) arc element, we get

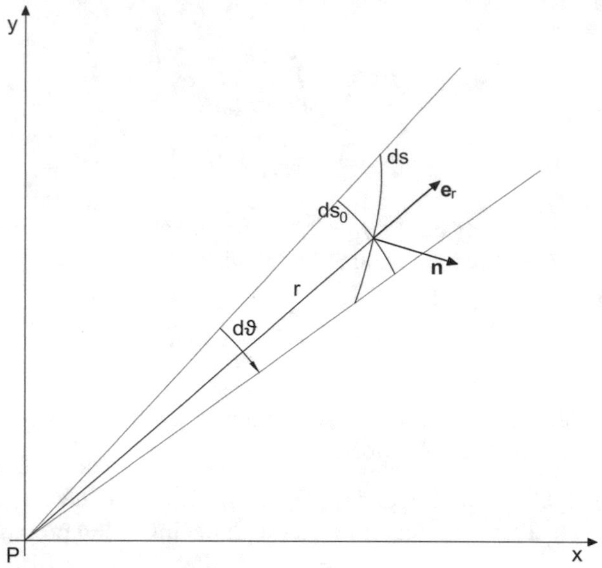

Fig. 2.5 The geometric proof of (2.18)

$$U(P) = \mu \int_{\mathscr{C}_0} \frac{\ell_{PQ}}{r} \boldsymbol{e}_r \cdot \boldsymbol{n}\, ds. \tag{2.17}$$

Finally, let us observe Fig. 2.5, where it is shown that

$$ds_0 = r d\vartheta = \boldsymbol{n} \cdot \boldsymbol{e}_r ds. \tag{2.18}$$

We note that if $\boldsymbol{n}$ would be on the opposite side with respect to that of Fig. 2.5, then $\boldsymbol{e}_r \cdot \boldsymbol{n}$ would change sign becoming negative, but in this case $d\vartheta$ would become negative too, so that (2.18) continues to hold.
By using (2.18), when $P \notin S_0$, in (2.17) we finally get the expression

$$U(P) = \mu \int_{\mathscr{C}_0} \sqrt{r_\vartheta^2 + H^2} d\vartheta, \tag{2.19}$$

where $r = r_\vartheta$ is the equation of $\mathscr{C}_0$ in polar coordinates.

Remark 2.3 Particular care should be taken, case by case, when one applies formula (2.19). For instance, looking at Fig. 2.6a, where P is outside S_0, we see that $\mathscr{C}_0$ has to be followed along $\mathscr{C}_{0+}$ with a positive $d\vartheta$, while the path along $\mathscr{C}_{0-}$ implies a negative $d\vartheta$. Therefore, if we want to work always with a positive $d\vartheta$, we are forced to write (2.19) as

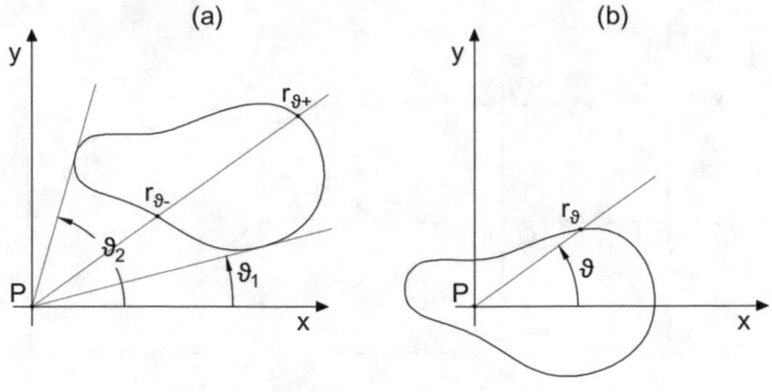

Fig. 2.6 Different limits for the integral (2.19)

$$U(P) = \mu \int_{\vartheta_1}^{\vartheta_2} \left(\sqrt{r_{\vartheta+}^2 + H^2} - \sqrt{r_{\vartheta-}^2 + H^2} \right) d\vartheta.$$

If, on the contrary, P is inside S_0, then $d\vartheta$, at least for the simple geometry in Fig. 2.6, is always positive and one has to write

$$U(P) = \mu \int_0^{2\pi} \sqrt{r_\vartheta^2 + H^2} d\vartheta. \tag{2.20}$$

However, in this case the application of the divergence theorem is less straightforward because the field $\frac{\ell_{PQ}}{r} \boldsymbol{e}_r$ is singular at P, i.e. for $r = 0$. So, instead of one integral only along $\mathscr{C}_0$, one has to isolate P with a small circle of radius ε, $\mathscr{C}_\varepsilon$, compute its contribution to $U(P)$, and then take the limit for $\varepsilon \to 0$. This is illustrated in Fig. 2.6b.
The result in this case is

$$\begin{aligned} P \in S_0, U(P) &= \mu \int_{\mathscr{C}_0} \sqrt{r_\vartheta^2 + H^2} d\vartheta - \mu \lim_{\varepsilon \to 0} \int_{\mathscr{C}_\varepsilon} \sqrt{\varepsilon^2 + H^2} d\vartheta = \\ &= \mu \int_{\mathscr{C}_0} \sqrt{r_\vartheta^2 + H^2} d\vartheta - 2\pi\mu H. \end{aligned} \tag{2.21}$$

Example 2.2 (The Rectangular Parallelepiped) We consider now a particular case of Example 2.1, when the projection of the cylinder on the (x, y)-plane is just a rectangle with vertices the $(x_0 \pm a, y_0 \pm b; a, b > 0)$; in this case therefore the cylinder is a rectangular parallelepiped.

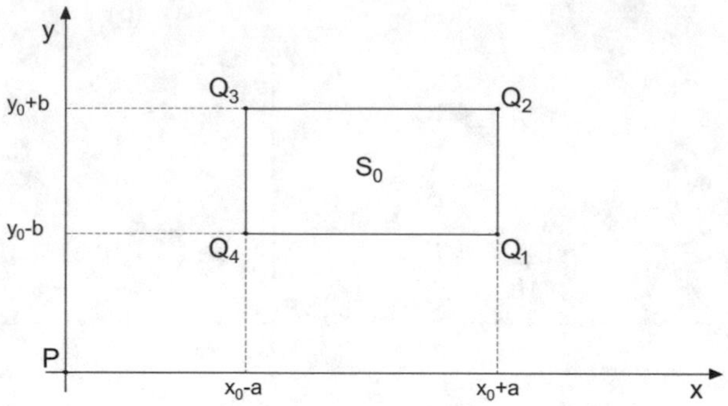

Fig. 2.7 The geometry of the calculation of the attraction of a parallelepiped

In view of the importance of this example, we present exercises (see Exercise 1, 2 and 3) at the end of the chapter that justify the result reported below.

We have to compute (2.18), so, referring to Fig. 2.7, we can start by using (2.20). We define

$$\ell_{Q_i} = \sqrt{r^2_{Q0i} + H^2}, \tag{2.22}$$

namely the distance from the origin of the vertices of a rectangle parallel to S_0, but translated at depth $Z = -H$.

We further define the functions

$$LN_1 = \log \frac{y_0 + b + \ell_{Q_2}}{y_0 - b + \ell_{Q_1}}, \quad LN_2 = \log \frac{x_0 + a + \ell_{Q_2}}{x_0 - a + \ell_{Q_3}},$$

$$LN_3 = \log \frac{y_0 + b + \ell_{Q_3}}{y_0 - b + \ell_{Q_4}}, \quad LN_4 = \log \frac{x_0 + a + \ell_{Q_1}}{x - a + \ell_{Q_4}},$$

$$AT_1 = \arctan \frac{H(y_0 + b)}{(x_0 + a)\ell_{Q_2}} - \arctan \frac{H(y_0 - b)}{(x_0 + a)\ell_{Q_1}},$$

$$AT_2 = \arctan \frac{H(x_0 + a)}{(y_0 + b)\ell_{Q_2}} - \arctan \frac{H(x_0 - a)}{(y_0 + b)\ell_{Q_3}},$$

$$AT_3 = \arctan \frac{H(y_0 + b)}{(x_0 - a)\ell_{Q_3}} - \arctan \frac{H(y_0 + b)}{(x_0 - a)\ell_{Q_4}},$$

$$AT_4 = \arctan \frac{H(x_0 + a)}{(y_0 - b)\ell_{Q_1}} - \arctan \frac{H(x_0 - a)}{(y_0 - b)\ell_{Q_4}}.$$

In terms of these functions, the solution $U(P)$ of (2.19) is given by

$$\begin{aligned} U(P) = \mu\{&(x_0 + a)LN_1 + H \cdot AT_1 \\ &+ (y_0 + b)LN_2 + H \cdot AT_2 \\ &- (x_0 - a)LN_3 - H \cdot AT_3 \\ &- (y_0 - b)LN_4 - H \cdot AT_4\}. \end{aligned} \tag{2.23}$$

Indeed, the expression (2.23) should be used once with $H = H_+$ and then with $H = H_-$ to compute $U_+(P)$ and $U_-(P)$ and finally, according to (2.10), to derive

$$g_Z(P) = -U_+(P) + U_-(P). \tag{2.24}$$

One may doubt that (2.23) does no longer hold when the evaluation point P is inside the rectangle S_0; however, this is not the case, as the reader is invited to prove in Exercise 1.

Example 2.3 (The Circular Cylinder) We consider a vertical cylinder when S_0 is just a circle, so that we are treating again a particular case of Example 2.1. Although one would be inclined to believe that, in this case, it should be easy to compute $g_Z(P)$, the authors are not aware of a simple proof for the closed expression which can be found in [88] and in papers cited therein.

So we shall content ourselves with finding here an algorithm for $g_Z(P)$, based on a series expansion.

We go back to (2.10) and (2.13), so that the problem is to compute

$$U(P) = \mu \int_{S_0} \frac{dS_0}{\sqrt{r_P^2 + r_Q^2 - 2r_P r_Q \cos\vartheta + H^2}} \tag{2.25}$$

and then to specialize the values of H.

Different from Fig. 2.4, we take as S_0 a circle centered at the origin and let R_0 denote its radius. So, looking at Fig. 2.8, calling s the radial integration variable and denoting $r_P = r$ to simplify notation, we have

$$U(r) = \mu \int_0^{R_0} s \int_0^{2\pi} \frac{d\vartheta ds}{\sqrt{s^2 + r^2 + H^2 - 2sr\cos\vartheta}}. \tag{2.26}$$

We put

$$L^2 = L_0^2 + s^2 = r^2 + H^2 + s^2, \quad m = \frac{2sr}{L^2} \tag{2.27}$$

and observe that (2.26) can be written in the form

$$U(r) = \mu \int_0^{R_0} \frac{s}{L} \int_0^{2\pi} \frac{d\vartheta ds}{\sqrt{1 - m\cos\vartheta}}. \tag{2.28}$$

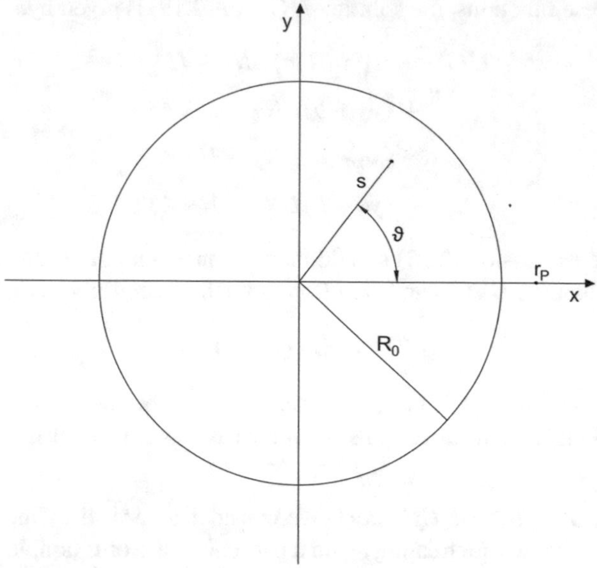

Fig. 2.8 Geometry of the calculation of (2.35)

The inner integral, due to the obvious relation

$$m < 1, \tag{2.29}$$

can be expanded in a binomial series

$$\frac{1}{\sqrt{1 - m\cos\vartheta}} = \sum_{k=0}^{+\infty} \binom{-1/2}{k} m^k \cos^k \vartheta,$$

so that (1.29) becomes

$$U(r) = \mu \sum_{k=0}^{+\infty} c_k \int_0^{R_0} \frac{s}{L} m^k ds, \tag{2.30}$$

with

$$c_k = \binom{-1/2}{k} \int_{k=0}^{2\pi} \cos^k \vartheta d\vartheta. \tag{2.31}$$

Indeed c_k is zero for all odd k; moreover, we have the recursive relation

$$a_\ell = \int_0^{2\pi} \cos^{2\ell} \vartheta d\vartheta = \frac{2\ell - 1}{2\ell} a_{\ell-1}, \ (a_1 = \pi) \tag{2.32}$$

which makes expedient the computation of

$$c_{2\ell} = \binom{-1/2}{2\ell} a_\ell. \tag{2.33}$$

Therefore, we are left with the problem of computing

$$J_\ell(R_0, L_0, r) = \int_0^{R_0} s \frac{s^{2\ell}(2r)^{2\ell}}{L^{4\ell+1}} ds.$$

One can verify that

$$J_\ell = L_0 \left(\frac{2r}{L_0}\right)^{2\ell} I_\ell \left(\frac{R_0}{L_0}\right), \tag{2.34}$$

$$I_\ell \left(\frac{R_0}{L_0}\right) = \sum_{k=0}^{\ell} \binom{\ell}{k} \frac{(-1)^{\ell-k}}{4\ell - 2k - 1} \left[1 - \frac{1}{\left(1 + \frac{R_0^2}{L_0^2}\right)^{2\ell-k-(1/2)}}\right]. \tag{2.35}$$

Summarizing, we get

$$U(r) = \mu \sum_{\ell=0}^{+\infty} c_{2\ell} L_0 \left(\frac{2r}{L_0}\right)^{2\ell} I_\ell \left(\frac{R_0}{L_0}\right), \tag{2.36}$$

with $c_{2\ell}$ given by (2.33) and I_ℓ given by (2.35) respectively.

This is certainly a "solution", though not very practical to work with, especially because of its slow convergence. There are alternative approaches, but none of them is more convenient, so we will not pursue them because later on we will be able to give a simple closed-form solution by Fourier methods.

Here we just give the first two terms of (2.36), which might provide a sufficient approximation when

$$H \gg R_0, \quad r < 2R_0,$$

namely

$$U(r) \cong \mu\pi \left\{2\left(\sqrt{R_0^2 + H^2 + r^2} - \sqrt{H^2 + r^2}\right) + \right.$$
$$\left. + \frac{r^2}{2}\left(\frac{2}{\sqrt{H^2 + r^2}} - \frac{3}{\sqrt{R_0^2 + H^2 + r^2}} + \frac{H^2 + r^2}{[R_0^2 + H^2 + r^2]^{3/2}}\right)\right]. \tag{2.37}$$

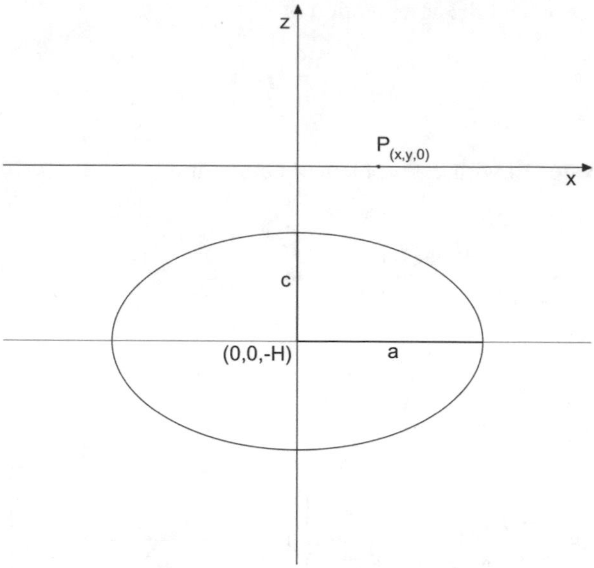

Fig. 2.9 A section of the ellipsoid in the (x, y)-plane

Example 2.4 (The Homogeneous Ellipsoid) The problem of the Newton field of a homogenous ellipsoid at external points has attracted considerable attention, starting with McLaurin's work in the eighteenth century, because of the almost ellipsoidal shape of the Earth. In this context we just wish to report the result of the attraction $g_Z(x, y, 0)$ of a homogenous ellipsoidal body buried under the (x, y)-plane, with the center at the point $(0, 0, -H)$ (see Fig. 2.9). The equation of the ellipsoidal surface is

$$\frac{x^2}{a^2} + \frac{y^2}{b^2} + \frac{(z + H)^2}{c^2} = 1, \tag{2.38}$$

i.e., the ellipsoid is referred to its axes. We will not reproduce here the quite involved proof, for which we send the interested reader to the book by M. D. McMillan [78].

The result, once we shift up the center of coordinates by H along the Z-axis, and supposing $a > b > c$, is

$$\begin{aligned} g_Z(x, y, 0) = & -\frac{4\pi \mu abcH}{(\sqrt{a^2 - c^2})(b^2 - c^2)} \times \\ & \times \left\{ E(\omega_k) - \sqrt{a^2 - c^2}\sqrt{\frac{b^2 + k}{(a^2 + k)(c^2 + k)}} \right\}, \end{aligned} \tag{2.39}$$

where

$$k = k(x, y)$$

is the largest root of the third-order algebraic equation

$$\frac{x^2}{a^2+k} + \frac{y^2}{b^2+k} + \frac{H^2}{c^2+k} = 1, \tag{2.40}$$

$$\omega_k = \arcsin\sqrt{\frac{a^2-c^2}{a^2+k}} \tag{2.41}$$

$$E(\omega_k) = \int_0^{\omega_k} \sqrt{1 - k_c^2 \sin^2\varphi}\, d\varphi = \text{(elliptic integral of second kind)}, \tag{2.42}$$

where $k_e^2 = \frac{a^2-b^2}{a^2-c^2}$.

Since third-order algebraic equations have closed-form solutions and the function $E(\omega)$ is computable by well-known algorithms (see, e.g., [111]), the formulas (2.39) through (2.42) provide the sought solution.

2.4 The Composition Lemma

When we compute the attraction of a body B as a Newton integral, if we decompose B into the union of a.e. non-overlapping subsets

$$B = \bigcap\nolimits_k B_k, \quad (B_k \cap B_j) = \emptyset,\ k \neq j$$

then one can exploit the identity

$$\begin{aligned} g_Z(P) &= -G\int_B \frac{\boldsymbol{e}_Z \cdot \boldsymbol{\ell}_{PQ}}{\ell_{PQ}^3}\rho\, dB_Q = \\ &= -G\sum_k \int_{B_k} \frac{\boldsymbol{e}_Z \cdot \boldsymbol{\ell}_{PQ}}{\ell_{PQ}^3}\rho\, dB_Q \end{aligned} \tag{2.43}$$

to formulate the following lemma.

Lemma 2.1 *The attraction along the Z-axis of the union of almost non-overlapping bodies is the sum of the attractions of the individual bodies.*

This elementary lemma allows to implement an approximate computation for a body of a general, maybe complicated, shape by decomposing it into parts that can be approximated by elementary bodies described in this section.

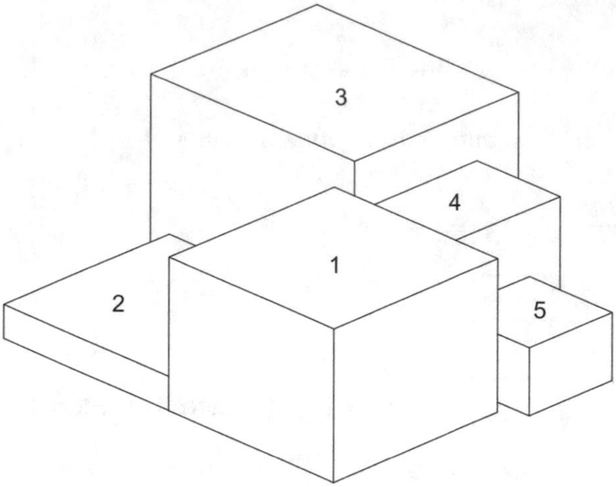

Fig. 2.10 A body B suitably approximated by 5 prisms

In particular, this can be done systematically by using rectangular prisms. Figure 2.10 shows a body that can be approximated by 5 prisms, for each of which we are able to compute $g_Z(P)$ at any point in space.

Sometimes a very easy to compute approximation is provided by a discrete set of point masses, or equivalently by homogenous spheres. This however becomes a very crude approximation if the computation point P is not far enough from the surface of the body B. In Exercise 3 we show how to express the attraction of a parallelepiped Π as a combination of one point mass and a higher-order multipole placed at the center of the parallelepiped itself. Such a formula can speed up the computation at points not too close to Π.

2.5 Stack of Laminas

In the effort of approximating a body B by a combination of elementary bodies with a known attraction formula, a theory was developed by Talwani [144] and subsequently perfected by Plouff [108], where the elementary bodies are vertical prisms with parallel polygonal top and bottom.

In Fig. 2.11 we show how a buried mountain can be approximated by a combination of such prisms, also called a *stack of laminas* by Talwani [144]. So our purpose now is to compute $g_Z(P)$ corresponding to a vertical prism that projects on the (x, y)-plane onto a polygon S_0, like that in Fig. 2.12. As we did before, the computation point P is placed at the origin. Moreover, since in Talwani's approach the order of integration matters, we return to computing $g_Z(P)$ by the Newton integral, according to

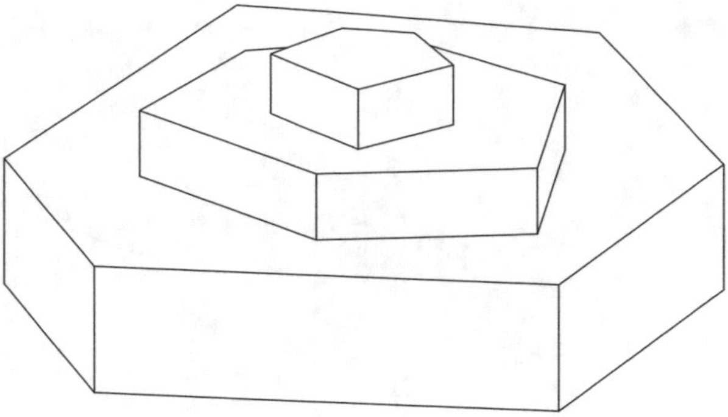

Fig. 2.11 A body decomposed into a stack of polygonal prisms

$$g_Z(P) = \mu \int_{H_-}^{H_+} \zeta \int_{S_0} \frac{dSd\zeta}{[r_{PQ}^2 + \zeta^2]^{3/2}}$$
$$(r_{PQ}^2 = (x_P - \xi)^2 + (y_P - \zeta)^2). \tag{2.44}$$

As we see in (2.39), the first step is to compute the inner integral over S_0, which represents the section of the prism with a horizontal plane at depth ζ. The key idea is that such an integral can be decomposed into the sum of integrals over triangles with the vertices in P, taken as positive when ϑ is increasing from P_{0k} to P_{0k+1} and negative in the opposite case.

In the example we easily see that one can write (see Fig. 2.12)

$$\int_{S_0} \cdot dS = \int_{T_1} \cdot dS + \int_{T_2} \cdot dS - \int_{T_3} \cdot dS - \int_{T_4} \cdot dS + \int_{T_5} \cdot dS. \tag{2.45}$$

With the above choice of sign we see that each integral over a specific triangle T has to be computed with increasing ϑ. So we are brought back to computing the integrals

$$\begin{aligned} g_T(P) &= \int_{\vartheta_1}^{\vartheta_2} \left(\int_0^{r_\vartheta} \frac{rdr}{[r^2 + \zeta^2]^{3/2}} \right) d\vartheta = \\ &= \int_{\vartheta_1}^{\vartheta_2} \left(\frac{1}{\sqrt{r_\vartheta^2 + \zeta^2}} - \frac{1}{|\zeta|} \right) d\vartheta = \\ &= I_T(\vartheta_1, \vartheta_2.\zeta) - \frac{(\vartheta_2 - \vartheta_1)}{|\zeta|}, \end{aligned} \tag{2.46}$$

where the symbols can be read out of Fig. 2.13.

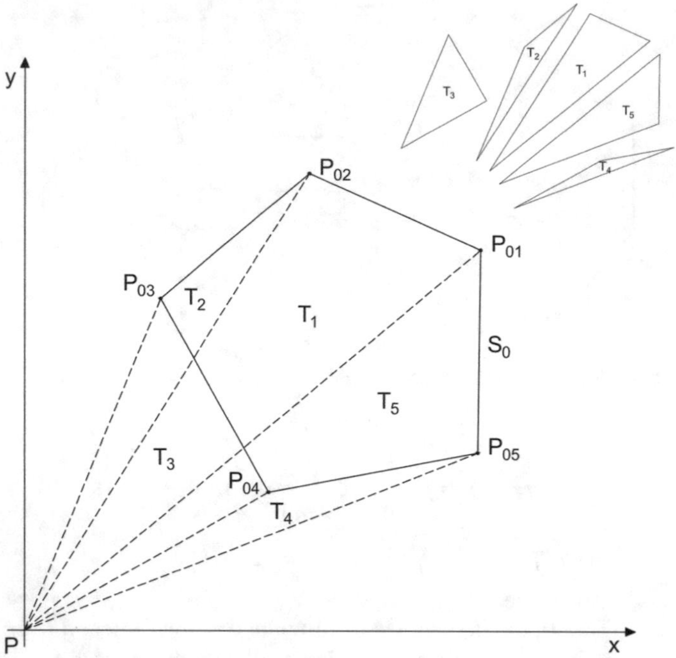

Fig. 2.12 The decomposition of the polygon S_0 into triangles with the vertices at the origin. Note that T_3 is the triangle (P, P_{04}, P_{03})

Indeed, the calculation of $g_Z(P)$ can be similarly decomposed into an algebraic sum of $g_T(P)$, the attraction of the various triangular prisms. Looking at Fig. 2.13, we have that

$$r_\vartheta \sin(\alpha - \vartheta) = D;$$

we also put

$$L^2 = D^2 + \zeta^2.$$

So upon substituting

$$\beta = \alpha - \vartheta, \quad \zeta \cos\beta = Lt$$

into (2.46), we get after some algebra

$$I_T(\vartheta_1, \vartheta_2; \zeta) = \frac{1}{\zeta}\int_{t_2}^{t_1} \frac{dt}{(1-t^2)^{1/2}} = \\ \frac{1}{\zeta}\arctan t_2 - \frac{1}{\zeta}\arcsin t_1, \tag{2.47}$$

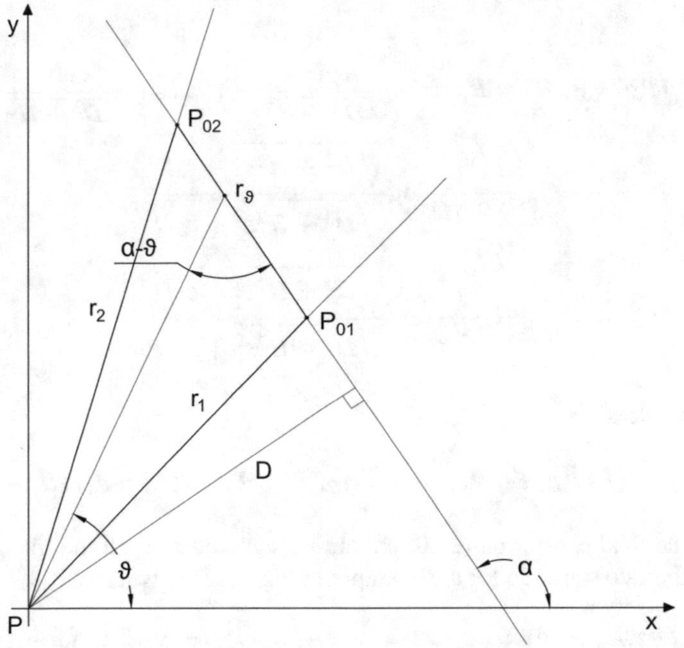

Fig. 2.13 The integration over T_1. Note that the coordinates of P_{01}, P_{02} define the anomalies ϑ_1, ϑ_2 as well as the angle α and the distance D of the origin to the straight line through P_{01}, P_{02}

where

$$t_i = \frac{\zeta}{L} c_i, \; c_i = \cos(\alpha - \vartheta_i) \quad (i = 1, 2). \tag{2.48}$$

Returning to (2.46) and then to (2.39), and observing that $\frac{\zeta}{|\zeta|} = -1$, we find that

$$\begin{aligned} g_T(P) = \mu \int_{-H_-}^{-H_+} \left(\arcsin \frac{\zeta}{\sqrt{D^2 + \zeta^2}} c_2 - \arcsin \frac{\zeta}{\sqrt{D^2 + \zeta^2}} c_1 \right) d\zeta + \\ + \mu (H_- - H_+)(\vartheta_2 - \vartheta_1). \end{aligned} \tag{2.49}$$

Up to here the work of Talwani, who proposed just to discretize the integral in (2.49). Then Plouff [108] realized that it was possible to compute the integral in (2.49) by parts (see also Exercise 4).

Letting

$$F(P_{01}, P_{02}, H) = H \arcsin \frac{c_2 H}{\sqrt{D^2 + H^2}} - H \arcsin \frac{c_1 H}{\sqrt{D^2 + H^2}} + \\ - \frac{1}{2c_2 D} \log \frac{\sqrt{D^2 + H^2 s_2^2} - c_2 D}{\sqrt{D^2 + H^2 s_2^2} + c_2 D} \\ + \frac{1}{2c_1 D} \log \frac{\sqrt{D^2 + H^2 s_1^2} - c_1 D}{\sqrt{D^2 + H^2 s_1^2} + c_1 D},$$

the final result is

$$g_T(P) = \mu \left[F(P_{02}, P_{01}, H_+) - F(P_{02}, P_{01}, H_-) + (\vartheta_2 - \vartheta_1)(H_- - H_+) \right]. \tag{2.50}$$

Then the final expression (2.50) should be summed over all the triangles, with their appropriate signs. So for the example in Fig. 2.12 we will have

$$g_Z(P) = g_{T_1}(P) + g_{T_2}(P) - g_{T_3}(P) - g_{T_4}(P) + g_{T_5}(P). \tag{2.51}$$

Let us remark that, in our example, the last terms in (2.50) will sum to zero because $\sum (\vartheta_{k+1} - \vartheta_k) = 0$, while $H_- - H_+$ is constant. This happens because P lies outside S_0. When on the contrary P is inside S_0, all the rotations $\vartheta_{k+1} - \vartheta_k$ sum to 2π and an additive term $2\pi(H_- - H_+)$ appears in (2.51).

Remark 2.4 (The Götze Paradigm) We start by noting that if we let $H \to \infty$ in the expression

$$F(P_{02}, P_{01}, H) - H(\vartheta_2 - \vartheta_1)$$

we get 0 in the limit. So letting $H_- \to \infty$ in (2.50) we find that a vertical triangular prism, unbounded below, produces in P the attraction

$$g_T(P) = F(P_{02}, P_{01}, H_+) - (\vartheta_2 - \vartheta_1) H_+. \tag{2.52}$$

On the other hand, we know from the general reasoning of Sect. 2.2 that $g_T(P)$ is basically the opposite of $U_T(P)$, where

$$U_T(P) = \mu \int_T \frac{\boldsymbol{n} \cdot \boldsymbol{e}_Z}{\ell_{PQ}} dS_Q; \tag{2.53}$$

the integral (2.53) is taken only over the upper triangular face of the prism, where in the case treated before in this section we have $\boldsymbol{n} \cdot \boldsymbol{e}_Z = 1$.

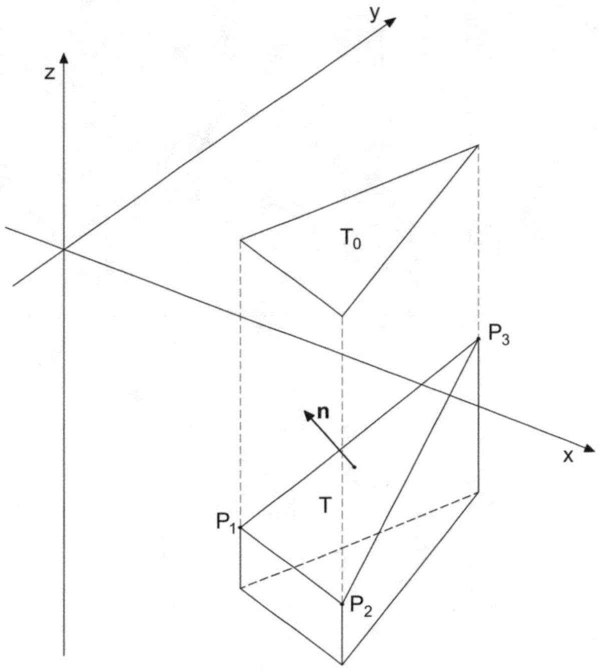

Fig. 2.14 A vertical triangular prism with a sloping upper face

Götze [56] considered the case where the face of the triangle is not parallel to the (x, y)-plane, as shown in Fig. 2.14.

The idea of Götze is to go back to (2.53) and take the constant $\boldsymbol{n} \cdot \boldsymbol{e}_Z$ out of the integral, so that

$$g_T(P) = -\mu U_T(P) = -\mu \boldsymbol{n} \cdot \boldsymbol{e}_Z \int_T \frac{dS}{\ell_{PQ}}. \tag{2.54}$$

Now we observe that

$$U_T'(P) = \mu \int_T \frac{dS}{\ell_{PQ}} \tag{2.55}$$

would be just what we should compute, e.g., by (2.52), if we had a new coordinate system (x', y', z') in which T is parallel to the (x', y')-plane. So we have only to rotate the reference system (x, y, z) keeping the origin fixed, in such a way that the z'-axis becomes parallel to $\boldsymbol{n}$, to be able to compute $U_T'(P)$ and then also

$$U_T(P) = \boldsymbol{n} \cdot \boldsymbol{e}_Z U_T'(P)$$

according to (2.54).

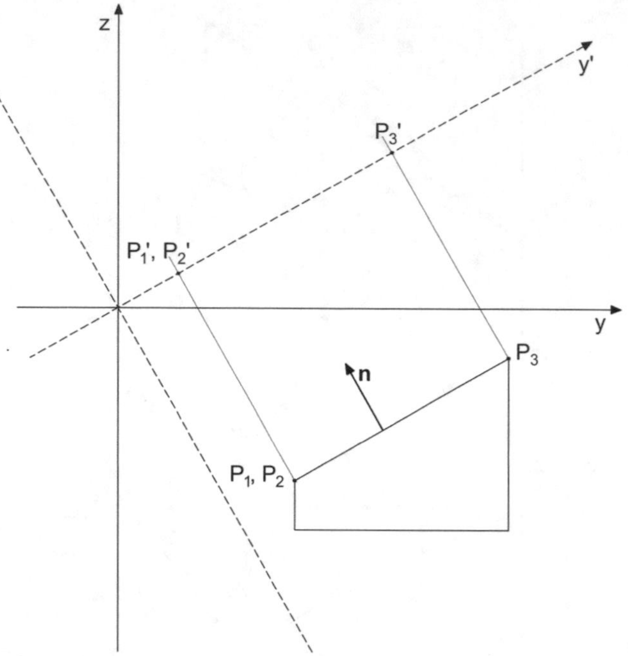

Fig. 2.15 A vertical triangular prism seen from the x-direction

The situation is presented, seen from the x-direction, in Fig. 2.15.

The problem of computing $g_Z(P)$ is reduced to: a) find the rotation from (x, y, z) to (x', y', z') in such a way that z' is parallel to $\boldsymbol{n}$, and b) find the (x', y', z') coordinates of P_1, P_2, P_3.

The solution of this problem is well known. Particulars can be found in literature (e.g., [56] and references therein). Here we observe only that this technique enables one to compute the attraction of any prismatic body by using the formulas seen in this section.

2.6 The Gravitational Signal for 2D Bodies

We call 2D body a body that has a constant density, a finite section in the (x, z)-plane and is invariant under translations in the y-direction (see Fig. 2.16). This is basically a horizontal cylinder with directrix parallel to the y-axis.

We observe that, for obvious symmetry reasons, the resulting vertical attraction g_Z on the (x, y)-plane will necessarily be a function of x only.

In fact, noting that the cylinder B will project onto a strip $I \otimes (-\infty, \infty)$ on the (x, y)-plane, recalling also Theorem 2.1 and formula (2.7), one has

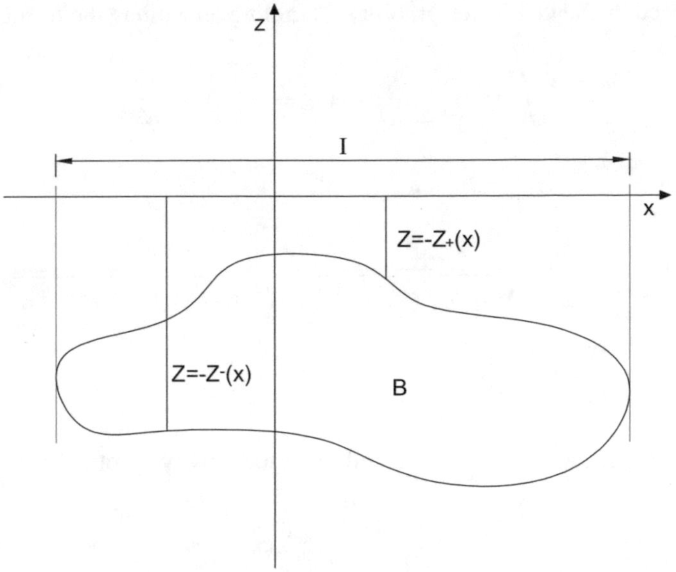

Fig. 2.16 Geometry of the (x, z)-section of a 2D body B. The upper and lower surfaces of B, namely $z = -Z_{\pm}(x)$, are function of x only

$$g_Z(x) = -\mu \int_I \int_{-\infty}^{+\infty} \left\{ \frac{1}{\sqrt{(x-\xi)^2 + \eta^2 + Z_+^2(\xi)}} - \frac{1}{\sqrt{(x-\xi)^2 + \eta^2 + Z_-^2(\xi)}} \right\} d\eta d\xi. \tag{2.56}$$

The point here is that the integral in $d\eta$ can be computed explicitly.

Before we proceed we note that, in general, the function $\{\eta^2 + A^2\}^{-1/2}$ does not have a finite integral over the whole axis. However, if we consider the integrand of (2.56), namely a function of the form

$$f(\eta) = \frac{1}{\sqrt{\eta^2 + A^2}} - \frac{1}{\sqrt{\eta^2 + B^2}},$$

then its asymptotic behaviour at infinity is

$$f(\eta) \sim \frac{1}{2} \frac{B^2 - A^2}{|\eta|^3},$$

which indeed guarantees its integrability. Therefore, exploiting the formula

$$\int_{-L}^{L} \frac{dy}{\sqrt{y^2 + A^2}} = \log \frac{\sqrt{A + L^2} + L}{A^2 + L^2 - L}, \tag{2.57}$$

we find that

$$\lim_{L \to \infty} \int_{-L}^{L} \left\{ \frac{1}{\sqrt{(x-\xi)^2 + Z_+(\xi)^2 + \eta^2}} - \frac{1}{\sqrt{(x-\xi) + Z_-^2(\xi) + \eta^2}} \right\} d\eta =$$

$$= \log \frac{(x-\xi)^2 + Z_-^2(\xi)}{(x-\xi)^2 + Z_+^2(\xi)}. \tag{2.58}$$

So, going back to (2.56), we conclude that the gravity "profile" $g_Z(x)$ is given by

$$g_Z(x) = -\mu \int_I \log \frac{(x-\xi)^2 + Z_-^2(\xi)}{(x-\xi)^2 + Z_+^2(\xi)} d\xi, \tag{2.59}$$

i.e., the problem is reduced to a simple quadrature.

Example 2.5 (The Horizontal Cylinder) This example is legitimately included in this section, but since we already computed the full vector signal of the horizontal cylinder in Example 1.12, it is here much more expedient to use that result instead of formula (2.59).

According to formula (1.114),

$$\boldsymbol{g}(P) = -2\pi\mu \frac{R^2}{r_{0p}} \boldsymbol{n}$$

(see Fig. 2.17), where

$$\boldsymbol{n} = \frac{1}{r_{0p}} (x\boldsymbol{e}_X + H_0 \boldsymbol{e}_Z).$$

Accordingly,

$$g_Z(x) = -2\pi\mu \frac{R^2}{r_{0p}^2} H_0 = -2\pi\mu R^2 \frac{H_0}{H_0^2 + x^2}. \tag{2.60}$$

Example 2.6 (The Trench) The geometry is illustrated, in the (x, z)-plane section in Fig. 2.18.

Obviously, according to our composition lemma (Sect. 2.4), we can compute the signal of the profile of Fig. 2.18 as the difference between a Bouguer plate of

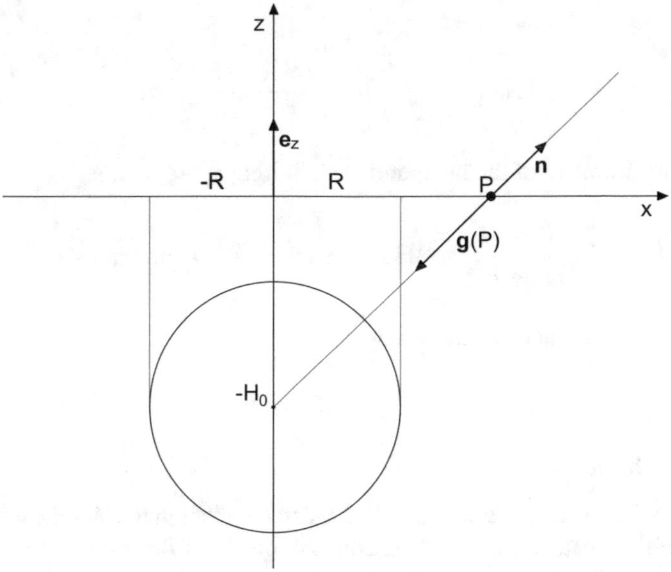

Fig. 2.17 The attraction of a horizontal cylinder with axis buried at depth $-H_0$, radius R and parallel to the y-axis

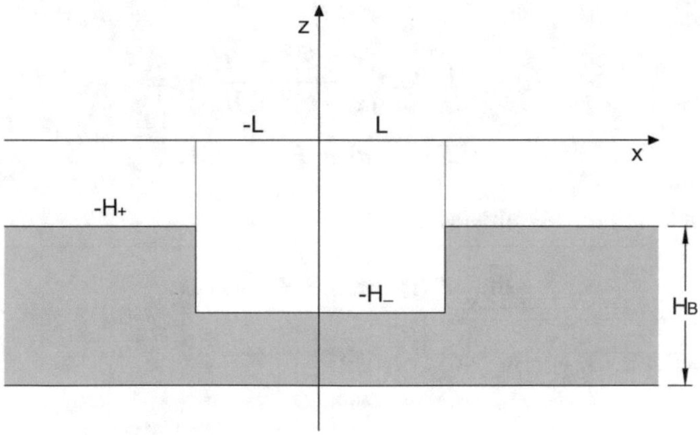

Fig. 2.18 The geometry of the trench

thickness H_B and a rectangular prism with sides $2L$ and $(H_- - H_+)$, namely (see Fig. 2.18 for symbols)

$$g_Z(x) = -2\pi\mu H_B + g_0(x), \tag{2.61}$$

where

$$g_0(x) = \mu \int_{-L}^{L} \log \frac{(x-\xi)^2 + H_-^2}{(x-\xi)^2 + H_+^2} d\xi. \tag{2.62}$$

Since the primitive of the integrand, substituting $x - \xi = t$, is

$$F(t) = t \log \frac{t^2 + H_-^2}{t^2 + H_+^2} + 2H_- \arctan \frac{t}{H_-} - 2H_+ \arctan \frac{t}{H_+} \tag{2.63}$$

the solution (2.62) can be expressed as

$$g_0 = \mu [F(x+L) - F(x-L)]. \tag{2.64}$$

This solves the problem.

Remark 2.5 The result of the Example 2.6 can be generalized significantly. In fact, let us note first that, if instead of placing the origin of the x-axis at the center of the prism, as in Fig. 2.18, we would allow it to range along the x-axis between two general values (a, b), formulas (2.62) and (2.64), that represent the negative of the Z-component of the attraction of the prism, could be written as

$$\begin{aligned} g_0 &= \mu \int_{a}^{b} \log \frac{(x-\xi)^2 + H_-^2}{(x-\xi)^2 + H_+^2} d\xi = \\ &= \mu [F(x-a) - F(x-b)]. \end{aligned} \tag{2.65}$$

Formula (2.63) readily implies that

$$\lim_{t \to \pm\infty} F(t) = \pm\pi (H_- - H_+). \tag{2.66}$$

So, if we let $a \to -\infty$ in (2.65), we get

$$g_0 = \pi\mu (H_- - H_+) - \mu F(x-b) \quad (a \to -\infty), \tag{2.67}$$

and if we let $b \to +\infty$ we obtain

$$g_0 = \pi\mu (H_- - H_+) + \pi\mu F(x-a) \quad (b \to +\infty). \tag{2.68}$$

The two formulas (2.67) and (2.68) give the negative of the g_Z signal of a semi-infinite slab, with the geometry displayed in Fig. 2.19. So with formulas (2.65), (2.67), and (2.68) we are able to compute the g_Z signal of semi-slabs and blocks. We note that these can always be combined to provide the g_Z signal of any 2D body composed of semi-slabs and blocks.

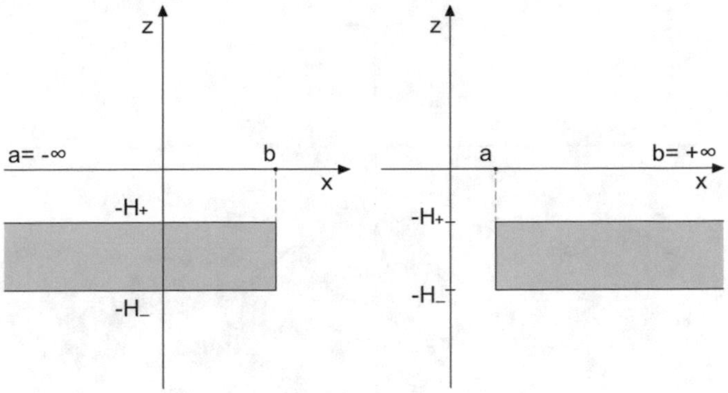

Fig. 2.19 Semi-slab infinite towards $x = -\infty$ (left); semi-slab infinite towards $x = +\infty$ (right)

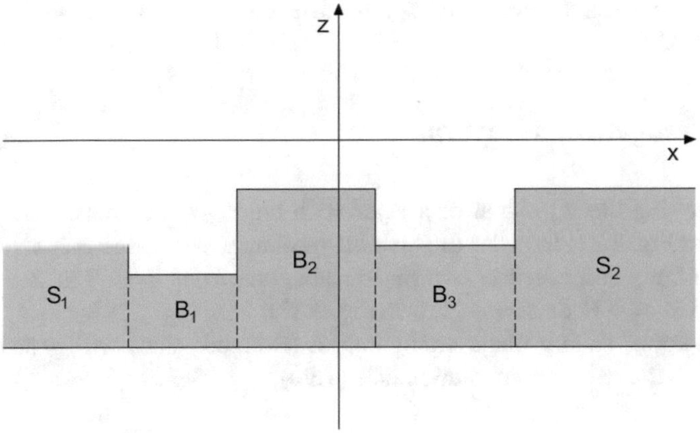

Fig. 2.20 An example of a 2D body with a stepwise upper surface

For instance, looking at Fig. 2.20 we see that g_Z of the whole body, ranging from $x = -\infty$ to $x = +\infty$, can be decomposed as

$$g_Z = g_Z(S_1) + g_Z(B_1) + g_Z(B_2) + g_Z(B_3) + g_Z(S_2)$$

and the above formulas provide a complete solution of the problem.

It goes without saying that the above decomposition is able to approximate a 2D body of arbitrary shape.

The same formulas as in Example 2.6 will be derived from a slightly different prospective in the next section.

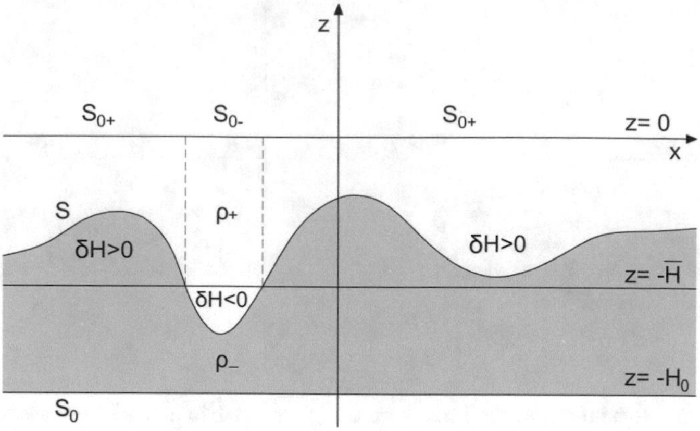

Fig. 2.21 The geometry of two layers: S_0 is the basement at depth H_0; $\rho_\pm$ are the densities of upper/lower layer

2.7 Homogeneous Layers

We are seeking the g_Z signal of a layer with homogeneous mass distribution, as depicted in Fig. 2.21. In order to solve our problem, we notice that we could first consider a body composed of two plane slabs, one from $Z = 0$ to $Z = -\overline{H}$, the other from $Z = -\overline{H}$ to $Z = -H_0$ (see Fig. 2.22).

As we know, such a plane configuration (Bouguer slab) will generate on the $Z = 0$ plane the constant gravitation (see (1.90))

$$g_{0Z} = -\left(2\pi G\rho_+\overline{H} + 2\pi G\rho_-\left(H_0 - \overline{H}\right)\right). \tag{2.69}$$

Now the actual two-layer body in Fig. 2.21 can be thought as the super-position of the two simple slabs considered above, plus a body with varying density $\rho(z)$, according to the following table (2.70), where we denote $\overline{\rho} = \rho_- - \rho_+$:

$$\rho(P) = \begin{cases} 0, & \text{if } -\overline{H} + \delta H < z < 0, \\ +\overline{\rho}, & \text{if } -\overline{H} < z < -\overline{H} + \delta H, \quad (\delta H > 0), \ (x, y) \in S_{0+} \\ 0, & \text{if } -H_0 < z < -\overline{H}, \end{cases}$$
$$\rho(P) = \begin{cases} 0, & \text{if } -\overline{H} < z < 0, \\ -\overline{\rho}, & \text{if } -\overline{H} + \delta H < z < -\overline{H}, \quad (\delta H < 0), \ (x, y) \in S_{0-} \\ 0, & \text{if } -H_0 < z < -\overline{H} + \delta H. \end{cases} \tag{2.70}$$

Such a body, with density 0 or $\pm\overline{\rho}$ for $-H_0 < Z < 0$, is represented in Fig. 2.22.

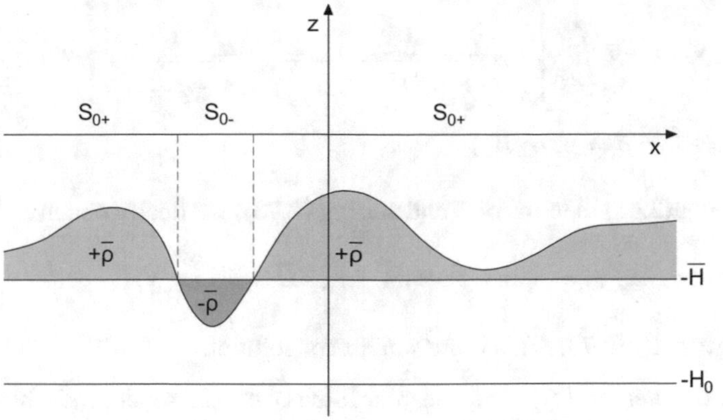

Fig. 2.22 The residual body obtained from Fig. 2.21 by subtracting the two Bouguer slabs

To make such a body compatible with the above description, one has indeed to assume that

$$0 < \overline{H} - \delta H < H_0, \tag{2.71}$$

implying that $\delta H(x, y)$ has to be a bounded function on the plane (x, y), in such a way that S never emerges above $Z = 0$, nor drops below $Z = -H_0$.

Now we can observe that in turn the body in Fig. 2.22 can be considered as the composition of two bodies, one with density $\overline{\rho}$ projecting onto S_{0+} on the (x, y)-plane, the other with density $-\overline{\rho}$, projecting onto S_{0-} on the (x, y)-plane. To each of them we can apply the theory developed in Sect. 2.2, in particular formulas (2.3), (2.5), and (2.7), yielding

$$\begin{aligned} \bar{g}_Z(x, y) = & -G\bar{\rho} \int_{S_{0+}} \left\{ \frac{1}{\sqrt{r_{PQ}^2 + \left(\bar{H} - \delta H\right)^2}} - \frac{1}{\sqrt{r_{PQ}^2 + \bar{H}^2}} \right\} d\xi d\eta + \\ & - G(-\bar{\rho}) \int_{S_{0-}} \left\{ \frac{1}{\sqrt{r_{PQ}^2 + \bar{H}^2}} - \frac{1}{\sqrt{r_{PQ}^2 + \left(\bar{H} - \delta H\right)^2}} \right\} d\xi d\eta \end{aligned} \tag{2.72}$$

where

$$\begin{aligned} & r_{PQ}^2 = (x - \xi)^2 + (y - \eta)^2 \quad (P \sim (x, y), Q \sim (\xi, \eta)), \\ & \delta H = \delta H(\xi, \eta). \end{aligned}$$

Since $S_{0+} \cup S_{0-}$ is the whole (x, y)-plane and due to the particular combination of signs, (2.72) can be recast as the simpler formula

$$\overline{g}_Z(x,y) = -\overline{\mu}\int_{R^2}\left\{\frac{1}{\sqrt{r_{PQ}^2 + \left(\overline{H}-\delta H\right)^2}} - \frac{1}{\sqrt{r_{pq}^2 + \overline{H}^2}}\right\} d\xi d\eta \tag{2.73}$$

$$(\overline{\mu} = G\overline{\rho}, \quad \overline{\rho} = \rho_- - \rho_+).$$

Rewriting (2.69) in terms of $\overline{\rho}$ and adding (2.73), we finally obtain

$$g_Z(x,y) = -2\pi G\rho_- H_0 + 2\pi g\overline{\rho}\overline{H} + \overline{g}_Z(x,y), \tag{2.74}$$

with $\overline{g}_Z$ given by (2.73). This is the sought for solution.

Remark 2.6 (Integrability) The meaningfulness of (2.74) depends heavily onto whether the integral (2.73) is converging or not. The point is that under the assumption (2.71) the integral (2.73) is always convergent. In fact, (2.71) implies that δH is necessarily bounded, for instance

$$|\delta H| < H_0. \tag{2.75}$$

So the integrand in (2.73) is certainly bounded and therefore integrable, on bounded sets of the (x, y)-plane and we need only to prove that (2.75) guarantees the integrability at infinity. To this purpose we note that, for P fixed and $r_{PQ} \to \infty$,

$$\left|\frac{1}{\sqrt{r_{PQ}^2 + \left(\overline{H}-\delta H\right)^2}} - \frac{1}{\sqrt{r_{PQ}^2 + \overline{H}^2}}\right| \sim \frac{1}{2}\frac{\left|\left(\overline{H}-\delta H\right)^2 - \overline{H}^2\right|}{r_{PQ}^3} = \tag{2.76}$$
$$= \frac{1}{2}\frac{|\delta H| \cdot \left|-2\overline{H} + \delta H\right|}{r_{PQ}^3} \le \frac{H_0^2}{r_{PQ}^3};$$

in this last step of (2.76) we have used both (2.71) and (2.75). Since (2.76) is an asymptotic relation when $r_{PQ} \to \infty$, we see that the integrand is integrable at infinity and (2.73) is finite on the (x, y)-plane.

More precise statements can be made under additional assumptions on δH. For instance, we can require that

$$\int_{\mathbb{R}^2} |\delta H| \, d\xi d\eta < +\infty, \tag{2.77}$$

a property that can be taken as the meaning of the statement: $S = \{z = -(\overline{H} + \delta H)\}$ is "supported" by the plane $\{z = -\overline{H}\}$. In addition, we can assume that $0 < a < H(x, y) = -\overline{H} - \delta H(x, y) < H_0$. Under these further conditions we could conclude that:

(i) $|g_Z(x, y)|$ is integrable on $\mathbb{R}^2$,

(ii) $g_Z(P) = g_Z(x, y)$ tends uniformly to 0 when $r_P \to \infty$.

In fact, if we use the Taylor formula for the integrand of (2.73), we find that

$$\left| \frac{1}{\sqrt{r_{PQ}^2 + (\overline{H} - \delta H)^2}} - \frac{1}{\sqrt{r_{PQ}^2 + \overline{H}^2}} \right| = \frac{\left|\overline{H} - \vartheta \delta H\right|}{\left[r_{pq}^2 + (\overline{H} - \vartheta \delta H)^2\right]^{3/2}} |\delta H(Q)| \le$$
$$\le \frac{H_0}{\left|r_{PQ}^2 + a^2\right|^{3/2}} |\delta H(Q)| \tag{2.78}$$

because

$$0 \le \vartheta = \vartheta(P, Q) \le 1.$$

Integrating (2.78) with respect to $d\xi d\eta$ we see that

$$\left|\overline{g}_Z(x, y)\right| \le \overline{\mu} H_0 \int_{\mathbb{R}^2} \frac{|\delta H(Q)|}{\left[r_{PQ}^2 + a^2\right]^{3/2}} d\xi d\eta; \tag{2.79}$$

integrating once more with respect to $dxdy$ and recalling that

$$\int_{\mathbb{R}^2} \frac{dxdy}{\left[r_{PQ}^2 + a^2\right]^{3/2}} = 2\pi \int_0^{+\infty} \frac{rdr}{(r^2 + a^2)^{3/2}} = \frac{2\pi}{a},$$

we conclude that

$$\int_{\mathbb{R}^2} |g_Z(x, y)| \, dxdy \le 2\pi \overline{\mu} \frac{H_0}{a} \int_{\mathbb{R}^2} |\delta H(\xi, \eta)| \, d\xi d\eta \tag{2.80}$$

and so (i) is proved.

As for (ii), we write, using (2.79),

$$|g_Z(x, y)| \le \bar{\mu} H_0 \int_{r_Q \leqslant R} \frac{|\delta H|}{\left(r_{PQ}^2 + a^2\right)^{3/2}} d\xi d\eta +$$
$$+ \bar{\mu} H_0 \int_{r_Q > R} \frac{|\delta H|}{\left(r_{PQ}^2 + a^2\right)^{3/2}} d\xi d\eta = \tag{2.81}$$
$$= \bar{\mu} H_0 (I_1 + I_2).$$

Now it is obvious that

$$I_2 \leq \frac{1}{a^3} \int_{r_q > R} |\delta H| \, d\xi \, d\eta$$

which can be made smaller than an arbitrarily fixed ε, by choosing a suitable fixed R.

Moreover, observing that $r_{PQ} \geq (r_P - R)$ when $r_P > R$ and $r_Q \leq R$, we have

$$I_1 \leq \frac{H_0}{\left[(r_P - R)^2 + a^2\right]^{3/2}} \int_{\mathbb{R}^2} |\delta H| \, d\xi \, d\eta,$$

so that

$$\lim_{r_P \to \infty} I_1 = 0.$$

Using this and the above bound on I_2 in (2.81) we find that, for fixed $\varepsilon > 0$, with a suitable R, we have

$$\overline{\lim_{r_P \to \infty}} \, |g_Z(x, y)| \leq \overline{\mu} H_0 \varepsilon,$$

which proves (ii), thanks to the arbitrariness of ε.

Remark 2.7 (Linearization) When the function $\delta H(\xi, \eta)$ has a general form, calculating the integral (2.73) explicitly becomes too difficult. So it is interesting to find an approximate expression that simplifies the calculation, under suitable assumptions.

Assume that

$$|\delta H| \ll \overline{H}. \tag{2.82}$$

Then we can expand (2.73) by the Taylor formula to the first order, namely, we put

$$\frac{1}{\sqrt{r_{PQ}^2 + \left(\overline{H} - \delta H\right)^2}} - \frac{1}{\sqrt{r_{PQ}^2 + \overline{H}^2}} \cong \frac{\overline{H}}{\left(r_{PQ}^2 + \overline{H}^2\right)^{3/2}} \cdot \delta H. \tag{2.83}$$

Accordingly, $g_Z(x, y)$ can be given by the approximate formula

$$g_Z(x, y) \cong -\overline{\mu} \overline{H} \int_{R^2} \frac{\delta H}{\left(r_{PQ}^2 + \overline{H}^2\right)^{3/2}} d\xi \, d\eta. \tag{2.84}$$

Just to place the above formula in a more geological context, we can think of $\overline{H}$ as the mean Moho depth (loosely speaking, the interface between crust and mantle) and for instance take $\overline{H} = 20\,\text{km}$. This gives a Bouguer plate signal of some 2 Gal, when ρ is the full mean density of the crust. This however is reduced to the order of 400 mGal when $\rho = \overline{\rho}$, the density contrast between crust and upper mantle.

So, if $\frac{|\delta H|}{H} < 10^{-1}$ (i.e., $|\delta H| < 2\,\text{km}$) we have the corresponding signal (2.84) of the order of 40 mGal; it is not difficult to verify, by computing the second-order term in the expansion (2.83), that the neglected part in using (2.84) is of the order of few mGal, which can be considered irrelevant in a first-order approximation.

Another remark about linearization is that the above formula could be applied not only when the support of S is a plane, of depth $\overline{H}$, but also if we think of S being supported by a smoother surface $\tilde{S} = \{z = -\overline{H}(x, y)\}$, always assuming that

$$|\delta H| \ll \overline{H}.$$

In this case formula (2.84) still holds, with $\overline{H} = \tilde{H}$, but now $\tilde{H}$ has to be taken under the integral because it is no longer a constant, i.e., we have

$$g_Z(x, y) \cong \overline{\mu} \int_{R^2} \frac{\tilde{H}\delta H}{\left(r_{PQ}^2 + \tilde{H}^2\right)^{3/2}} d\xi d\eta. \tag{2.85}$$

We close this remark by observing that (2.84) has the mathematical form of a convolution, namely an integral transform with a kernel depending only on $x - \xi$, $y - \eta$, a fact this that will be of importance in the next chapter. On the contrary, the integral transform (2.85) looses this property because there $\tilde{H}$ is a function of (ξ, η).

2.8 Layers in 2D

This is just a specialization of the previous section to the case where

$$\delta H = \delta H(\xi),$$

i.e., the interface S between the layers is constant in the y-direction. In this case the integral (2.73) can be reduced by integrating with respect to $d\eta$, which is performed by using the relation

$$\int \frac{d\eta}{\sqrt{A^2 + \eta^2}} = \log \eta + \sqrt{A + \eta^2} + C, \tag{2.86}$$

in the inner integral of (2.73). This yields

$$\lim_{L\to\infty}\int_{-L}^{L}\left\{\frac{1}{\sqrt{(x-\xi)^2+\left(\overline{H}-\delta H\right)^2+\eta^2}}-\frac{1}{\sqrt{(x-\xi)^2+\overline{H}^2+\eta^2}}\right\}d\eta=$$
$$=\lim_{L\to\infty}\log\frac{\sqrt{(x-\xi)^2+\left(\overline{H}-\delta H\right)^2+L^2}+L}{\sqrt{(x-\xi)^2+\left(\overline{H}-\delta H\right)^2+L^2}-L}\cdot\frac{\sqrt{(x-\xi)^2+\overline{H}^2+L^2}-L}{\sqrt{(x-\xi)^2+\overline{H}^2+L^2}+L}$$
$$=\log\frac{(x-\xi)^2+\overline{H}^2}{(x-\xi)^2+\left(\overline{H}-\delta H\right)^2}. \tag{2.87}$$

Therefore, considering that in the present case $\overline{g}_Z$ is function of x only,

$$\overline{g}_Z(x)=-\overline{\mu}\int_{-\infty}^{+\infty}\log\frac{(x-\xi)^2+\overline{H}^2}{(x-\xi)^2+\left(\overline{H}-\delta H\right)^2}d\xi. \tag{2.88}$$

As we see, we are back to the formula (2.59), when I becomes the full x-axis. This means that we have indeed computed the same $g_Z(x)$, only following a different order of integration.

Example 2.7 (The Ramp) We have already computed in Remark 2.5 the signal of a semi-infinite slab, so we just reproduce this formula, derived from (2.67) with a different reasoning:

$$g_Z(x;b)=-\overline{\mu}\pi\left(H_- - H_+\right)+\overline{\mu}(x-b)\log\frac{(x-b)^2+H_-^2}{(x-b)^2+H_+^2}+$$
$$+2\overline{\mu}H_-\arctan\frac{x-b}{H_-}-2\overline{\mu}H_+\arctan\frac{x-b}{H_+}. \tag{2.89}$$

If we put

$$H_-=H,\ ;H_+=H+dH,\ b=\xi$$

and compute only the first-order differential, then from (2.89) we get the contribution of a semi-infinite slab of infinitesimal width, terminating at ξ (see Fig. 2.23), and after some calculations we get

$$dg_Z(x)=-\overline{\mu}\pi dH+\overline{\mu}2dH\ \arctan\frac{x-\xi}{H}. \tag{2.90}$$

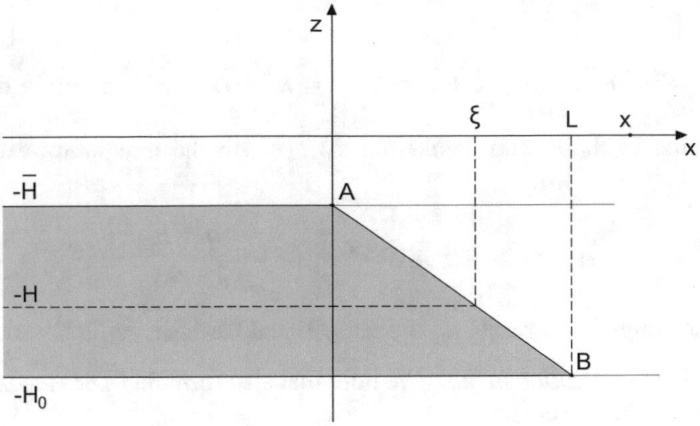

Fig. 2.23 Decomposition of the semi-infinite slab with ramp AB into infinitesimal semi-infinite slabs

Looking at Fig. 2.23, we see that the equation of the ramp can be written as

$$z = -H = -\overline{H} - m\xi \\ \left(m \geq 0,\ \ 0 \leq \xi \leq L,\ \ \overline{H} \leq H \leq H_0 = \overline{H} + mL\right). \tag{2.91}$$

Therefore,

$$dH = md\xi,$$

so (2.90) becomes

$$dg_Z\,(x) = -2\overline{\mu}md\xi\left(\frac{\pi}{2} - \arctan\frac{x-\xi}{H}\right) = \\ = -2\overline{\mu}md\xi\,\arctan\frac{\overline{H}+m\xi}{x-\xi}; \tag{2.92}$$

this relation has now to be integrated in $d\xi$ from O to L to get the full gravitation $g_Z(x)$. This calculation is laborious, however it can be done by parts, yielding

$$g_Z\,(x) = -2\overline{\mu}mL\,\arctan\frac{H_0}{x-L} + \\ + \frac{\overline{\mu}m}{\alpha}\log\frac{\alpha L^2 - 2\beta L + \gamma}{\gamma} + \frac{2\overline{\mu}m}{\alpha\Delta}\arctan\frac{\Delta L}{\gamma - \beta L} \tag{2.93}$$

where

$$\alpha = 1 + m^2, \quad \beta = x + \overline{H}m, \quad \gamma = x^2 + \overline{H}^2, \quad \Delta^2 = \alpha\gamma - \beta^2.$$

A warning to those who would like to perform the integration exercise: the relation

$$\arctan u + \arctan v = \arctan \frac{u + v}{1 - uv}$$

was used to simplify the result. So the ramp signal has been explicitly found.

Remark 2.8 (Linearization in 2D*)* We note that also formula (2.88) lends itself to a linearization in δH when

$$|\delta H| \ll \overline{H};$$

the differentiation work is easy and yields

$$g_Z(x) \cong -2\overline{\mu}\overline{H} \int_{+\infty}^{-\infty} \frac{\delta H(\xi)}{(x - \xi)^2 + \overline{H}^2} d\xi. \tag{2.94}$$

2.9 Exercises

Exercise 1 Prove that Eq. (2.23) holds even if the point P is inside the S_0 rectangle.

Solution We begin by generalizing (2.23) by considering a generic P of coordinates (x_P, y_P, z_P). This can be done by applying a translation of the reference system by a quantity $-x_P$, $-y_P$ and $-z_P$ in the x, y, z direction respectively, i.e., by replacing in (2.22) and (2.23) $x_0 - x_P$ by x_0, $y_0 - y_P$ by y_0 and $H - z_P$ by H.

In order to prove (2.23) we have to take P such that $x_0 - a < x_P < x_0 + a$ and $y_0 - b < y_P < y_0 + b$ and compute the integral (2.19). This can be done by moving to a Cartesian reference system and considering that

$$\theta = \begin{cases} \arctan\left(\frac{y}{x}\right) \text{ if } x > 0 \text{ and } y > 0, \\ \arctan\left(\frac{y}{x}\right) + 2\pi \text{ if } x > 0 \text{ and } y < 0, \\ \arctan\left(\frac{y}{x}\right) + \pi \text{ if } x < 0, \end{cases}$$

and therefore

$$d\theta = -\frac{y}{x^2 + y^2} dx + \frac{x}{x^2 + y^2} dy.$$

The integral can be split into four parts according to the angle subtended by each segment which forms the rectangle. So considering for instance the angle θ_1 subtended by the segment $\overline{Q_{04}Q_{01}}$ we have that $dy = 0$ and therefore we have to calculate

$$U(P)_{\overline{Q_{04}Q_{01}}} = -\mu y \int_{Q_{04}}^{Q_{01}} \frac{\sqrt{H^2 + x^2 + y^2}}{\left(x^2 + y^2\right)} dx$$

Substituting $u = \arctan \frac{x}{\sqrt{H^2+y^2}}$ we obtain:

$$U(P)_{Q_{04}Q_{01}} = -\mu \left(H^2 y + y^3\right) \int_{Q_{04}}^{Q_{01}} \frac{\cos(u)}{\left[\sin^2(u) - 1\right] \left[H^2 \sin^2(u) + y^2\right]} du.$$

Substituting again $s = \sin(u)$ we obtain

$$U(P)_{\overline{Q_{04}Q_{01}}} = \mu \left(H^2 y + y^3\right) \int_{Q_{04}}^{Q_{01}} \frac{1}{\left(s^2 - 1\right) \left(H^2 s^2 + y^2\right)} ds.$$

The above integral can be decomposed by means of partial fractions in the following way:

$$U(P)_{\overline{Q_{04}Q_{01}}} = I_1 + I_2 + I_3,$$

where

$$I_1 = -\mu \frac{H^4 y + H^2 y^3}{y^2 \left(H^2 + y^2\right)} \int_{Q_{04}}^{Q_{01}} \frac{1}{\frac{H^2 s^2}{y^2} + 1} ds,$$

$$I_2 = -\mu \left(\frac{H^2 y}{2} + \frac{y^3}{2}\right) \int_{Q_{04}}^{Q_{01}} \frac{1}{s\left(H^2 + y^2\right) + H^2 + y^3} ds,$$

$$I_3 = \mu \left(\frac{H^2 y}{2} + \frac{y^3}{2}\right) \int_{Q_{04}}^{Q_{01}} \frac{1}{s\left(H^2 + y^2\right) - H^2 - y^2} ds.$$

Recalling that $\int \frac{1}{x^2+1} dx = \arctan x$ and $\int \frac{1}{x} dx = \log x$, we obtain

$$I_1 = -\mu \frac{H^4 + H^2 y^2}{y\left(H^2 + y^2\right)} \arctan\left(\frac{Hs}{y}\right)\Bigg|_{Q_{04}}^{Q_{01}},$$

$$I_2 = -\mu \left(\frac{H^2 y}{2} + \frac{y^3}{2}\right) \log\left(s\left(H^2 + y^2\right) + H^2 + y^2\right)\Bigg|_{Q_{04}}^{Q_{01}}$$

and

$$I_3 = \mu \left(\frac{H^2 y}{2} + \frac{y^3}{2} \right) \log \left(s \left(H^2 + y^2 \right) - H^2 - y^2 \right) . \Bigg|_{Q_{04}}^{Q_{01}} .$$

Substituting back u and x we finally obtain:

$$U(P)_{\overline{Q_{04} Q_{01}}} = -y \log \left(\sqrt{H^2 + x^2 + y^2} + x \right) H \arctan \frac{Hx}{\left(y\sqrt{H^2 + x^2 + y^2} \right)} \Bigg|_{Q_{04}}^{Q_{01}} .$$

Proceeding in a similar way with the three remaining sides of the rectangle we obtain relation (2.23).

Note that the above solution is valid everywhere except when the point P lies on a corner of the rectangle, i.e., when both H and one among x or y is zero. In that case we have that the argument of arctan is $\frac{0}{0}$ and therefore in principle the proper limit should be considered. However, when dealing with real-life problems it is easier, from the computational point of view, to add a suitable small value ε (e.g., 10^{-5} m) thus avoiding the instability at the rectangle corner and keeping a unique formula for the computation.

Exercise 2 Compute the gravitational field at the point of coordinates (0,0,0) of a buried ellipsoidal salt dome with density equal to 2200 kg/m^3 with respect to a background sedimentary layer with constant density equal to 2670 kg/m^3 supposing the following parameters:

- $a = 102$ m;
- $b = 101.1$ m;
- $c = 101$ m;
- Coordinates of the ellipsoid center $(0, 0, -104)$;
- $\Delta\rho = 2670 - 2200\,\text{kg/m}^3$.

Compare the result with the one for an homogeneous sphere with the same mass and the same density.

Solution We begin by simplifying (2.40), taking $x = y = 0$. In this simple case we get

$$\frac{H^2}{c^2 + k} = 1,$$

which gives

$$k = H^2 - c^2 = 615.$$

With this value we compute ω_k by means of Eq. (2.41), which gives 0.1362. Computing now the elliptic integral of second kind we obtain $E(\omega_k) = 0.1358$ and finally a $g_Z(0, 0, 0) = 1 : 26$ mGal. The total mass anomaly of the above ellipsoid with respect to the background is given by $M = \frac{4}{3}\pi abc\Delta\rho$, therefore the gravitational effect of a uniform sphere with the same mass gives $g_{sZ} = \frac{GM}{104^2} =$ 1.27 mGal.

```
%Matlab code for Exercise 2
a = 102; b = 101.1; c = 101; %ellipsoid axis [m]
H = a + 2; %ellipsoid center depth [m]
x = 0; y = 0; %planimetric coordinates of P [m];

G = 6.7e-11; % Universal gravitational constant [m3kg-1 s-2];
mu_b = 2670; % background density [kgm-3];
mu_s = 2200; % background density [kgm-3];
dmu = 2670 - 2200; %density contrast [kgm-3];

dM_e = dmu.*4 3 * pi * 101 * 101.1 * 102; % total mass of the ellipsoidal
body [kg];

k = H.*H - c.*c; k2=(a.*a-b.*b)./(a.*a-c.*c);
wk = asin( sqrt(a.*a-c.*c)./(a.*a+k));
[~, a.*a-c.*c,~] = elliptic12(wk, k2); %elliptic integral

g_z=-4.*pi.*dmu.*G.*a.*b.*c.*H.(sqrt(a.*a-c.*c).*(b.*b-c.*c)).*(E-sqrt(a.*a-c.*c).
*sqrt((b.*b+k).((a.*a+k).*(c.*c+k)))).*1e5 %[mGal]
g_sz = G .* dM_e . *H.*H*1e5 %[mGal]
```

Exercise 3 Compute the approximate formula for the gravitational potential of a right rectangular prism given as a combination of one point mass and a higher–order multipole placed at the centered of the prism.

Compare the gravitational field at point P obtained by the exact analytical solution and with that obtained by the approximated formulas of a prism with the following characteristics:

$\Delta x = \Delta y = 100$ m;
$\Delta z = 500$ m;
coordinates of the center of the prism: (0,0,250);
$P = (0, 101, 500)$ and $P = (0, 500, 500)$.

Solution

- In order to compute the gravitational acceleration we start form recalling the expression for the gravitational potential U of the right rectangular prism of homogeneous mass density at a point P of coordinates (x, y, z) as given by the Newton law in Cartesian coordinates:

$$U(P) = \mu \int_{z_1}^{z_1} \int_{y_1}^{y_1} \int_{x_1}^{x_1} \frac{dxdydz}{\ell_{PQ}},$$

where x_1, x_2, y_1, y_2, and z_1, z_2 define the vertices of the prism. A simpler integral can be obtained by replacing the above integrand, i.e., $\frac{1}{\ell_{PQ}}$, with its Taylor expansion about the geometric center C of the prism given by $x_C = \frac{x_1+x_2}{2}$, $y_C = \frac{y_1+y_2}{2}$, and $z_C = \frac{z_1+z_2}{2}$, i.e., with

$$\frac{1}{\ell_{PQ}} = \sum_{i,j,k} I(P) \left[x(Q) - x_C\right]^i \left[y(Q) - y_C\right]^j \left[z(Q) - z_C\right]^k,$$

and $I(P)$ formally given by

$$I(P) = \frac{1}{(i+j+k)!} \frac{\partial^{i+j+k}}{\partial x^i \partial y^i \partial z^k} \frac{1}{\ell_{PC}}.$$

Inserting the above Taylor expansion in Newton's law we have:

$$U(P) = \mu \int_{z_1}^{z_1} \int_{y_1}^{y_1} \int_{x_1}^{x_1} \sum_{i,j,k} I \left[x(Q) - x_C\right]^i \left[y(Q) - y_C\right]^j \left[z(Q) - z_C\right]^k dxdydz.$$

We see that the computation of the potential $U(P)$ reduces to the computation of a set of integrals of the type $\int_{x_1}^{x_2} [x(Q) - x_C]^i \, dx$, which gives 0 if i is odd and $\frac{(x_2-x_1)^{i+1}}{2^i(i+1)}$ if i is even. Therefore the gravitational potential of the prism, neglecting terms of order higher than 2, is

$$U(P) \approx \mu \Delta x \Delta y \Delta z \left[\frac{1}{\ell_{PC}} + \frac{3(x_C - x)^2 - \ell_{PC}^2}{24\ell_{PC}^5} \Delta x^2 \right.$$
$$\left. + \frac{3(y_C - y)^2 - \ell_{PC}^2}{24\ell_{PC}^2} \Delta y^2 + \frac{3(z_C - z)^2 - \ell_{PC}^2}{24\ell_{PC}^2} \Delta z^2 \right].$$

First of all, note that $\mu \Delta x \Delta y \Delta z = GM$, where M is the total mass of the prism, given by its density multiplied by its volume. As we have already seen several times, the derivative of the first term

$$\frac{\partial}{\partial z} \frac{Gm}{\ell_{PC}} = GM \frac{z - z_C}{\ell_{PC}^3},$$

which is equivalent to the effect of a point mass placed in the geometric center of the prism. The derivative of the second term:

$$GM\Delta x^2 \frac{\partial}{\partial z}\frac{3\,(x_C-x)^2-\ell_{PC}^2}{24\ell_{PC}^5} = GM\Delta x^2 \frac{(z_C-z)[15(x_C-x)^2+4\ell_{PC}]}{24\ell_{PC}^7}.$$

The effect of the third term can be found in a similar way, while the derivative of the last term is

$$(z_C-z)\,\frac{2\,(z_C-z)^2-3\left(x^2+y^2\right)}{8\ell_{PC}^7},$$

thus giving the sought result.

- Substituting the provided values in the prism formula and the approximate formula the following results are obtained:
  $\Delta g\,(0, 101, 500) = -1.481\,\text{mGal}$;
  $\Delta\,\tilde{g}\,(0, 101, 500) = -1.756\,\text{mGal}$;
  $\Delta\,\tilde{g}_S\,(0, 101, 500) = -1.136\,\text{mGal}$;
  $\Delta\,g\,(0, 500, 500) = -0.105\,\text{mGal}$;
  $\Delta\,\tilde{g}\,(0, 500, 500) = -0.100\,\text{mGal}$;
  $\Delta\,\tilde{g}_S\,(0, 500, 500) = -0.127\,\text{mGal}$;
  where $\Delta\,g$ is the result obtained with the complete prism formula, $\Delta\,\tilde{g}$ is the result obtained with the approximate formula presented above and $\Delta\,\tilde{g}_S$ is the result obtained by concentrating the whole prism mass in its geometrical center. The results show that the multipole expansion drastically improves the accuracy with respect to the point mass approximation, reducing the relative error from about 23% to 18% for the observation point close to the prism and from 21% to just 4% for the others point. Note also how the relative accuracy of the approximate formula increase when the distance between the prism and P increases.

Exercise 4 Prove the solution of integral in Eq. (2.49).

Solution To this end basically we have to compute

$$I = \int_b^a \arcsin\frac{\xi}{\sqrt{D^2+\xi^2}}d\xi.$$

This can be done by applying the well-known formula of integration by parts $\int f dg = fg - \int g d\,f$, taking

$$f = \arcsin\frac{\xi}{\sqrt{D^2+\xi^2}},$$

$dg = d\xi$, and

$$d\,f = \frac{D}{D^2+\xi^2}d\xi.$$

This yields

$$I = \xi \arcsin \frac{\xi}{\sqrt{D^2 + \xi^2}} - \int \xi \frac{D}{D^2 + \xi^2} d\xi.$$

Denote

$$I_1 = \int \xi \frac{D}{D^2 + \xi^2} d\xi.$$

Passing to the variable $u = D^2 + \xi^2$, we have

$$I_1 = \frac{D}{2} \int \frac{1}{u} du$$

which gives simply $\log u$. Substituting back ξ, we have

$$I = \xi \arcsin \frac{\xi}{D^2 + \xi^2} - \frac{D}{2} \log \left(D^2 + \xi^2 \right)$$

The solution of Eq. (2.49) can be easily obtained by computing all the terms of the integral between $-H_-$ and $-H_+$.

Exercise 5 Show that the attraction along the X-axis of an infinite semi-cylinder with the flat face in the $(x,\ y)$-plane can be expressed in terms of elementary functions.

Solution First, observe that, by symmetry, we only need to compute the attraction for $x > 0$. Recalling back Eq. (2.59), we have to compute the integral

$$g_Z(x) = -\mu \int_I \log \frac{(x - \xi)^2 + Z_-^2(\xi)}{(x - \xi)^2 + Z_+^2(\xi)} d\xi$$

with $Z_-(\xi) = \sqrt{R^2 + \xi^2}$ and $Z_+(\xi) = 0$, i.e.,

$$g_Z(x) = -\mu \int_{-R}^{R} \log \frac{(x - \xi)^2 + R^2 + \xi^2}{(x - \xi)^2} d\xi = -\mu \int_{-R}^{R} \log \frac{x^2 + R^2 - 2x\xi}{(x - \xi)^2} d\xi.$$

This in turn is equal to

$$g_Z(x) = -\mu \left(\int_{-R}^{R} \log \left(x^2 + R^2 - 2x\xi \right) d\xi - \int_{-R}^{R} \log (x - \xi)^2 d\xi \right).$$

For the first integral we note that $x^2 + R^2 - 2x\xi \geq x^2 + R^2 - 2Rx = (R - x)^2 \geq 0$, and so the log function in the integral is well defined.

Substituting in the first integral $u = x^2 + R^2 - 2x\xi$ and considering that $du = -2x d\xi$, we have to calculate:

$$\frac{1}{2x}\int_{-R}^{R} \log(u)\, du.$$

Integrating $\log(u)$ by part we obtain:

$$\frac{1}{2x}\int \log(u)\, du = \frac{-u\log(u)}{2x} + \frac{u}{2x}.$$

Substituting back $x^2 + R^2 - 2x\xi$ we conclude that

$$\int_{-R}^{R} \log\left(x^2 + R^2 - 2x\xi\right) d\xi = -\frac{(R-x)^2}{2x}\log(R-x)^2 + \frac{(R+x)^2}{2x}\log(R+x)^2 - 2R.$$

The second integral can be calculated similarly, obtaining:

$$\int_{-R}^{R} \log(x-\xi)^2 d\xi = 2\int_{-R}^{R} \log|x-\xi|\, d\xi = (R-x)\log(R-x)^2 + (R+x)\log(R+x)^2 - 4R.$$

Since both integrals are regular at $x = 0$ as at $x = R$, the final results is

$$g_Z(x) = \frac{\mu}{2}\left[\frac{R^2 - x^2}{x}\log\frac{(R+x)^2}{(R-x)^2} + 4R\right].$$

Exercise 6 Compute the gravitational attraction with respect to the vertical axis of a vertical cylinder of height H and radius R, at a point P with height h on the symmetry axis of the cylinder. Linearise the attraction at $h = H$.

Apply the exact and approximate formula for $H = 100\,\text{m}$, $R = 100\,\text{m}$, $h = 110\,\text{m}$ and $\rho = 2670\,\text{kgm}^{-3}$.

Repeat the exercise supposing $h < H$ (e.g., $h = 90\,\text{m}$).

Solution

- Suppose first that $h > H$. Recalling Eq. (2.24), we have

$$g_Z(h) = -U_+(h) + U_-(h).$$

In a cylindrical reference frame with the origin at the symmetry axis of the cylinder we have that:

$$U_+(h) = \mu \int_0^{2\pi} d\alpha \int_0^R \frac{r dr}{\sqrt{r^2 + (h-H)^2}} = \pi\mu \int_0^{R^2} \frac{dt}{\sqrt{t + (h-H)^2}},$$

where $t = r^2$. The above integral simply gives:

$$U_+(h) = 2\pi\mu \sqrt{t + (h-H)^2}\Big|_0^{R^2} = 2\pi\mu \left(\sqrt{R^2 + (h-H)^2} - h - H\right).$$

In a similar way one computes also $U_-(h)$, obtaining

$$U_-(h) = 2\pi\mu \left(\sqrt{R^2 + h^2} - h\right).$$

As a consequence we can compute $g_Z(h) = -U_+(h) + U_-(h)$ as:

$$g_Z(h) = -2\pi\mu \left(\sqrt{R^2 + (h-H)^2} - \sqrt{R^2 + h^2} + H\right).$$

The variation of $g_Z(h)$ with respect to h can be computed as $\frac{\partial}{\partial h} g_Z(h)$:

$$g'_Z(h) = -2\pi\mu \frac{h-H}{\sqrt{R^2 + (h-H)^2}} + 2\pi\mu \frac{h}{\sqrt{R^2 + h^2}}.$$

The linearized formula for a point P close to the cylinder surface is given as $g_Z(h) \approx g_Z(H) + g'_Z(H)(h-H)$, i.e:

$$g_Z(h) \approx -2\pi\mu \left(R + H\sqrt{R^2 + H^2}\right) + 2\pi\mu \frac{H}{\sqrt{R^2 + H^2}} (h-H).$$

- Considering the exact solution, we have

$$g_Z(h) = -2\pi\mu \left(\sqrt{100^2 + (110-100)^2} - \sqrt{100^2 + 110^2} + 100\right) = -5.83\,\text{mGal}.$$

The approximate solution is

$$\tilde{g}_Z(h) = -2\pi\mu * \left(100 + 100\sqrt{100^2 + 100^2}\right) + 2\pi\mu \frac{100}{\sqrt{100^2 + 100^2}} (10)$$
$$= -5.79\,\text{mGal}.$$

- If we now consider $h < H$ we basically have only to change the signs of $U_+(h)$ and of $U_-(h)$:

$$g_Z(h) = -2\pi\mu\left(R + h - \sqrt{R^2 + h^2}\right) + 2\pi\mu\left(R + H - h - \sqrt{R^2\,(H-h)^2}\right)$$

i.e.,

$$g_Z(h) = -2\pi\mu\left[2h - H - \sqrt{R^2 + h^2} + \sqrt{R^2 + (H-h)^2}\right].$$

We can observe from this formula that $g_Z\,(H_-) = g_Z\,(H_+)$. Considering the approximate formula, the derivative of $g_Z(h)$ with respect to h is,

$$g'_Z(h) = -2\pi\mu\left(2 - \frac{h}{\sqrt{R^2 + h^2}} - \frac{H-h}{\sqrt{R^2 + (H-h)^2}}\right).$$

The derivative in H is

$$g'_Z\,(h) = -2\pi\mu\left(2 - \frac{H}{R^2 + H^2}\right).$$

Note that $g'_Z\,(H_+) - g'_Z\,(H_-) = -4\pi\mu$. If we substitute the numerical values with $h = 90\,\text{m}$ in the above expressions we obtain

$$\begin{aligned} g_Z\,(h) &= -2\pi\mu\left[180 - 100 - \sqrt{100^2 + 90^2} + \sqrt{100^2 + (100-90)^2}\right] \\ &= -5.16\,\text{mGal} \end{aligned}$$

and

$$\begin{aligned} \tilde{g}_Z\,(h) &= -2\pi\mu\left[\left(100 + 100 - \sqrt{100^2 + 100^2}\right) + \left(2 + \frac{90}{100^2 + 100^2}\right)\right] \\ &= -6.81\,\text{mGal} \end{aligned}$$

Chapter 3
Fourier Methods

3.1 Outline of the Chapter

The fundamental kernel of Newton's gravitational theory, namely ℓ_{PQ}^{-1}, is invariant, together with all its derivatives, under translations of the coordinate system.

This endows its integral with special properties in terms of Fourier transforms, making Fourier theory intimately related to the geophysical interpretation of the gravity field. Indeed, Fourier theory has wide applications in mathematical analysis and its full understanding requires a deeper background than the one required from the readers of this book. So we need to introduce the basics of Fourier methods to get acquainted with a fundamental tool of gravity interpretation, to obtain theoretical as well as numerical results, and then we shall go through the statement of the most important properties of Fourier transforms, keeping a rigorous approach, yet without entering into most of the proofs.

This program will be pursued in the first three sections. Then in Sect. 3.5 we shall establish the link between Fourier and gravitation theories. In the next two sections we will explore applications to some of the specific cases described in Chap. 2.

Then a section follows, where the Fourier transform is discretized in a way that makes it useful for numerical applications, giving rise to the Discrete Fourier Transform (DFT) theory, which nowadays, is probably the most effective tool for the numerical analysis of linear problems in signal theory.

In the last section we give some practical considerations on the sensitivity of the gravitational signal to power, spatial resolution and depth of its density source.

3.2 Fourier Transform in $\mathbb{R}^1$: Definition and First Properties

Let $f(x)$ be a function defined on $\mathbb{R}^1$. We assume that $f(x)$ is Lebesgue integrable on $\mathbb{R}^1$, namely that $f \in L(\mathbb{R}) \equiv L^1$. In this case the following integral is convergent for all $p \in \mathbb{R}^1$ and defines a complex function

F. Sansò, D. Sampietro, *Analysis of the Gravity Field*, Lecture Notes in Geosystems Mathematics and Computing, https://doi.org/10.1007/978-3-030-74353-6_3

$$\hat{f}(p) = \int_{-\infty}^{+\infty} e^{i2\pi px} f(x)\, dx \tag{3.1}$$

which by definition is the Fourier transform of $f(x)$. When the variable x has the meaning of time, p has the dimension of the inverse of time and it is called frequency; in this case p is often substituted by the letter f. When x is a length (e.g., a spatial Cartesian coordinate), p has the dimension of the inverse of a length and it is sometimes called a wave number, i.e., p^{-1} is a wavelength.

At the level of notation, we shall write (3.1) also in the concise form as

$$\hat{f}(p) = \mathscr{F}\{f(x)\}(p), \tag{3.2}$$

or, when no ambiguity on the independent variables arises, we will use the even shorter form

$$\hat{f} = \mathscr{F}\{f\}. \tag{3.3}$$

The notation (3.3) emphasizes that $\mathscr{F}$ is in fact an integral operator, with kernel $e^{i2\pi pq}$, acting between L^1 and a space of functions that, as we shall show later, is univocally defined by $\mathscr{F}$.

Remark 3.1 The choice of using the factor 2π in the exponent is not mandatory: some authors prefer to use a variable $q = 2\pi p$ instead of p. Yet our choice simplifies formulas, avoiding the use of normalization factors in front of the integral in (3.1). This is particularly useful when we go to higher dimensions. Another warning is that in literature one often finds the operator $\mathscr{F}$ defined by the kernel $e^{-i2\pi px}$; we shall rather use the symbol $\mathscr{F}^*$ for the operator corresponding to such a kernel.

Since the theory is perfectly symmetrical with respect to this change of sign, the choice is a pure matter of taste.

Symmetry Properties. We first observe that when $f(x)$ is real, the relation (3.1) can be written as

$$\begin{aligned} \hat{f}(p) = \hat{c}(p) + i\hat{s}(p) = &\int_{\infty}^{+\infty} \cos(2\pi px) f(x)\, dx + \\ &+ i\int_{-\infty}^{+\infty} \sin(2\pi px) f(x)\, dx \end{aligned} \tag{3.4}$$

Further, since $\cos(2\pi px)$ is an even function and $\sin(2\pi px)$ is odd, we see that

$$\begin{cases} \hat{c}(-p) = \hat{c}(p) \\ \hat{s}(-p) = -\hat{s}(p) \end{cases}$$

and $\hat{c}$, $\hat{s}$ depend respectively on the even and odd components of $f(x)$. The above relations can be summarized in complex form as

$$\hat{f}(-p) = \hat{f}^*(p). \tag{3.5}$$

Moreover, when f is a real function, we have also

$$\begin{aligned} \mathscr{F}\{f(-x)\} &= \int_{-\infty}^{+\infty} e^{i2\pi px} f(-x)\,dx = \\ &= \int_{-\infty}^{+\infty} e^{-i2\pi pt} f(t)\,dt = \mathscr{F}^*\{f(x)\} = \hat{f}^*(p) \end{aligned} \tag{3.6}$$

Translation and Change of Scale Properties. Let us define the (forward) translation operator T_t as

$$F(x) = T_t\{f(x)\} = f(x-t). \tag{3.7}$$

The geometrical meaning of (3.7) is illustrated in Fig. 3.1

One has

$$\begin{aligned} &\mathscr{F}\{T_t(f)\} = \mathscr{F}\{f(x-t)\} = \int_{-\infty}^{+\infty} e^{i2\pi px} f(x-t)\,dx = \\ &\int_{-\infty}^{+\infty} e^{-i2\pi p(y+t)} f(y)\,dy = e^{i2\pi pt}\hat{f}(p) \end{aligned} \tag{3.8}$$

Moreover, let us consider a change of scale of the original variable x, i.e., define

$$f_\lambda(x) = f(\lambda x).$$

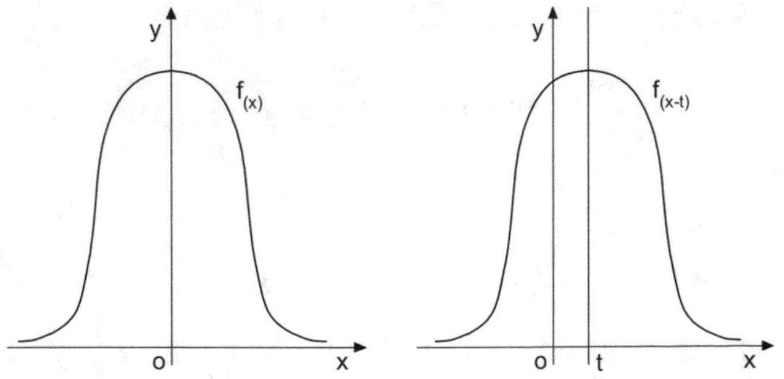

Fig. 3.1 A function f and its translate

Passing to the Fourier transform, we find, with the change of variable $\lambda x = t$, that

$$\begin{aligned}\hat{f}_\lambda &= \int_{-\infty}^{+\infty} e^{i2\pi px} f(\lambda x)\, dx = \\ &= \frac{1}{\lambda}\int_{-\infty}^{+\infty} e^{i2\pi \frac{p}{\lambda} t} f(t)\, dt \\ &= \frac{1}{\lambda}\hat{f}\left(\frac{p}{\lambda}\right).\end{aligned} \tag{3.9}$$

The Convolution of L^1 Functions.
Let us recall that the convolution of two functions $f, g \in L^1$ is defined as

$$h(x) = (f * g)(x) = \int_{-\infty}^{+\infty} f(x-y)\, g(y)\, dy. \tag{3.10}$$

Thanks to the inequality

$$\begin{aligned}\int_{-\infty}^{+\infty} |h(x)|\, dx &\le \int_{-\infty}^{+\infty}\int_{-\infty}^{+\infty} |f(x-y)|\, |g(y)|\, dydx = \\ &= \int_{-\infty}^{+\infty} |f(t)|\, dt \int_{-\infty}^{+\infty} |g(y)|\, dy\end{aligned} \tag{3.11}$$

one verifies that

$$f, g \in L^1 \Rightarrow h \in L^1,$$

so that we are allowed to speak of the Fourier transform of all the three functions. A direct computation and use of (3.10) yields

$$\begin{aligned}\hat{h}(p) = \mathscr{F}\{h\} &= \int_{-\infty}^{+\infty} \mathscr{F}\{f(x-y)\}\, g(y)\, dy = \\ &= \int_{-\infty}^{+\infty} e^{i2\pi py} \hat{f}(p)\, g(y)\, dy = \hat{f}(p)\, \hat{g}(p),\end{aligned} \tag{3.12}$$

that is

$$\mathscr{F}\{f * g\} = \mathscr{F}\{f\}\mathscr{F}\{g\}\}. \tag{3.13}$$

Theorem 3.1 (Convolution Theorem) *Relation* (3.13) *holds for any two functions* $f, g \in L^1$.

Regularity and Asymptotic Properties.

The following three properties are common to all Fourier transforms of L^1 functions (see [24])

(a) $\hat{f}(p)$ is bounded, in fact

$$\left|\hat{f}(p)\right| \leqslant \int_{-\infty}^{+\infty} |f(x)|\,dx; \tag{3.14}$$

(b) $\hat{f}(p)$ is continuous for all $p \in \mathbb{R}^1$:

$$\lim_{p \to p_0} \hat{f}(p) = \hat{f}(p_0), \tag{3.15}$$

and in fact uniformly continuous over the whole real axis;

(c) one has

$$\lim_{p \to \pm\infty} \hat{f}(p) = 0. \tag{3.16}$$

The first two statements are easy to prove; the third is also known as *Riemann's Theorem*.

Now if we add more conditions on $f(x)$ we can obtain more refined statements on the regularity of $\hat{f}(p)$. For instance, assume that $f(x)$ and its derivatives $D^k f(x)$ $(k = 1, \ldots, m)$ belong to L^1, and furthermore that

$$\lim_{|x| \to \infty} f(x) = 0. \tag{3.17}$$

Then:

(d) the Fourier transforms of the derivatives are related to one another by the formula

$$\mathscr{F}\left\{D^k f\right\} = (-2\pi i p)^k \mathscr{F}\{f\}. \tag{3.18}$$

Conversely, assume that $f(x)$ as well as $|x|^m f(x)$ are in L^1. Then the function $\hat{f}(p)$ is differentiable m times and one has

$$D^k \hat{f}(p) = \mathscr{F}\left\{(2\pi i x)^k f(x)\right\} \quad k = 1, 2, \ldots, m. \tag{3.19}$$

To proceed further in presenting Fourier theory, we need to extend it to other classes of functions, particularly to square integrable functions f, i.e.,

$$f \in L^2 \Longleftrightarrow \int_{-\infty}^{+\infty} f^2(x)\,dx < +\infty, \tag{3.20}$$

since it is in this space, L^2, that it acquires its simpler and more natural form.

3.3 Inversion Formula and Extensions

When we come to the matter of the inversion of the Fourier transform, i.e., a formula that allows to go back from $\hat{u}(p)$ to $u(x)$, this appears on the same time very simple and very complicated.

The answer depends on the choice of initially defining $\mathscr{F}$ under very natural conditions on u, namely that $u \in L^1$, which guarantees the existence of the integral (3.1) without any other requirements. In this case in fact we know (thanks to (3.15) and (3.16)), that when $u \in L^1$, then $\hat{u}(p)$ is continuous, bounded and tending to zero when $p \to \pm\infty$. Yet, these necessary conditions are not enough to maintain that $\hat{u} = \mathscr{F} u \in L^1$. In particular, it is not guaranteed that one could apply again $\mathscr{F}$, or better $\mathscr{F}^*$, to such a $\hat{u}$. Would this be the case, we could propose the formula

$$\mathscr{F}^*(\hat{u}) = \int_{-\infty}^{+\infty} e^{-i2\pi px}\hat{u}(p)dp = u(x) \tag{3.21}$$

as the solution of our problem. In fact, there are many couples of functions that do satisfy the equation (3.21), yet this is not true for all $u \in L^1$.

Here we propose some examples to clarify the above statement and also to prepare formulas that will prove useful in the sequel.

Example 3.1 Let us consider the function

$$\psi(x) = e^{-\pi x^2}. \tag{3.22}$$

We want to show that

$$\mathscr{F}\left\{e^{-\pi x^2}\right\} = e^{-\pi p^2}, \tag{3.23}$$

namely

$$\mathscr{F}\{\psi\} = \hat{\psi}(p) \equiv \psi(p). \tag{3.24}$$

In fact

$$\begin{aligned}\hat{\psi}(p) &= \int_{-\infty}^{+\infty} e^{-\pi x^2+2\pi ipx}dx = e^{-\pi p^2}\int_{-\infty}^{+\infty} e^{-\pi(x-ip)^2}dx \\ &\equiv e^{-\pi p^2}I(p),\end{aligned} \tag{3.25}$$

as one easily verifies by expanding the square of the binomial in the exponent. Now we observe that

$$I(p) = D_p\int_{-\infty}^{+\infty} e^{-\pi(x-ip)^2}dx = -i\int_{-\infty}^{+\infty} D_x e^{-\pi(x-ip)^2}dx = 0,$$

so that

$$I(p) = I(0) = \int_{-\infty}^{+\infty} e^{-\pi x^2} dx = 1,$$

as it is well-known from the integral of a Gaussian distribution.

So returning to (3.25) we get (3.23). Since the above reasoning does not change when one changes the sign of p, we see that we also have

$$\mathscr{F}^* \left\{ e^{-\pi p^2} \right\} = \psi(x),$$

i.e.,

$$\mathscr{F}^* \{\mathscr{F}\{\psi\}\} = \psi. \tag{3.26}$$

Example 3.2 We consider the function

$$u(x) = e^{-2\pi|x|}; \tag{3.27}$$

we claim that

$$\hat{u}(p) = \mathscr{F}\{u\} = \frac{1}{\pi} \frac{1}{1+p^2} \tag{3.28}$$

and

$$\mathscr{F}^* \left\{ \frac{1}{\pi} \frac{1}{1+p^2} \right\} = e^{-2\pi|x|}. \tag{3.29}$$

The first relation is easy to prove and we will do it here; the second requires the *Residue Theorem* of the theory of functions of a complex variable and we send the interested reader to the literature.

Coming to (3.28), we have

$$\begin{aligned} \hat{u}(p) &= \int_{-\infty}^{+\infty} e^{-2\pi|x|+2\pi i p x} dx = \\ &= \int_{-\infty}^{0} e^{+2\pi x(1+ip)} dx + \int_{0}^{+\infty} e^{-2\pi x(1-ip)} dx \\ &= \frac{1}{2\pi} \left[\frac{1}{1+ip} - \frac{1}{i-ip} \right] = \frac{1}{\pi} \frac{1}{1+p^2} \end{aligned}$$

as claimed.

As for (3.29), we limit ourselves to observe that $\frac{1}{\pi}\frac{1}{1+p^2}$ is indeed an L^1 function, so that the operator $\mathscr{F}^*$ can be properly applied to it. Moreover, in contrast with the next example, we have

$$\frac{1}{\pi}\int_{-\infty}^{+\infty} e^{-i2\pi px}\frac{1}{1+p^2}\,dx = \lim_{A\to\infty}\frac{1}{\pi}\int_{-A}^{+A} e^{-i2\pi px}\frac{1}{1+p^2}\,dp,$$

which is useful for carrying out the integration along a suitable path in the complex plane [18].

Example 3.3 Now we consider the function

$$u(x) = \begin{cases} e^{-2\pi x}, & \text{if } x \geq 0, \\ 0, & \text{if } x < 0; \end{cases}$$

as we see, $u \in L^1$ and it is not symmetric, so we can expect that its Fourier transform is complex. In fact

$$\hat{y}(p) = \int_{-\infty}^{+\infty} e^{-2\pi x(1-ip)}dx = \frac{1}{(2\pi)(1-ip)}. \tag{3.30}$$

So $\hat{u}(p)$ is not an L^1 function and we are not permitted to define directly its Fourier transform. Yet, resorting again to the *Residue Theorem*, one can prove that the following limit exists:

$$\lim_{A\to\infty}\int_{-A}^{+A}\frac{e^{-i2\pi px}}{2\pi(1-ip)}dp = u(x) \tag{3.31}$$

So, in a sense, we can again apply the operator $\mathscr{F}^*$, but we need to somehow regularize before the function $\hat{u}(p)$.

The result of Example 3.3 suggests that we need to generalize in some sense the definition of the operators $\mathscr{F}$, $\mathscr{F}^*$ in order to ensure, with more generality, the validity of the inversion of formula (3.21). This can be (and has been) done in several ways; we follow here the approach that seems the simplest to us.

Definition of Generalized Fourier Transform. We define the two generalized operators $\overline{\mathscr{F}}$, $\overline{\mathscr{F}}^*$ (which for the moment we denote with an overbar to distinguish them from $\mathscr{F}$, $\mathscr{F}^*$) as

$$\overline{\mathscr{F}}\{u\} = \lim_{\varepsilon\to 0}\int_{-\infty}^{+\infty} e^{i2\pi px}\psi(\varepsilon x)\,u(x)dx \tag{3.32}$$

$$\overline{\mathscr{F}}^*\{\hat{u}\} = \lim_{\varepsilon\to 0}\int_{+\infty}^{-\infty} e^{-i2\pi px}\psi(\varepsilon p)\,\hat{u}(p)dp, \tag{3.33}$$

where $\psi(x)$ is the same function as in Example 3.1.

By using the generalized FT operators, we can now provide the following nice theorem for functions $u \in L^1$.

Theorem 3.2 *For any $u \in L^1$ we have*

$$\text{(a)} \qquad \hat{u}(p) = \overline{\mathscr{F}}(u) = \mathscr{F}(u), \tag{3.34}$$

i.e., $\overline{\mathscr{F}}$ and $\mathscr{F}$ provide the same results on L^1; moreover,

$$\text{(b)} \qquad \overline{\mathscr{F}}^*\{\hat{u}\} = u(x), \tag{3.35}$$

where the limit, implicit in (3.35)*, can be understood either in the L^1 norm, namely*

$$\int_{-\infty}^{+\infty} \left| \int_{-\infty}^{+\infty} \psi\,(\varepsilon p)\, \hat{u}\,(p) \cdot e^{-2\pi i p x} d\,p - u\,(x) \right| dx \underset{(\varepsilon \to 0)}{\longrightarrow} 0 \tag{3.36}$$

or almost everywhere, i.e., for all the real x except for a set of measure zero.

Proof Item (a) is a simple consequence of the dominated convergence theorem of Lebesgue (see [117]). In fact

$$\left| e^{i2\pi p x} \psi(\varepsilon x) u(x) \right| < |u\,(x)|\,,$$

because $\psi\,(\varepsilon x) < 1$ and the second member is integrable, we can pass to the limit in (3.32) under the integral sign and since

$$\lim_{\varepsilon \to 0} \int_{-\infty}^{+\infty} e^{i2\pi p x} \psi(\varepsilon x) u(x) dx = \int_{-\infty}^{+\infty} e^{i2\pi p x} u(x) dx,$$

we obtain (3.34).

For item (b), denote

$$F\,(x, \varepsilon) = \int_{-\infty}^{+\infty} e^{-i2\pi p x} \hat{u}(p) \psi(\varepsilon p) dp \tag{3.37}$$

and note that, by using the definition of $\hat{u}(p)$ and making the substitution $q = \varepsilon p$, we get

$$\begin{aligned} F\,(x, \varepsilon) &= \int_{-\infty}^{+\infty} e^{-i2\pi p x} \psi(\varepsilon p) \left(\int_{-\infty}^{+\infty} e^{i2\pi p y} u(y) dy \right) dp = \\ &= \int_{-\infty}^{+\infty} u(y) \left(\int_{-\infty}^{+\infty} \frac{1}{\varepsilon} e^{i2\pi q \frac{(y-x)}{\varepsilon}} \psi(q) dq \right) dy; \end{aligned} \tag{3.38}$$

Since from Example 3.1 it follows that

$$\hat{\psi}(x) = \psi(x),$$

(3.38) becomes

$$\begin{aligned} F(x,\varepsilon) &= \int_{-\infty}^{+\infty} u(y)\,\psi_\varepsilon(y-x)\,dy \\ &= \int_{-\infty}^{+\infty} \psi_\varepsilon(x-y)\,u(y)dy = \int_{-\infty}^{+\infty} \psi_\varepsilon(y)u(x-y)\,dy, \end{aligned} \tag{3.39}$$

where we have put

$$\psi_\varepsilon(x-y) = \frac{1}{\varepsilon}\psi\left(\frac{x-y}{\varepsilon}\right), \tag{3.40}$$

which is indeed an even function of $x-y$.
Let us note that, for every ε,

$$\int_{-\infty}^{+\infty} \psi_\varepsilon(y)dx = \int_{-\infty}^{+\infty} \psi(x)dx = 1.$$

Moreover, for every $\omega > 0$, $\int_{-\omega}^{+\omega} \psi_\varepsilon(x)dx = \int_{-\frac{\omega}{\varepsilon}}^{\frac{\omega}{\varepsilon}} \psi(x)dx$, so that when $\varepsilon \to 0$, this integral tends to 1. Consequently, $\lim_{\varepsilon\to 0} \int_{|x|>\omega} \psi_\varepsilon(x)\,dx = 0$ for all ω.

Since we want to prove (3.36), we consider

$$\begin{aligned} &\int_{-\infty}^{+\infty} |F(x,\varepsilon) - u(x)|\,dx = \\ &= \int_{-\infty}^{+\infty} \left|\int_{-\infty}^{+\infty} \psi_\varepsilon(y)u(x-y)\,dy - u(x)\int_{-\infty}^{+\infty} \psi_\varepsilon(y)dy\right| dx \le \\ &\le \int_{-\infty}^{+\infty} \psi_\varepsilon(y) \int_{-\infty}^{+\infty} |u(x-y) - u(x)|\,dxdy. \end{aligned} \tag{3.41}$$

Now denote

$$\Delta_u(y) = \int_{-\infty}^{+\infty} |u(x-y) - u(x)|\,dx \tag{3.42}$$

which is known as the L^1 modulus of continuity of $u(x)$. It is clear that

$$0 \le \Delta_u(y) \le 2\,\|u\|_{L^1}.$$

Moreover, we know that for any $u \in L^1$ there is a continuous function v with compact support, such that $\|u - v\|_{L^1} < \delta$, however $\delta > 0$ is chosen. For a v of this kind we have clearly, uniformly in y

$$\Delta_v(y) \underset{(y \to 0)}{\longrightarrow} 0 \qquad \forall y \in R. \tag{3.43}$$

On the other hand,

$$\Delta_u(y) \le \Delta_{u-v}(y) + \Delta_v(y). \tag{3.44}$$

But

$$\Delta_{u-v}(y) \le \int_{-\infty}^{+\infty} |(u-v)(x-y)|\,dx + \int_{-\infty}^{+\infty} |(u-v)(x)|\,dx \le 2\delta,$$

so that

$$\overline{\lim_{y \to 0}}\, \Delta_u(y) \le 2\delta + \overline{\lim}\, \Delta_v(y) = 2\delta,$$

and, in view of the arbitrariness of δ,

$$\lim_{y \to 0} \Delta_u(y) = 0.$$

So fix any $\eta > 0$, we can find ω such that

$$\Delta_u(y) < \eta, \qquad \forall\, |y| < \omega;$$

one can write then, according to (3.41) and (3.42) and previous remarks,

$$\int_{-\infty}^{+\infty} |F(x, \varepsilon) - u(x)|\,dx \le \int_{|y|<\omega} \psi_\varepsilon(y)\Delta_u(y)dy + \int_{|y|>\omega} \psi_\varepsilon(y)\Delta_u(y)dy \le$$
$$\le \eta \int_{|y|<\omega} \psi_\varepsilon(y)dy + 2\,\|u\|_{L_1} \int_{|y|>\omega} \psi_\varepsilon(y)dy.$$

As we have already observed, when $\varepsilon \to 0$,

$$\int_{|y|<\omega} \psi_\varepsilon(y)dy \to 1, \int_{|y|>\varepsilon} \psi_\varepsilon(y)dy \to 0,$$

so we get

$$\overline{\lim_{\varepsilon \to 0}} \int_{-\infty}^{+\infty} |F(x, \varepsilon) - u(x)|\,dx \le \eta,$$

i.e. (3.36) is proved.

As for the last point of the almost everywhere convergence we send the reader to [142, Th. 2, §2.2.] □

An important consequence of Theorem 3.2 is illustrated by

Corollary 3.1 *Let $u \in C_0(\mathbb{R})$, i.e., u is continuous on $\mathbb{R}$ and equal to zero outside some finite interval; then we have*

$$\left(\hat{u}(p), \hat{u}(p)\right)_{L^2} = \left(u\left(x\right), u\left(x\right)\right)_{L^2}, \tag{3.45}$$

which is known as the Parseval identity.

Proof Recall that the $L^2(\mathbb{R})$ scalar product in (3.45) is given, for any two complex square-integrable functions u and v, by

$$(u, v)_{L^2} = \int_{-\infty}^{+\infty} u^*(x)v(x)dx. \tag{3.46}$$

Note also that

$$u \in C_0(\mathbb{R}) \Rightarrow u \in L^1; u \in L^2.$$

Hence, by Theorem 3.2,

$$\overline{\mathscr{F}}^*\{\hat{u}(p)\} = \lim_{\varepsilon \to 0} \int e^{-i2\pi px}\psi(\varepsilon p)\hat{u}(p)dp = u(x), \tag{3.47}$$

the limit taking place in L^1 and almost everywhere. We take the L^2 scalar product of (3.47) by $u(x)$ and observe that, since $u(x)$ is of class C_0, the corresponding integral in the definition of the product is over a finite interval and we can then interchange limit and integration. In this way we get

$$\lim_{\varepsilon \to 0} \int_{-\infty}^{+\infty} \left(\int_{-\infty}^{+\infty} e^{-i2\pi px}u(x)dx\right) \psi(\varepsilon p)\hat{u}(p)dp = (u, u)_{L^2} \tag{3.48}$$

and, recalling the definition of the FT,

$$\begin{aligned} &\lim_{\varepsilon \to 0} \int_{-\infty}^{+\infty} (\psi(\varepsilon p)\hat{u}^*(p)\hat{u}(p)dp = \\ &= \lim_{\varepsilon \to 0} \int_{-\infty}^{+\infty} \psi(\varepsilon p)|\hat{u}(p)|^2 dp = (u, u)_{L^2}. \end{aligned} \tag{3.49}$$

Now we observe that $\psi(\varepsilon p)|\hat{u}(p)|^2$ is indeed positive, increasing when $\varepsilon \to 0$, its integral is uniformly bounded, by a positive constant, and finally

$$\lim_{\varepsilon \to 0} \psi(\varepsilon p)|\hat{u}(p)|^2 = |\hat{u}(p)|^2, \ \forall p. \tag{3.50}$$

Therefore, by the Beppo Levi Theorem (see Riesz–Nagy [117], and Proposition 0.9), $|\hat{u}(p)|^2$ is integrable, i.e., $\hat{u} \in L^2$, and

$$(\hat{u}, \hat{u})_{L^2} = (u, u)_{L^2}. \tag{3.51}$$

□

Remark 3.2 If we multiply (3.47) by $v(x) \in C_0$, one can modify the proof of the Corollary so as to get the relation

$$(\hat{v}, \hat{u})_{L^2} = (u, v)_{L^2}, \tag{3.52}$$

which is known in analysis as the *Plancherel Theorem.*

Extension of the FT to L^2. Note first that when $u \in C_0$, then thanks to Theorem 3.2 we can assert that

$$\overline{\mathscr{F}}^* \{\overline{\mathscr{F}}(u)\} = \overline{\mathscr{F}}^* \{\hat{u}\} = u. \tag{3.53}$$

Furthermore, relation (3.51) says that, at least when $u \in C_0$,

$$||\overline{\mathscr{F}}(u)||^2_{L^2} = ||\hat{u}||^2_{L^2} = ||u||^2_{L^2}, \tag{3.54}$$

showing that:

(a) restricted to the class C_0, the operator $\overline{\mathscr{F}}$ is a unitary operator, in the topology of L^2, namely it preserves the L^2 norm,
(b) one has, at least for functions u in C_0,

$$\overline{\mathscr{F}}^* = \overline{\mathscr{F}}^{-1}. \tag{3.55}$$

In particular, the above property (a) implies that if $\{u_n\}$ is an L^2 Cauchy sequence, so is its image $\{\hat{u}_n\}$; this opens the way to extend $\overline{\mathscr{F}}$ to the whole of L^2. It is in fact known that for any function $u \in L^2$ we can find a sequence $\{u_n\} \in C_0$ such that $||u - u_n||_{L^2} \to 0$ as $n \to \infty$. But this implies that $\{u_n\}$ satisfies Cauchy's criterion and so the same must hold for $\{\hat{u}_n\}$. Therefore, there is a $\hat{u} \in L^2$ such that

$$\hat{u} \underset{(L^2)}{=} \lim_{n \to \infty} \hat{u}_n. \tag{3.56}$$

Accordingly, we can extend $\overline{\mathscr{F}}$ defining an operator $\overline{\overline{\mathscr{F}}}$ such that

$$\hat{u} = \overline{\overline{\mathscr{F}}} u. \tag{3.57}$$

In this way we can extend the FT to the whole L^2. One can prove that, like in the L^1 case, one has

$$\lim_{\varepsilon\to 0}\int e^{+i2\pi px}\psi(\varepsilon x)u(x)dx = \hat{u}(p) \tag{3.58}$$

and

$$\lim_{\varepsilon\to 0}\int e^{-i2\pi px}\psi(\varepsilon x)\hat{u}(p)dp = u(x), \tag{3.59}$$

with the limits above, understood both in L^2 and almost everywhere.

Having achieved the above extension, let us stipulate that from now on we shall no longer distinguish, at the level of notation, between $\mathscr{F}, \overline{\mathscr{F}}, \overline{\overline{\mathscr{F}}}$, assuming that each time the right operator is applied, according to the circumstances.

Remark 3.3 Let us observe that our presentation of the Fourier transform theory was based on the regularizing kernel $\psi(\varepsilon x)$; this however is by no means the only way to treat the matter and a whole family of suitable kernels could be used instead of ψ. For instance, the L^2 extension of FT could be based on the formulas

$$\hat{u} = \mathscr{F}(u) = \lim_{A\to\infty}\int_{-A}^{A} e^{i2\pi px}u(x)dx, \tag{3.60}$$

$$u = \mathscr{F}^*(\hat{u}) = \lim_{A\to\infty}\int_{-A}^{A} e^{-i2\pi px}\hat{u}(p)dp, \tag{3.61}$$

where the regularizing kernel is clearly the function

$$\chi_A(x) = \begin{cases} 1, & \text{if } |x| \le A, \\ 0, & \text{if } |x| < A. \end{cases}$$

Remark 3.4 There is still another extension of the FT operator that is somehow useful in the present context, concerning the so-called *tempered distributions*. We shall only briefly outline the matter without too many details, focusing only on the FT of Dirac's $\delta(x)$ functional and on related formulas.

We start by defining the class of "good" functions $\mathscr{S}$ as

$$\mathscr{S} = \{\varphi;\, x^m D_x^n\varphi \in L^2 \quad \forall m, n \ge 0\}. \tag{3.62}$$

It is easy to see that if $\varphi \in \mathscr{S}$, then φ together with all its derivatives are continuous and tend to zero at infinity faster than any negative power of x.

Thanks to the differentiation formulas (3.18) and (3.19) and to the fact that $\mathcal{F}(L^2) = L^2$, we see that

$$\varphi \in \mathcal{S} \Rightarrow \hat{\varphi} \in \mathcal{S}, \tag{3.63}$$

i.e.,

$$\mathcal{F}(\mathcal{S}) \equiv \mathcal{S}.$$

Since $\mathcal{F}^* \equiv \mathcal{F}^{-1}$, we must have as well $\mathcal{F}^*(\mathcal{S}) \equiv \mathcal{S}$.

Furthermore, as $\mathcal{S} \subset L^2$ (in fact $\mathcal{S}$ is dense in L^2 with respect to the L^2-topology), for any φ, $\chi \in \mathcal{S}$ we can write the Plancherel relation

$$(\hat{\chi}, \hat{\varphi})_{L^2} = (\chi, \varphi)_{L^2}. \tag{3.64}$$

First of all it is clear that through the integral

$$(\chi, \varphi)_{L^2} = \int_{-\infty}^{+\infty} \chi(x)^* \varphi(x) dx \tag{3.65}$$

we can identify $\chi(x)$ with a continuous linear functional on $\mathcal{S}$, in the sense that (χ, φ) is linear in φ and if

$$\varphi_k \to \bar{\varphi} \text{ in } \mathcal{S} \text{ (i.e., } x^m D_x^n \varphi_k \to x^m D_x^n \varphi \text{ in } L^2 \quad \forall n, m),$$

then

$$(\chi, \varphi_k)_{L^2} \to (\chi, \bar{\varphi})_{L^2}.$$

Now one can prove that the dual $\mathcal{S}^*$ of $\mathcal{S}$, namely the space of continuous linear functionals on $\mathcal{S}$, not only contains $\mathcal{S}$ via the representation (3.65), but also it contains $\mathcal{S}$ densely, namely, given $T \in \mathcal{S}^*$ one can find a sequence $\chi_n \in \mathcal{S}$ such that $\chi_n \to T$ in $\mathcal{S}$ (i.e., $(\chi_n, \varphi) \to T(\varphi)$, $\forall \varphi \in \mathcal{S}$).

Consider then the sequence of the **FT**, $\hat{\chi}_n$, which also belongs to $\mathcal{S}$. Seen as elements of $\mathcal{S}^*$ one has

$$(\hat{\chi}_n, \varphi)_{L^2} = (\chi_n, \mathcal{F}^*(\varphi)) \to T[\mathcal{F}^*(\varphi)] \text{ as } n \to \infty.$$

It is readily verified that $T[\mathcal{F}^*(\varphi)]$ is also a continuous linear functional on $\mathcal{S}$, so we put by definition

$$\hat{T}[\varphi] = T[\mathcal{F}^*(\varphi)] \tag{3.66}$$

and we say that $\hat{T}$ is the distributional FT of T.

Example 3.4 (Dirac's δ-Functional). The functional

$$\delta[\varphi] \equiv \varphi(0) \tag{3.67}$$

is called the *Dirac functional*. It is clear that $\delta[\cdot] \in \mathscr{S}^*$. By extending the representation (3.65), formula (3.67) is also written as

$$\delta[\varphi] = \int_{-\infty}^{+\infty} \delta(x)\varphi(x)dx = \varphi(0). \tag{3.68}$$

although it is clear that the symbols in (3.68) cannot represent any kind of integral, exactly as $\delta(x)$ cannot represent any function, which in this case should be zero everywhere outside the origin and have an infinite spike at $x = 0$. Rather, $\delta(x)$ represents the limit of a sequence of good functions, directly written under the integral sign. For instance, if one takes the kernels $\psi_\varepsilon(x)$, one sees that

$$\lim_{\varepsilon \to 0} \int_{-\infty}^{+\infty} \psi_\varepsilon(x)\varphi(x)dx = \varphi(0) = \int_{-\infty}^{+\infty} \delta(x)\varphi(x)dx. \tag{3.69}$$

On the other hand, we know that

$$\hat{\psi}_\varepsilon(x) = \psi(\varepsilon p),$$

so that

$$\begin{aligned} \lim_{\varepsilon \to 0} \int_{-\infty}^{+\infty} \psi_\varepsilon(x)\widehat{\varphi}(x)dx &= \lim_{\varepsilon \to 0} \int_{-\infty}^{+\infty} \psi(\varepsilon p)\varphi(p)dp \\ &= \int_{-\infty}^{+\infty} \varphi(p)dp; \end{aligned} \tag{3.70}$$

comparing with (3.66) one finds that

$$\hat{\delta}(p) \equiv 1.$$

This can also be written symbolically as

$$\int_{-\infty}^{+\infty} \delta(x)e^{i2\pi px}dx = 1. \tag{3.71}$$

Similarly, one can write the inverse relation

$$\int_{-\infty}^{+\infty} e^{-i2\pi px}dp = \delta(x). \tag{3.72}$$

Since $\delta(x)$ is obviously a real function, the complex conjugate of (3.72) has to hold too, namely

$$\int_{-\infty}^{+\infty} e^{i2\pi px} dp = \delta(x). \tag{3.73}$$

We note that shifting the variable x we can alternatively write

$$\int_{-\infty}^{+\infty} e^{i2\pi(x-y)} dp = \delta(x-y),$$

so that for any $\varphi(x) \in \mathscr{S}$ one has

$$\int_{-\infty}^{+\infty} \varphi(x) \int_{-\infty}^{+\infty} e^{i2\pi(x-y)} dp dx = \int_{-\infty}^{+\infty} \delta(x-y)\varphi(x) dx \equiv \varphi(y),$$

or

$$\int_{-\infty}^{+\infty} e^{-i2\pi py} \int_{-\infty}^{+\infty} e^{i2\pi px} \varphi(x) dx dp \equiv \varphi(y).$$

Thus, were covered the inversion formula of the FT for functions $\varphi \subset \mathscr{S}$.

As a last result, and exploiting the formalism of Example 3.4, we can state the following result.

Theorem 3.3 (Converse of the Convolution Theorem) *Let $\hat{f}, \hat{g} \in L^2$, satisfying condition* (3.3)*, so that we can assume that*

$$\hat{f} = \mathscr{F}\{f\}, \quad \hat{g} = \mathscr{F}(g), \text{ with } f, g \in L^2 \textit{ and real}. \tag{3.74}$$

Then the inverse of (3.74) *holds, namely*

$$\mathscr{F}^*\{\hat{f}\hat{g}\} \equiv f * g. \tag{3.75}$$

Proof We first prove (3.75), when $\hat{f}, \hat{g} \in \mathscr{S}$. The proof is then completed by an approximation process, taking $\hat{f}_n, \hat{g}_n \in \mathscr{S}$ and $\hat{f}_n \underset{L^2}{\to} \hat{f}$, $\hat{g}_n \underset{L^2}{\to} \hat{g}$, observing that in this case we have also $\hat{f}_n \hat{g}_n \underset{L^1}{\to} \hat{f}\hat{g}$ and $f_n * g_n \to f * g$ for all $x \in \mathbb{R}$.

So assume $\hat{f}, \hat{g} \in \mathscr{S}$; thanks to (3.63), we know that $f, g \in \mathscr{S}$ as well. As a matter of fact, it is readily verified that $\hat{f}\hat{g}$ and $f * g$ belong to $\mathscr{S}$ too, so that one can apply FT's in their original form.

We have, recalling (3.73),

$$\begin{aligned}
\mathscr{F}^*\{\hat{f}\hat{g}\} &= \int_{-\infty}^{+\infty} e^{-i2\pi px}\hat{f}(p)\hat{g}(p)dp = \\
&= \int_{-\infty}^{+\infty} e^{-i2\pi px}\int_{-\infty}^{+\infty} e^{i2\pi py}f(y)\int_{-\infty}^{+\infty} e^{i2\pi pz}g(z)dzdydp = \\
&= \int_{-\infty}^{+\infty}\int_{-\infty}^{+\infty} f(y)g(z)\int_{-\infty}^{+\infty} e^{i2\pi p(y+z-x)}dpdzdy = \\
&= \int_{-\infty}^{+\infty}\int_{-\infty}^{+\infty} f(y)g(z)\delta(y+z-x)dzdy \\
&= \int_{-\infty}^{+\infty} f(y)g(x-y)dy \equiv f * g.
\end{aligned}$$

The inversion of the integrals in the proof is justified once they are interpreted as the dual pairing between $\mathscr{S}$ and $\mathscr{S}^*$. □

3.4 The Fourier Transform in $\mathbb{R}^2$ and $\mathbb{R}^3$

The concept of Fourier Transform is readily generalized to functions of $x \in \mathbb{R}^n$; here we report only the cases $x \in \mathbb{R}^2$ and $x \in \mathbb{R}^3$ that are relevant to our subject.

We start with the basic case of $\mathbb{R}^2$, since then treating the higher dimensional case is just a matter of notation.

So let

$$\boldsymbol{x} = \begin{pmatrix} x_1 \\ x_2 \end{pmatrix}; \; f(\boldsymbol{x}) = f(x_1, x_2) \in L^1 \equiv L^1(\mathbb{R}^2);$$

the basic tool to generalize the FT is a combination of the theorems of Tonelli and Fubini, according to which

(a) for almost any fixed x_1, $f(x_1, x_2) \in L^1(\mathbb{R}^1)$ as a function of x_2; the same is then true for any p_2 for $f(x_1, x_2)^{ei2\pi p_2 x_2}$, so that we can define

$$\hat{f}(x_1, p_2) = \int_{-\infty}^{+\infty} f(x_1, x_2)e^{i2\pi p_2 x_2}dx_2, \tag{3.76}$$

(b) the function $\hat{f}(x_1, p_2)$ is then again in $L^1(\mathbb{R}^1)$ as a function of x_1, so that we can define for any $p_1 \in \mathbb{R}$

$$\hat{\hat{f}}(p_1, p_2) = \int_{-\infty}^{+\infty} f(x_1, p_2)e^{i2\pi p_1 x_1}dx_1 \tag{3.77}$$

(c) the iterated integral obtained by combining (3.76) and (3.77) gives the same result as a double integral over $\mathbb{R}^2$, and since

$$e^{i2\pi p_1 x_1} e^{i2\pi p_2 x_2} = e^{i2\pi(p_1 x_1 + p_2 x_2)} = e^{i2\pi \boldsymbol{p}\cdot\boldsymbol{x}},$$

where

$$\boldsymbol{p} = \begin{pmatrix} p_1 \\ p_2 \end{pmatrix},$$

one can write

$$\begin{aligned} \widehat{f}(\boldsymbol{p}) =& \widehat{\widehat{f}}(p_1, p_2) = \int_{-\infty}^{+\infty} \int_{-\infty}^{+\infty} e^{i2\pi \boldsymbol{p}\cdot\boldsymbol{x}} f(x_1, x_2) dx_1 dx_2 = \\ =& \int_{\mathbb{R}^2} e^{i2\pi \boldsymbol{p}\cdot\boldsymbol{x}} f(x) d_2 x \end{aligned} \tag{3.78}$$

which can be taken as the definition of the FT in $\mathbb{R}^2$, when $f \in L^1$.

Let us note that indeed in general $\hat{f}(\boldsymbol{p})$ fails to be in L^1. Yet we can repeat the reasonings of Sect. 3.3, in particular defining in $\mathbb{R}^2$ the regularizing kernel

$$\psi(x) = e^{-\pi(x_1^2 + x_2^2)} = \psi(x_1)\psi(x_2),$$

and then the inverse FT as

$$\mathscr{F}^*\{\hat{f}(\boldsymbol{p})\} = \lim_{\varepsilon\to 0} \int_{\mathbb{R}^2} e^{-i2\pi \boldsymbol{p}\cdot\boldsymbol{x}} \psi(\varepsilon x) \hat{f}(\boldsymbol{p}) d_2 p. \tag{3.79}$$

An iterated application of Theorem 3.2 leads to the inversion result

$$\mathscr{F}^*\{\hat{f}(p)\} = f(x) \quad \text{a.e.} \tag{3.80}$$

The above result is then extended to $f \in L^2(\mathbb{R}^2)$, with the help of the kernel $\psi(x)$. So we arrive at the following proposition.

Proposition 3.1 *Let $\mathscr{S}$, $\mathscr{S}^*$ be defined by*

$$\hat{f} = \mathscr{F}\{f\} = \lim_{\varepsilon\to 0} \int_{\mathbb{R}^2} e^{i2\pi \boldsymbol{p}\cdot\boldsymbol{x}} f(x)\psi(\varepsilon x) d_2 x, \tag{3.81}$$

$$\mathscr{F}^*\{\hat{f}\} = \lim_{\varepsilon\to 0} \int_{\mathbb{R}^2} e^{-i2\pi \boldsymbol{p}\cdot\boldsymbol{x}} \hat{f}(p)\psi(\varepsilon p) d_2 p. \tag{3.82}$$

Then $\forall f \in L^1$ and $\forall f \in L^2$ one has

$$\hat{f} = \mathscr{F}\{f\}, f = \mathscr{F}^*\{\hat{f}\}. \tag{3.83}$$

Remark 3.5 Also on $\mathbb{R}^2$ one can define the space $\mathscr{S}$ of functions that are infinitely differentiable and rapidly decreasing at infinity. Accordingly one can define the dual space $\mathscr{S}^*$ and the Dirac functional $\delta(x)$ as

$$\delta[\varphi] = \varphi(0). \tag{3.84}$$

We notice that the possibility of calculating a double integral as two iterated one-dimensional integrals, suggests that one should have

$$\delta(\boldsymbol{x}) = \delta(x_1)\delta(x_2); \tag{3.85}$$

such a formula can in fact be made rigorous, with the caution that the first δ is two-dimensional, while in the right-hand side we have two one-dimensional δ's. In any event, the formulas

$$\widehat{\delta}(\boldsymbol{p}) = \mathscr{F}\{\delta(x)\} = \int_{R^2} \delta(\boldsymbol{x}) e^{i2\pi \boldsymbol{p}\cdot\boldsymbol{x}} d_2x \equiv 1 \tag{3.86}$$

$$\delta(\boldsymbol{x}) = \mathscr{F}^*\{\widehat{\delta}(p)\} = \int_{R^2} e^{i2\pi \boldsymbol{p}\cdot\boldsymbol{x}} d_2p \tag{3.87}$$

continue to hold as in the Example 3.4.

As one can deduce from the preceding discussion, the main properties of the 1D FT carry over to $\mathbb{R}^2$; in particular, this is true for the convolution theorem and its converse.

Let us recall that in 2D the convolution of two functions, either in L^1 or in L^2, is defined as

$$h(\boldsymbol{x}) = \int_{R^2} f(\boldsymbol{x} - \boldsymbol{y}) g(\boldsymbol{y}) d_2y = f * g. \tag{3.88}$$

Proposition 3.2 *Let f, g be either in L^1 or in L^2. Then one has*

$$F\{f * g\} = \hat{f}(p)\hat{g}(p), \tag{3.89}$$

$$\mathscr{F}^*\{\hat{f}\hat{g}\} = f * g. \tag{3.90}$$

Remark 3.6 (The Bessel Transform) There is one further property of the FT in $\mathbb{R}^2$ that will be useful when we will come to stochastic prediction methods applied in gravimetry: this is the particular form taken by the FT of a function $f(\boldsymbol{x})$ that depends on $|\boldsymbol{x}| = r$ only. With a small abuse of notation we shall put

$$f(\boldsymbol{x}) = f(|\boldsymbol{x}|) = f(r). \tag{3.91}$$

We note that in this case

$$\hat{f}(\boldsymbol{p}) = \int_{R^2} e^{i2\pi \boldsymbol{p}\cdot\boldsymbol{x}} f(r) d_2x;$$

switching to polar coordinates in the plane and calling ϑ the angle between $\boldsymbol{p}$ (which is held fixed in the integration process) and $\boldsymbol{x}$ one has

$$\hat{f}(\boldsymbol{p}) = \int_{-\infty}^{+\infty} rf(r) \int_0^{2\pi} e^{i2\pi pr\cos\vartheta} d\vartheta dr. \tag{3.92}$$

The function

$$J_0(2\pi s) = \frac{1}{2\pi} \int_0^{2\pi} e^{i2\pi s\cos\vartheta} d\vartheta, \; s \geq 0, \tag{3.93}$$

is well-known in mathematics [1] and designated as Bessel function of order 0. It is easy to verify that J_0 is a real function, so that

$$2\pi \, J_0(2\pi s) = \int_0^{2\pi} \cos(2\pi s \cos\vartheta) d\vartheta = \int_0^{2\pi} e^{-i2\pi s\cos\vartheta} d\vartheta. \tag{3.94}$$

Note also that $J_0(0) = 1$, which is the maximum of J_0; furthermore, $J_0(x)$ goes to zero slowly, oscillating so that the following asymptotic formula holds [1]:

$$J_0(x) \sim \sqrt{\frac{2}{\pi x}} \cos\left(x - \frac{\pi}{x}\right) + O\left(\frac{1}{x}\right). \tag{3.95}$$

We provide also a plot of $J_0(x)$ because of the importance of this function in our context (Fig. 3.2).

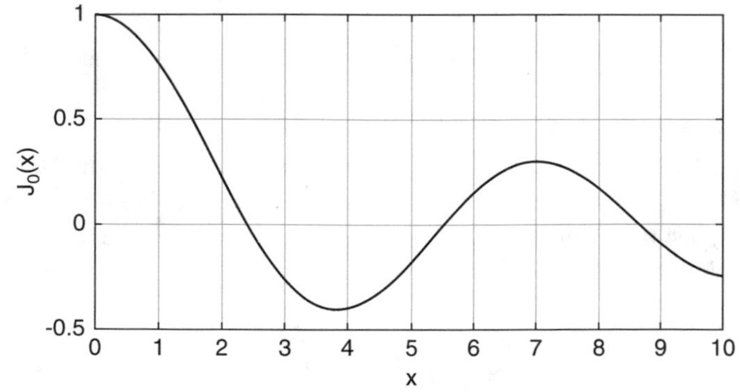

Fig. 3.2 The function $J_0(x)$; the first two zeros of J_0 are $x_1 = 2.41$, $x_2 = 5.52$

With the help of the above remarks we find that

$$\widehat{f}(\boldsymbol{p}) = 2\pi \int_0^{+\infty} f(r) J_0(2\pi pr) r dr; \tag{3.96}$$

therefore, the FT $\widehat{f}(\boldsymbol{p})$ is in fact a real function of $p \equiv |\boldsymbol{p}|$ only, that again abusing notation we call $\widehat{f}(p)$.

As we can see, (3.96) can be taken another transform of $f(x)$, as a function defined on $\mathbb{R}_+ = [0, +\infty]$, into another function $\widehat{f}(p)$ again defined on $\mathbb{R}_+$. If we assume that $f \in L^2(\mathbb{R}^2)$, this translates into the condition

$$\int_0^{+\infty} f^2(r) r dr < +\infty, \tag{3.97}$$

then similarly, since $\widehat{f}(p)$ has to be in $L^2(\mathbb{R}^2)$ as a function of p, we see that

$$\int_0^{+\infty} \widehat{f}(p)^2 p dp < +\infty \tag{3.98}$$

So we can consider that (3.96) defines a transform of the space $L^2(\mathbb{R}_+, r)$ (i.e., functions satisfying (3.97)) into itself; this transform is called the Bessel transform

$$\widehat{f}(p) = \mathscr{B}\{f(r)\} = 2\pi \int_0^{+\infty} f(r) J_0(2\pi pr) r dr \tag{3.99}$$

and it is easy to verify, by applying the inverse FT to (3.99), that

$$f(r) = \mathscr{B}\{\widehat{f}(p)\} = 2\pi \int_0^{+\infty} \widehat{f}(p) J_0(2\pi pr) p dp. \tag{3.100}$$

This implies that $\mathscr{B}$ is a unitary transformation of $L^2(\mathbb{R}_+, r)$ onto itself, i.e., in particular,

$$\mathscr{B}^2 = I,$$

with I the identity operator in $L^2(\mathbb{R}_+, r)$.

Since we have purposely adopted a vector notation to define the FT in $\mathbb{R}^2$ (see (3.81) and (3.82)) we can extend our reasonings from $\mathbb{R}^2$ to $\mathbb{R}^3$ (in fact even to $\mathbb{R}^n$ and beyond) just by stipulating that integrals are now over $\mathbb{R}^3$, $d_3x = dx_1, dx_2, dx_3$ is the $\mathbb{R}^3$ volume element and

$$p = \begin{pmatrix} p_1 \\ p_2 \\ p_3 \end{pmatrix}, x = \begin{pmatrix} x_1 \\ x_2 \\ x_3 \end{pmatrix}.$$

All the statements of the FT in $\mathbb{R}^2$ are then translated without changes into the analogous statements on FT in $\mathbb{R}^3$; only when we came to the matter of Remark 3.6 we have to take into account the different dimension of the underlying space.

In fact, if we apply the definition of FT to a function $f = f(r)$ by using 3D polar coordinates we obtain the formula

$$\widehat{f}(\boldsymbol{p}) = \int r^2 f(r)\left(\int_\sigma e^{i2\pi pr\cos\vartheta} d\vartheta\right)dr \tag{3.101}$$

where σ is the unit sphere and ϑ is the angle between $\boldsymbol{p}$ (a fixed vector) and $\boldsymbol{x}$. If we take spherical coordinates (ϑ, α) on the unit sphere, we know that

$$d\sigma = \sin\vartheta\, d\vartheta\, d\alpha,\ 0 \le \vartheta \le \pi,\ 0 \le \alpha < 2\pi,$$

so the kernel of (3.101) becomes

$$\int_0^{2\pi}\int_0^{2\pi} d\vartheta \sin\vartheta\, e^{i2\pi pr\cos\vartheta} d\alpha = \frac{2\sin 2\pi pr}{pr}.$$

Therefore, by using the function $\operatorname{sinc} x = \sin x/x$, we obtain

$$\widehat{f}(\boldsymbol{p}) = \widehat{f}(p) = \int_0^{+\infty} r^2 f(r)(4\pi \operatorname{sinc} 2\pi pr)dr. \tag{3.102}$$

Once again we find that the FT is a real function of $p = |\boldsymbol{p}|$ only.

Formula (3.102), too, defines a transform of $L^2(R_+, r^2)$ into itself and the same kernel can be used to invert it, i.e., to go back from $\widehat{f}(p)$ to $f(r)$.

3.5 Fourier Gravitation Formulas for a Homogeneous Body

The context in which we move in the present section is the same as outlined in §2.1 and developed in Sects. 2.2 and 2.3, only here we shall introduce, whenever possible and convenient, the FT formalism.

The basic tool we will use is contained in the following Proposition 3.3, of which, given its importance, we shall provide a proof.

Proposition 3.3 *Denote*

$$H = |z| \ge 0. \tag{3.103}$$

Then the two functions

$$f(r) = \frac{H}{(r^2 + H^2)^{3/2}} \quad (r = |x|,\ x \in \mathbb{R}^2), \tag{3.104}$$

$$\widehat{f}(p) = 2\pi e^{-2\pi p H} \quad (p = |p|,\ p \in \mathbb{R}^2) \tag{3.105}$$

form a Fourier pair, i.e.,

$$\widehat{f}(p) = \mathscr{F}\{f(r)\},\ f(r) = \mathscr{F}^*\{\widehat{f}(p)\}. \tag{3.106}$$

Proof We first of all observe that since both f and $\widehat{f}$ are real, the roles of $\mathscr{F}$ and $\mathscr{F}^*$ in (3.106) can be switched. Furthermore since $f, \widehat{f} \in L^1(\mathbb{R}^2)$ we can simply prove one of the two, the other being then a consequence of Theorem 3.2.

Expressing the transform in polar coordinates, we have

$$\begin{aligned} F(r) &= \mathscr{F}^*\{2\pi e^{-2\pi p H}\} = 2\pi \int_0^{2\pi}\int_0^{+\infty} e^{-2\pi p(H+ir\cos\vartheta)} pd\, pd\vartheta \\ &= -\frac{\partial}{\partial H}\int_0^{2\pi}\int_0^{+\infty} e^{-2\pi p(H+ir\cos\vartheta)} d\, pd\vartheta = \\ &= -\frac{\partial}{\partial H}\int_0^{2\pi} \frac{d\vartheta}{2\pi(H+ir\cos\vartheta)} = -\frac{1}{2\pi}\frac{\partial}{\partial H}\int_0^{2\pi}\frac{H-ir\cos\vartheta}{H^2+r^2\cos^2\vartheta}d\vartheta. \end{aligned} \tag{3.107}$$

For symmetry reasons it is obvious that

$$\int_0^{2\pi}\frac{\cos\vartheta}{H^2+r^2\cos^2\vartheta}d\vartheta = 0,$$

so that (3.107) becomes

$$F(r) = -\frac{\partial}{\partial H}\frac{1}{2\pi}\int_0^{2\pi}\frac{H}{H^2+r^2\cos^2\vartheta}d\vartheta = -\frac{\partial}{\partial H}\frac{4H}{2\pi}\int_0^{\frac{\pi}{2}}\frac{d\vartheta}{H^2+r^2\cos^2\vartheta}.$$

Upon substitution $t = \tan\vartheta$, we receive

$$\begin{aligned} F(r) &= -\frac{2}{\pi}\frac{\partial}{\partial H}\int_0^{+\infty}\frac{H dt}{H^2t^2+H^2+r^2} = -\frac{2}{\pi}\frac{\partial}{\partial H}\frac{\frac{\pi}{2}}{\sqrt{H^2+r^2}} = \\ &= \frac{H}{(H^2+r^2)^{3/2}} = f(r) \end{aligned}$$

as it was to be proved. □

Now let us return to the formula of the gravitation g_Z at a point P (in the $z = 0$ plane) generated by a body B of constant density ρ placed below the $z = 0$ plane, namely to formula (2.3).

Let us stipulate that $\boldsymbol{x} = (x_1, x_2)$ is the position vector of P on the horizontal plane, $\boldsymbol{y} \equiv (y_1, y_2)$ the position vector of the projection of Q on $\{z = 0\}$, and $(\boldsymbol{y}, \zeta)$ the 3D position vector of Q (implying $\zeta < 0$), and that the body B is identified by the two surfaces $S_+ \equiv \{\zeta = -Z_+(\boldsymbol{y})\}$ and $S_- \equiv \{\zeta = -Z_-(\boldsymbol{y})\}$. Denoting

$$r = |\boldsymbol{x} - \boldsymbol{y}|,$$

we write (2.3) in the form

$$g_Z(x) = \mu \int_B \frac{\zeta}{(r^2 + \zeta^2)^{3/2}} dB = \tag{3.108}$$

$$= \mu \int_{S_0} \int_{-Z_-(y)}^{-Z_+(y)} \frac{\zeta}{(r^2 + \zeta^2)^{3/2}} d\zeta \, d_2 y, \tag{3.109}$$

where S_0 is the projection of B on the plane $z = 0$. Now, using Proposition 3.3 with $H = -\zeta$, we write (3.108) in the form

$$\begin{aligned} g_Z(\boldsymbol{x}) &= -\mu \int_{S_0} \int_{-Z_-(y)}^{-Z_+(y)} \int_{\mathbb{R}^2} 2\pi e^{2\pi p\zeta} e^{-i2\pi \boldsymbol{p}\cdot(\boldsymbol{x}-\boldsymbol{y})} d_2 p \, d\zeta \, d_2 y \\ &= \mu \int_{R^2} e^{-i2\pi \boldsymbol{p}\cdot\boldsymbol{x}} \left(- \int_{S_0} e^{i2\pi \boldsymbol{p}\cdot\boldsymbol{y}} \frac{e^{-2\pi p Z_+(\boldsymbol{y})} - e^{-2\pi p Z_-(\boldsymbol{y})}}{p} d_2 y \right) d_2 p \end{aligned} \tag{3.110}$$

This formula clearly shows that $g_Z(\boldsymbol{x})$ is the inverse transform, $\mathscr{F}^{-1}$, of the function in parentheses, so that we can write

$$\widehat{g}_Z(\boldsymbol{p}) = \mathscr{F}\{g_Z(\boldsymbol{x})\} = \mu \int_{S_0} e^{i2\pi \boldsymbol{p}\cdot\boldsymbol{y}} \frac{e^{-2\pi Z_-(\boldsymbol{y})} - e^{-2\pi p Z_+(\boldsymbol{y})}}{p} d_2 y. \tag{3.111}$$

Such a formula is deceivingly simple and cannot be further elaborated unless we restrict the shape of the body B, for instance assuming that B is a general cylinder (see Example 2.1), i.e.,

$$Z_\pm(\boldsymbol{y}) = H_\pm(\text{constant}) = \bar{H} \mp \Delta. \tag{3.112}$$

In this case (3.111) factors as

$$\widehat{g}_Z(\boldsymbol{p}) = -\mu e^{-2\pi p \bar{H}} \frac{2Sh2\pi p\Delta}{p} \cdot \widehat{\chi}_{S_0}(\boldsymbol{p}), \tag{3.113}$$

where

$$\widehat{\chi}_{S_0}(\boldsymbol{p}) = \int_{R^2} \widehat{\chi}_{S_0}(\boldsymbol{x}) \, e^{i2\pi \boldsymbol{p}\cdot\boldsymbol{x}} d_2 x = \int_{S_0} e^{i2\pi \boldsymbol{p}\cdot\boldsymbol{x}} d_2 x. \tag{3.114}$$

Before going to see examples where (3.113) is specified by the choice of S_0, we want to emphasize that $\widehat{g}_Z(\boldsymbol{p})$ in (3.113) is certainly a smooth function when the body B is really buried, i.e., $H_+ = \bar{H} - \Delta > 0$; in fact, $\widehat{\chi}_{S_0}(\boldsymbol{p})$ is very regular and bounded
and

$$\frac{2Sh2\pi p\Delta}{p} \to 2\pi(2\Delta) \quad \text{when} \quad p \to 0$$

while

$$e^{-2\pi p\bar{H}}\frac{2Sh2\pi p\Delta}{p} \sim \frac{e^{-2\pi pH_+}}{p} \quad \text{when} \quad p \to \infty.$$

We further observe that we consider the implementation of (3.113) as a "solution" of the direct calculation of the gravitation of a given cylinder, because the inversion of $\widehat{g}_Z(p)$, to return to $g_Z(\boldsymbol{x})$, can be handled in a purely numerical manner, as we shall see in Sect. 3.7.

Example 3.5 (The Parallelepiped) We take up again the Example 2.2, referring to Fig. 2.7 for notation. Notice that here we use Cartesian coordinates (x, y), instead of (x_1, x_2), on the plane $\{z = 0\}$. To implement (3.113) we need only to compute

$$\begin{aligned}\widehat{\chi}_{S_0}(\boldsymbol{p}) &= \int_{S_0} e^{i2\pi\,\boldsymbol{p}\cdot\boldsymbol{x}} d_2x = \int_{x_0-a}^{x_0+a}\int_{y_0-b}^{y_0+b} e^{i2\pi(px+qy)} dy dx \\ &= e^{i2\pi(px_0+qy_0)}\int_{-a}^{a} e^{i2\pi px} dx \int_{-b}^{+b} e^{i2\pi py} dy = \\ &= e^{i2\pi\,\boldsymbol{p}\cdot\boldsymbol{x}_0}\,\mathrm{sinc}\,(2\pi pa)\ \mathrm{sinc}\,(2\pi qb)\ (2a2b);\end{aligned}$$

here $\boldsymbol{p} = \begin{pmatrix} p \\ q \end{pmatrix}$ and $\boldsymbol{x}_0 = \begin{pmatrix} x_0 \\ y_0 \end{pmatrix}$.

Therefore, (3.113) takes on the form

$$\widehat{g}_Z(\boldsymbol{p}) = -\mu e^{2\pi(-\bar{H}p+i\boldsymbol{p}\cdot\boldsymbol{x_0})}.\frac{Sh2\pi p\Delta}{p\Delta}(\mathrm{sinc}2\pi pa)(\mathrm{sinc}2\pi gb)\cdot(8ab\Delta) \qquad (3.115)$$

Referring to Example 2.2, we can assert that the gravitation of a parallelepiped has a much simpler form in the Fourier domain than in geometric space.

Moreover, we see that $\widehat{g}_Z(\boldsymbol{p})$ in (3.115) is decomposed into a position factor

$$F_B(\bar{H}, \boldsymbol{x_0}) = e^{2\pi(-\bar{H}p+i\boldsymbol{p}\cdot\boldsymbol{x_0})} \qquad (3.116)$$

that depends only on the position of the barycenter of the parallelepiped, and a shape factor

$$F_S(\boldsymbol{p}) = \mu \frac{Sh2\pi p\Delta}{p\Delta}(\text{sinc}2\pi pa)(\text{sinc}2\pi qb) \cdot (8ab\Delta) \tag{3.117}$$

that depends only on the shape of the parallelepiped, namely a, b, Δ. Note also that the last factor in (3.117) is just the volume of the parallelepiped.

Such a factorization becomes useful if one wants to compute the gravitation of a body of complicated geometry by (approximately) decomposing it into blocks of identical size, i.e., having the same shape factor, and then adding the effects according to the decomposition Lemma Sect. 2.4.

A final remark is that once $\widehat{g}_Z(\boldsymbol{p})$ is achieved, numerical methods for the computation of $g_Z(\boldsymbol{x})$ will provide the value of gravitation all over the plane $\{z = 0\}$.

Example 3.6 (The Circular Cylinder) We take up again Example 3.3 and demonstrate that in the Fourier domain we are able to reduce it to a closed form. For the sake of generality we assume that the circle S_0 is in an arbitrary position in the plane, as in Fig. 3.3.

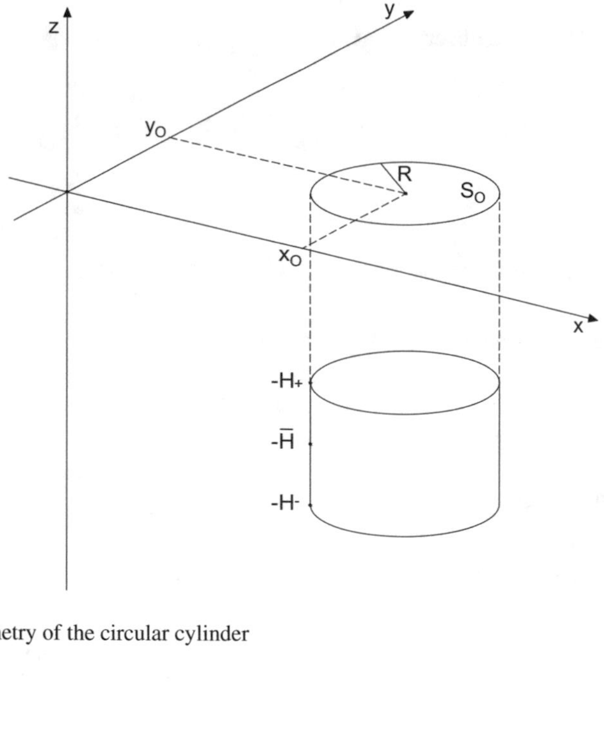

Fig. 3.3 Geometry of the circular cylinder

Our task is to compute

$$\widehat{\chi}_{S_0}(\boldsymbol{p}) = \int_{S_0} e^{i2\pi\,\boldsymbol{p}\cdot\boldsymbol{x}} d_2x = e^{i2\pi\,\boldsymbol{p}\cdot\boldsymbol{x}_0} \int_{|\xi|\le R} e^{i2\pi\,\boldsymbol{p}\cdot\boldsymbol{\xi}} d_2\xi =$$

$$= e^{i2\pi\,\boldsymbol{p}\cdot\boldsymbol{x}_0} \int_0^{2\pi}\int_0^R e^{i2\pi pr\cos\vartheta} r dr d\vartheta .$$

Recalling the definition (3.93) of the function $J_0(2\pi p)$, we have

$$\widehat{\chi}_{S_0}(\boldsymbol{p}) = 2\pi e^{i2\pi\,\boldsymbol{p}\cdot\boldsymbol{x}_0} \int_0^R J_0(2\pi pr) r dr. \tag{3.118}$$

To calculate (3.118) we need the Bessel function of order 1, $J_1(x)$, which is also well-known and tabulated [1]. What is of interest to us is the relation

$$J_0(x) = J_1'(x) + \frac{J_1(x)}{x},$$

that we rewrite as

$$x J_0(x) = D_x[x J_1(x)].$$

If we put $x = 2\pi pr$ and observe that $D_x = \frac{1}{2\pi p} D_r$, the above relation becomes

$$2\pi pr\, J_0(2\pi pr) = \frac{1}{2\pi p} D_r[2\pi pr\, J_1(2\pi pr)],$$

or

$$2\pi r\, J_0(2\pi pr) = \frac{1}{2\pi p^2} D_r[2\pi pr\, J_1(2\pi pr)].$$

Inserting this expression in (3.118) we obtain

$$\widehat{\chi}_{S_0}(\boldsymbol{p}) = e^{i2\pi\,\boldsymbol{p}\cdot\boldsymbol{x}_0} \frac{J_1(2\pi pR)}{p} \cdot R,$$

which combined with (3.113) gives

$$\widehat{g}_Z(\boldsymbol{p}) = -\mu e^{2\pi(-p\bar{H}+i\boldsymbol{p}\cdot\boldsymbol{x}_0)} 2\frac{Sh2\pi r\Delta}{p}\frac{J_1(2\pi pR)}{p} R. \tag{3.119}$$

Once more the decomposition into the product of a position factor and a shape factor is quite evident.

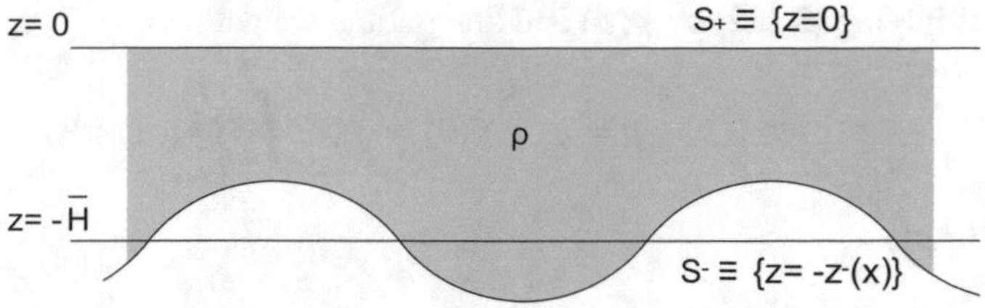

Fig. 3.4 A layer with density contrast ρ with respect to the background

3.6 Fourier Gravitation Formulas for a Two-Layers Configuration in 2D and 3D

In this section we will use the FT method for the cases treated in Sects. 2.7 and 2.8. We will basically apply the same approach as in the previous section, only to a body B which now, after reductions described in Sect. 2.7, has an upper surface which is the $\{z = 0\}$ plane and a lower surface at

$$z = -Z_{-}(x) = -(\bar{H} - \delta H(x))$$

as illustrated in Fig. 3.4

If we start from (3.108) with $S_0 \equiv \mathbb{R}^2$ and $Z_+ = 0$, $Z_- = (\bar{H} - \delta H)$ we arrive at (3.111) in the form

$$\begin{aligned} \widehat{g}_Z(\boldsymbol{p}) &= \mu \int_{\mathbb{R}^2} e^{i2\pi \boldsymbol{p}\cdot\boldsymbol{x}} \frac{1 - e^{2\pi p \bar{H}}}{p} d_2 y + \\ &+ \mu \int_{\mathbb{R}^2} e^{i2\pi \boldsymbol{p}\cdot\boldsymbol{x}} \frac{e^{-2\pi p \bar{H}} - e^{-2\pi p(\bar{H} - \delta H)}}{p} d_2 y = \\ &= \widehat{g}_B(\boldsymbol{p}) + \delta\widehat{g}_Z(\boldsymbol{p}). \end{aligned} \tag{3.120}$$

We have

$$\begin{aligned} \widehat{g}_B(\boldsymbol{p}) &= -\mu \frac{1 - e^{-2\pi p \bar{H}}}{p} \int_{\mathbb{R}^2} e^{i2\pi \boldsymbol{p}\cdot\boldsymbol{x}} d_2 y = \\ &= -\mu \frac{1 - e^{-2\pi \bar{H} p}}{p} \delta(\boldsymbol{p}) \end{aligned} \tag{3.121}$$

where we used the $\mathbb{R}^2$ analogoue of (3.72).

The following formula holds for any sufficiently regular $g(\boldsymbol{p})$:

$$g(\boldsymbol{p})\delta(\boldsymbol{p}) = g(0)\delta(\boldsymbol{p}); \tag{3.122}$$

in fact, multiplying (3.122) by $\varphi(\boldsymbol{p})$ and integrating, we get

$$\int \varphi(\boldsymbol{p}) f(p)\delta(\boldsymbol{p}) d_2 p = \varphi(0) f(0) = f(0) \int \delta(\boldsymbol{p})\varphi(p) d_2 p.$$

Since in (3.121)

$$\lim_{p\to 0} \frac{1 - e^{-2\pi p\bar{H}}}{p} = 2\pi \bar{H},$$

we see that (3.121) can be written as

$$\widehat{g}_B(\boldsymbol{p}) = -\mu 2\pi \bar{H} \delta(\boldsymbol{p}). \tag{3.123}$$

We recognize in (3.123) the FT of the attraction of a Bouguer plate of width $\bar{H}$. Coming to the attraction $\delta\widehat{g}_Z(\boldsymbol{p})$ of the interface between the two layers, we can further elaborate (3.120) as

$$\delta\widehat{g}_Z(\boldsymbol{p}) = -\mu \int_{R^2} e^{2\pi(-p\bar{H} + i\boldsymbol{p}\cdot\boldsymbol{y})} \frac{1 - e^{2\pi p \delta H}}{p} d_2 y. \tag{3.124}$$

From the expression (3.124) we can draw only some general statement on the nature of d $\delta\widehat{g}_Z(\boldsymbol{p})$. For instance, if we assume that $\delta H \in L^1(\mathbb{R}^2)$ and at the same time

$$|\delta H(\boldsymbol{y})| < \alpha \bar{H} \quad (\alpha < 1),$$

so that S_- is always below the reference horizontal pane, then by employing the obvious inequality

$$\left|\frac{e^x - 1}{x}\right| = |e^{\vartheta x}| \le e^{|x|} \quad \text{for} \quad |x| \le \bar{x} \quad (|\vartheta| < 1)$$

we find that

$$|\delta\widehat{g}_Z(\boldsymbol{p})| \le \mu 2\pi e^{-2\pi(1-\alpha)p\bar{H}} \int_{R^2} |\delta H| \, d_2 y. \tag{3.125}$$

This implies that $\delta\widehat{g}_Z(\boldsymbol{p})$ is not only bounded but it is also in L^1 and in L^2 too, so that the FT theory and its inversion can be readily applied.

If we want to go further, we need to simplify (3.124), for instance by linearizing with respect to δH, namely

$$\delta\widehat{g}_Z(\boldsymbol{p}) \cong \mu \int_{R^2} e^{2\pi(-p\bar{H}+i\boldsymbol{p}\cdot\boldsymbol{y})} 2\pi\delta H(\boldsymbol{y})d_2y \equiv$$

$$\equiv 2\pi\mu e^{-2\pi p\bar{H}} \int e^{i2\pi\boldsymbol{p}\cdot\boldsymbol{y}}\delta H(\boldsymbol{y})d_2y;$$

in this way we obtain the simple relation

$$\delta\widehat{g}_Z(\boldsymbol{p}) \equiv F\{\delta g_Z(\boldsymbol{x})\} = 2\pi\mu e^{-2\pi p\bar{H}}\mathscr{F}\{\delta H\}. \tag{3.126}$$

Comparing (3.126) to (2.84) we recognize, by applying the convolution theorem, that the former is just the FT of the latter, because

$$\mathscr{F}\left\{\frac{\bar{H}}{(r^2+\bar{H}^2)^{3/2}}\right\} = 2\pi e^{-2\pi p\bar{H}},$$

according to Proposition 3.3.

The difference in sign in this case is due to the fact that in (3.126) $\mu = G\rho$, with ρ the density contrast of the upper layer minus the lower layer, which was the opposite in (2.84).

In any case, (3.126) constitutes a quite simple recipe to compute δg from δH showing the power of FT theory when convolutions have to be computed.

Remark 3.7 Wishing to complete the analysis of a two-layer body we go to the 2D case, namely, when the interface δH is a function of only one of the plane coordinates: $\delta H = \delta H(x)$.Without too much discussion, we go directly to the linearized formula (2.94), only taking into account that with the convention of this section, $\mu = G\rho$, with ρ the difference of density between upper and lower layers, is just the opposite of $\bar{\mu}$ in (2.94). So we have here

$$\delta g_Z(x) = 2\mu\bar{H}\int_{-\infty}^{+\infty}\frac{\delta H(\xi)}{(x-\xi)^2+\bar{H}^2}d\xi \tag{3.127}$$

As we see, (3.127) has the form of a 1D convolution product, namely

$$\delta g_Z(x) = \mu f * \delta H,$$

with

$$f(x) = \frac{2\bar{H}}{(x^2+\bar{H}^2)}. \tag{3.128}$$

Hence, by the convolution theorem,

$$\delta\widehat{g}_Z(\boldsymbol{p}) = \mu\widehat{f}(p)\delta\widehat{H}(p). \tag{3.129}$$

We show that

$$\widehat{f}(\boldsymbol{p}) = 2\pi e^{-2\pi|p|\bar{H}} \tag{3.130}$$

by using the inverse transform formula. In fact

$$\begin{aligned} F(\xi) &= 2\pi \int_0^{+\infty} e^{-2\pi p(\bar{H}+i\xi)} dp + 2\pi \int_{-\infty}^0 e^{2\pi p(\bar{H}+i\xi)} dp \\ &= \frac{1}{\bar{H}-i\xi} + \frac{1}{\bar{H}+i\xi} = \frac{2\bar{H}}{\bar{H}^2+\xi^2}. \end{aligned}$$

So, combining (3.130) and (3.129) we get

$$\delta\widehat{g}_Z(\boldsymbol{p}) = 2\pi\mu e^{-2\pi|p|\bar{H}} \delta\widehat{H}(p), \tag{3.131}$$

which is very similar to (3.126), though not identical, considering that in (3.131) p is a 1D variable running from $-\infty$ to $+\infty$, while in (3.126) $p = |\boldsymbol{p}|$.

Example 3.7 Assume that the 2-layer interface is

$$z = (\bar{H} - \delta H)$$

where

$$\delta H(x) = A \frac{\sin 2\pi a x}{2\pi a x} = A\mathrm{sinc} 2\pi a x.$$

When $A > 0$ we have a lack of mass with density ρ (Fig. 3.5) under the origin, an excess of mass for the two-side lobes and so forth, with bumps following a damped oscillatory profile.

So we still have

$$\begin{aligned} \widehat{g}_Z(\boldsymbol{p}) &= -\mu 2\pi \bar{H} \delta(p) + \delta\widehat{g}_Z(\boldsymbol{p}) \\ &= -\mu 2\pi \bar{H} \delta(p) + \mu 2\pi e^{-2\pi|p|\bar{H}} \delta\widehat{H}(p). \end{aligned}$$

where the first term on the right-hand side represents the Bouguer plate and the second term the interface effect.

Let

$$\frac{1}{2a}\chi_a(p) = \frac{1}{2a} \begin{cases} 1, & \text{if } |p| \leq a, \\ 0, & \text{if } |p| > a. \end{cases}$$

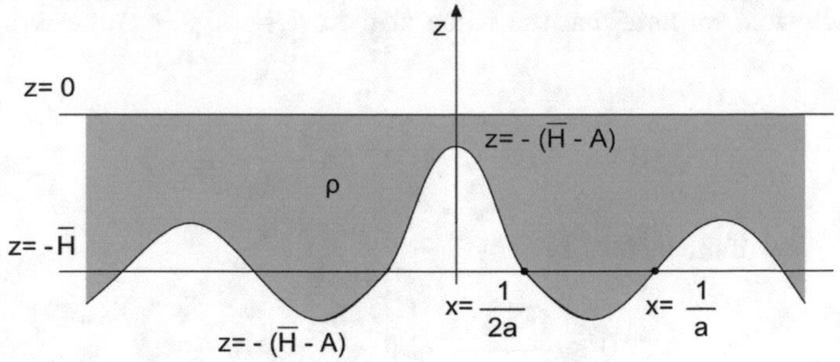

Fig. 3.5 The 2D layer of Example 3.7

Then

$$\mathscr{F}^*\left\{\frac{1}{2a}\chi_a(p)\right\} = \frac{1}{2a}\int_{-a}^{a} e^{-i2\pi px}dp = \operatorname{sinc} 2\pi ax,$$

and so

$$\delta\widehat{H}(p) = \frac{A}{2a}\chi_a(p).$$

It is remarkable that in this elementary example we are able to apply the inverse FT and find the explicit form of $g_Z(x)$. To accomplish this, we need the transform formula

$$\begin{aligned}
\mathscr{F}^*\{e^{-2\pi|p|\overline{H}}\chi_a(p)\} &= \int_{-a}^{a} e^{-2\pi(|p|\overline{H}+ipx)}dp = \\
&= 2\mathrm{Re}\int_0^a e^{-2\pi p(\overline{H}+ix)}dx = \\
&= \frac{1}{\pi}\mathrm{Re}\left\{\frac{1}{\overline{H}+ix}\cdot\left[1-e^{-2\pi a(\overline{H}+ix)}\right]\right\} = \\
&= \frac{1}{\pi}\frac{1}{\overline{H}^2+x^2}\Big[\overline{H}(1-e^{-2\pi a\overline{H}}\cos 2\pi ax) \\
&\quad +xe^{-2\pi a\overline{H}}\sin 2\pi ax\Big].
\end{aligned}$$

So the final result is

$$g_Z(x) = -\mu 2\pi\overline{H} + \mu\frac{A}{a}\frac{1}{\overline{H}^2+x^2}\left[\overline{H}(1-e^{-2\pi a\overline{H}}\cos 2\pi ax) + x\sin 2\pi ax\right].$$

In conclusion, we note that the value at the origin of g_Z can be written in the form

$$g_Z(0) = -\mu 2\pi \overline{H} + \mu 2\pi A \frac{1 - e^{-2\pi a \overline{H}}}{2\pi a \overline{H}}$$

so that, in view of the inequality

$$0 \leq \frac{1 - e^{-x}}{x} < 1, \quad \forall x > 0,$$

we see that, for $A < \overline{H}$, $g_Z(0)$ is less negative than the pure Bouguer factor, i.e., the attraction at the origin is less intense than that of the Bouguer plate, because we have major lack of mass exactly below $x = 0$. This confirms our geometric and physical intuition.

3.7 Numerical Issues: Fourier Series and the Discrete Fourier Transform

Up to this point, we have introduced the FT theory and studied its application to linear (and nonlinear) problems in gravitation formulas: as we have seen, FT is mostly effective when we apply it to convolutions by using Theorems 3.1 and 3.3. Yet such an application of FT requires the knowledge of the FT of various functions, like the interface ΔH between two layers, which, beyond general characteristics regarding their regularity can be quite complicated; and certainly we need generally to use transforms, that go far beyond the typical list of those for which a closed form is known in the literature.

The answer is provided by suitable pieces of numerical analysis and signal theory. The idea is first of all that we will use functions which not only are in L^1 and L^2 (i.e., $f \in L^2 \cap L^1$), but also tend to zero in geometric space (i.e., for $|\boldsymbol{x}| \to 0$), as well as possessing FT tending to zero in the frequency domain (i.e., for $|\boldsymbol{p}| \to \infty$). The focus of the section will be on the numerical approximation of the calculations of convolutions; we will do that for the 1D case for the sake of simplicity and because the generalization to higher dimensions is straightforward.

So let us start with what we mean by *numerical approximation* of a smooth function $f(x) \in L^1 \cap L^2$, tending to zero at infinity and such that $\widehat{f}(p) \in L^1 \cap L^2$.

We achieve this approximation in two steps:

(a) First of all, we identify an interval, $\left\{|x| \leq \frac{L}{2}\right\}$, such that the truncated function

$$f_T(x) = \begin{cases} f(x), & \text{if } |x| \leq \frac{L}{2}, \\ 0, & \text{if } |x| > \frac{L}{2} \end{cases} \tag{3.132}$$

can be considered a good approximation in $L^2 \cap L^1$ to $f(x)$, i.e., a norm of the difference, for instance

$$\begin{aligned} \|f - f_T\|_{L^2} &= \left\{ \int_{-\infty}^{+\infty} |f - f_T|^2 \, dx \right\}^{1/2} \\ &= \left\{ \int_{|x|>\frac{L}{2}} |f|^2 \, dx \right\}^{1/2}, \end{aligned} \tag{3.133}$$

is smaller than a preassigned value; for functions of this kind the Fourier theory is expressed in terms of Fourier series rather than Fourier Transform. We will also show that a suitable convolution theorem holds for such functions.

(b) We shall then discretize the function $f(x)$ in the relevant interval; for reasons that will be soon clarified, we consider $f_T(x)$ on the full interval $[-L, L]$, extending it by zero to the two intervals $[-L, -L/2]$, $[L/2, L]$. The interval $[-L, L]$ is then split into $2M$ regular subintervals of width $\Delta = \frac{2L}{2M}$ and with endpoints

$$x_k = k\Delta, \quad k = -M, \ldots, 0, \ldots M; \tag{3.134}$$

Notice that the total number of points (3.134) is $N = 2M + 1$. Finally, we collect the values

$$f_k = f(k\Delta)k = -M, \ldots M \tag{3.135}$$

into a vector, for which we know that at the beginning and at the end there are a number of zeros for about half of the components.

Let us observe that, once Δ (or M or N) are chosen, we can always interpolate the discrete values $\{f_k\}$, e.g., with linear splines in $\mathbb{R}^1$ (or bilinear splines in $\mathbb{R}^2$) to obtain f_S, a "spline version" of f_T. We can then compute a "sampling error" index like $\varepsilon_S = \|f_T - f_S\|_{L^2}$; indeed, we expect that ε_S tends to zero when $\Delta \to 0$. If we impose a bound to such an error, we have a guide for choosing Δ.

Let us notice as well that the choice of sampling f at a regular grid points (3.134) is not mandatory at this level, but it simplifies a lot further developments. So the combination of the two steps a) and b) associates to $f(x)$ an N-dimensional real vector $\boldsymbol{f}$ of components (3.135). The process, that depends on the two parameters L and Δ (or N), is illustrated in Fig. 3.6.

From the vector $\boldsymbol{f}$ we can reconstruct by interpolation $f_S(x)$, which is an approximation to $f_T(x)$, which in turn is an approximation to $f(x)$. Armed with these tools, we pass now to present Fourier theories and convolution theorems adapted to the points (a) and (b) above. Our purpose is to approximate the calculus of

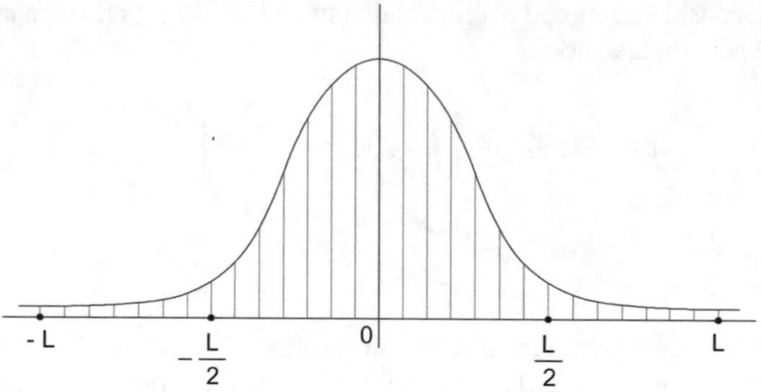

Fig. 3.6 The digitization or sampling process for $f(x)$ with $L = 14$, $\Delta = 1$, $M = 7$, $N = 29$; note that in the processing the first 7 and last 7 values are set to zero

$$h * f = \int_{-\infty}^{+\infty} h(x-y)f(y)dy = g(x). \tag{3.136}$$

(a) Truncation and Fourier Series. According to our program at point (a), we want to replace $h(x)$ and $f(x)$ by $h_T(x)$ and $f_T(x)$ in (3.136). The first question then is: will this provide a good approximation to $h * f$? The answer is in the affirmative, because:

$$\|f_T\|_{L^1} = \int_{-L}^{+L} |f(x)|\, dx \leq \int_{-\infty}^{+\infty} |f(x)|\, dx = \|f\|_{L^1} \tag{3.137}$$

and

$$\begin{aligned} \|h * f\|_{L^1} &= \int_{-\infty}^{+\infty} \left| \int_{-\infty}^{+\infty} h(x-y)f(y)dy \right| dx \leq \\ &\leq \int_{-\infty}^{+\infty} \int_{-\infty}^{+\infty} |h(x-y)|\, |f(y)|\, dydx = \\ &= \int_{-\infty}^{+\infty} |f(y)| \int_{-\infty}^{+\infty} |h(x-y)|\, dxdy = \|h\|_{L^1} \|f\|_{L^1} . \end{aligned} \tag{3.138}$$

Therefore

$$\begin{aligned} \|h * f - h_T * f_T\|_{L^1} &= \|(h - h_T) * f + h_T * (f - f_T)\|_{L^1} \leq \\ &\leq \|h - h_T\|_{L^1} \|f\|_{L^1} + \|h\|_{L^1} \|f - f_T\|_{L^1} , \end{aligned}$$

which clearly tends to zero when $L \to \infty$.

Then we can replace (3.136) by

$$h_T * f_T = g_T(x) \quad \forall x \in \mathbb{R}^1. \tag{3.139}$$

On the other hand, both h_T and f_T have support (i.e., they are $\neq 0$) in $[-L/2, L/2]$. It follows that $g_T\ (x)$ will have support in $[-L, L]$, because for $|x| > L$ the supports of $h_T(x-y)$ and $f_T\ (y)$ are disjoint. So (3.133) can be written as

$$\int_{-L}^{L} h_T(x-y) f_T(y)dy = g_T(x) \quad \forall x \in [-L,\ L]; \tag{3.140}$$

note that the integral is this a formula could be restricted to the interval $[-L/2, L/2]$, yet the form (3.140) has the great advantage that all three functions $h_T,\ f_T,\ g_T$ can now be considered as defined on the *same* interval $[-L, L]$. A careful consideration of the supports of $h_T(x-y)$ and $f_T\ (y)$ in (3.140) shows that g_T not only lies in L^1, but it is also bounded, implying $g_T \in L^2$.

The next step is to establish the Fourier series representation of any $f \in L^2([-L,\ L])$, namely

$$f(x) = \sum_{k=-\infty}^{+\infty} c_k e^{i2\pi k\frac{x}{2L}}, \quad x \in [-L,\ L], \tag{3.141}$$

$$c_k = \frac{1}{2L}\int_{-L}^{+L} f(x) e^{-i2\pi k\frac{x}{2L}} dx. \tag{3.142}$$

The series in (3.141) converges in the $L^2\ ([-L, L])$ topology and equality holds a.e. in $[-L, L]$. That the sequence $\left\{\frac{e^{i2\pi k\frac{x}{2L}}}{\sqrt{2L}}\right\}$ is orthonormal in L^2 is a consequence of the readily proved relation

$$\frac{1}{2L}\left(e^{i2\pi\ell\frac{x}{2L}}, e^{i2\pi k\frac{x}{2L}}\right)_{L^2} = \frac{1}{2L}\int_{-L}^{L} e^{i2\pi(k-\ell)\frac{x}{2L}} dx = \delta_{k\ell}. \tag{3.143}$$

That this sequence is also complete in $L^2\ ([-L, L])$ is a theorem for which we send the reader to literature [24]. The conclusion is that (3.142), or the condensed formula

$$f(x) = \sum_{-\infty}^{+\infty}\left(\frac{e^{i2\pi k\frac{x}{2L}}}{\sqrt{2L}}, f(x)\right)_{L^2} \frac{e^{i2\pi k\frac{x}{2L}}}{\sqrt{2L}},$$

hold true. Then, considering that the three functions $h_T,\ f_T,\ g_T$ are all in L^2, we can use the existing Fourier Series representations

$$h_T(x) = \sum_{k=-\infty}^{+\infty} a_k e^{i2\pi k\frac{x}{2L}}, \quad a_k = \frac{1}{2L}\int_{-L}^{L} e^{-i2\pi k\frac{x}{2L}} h(x)dx$$

$$f_T(x) = \sum_{k=-\infty}^{+\infty} b_k e^{i2\pi k\frac{x}{2L}}, \quad b_k = \frac{1}{2L}\int_{-L}^{L} e^{-i2\pi k\frac{x}{2L}} f(x)dx$$

$$g_T(x) = \sum_{k=-\infty}^{+\infty} c_k e^{i2\pi k\frac{x}{2L}}, \quad c_k = \frac{1}{2L}\int_{-L}^{L} e^{-i2\pi k\frac{x}{2L}} g(x)dx$$

and insert them into (3.140) to obtain, with the help of (3.143),

$$\begin{aligned}&\int_{-L}^{L} \sum_{k=-\infty}^{+\infty} a_k e^{i2\pi k\frac{x-y}{2L}} \sum_{\ell=-\infty}^{+\infty} b_\ell e^{i2\pi \ell\frac{y}{2L}} dy = \\ &= \sum_{k=-\infty}^{+\infty} a_k b_k 2L\, e^{i2\pi k\frac{x}{2L}} \equiv \sum_{k=-\infty}^{+\infty} c_k e^{i2\pi k\frac{x}{2L}}.\end{aligned} \tag{3.144}$$

Since $\left\{e^{i2\pi k\frac{x}{2L}}\right\}$ is a complete orthogonal system in L^2, the above relation implies

$$2La_k b_k = c_k \quad (-\infty < k < +\infty), \tag{3.145}$$

which is nothing but the convolution theorem for Fourier series.

Remark 3.8 A word of caution is necessary to understand the proof of (3.145).

In fact, note that when y runs from $-L$ to L in the convolution integral, $x - y$ runs from $x - L$ to $x + L$, i.e., $h_T(x - y)$ is evaluated, for instance when $x > 0$, at points to the right of the interval $[-L, L]$.

Since in formula (3.144) $h_T(x - y)$ is defined in terms of its Fourier series, it is this function that we need to evaluate outside $[-L, L]$. To understand what happens in (3.144), we define the extended function

$$\overline{h}(t) = \sum_{k=-\infty}^{+\infty} a_k e^{i2\pi k\frac{t}{2L}}, \quad \forall t \in R^1$$

and we observe that indeed

$$\overline{h}(t) = h_T(t) \quad -L \le t \le L.$$

However, since $e^{i2\pi k\frac{t}{2L}}$ is periodic of period $2L$, whatever is k, we see that $\bar{h}(t)$ is jut the periodic repetition of $h_T\,(t)$ over the whole axis, as shown in Fig. 3.7.

Figure 3.8 shows the relative position of the supports of $\bar{h}(x-y)$ and $f\,(y)$, when $\frac{L}{2} < x < L$; we realize then that the support of the product is just $\left[\left(x - \frac{L}{2}\right), \frac{L}{2}\right]$

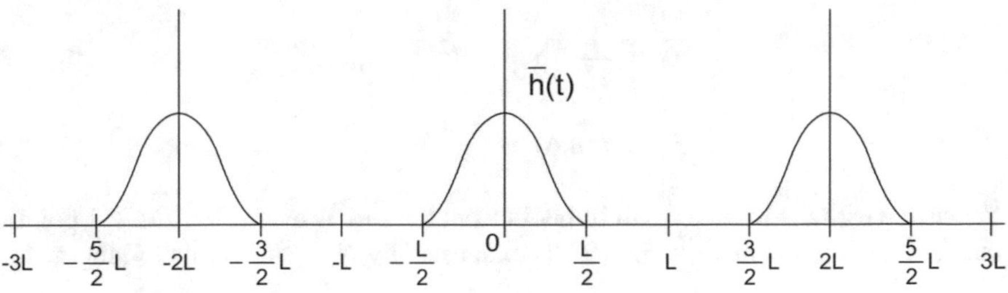

Fig. 3.7 The plot of the periodic $\bar{h}(t)$

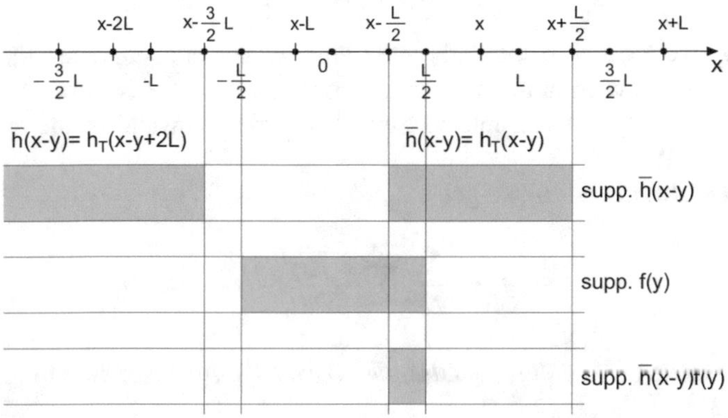

Fig. 3.8 The supports of $\bar{h}(x - y)$, $f(y)$ and $\bar{h}(x - y)\ f(y)$. Note that on supp $\bar{h}(x - y)\ f(y)$ we have $\bar{h}(x - y) = h_T(x - y) (= h(x - y))$; note also that when $x \to L$, the right endpoint of supp $h_T(x - y + 2L)$ tends to $-L/2$ from below, never overlapping the supp of $f\ (y)$; the situation is symmetric when $x < 0$

where $h(x - y) \equiv h_T(x - y)$, so that the integral (3.144) is in fact identical to the integral (3.140) that we want to compute. This is an effect of our choice of padding $\left[-\frac{L}{2},\ \frac{L}{2}\right]$ with two intervals, namely $\left[-L,\ -\frac{L}{2}\right]$, $\left[\frac{L}{2},\ L\right]$, of zero values. Although not strictly necessary, such a choice makes our discretization process more transparent. By the way, should the zero padding be longer than $2\left(\frac{L}{2}\right) = L$, nothing would change in the reasoning of this Remark.

Now while the knowledge of a certain number of coefficients c_k could be considered as a good information to approximate $g_T\ (x)$ by a finite Fourier series, the computation of a_k, b_k bears more or less the same difficulty that we have in computing Fourier Transform. So, in order to accomplish our approximation program, we need to go to step (b).

(b) Sampling and the Discrete Fourier Transform (DFT). The sampling of f_T on $[-L, L]$ has already been defined in (3.134), (3.135) and Fig. 3.6. Here we just consider the discretization of the integral (3.142) for the calculus of Fourier series coefficients. The idea is to replace (3.142) by

$$\widehat{f_k} = \frac{1}{2L} \sum_{n=-M}^{M} e^{-i2\pi k \frac{n\Delta}{2L}} f_n \Delta, \tag{3.146}$$

$$f_n = f(n\Delta). \tag{3.147}$$

Yet such a formula hides a small imperfection; the sum is over $N = 2M + 1$ points, but the interval Δ is just $\frac{2L}{2M}$, i.e., $2L$ is exhausted by $N - 1$ intervals of length Δ.

To correct for this effect we modify Δ slightly by putting

$$2L = N\Delta; \tag{3.148}$$

In this way we have N Δ-intervals in $2L$ and we can accommodate the points x_k in such a way that each of them belongs to one Δ-interval only. Indeed, x_k will not be at the center of the small interval, but this does not invalidate the approximate formula (3.146).

Using (3.148) in (3.146) we get

$$\widehat{f_k} = \frac{1}{N} \sum_{n=-M}^{M} e^{-i2\pi \frac{kn}{N}} f_n, \tag{3.149}$$

which we take as the *definition of Discrete Fourier Transform* of the sample $\{f_n; n = -M, \dots, M\}$.

It is remarkable that in (3.149) the only place where Δ appears implicitly is in the components $f_n = f(n\Delta)$, so that once the vector $\boldsymbol{f}$ is formed, the shape of the DFT is fixed by one parameter only, N. Another remark is that the mapping of $\boldsymbol{f}$ into $\hat{f} = \{f_k; k = -M, \dots, M\}$ seems to multiply by a factor of 2 the number of involved parameters, because the components of $\boldsymbol{f}$ are real, while the components of $\hat{f}$ are complex. Yet this is not the case since: first of all

$$\widehat{f_0} = \frac{1}{N} \sum_{n=-M}^{M} f_n$$

is indeed always real, i.e., represented by one parameter; furthermore,

$$\widehat{f_{-n}} = \widehat{f_n^*},$$

hence the number of independent parameters in a vector $\widehat{\boldsymbol{f}} \in \mathbb{C}^N$, which is the DFT of a real vector $\boldsymbol{f} \in \mathbb{R}^N$, is again $2M + 1 = N$. It is then natural to ask whether the operator (in this case the matrix) that implements the DFT, namely

$$\Phi \equiv \left\{ e^{-i2\pi \frac{kn}{N}} \right\}, \quad (k = -M, \dots, M, n = -M, \dots, M) \tag{3.150}$$

is invertible or not.

Before we answer the question, let us note that with the notation (3.150), formula (3.149) takes on the vector form

$$\widehat{f} = \frac{1}{N}\Phi f. \tag{3.151}$$

Moreover, for a matrix, like Φ, with complex entries, there is an important related matrix called the adjoint of Φ, denoted by $\Phi^\dagger$, and defined as

$$\Phi^\dagger = (\Phi^\top)^*, \tag{3.152}$$

where $\top$ stands for transpose and (*) stands for complex conjugation. So when Φ is (3.150)

$$(\Phi^\dagger)_{kn} = \left\{e^{i2\pi\frac{nk}{N}}\right\}, \quad (k = -M, \ldots, M; n = -M, \ldots, M). \tag{3.153}$$

As we see, transposition has little effect on Φ because this is symmetric, while conjugating changes the sign of the exponentials.

The inversion formula of (3.151) is a consequence of the following proposition.

Proposition 3.4 *With the notation of* (3.150) *and* (3.152) *one has*

$$\Phi^\dagger\Phi = N I. \tag{3.154}$$

Proof We have to check that

$$(\Phi^\dagger\Phi)_{\ell n} - \sum_{k=-M}^{M} (\Phi^\dagger)_{\ell k}(\Phi)_{kn} = N\delta_{\ell n}. \tag{3.155}$$

Since

$$(\Phi^\dagger\Phi)_{\ell n} = \sum_{k=-M}^{M} e^{i2\pi\frac{k(\ell-n)}{N}}, \tag{3.156}$$

we immediately see that, for $n = \ell$,

$$(\Phi^\dagger\Phi)_{nn} = N,$$

showing that the diagonal part of (3.155) is true. For $n \neq \ell$ put

$$\xi = e^{i2\pi\frac{\ell-n}{N}};$$

note that since both n, ℓ are between $-M$ and M, for $n \neq \ell$ we necessarily have

$$\xi \neq 1. \tag{3.157}$$

Then we find

$$\sum_{k=-M}^{M} \xi^k = \xi^{-M} \sum_{k=0}^{2M} \xi^k = \xi^{-M} \frac{1 - \xi^{2M+1}}{1 - \xi}; \tag{3.158}$$

in view of (3.157) and (3.158) is perfectly meaningful.

On the other hand

$$\xi^{2M+1} = \xi^N = \left[e^{i2\pi \frac{\ell - n}{N}}\right]^N = e^{i2\pi(\ell - n)} = 1, \tag{3.159}$$

because $\ell - n$ is always an integer.

So, from (3.159) and (3.158)

$$\ell \neq n, \ (\Phi^\dagger \Phi)_{\ell n} = \sum_{k=-M}^{M} \xi^k = 0,$$

and the off-diagonal part of (3.155) is proved too. □

With the help of Proposition 3.4, we easily invert formula (3.151), namely

$$\boldsymbol{f} = \Phi^\dagger \widehat{\boldsymbol{f}}, \tag{3.160}$$

or, in components,

$$f_n = \sum_{k=-M}^{M} e^{i2\pi \frac{nk}{N}} \widehat{f}_k. \tag{3.161}$$

Formula (3.161) defines the inverse DFT.

Remark 3.9 There are other ways to express (3.154), namely one says that the matrix

$$DF = \frac{1}{\sqrt{N}} \Phi$$

is unitary, or that its column vectors are orthonormal with respect to the Euclidean scalar product in $\mathbb{C}^N$, defined for $\boldsymbol{a}, \boldsymbol{b} \in \mathbb{C}^N$ by

$$\langle \boldsymbol{a}, \boldsymbol{b} \rangle = \boldsymbol{a}^\dagger \boldsymbol{b}.$$

We can pass now to the discretization of the convolution formula (3.140). Let us first discretize the integral, replacing (3.140) by

$$\sum_{k=-M}^{M} h_T(x - k\Delta) f_T(k\Delta)\Delta = g_T(x), \quad -L \leq x \leq L.$$

However, since we want to use only the sampled versions of h_T, f_T, g_T, we require that the above relation holds not on the whole interval $[-L, L]$, but only at the grid points $x_n = n\Delta$, namely we write:

$$\sum_{k=-M}^{M} h_T((n-k)\Delta) f_T(k\Delta)\Delta = g_T(n\Delta),$$

or

$$\sum_{k=-M}^{M} h_{n-k} f_k \Delta = g_n, \qquad n = -M, \ldots, M. \tag{3.162}$$

This formula is the discretized version of (3.140).

In (3.162) we substitute

$$h_{n-k} = \sum_{\ell=-M}^{M} \widehat{h}_\ell e^{i2\pi\ell\frac{n-k}{N}}. \tag{3.163}$$

Notice that by (3.163) we can define in fact a function of $n - k$ that extends periodically to the right of M, or to the left of M, as we have already observed in Remark 3.8.

Such an extension though coincides with h_{n-k} at the integers contained in the interval $\left[-\frac{M}{2}, \frac{M}{2}\right]$, where f_n is different from zero.

So we use (3.163), together with

$$f_k = \sum_{j=-M}^{M} \widehat{f}_j e^{i2\pi j\frac{k}{N}},$$

to obtain, recalling (3.155),

$$\begin{aligned} g_n &= \sum_{k=-M}^{M} \left(\sum_{\ell=-M}^{M} \widehat{h}_\ell e^{i2\pi\ell\frac{n-k}{N}} \right) \sum_{k=-M}^{M} \left(\sum_{j=-M}^{M} \widehat{f}_j e^{i2\pi j\frac{k}{N}} \right) \Delta = \\ &= \sum_{\ell,j=-M}^{M} \widehat{h}_\ell \widehat{f}_j e^{i2\pi\ell\frac{n}{N}} \sum_{k=-M}^{M} e^{i2\pi k\frac{\ell-j}{N}} \Delta = \\ &= \sum_{\ell=-M}^{M} (\widehat{h}_\ell \widehat{f}_\ell N\Delta) e^{i2\pi\ell\frac{n}{N}}. \end{aligned} \tag{3.164}$$

In view of the uniqueness of the DFT and noting that $N\Delta = 2L$, (3.164) yields

$$2L\widehat{h}_\ell \widehat{f}_\ell = \widehat{g}_\ell \ (\ell = -M, \ldots, M) \tag{3.165}$$

which is the *convolution theorem for DFT.*

Note the similarity of (3.165) with (3.145), the difference being in the range swept by the respective indexes.

As we see, formula (3.165) is the core of our solution; namely, from $h(x)$, $f(x)$ by sampling we compute the vectors $\{h_n\}$, $\{f_n\}$, then we apply the DFT and by (3.165) we get the DFT of $g = h * f$; the inverse DFT (3.161) gives us $g(n\Delta)$ in the range $-M \leq n \leq M$, where we know that $g(x)$ is significantly different from zero. A final interpolation, if needed, gives us a continuous (approximation) of $g(x)$ on $[-L, L]$.

Remark 3.10 One of the reasons why the DFT algorithm is so powerful numerically, is that when $N = 2^m$, i.e., N is a power of 2, one can apply an algorithm, known as *Fast Fourier Transform* (FFT), which computes the DFT in an extremely short time. For instance, the algorithm implemented in C can compute the DFT of a vector of $1,048,576 = 2^{20}$ data in less than 0.1 second on a standard desktop PC. When we can choose N we can always satisfy the above condition; however, if N is fixed, for instance because the values of one or more functions are derived from measurements, we can always pass to a vector of length 2^m, for some m, by further padding the data with zero components, which does not invalidate the reasoning of this section. The interested readers can consult [48, 111].

Before passing to the next item, we still want to analyse the relation of the DFT $\widehat{\boldsymbol{f}}$ with the actual FT of $f(x)$. This will allow us to clarify the phenomenon of aliasing of $\widehat{f}(p)$ due to discretization, known in signal theory as the *Nyquist Theorem* [99]. We start observing that by truncating and (regularly) sampling we reduce the information on $f(x)$ to the vector

$$\boldsymbol{f} \equiv \{f_n\} \equiv \{f(n\Delta)\}, \quad n = -M, \ldots, M.$$

But then we can write

$$f_n = \int_{-\infty}^{+\infty} e^{-i2\pi n\Delta p} \widehat{f}(p) dp. \tag{3.166}$$

First of all let us observe that the functions

$$e^{-i2\pi n\Delta p} = e^{-i2\pi n \frac{p}{\overline{p}}}$$

share the same period $\overline{p} = 1/\Delta$ for all n, namely

$$e^{-i2\pi n \frac{(p+j\bar{p})}{\bar{p}}} = e^{-i2\pi n \frac{p}{\overline{p}}}. \tag{3.167}$$

Now we write (3.166) in the form

$$
\begin{aligned}
f_n &= \sum_{j=-\infty}^{+\infty} \int_{j-\frac{1}{2}\bar{p}}^{j+\frac{1}{2}\bar{p}} e^{-i2\pi n \frac{p}{\bar{p}}} \widehat{f}(p)dp = \\
&= \sum_{j=-\infty}^{+\infty} \int_{-\frac{1}{2}\bar{p}}^{\frac{1}{2}\bar{p}} e^{-i2\pi n \frac{j\bar{p}+q}{\bar{p}}} \widehat{f}(j\bar{p}+q)dq = \\
&= \int_{-\frac{1}{2}\bar{p}}^{\frac{1}{2}\bar{p}} e^{-i2\pi n \frac{q}{\bar{p}}} \sum_{j=-\infty}^{+\infty} \widehat{f}(j\bar{p}+q)dq.
\end{aligned}
\tag{3.168}
$$

We put

$$
-\frac{1}{2}\bar{p} \le q \le \frac{1}{2}\bar{p}, \; F(q) = \sum_{j=-\infty}^{+\infty} \widehat{f}(j\bar{p}+q) \tag{3.169}
$$

in (3.168) and we further discretize the integral, with a basic interval of width $p_0 = \frac{\bar{p}}{N} = \frac{1}{N\Delta}$, yielding

$$
\begin{aligned}
f_n &\cong \sum_{k=-M}^{M} e^{-i2\pi n \frac{kp_0}{\bar{p}}} F(kp_0) \cdot p_0 = \\
&= \sum_{k=-M}^{M} e^{-i2\pi \frac{nk}{\bar{p}}} F(kp_0)p_0.
\end{aligned}
$$

Comparing with (3.161) we get

$$
\begin{aligned}
\widehat{f}_k &= F(kp_0)p_0 = \\
&= \sum_{j=-\infty}^{+\infty} \widehat{f}(j\bar{p}+kp_0)p_0 = \\
&= \sum_{j=-\infty}^{+\infty} \widehat{f}\left[(jN+k)\,p_0\right]p_0 = \\
&= \sum_{j=-\infty}^{+\infty} \widehat{f}\left[\left(j+\frac{k}{N}\right)\bar{p}\right]p_0, \quad k=-M,\ldots,M.
\end{aligned}
\tag{3.170}
$$

Formula (3.170) can be read as

$$
\widehat{f}_k = \widehat{f}(kp_0)p_0 + \widehat{f}(\bar{p}+kp_0)p_0 + \widehat{f}(-\bar{p}+kp_0)p_0 + \cdots,
$$

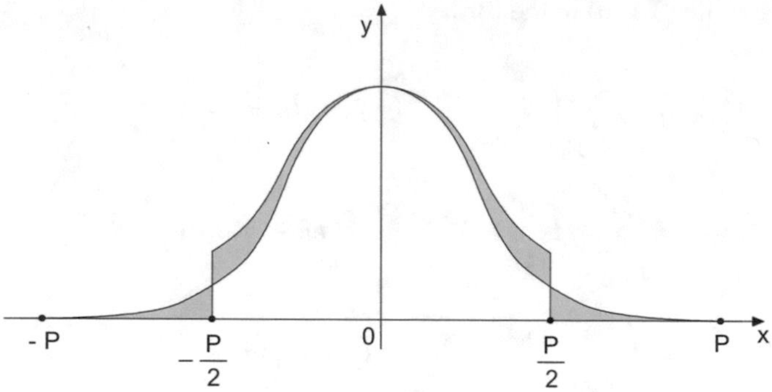

Fig. 3.9 Illustration of the aliasing phenomenon

which shows that $\widehat{f_k}$ is the sampling value of the spectrum at $kp_0 = \frac{k}{N}\bar{p}$ plus the values at $\pm\bar{p} + \frac{k}{N}\bar{p}, \pm 2\bar{p} + \frac{k}{N}\bar{p}, \ldots$ folded into the interval $\left[-\frac{M}{N}\bar{p}, \frac{M}{N}\bar{p}\right]$. This folding of the Fourier Transform $f\,(p)$, represented in Fig. 3.9, is called *spectral aliasing*.

On the one hand the aliasing phenomenon shows that no information on $\widehat{f}(p)$ is given by $\widehat{f_k}$ above the *Nyquist frequency*

$$|p| \leq p_{N_y} = \frac{M}{N}\bar{p} \cong \frac{1}{2\Delta}; \tag{3.171}$$

on the other hand, we find that $\widehat{f_k}$ is a reliable estimate of $\widehat{f}(p_k) = \widehat{f}(kp_0) = \widehat{f}(\frac{k}{N}\bar{p})$ only if the content of $\widehat{f}(p)$ above the Nyquist frequency p_{Ny} is really negligible and rapidly decaying to zero at infinity.

3.8 Some Considerations About Sensitivity of the Gravitational Signal

In this first part of the book we are studying the effects of anomalous mass densities on the vertical component of the gravity vector, in particular considering their geometric distribution.

So what we treated are isolated anomalies that give a relatively small contribution to a signal, the modulus of the gravity vector, which has a very large value $g \sim 983$ gal, a large part of which however is accounted for by the ellipsoidal shape of the Earth and by its rotation. The remaining part, that we will call gravity anomaly, can amount to a maximum of say 200 mGal. Most of this is explained by the influence of topographic masses and by geophysical features with a scale (wavelength) of several thousands down to some dozens of kilometres.

It is our purpose to roughly illustrate here how local mass anomalies with horizontal extent of less than 100 km and with a depth of less than 10 km and a density contrast of 0.3 gm^{-3} at most, can produce a signal up to some 20/30 mGal.

The choice of $\rho = 0.3$ g cm^{-3} is large but not unrealistic, so we shall use it in this context, very much aware that in any event the resulting values of g_Z are just proportional to ρ, as far as we are treating homogeneous bodies.

The accuracy of the observations of gravity can go down to a few μGal, yet there are many local accidental phenomena that can easily produce a signal of some tenth of a mGal. So, just to fix ideas, we will focus on signals in the range between 0.1 and 20 mGal. In particular, we are interested in getting acquainted with the sensitivity of the gravity signal to the resolution of the source, i.e., the characteristic dimension of a mass anomaly that can produce a perceivable (and later on interpretable) signal.

We first of all do that for a 2-layer situation in both 2D and 3D, by applying the linearized formulas (3.131) and (3.126), respectively.

Let us start with the easier, 2D case. We assume that the interface anomaly δH has the shape

$$\delta H(x) = A \cos 2\pi \bar{p} x \equiv A \cos 2\pi \frac{x}{L}, \tag{3.172}$$

where A is the amplitude of the source signal and $L = \frac{1}{\bar{p}}$ is the resolution parameter. Notice that, strictly speaking, the function (3.172) does not have a Fourier Transform as a regular function, yet, by recalling (3.72) and (3.73) and the Euler formula

$$\cos 2\pi \bar{x} = \frac{1}{2}\left(e^{i2\pi \bar{p}} + e^{-i2\pi \bar{p} x}\right),$$

we immediately find that $\mathscr{F}\{\delta H\}$ is a distribution, namely

$$\mathscr{F}\{\delta H\} = \frac{A}{2}[\delta(p + \bar{p}) + \delta(p - \bar{p})].$$

If we use this in (3.131) and apply the inverse FT, we get

$$\begin{aligned} g_Z(x) &= 2\pi \mu A \int_{-\infty}^{+\infty} e^{-2\pi |p| \bar{H} + 2\pi i p x} \frac{1}{2} [\delta(p + \bar{p}) + \delta(p - \bar{p})]\, dp \\ &= 2\pi \mu A e^{-2\pi \bar{p} \bar{H}} \cos 2\pi \bar{p} x. \end{aligned} \tag{3.173}$$

Computing (3.173) at $x = 0$, as representative value, we get

$$g_Z(0) = 2\pi \mu A e^{-2\pi \frac{\bar{H}}{L}}. \tag{3.174}$$

In (3.174) the factor $2\pi\mu = 2\pi G\rho$ is fixed by our choice of $\rho = 0.3$ g cm^{-3} and it turns out that $2\pi\mu = 12.6$ mGal km^{-1}. Given that, $g_Z(0)$ is a function of the three length parameters A, L, H.

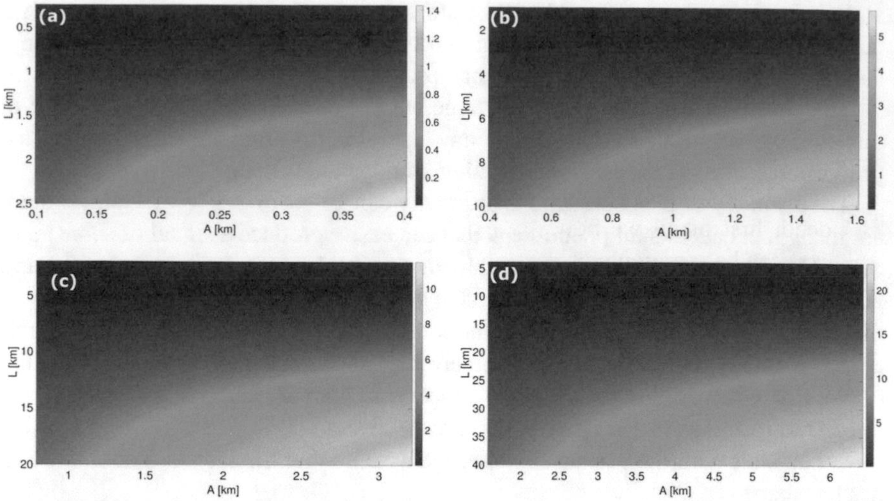

Fig. 3.10 Values of $g_Z(0)$ corresponding to (**a**) $\bar{H} = 0.5$ km, (**b**) $\bar{H} = 2$ km, (**c**) $\bar{H} = 4$ km, and (**d**) $\bar{H} = 8$ km for different values of A and L, in km. $g_Z(0)$ values are in mGal

Table 3.1 Values of $\bar{H}$ investigated in the (a), (b, (c) and (d) scenarios

a	b	c	d	
$\bar{H} = 0.5$	2	4	8	(km)

Table 3.2 $g_Z(0)$ for different values of the L parameter

L	2	4	6	8	(km)
$g_Z(0)$	0.02	0.4	1.2	2.1	(mGal)

The values of the function are presented in Fig. 3.10a–d. In each panel we fix H, according to Table 3.1. Moreover, the range of A is taken between 0.2 $\bar{H}$ and 0.8 $\bar{H}$, recalling that we must always have $0 < A < \bar{H}$. The range of L, on the contrary, is chosen between 0.5 $\bar{H}$ and 5 $\bar{H}$, so that L is never larger than half the diameter of the area in which we perform our analysis.

Just to analyze some numbers, take $\bar{H} = 2$ km, $A = 0.8$ km and $L = 2, 4, 6, 8$ km, as in Table 3.2. As we see when the resolution is $L = 2$ km $= \bar{H}$ we fall below our lower threshold of the signal, as one can also see in Fig. 3.10b.

At this point we wish to use the same example in a 3D context, on account of the fact that the transfer function of (3.126) $2\pi\mu e^{-2\pi p\bar{H}}$ is formally the same as that in (3.131). We only have to understand what is the interface $\delta H(x)$ that can produce the same result.

We claim that this is the case if

$$\delta H(x) = A J_0(2\pi \bar{p} r)$$
$$\left(\bar{p} = \frac{1}{L}, r = |x|\right); \tag{3.175}$$

see Remark 3.6 for definition of J_0.

Indeed, first of all we notice that the identity operator I in $L^2(0, +\infty)$ with weight p can be represented by the distribution

$$I \sim \frac{\delta(p-q)}{q},$$

because

$$\int_0^{+\infty} \frac{\delta(p-q)}{q} f(q) \cdot q dq = f(p).$$

Therefore we consider that the identity

$$B^2 = I,$$

established in Remark 3.6, can be written in the form (cf. (3.99) for the definition of B)

$$4\pi^2 \int_0^{+\infty} J_0(2\pi pr) J_0(2\pi qr) r dr = \frac{\delta(p-q)}{q}. \tag{3.176}$$

Finally we recall that, see (3.96), when a function f depends on $r = |\boldsymbol{x}|$, then

$$\widehat{f}(p) = \mathscr{F}\{f\} = B\{f\} = 2\pi \int_0^{+\infty} f(r) J_0(2\pi pr) r dr. \tag{3.177}$$

Putting $f(r) = \delta H(r) = A J_0(2\pi \bar{r})$ in (3.177) and using (3.176), we see that

$$\delta \widehat{H}(q) = \frac{A}{2\pi} \frac{\delta(\bar{p}-q)}{q}.$$

Therefore from (3.126), after applying the inverse FT, we obtain

$$\begin{aligned} \delta g_Z(r) &= 2\pi \int_0^{+\infty} J_0(2\pi qr) \delta \widehat{g}_z(q) q dq = \\ &= 4\pi^2 \mu A \int_0^{+\infty} J_0(2\pi qr) e^{-2\pi q \bar{H}} \frac{1}{2\pi} \frac{\delta(\bar{p}-q)}{q} q dq = \\ &= 2\pi \mu A J_0(2\pi \bar{p} r) e^{-2\pi \bar{p} H}. \end{aligned} \tag{3.178}$$

By taking $r = 0$ in (3.178) we get

$$\delta g_Z(0) = 2\pi \mu A e^{-2\pi \frac{\bar{H}}{L}}$$

which is formally identical to (3.174), with the only difference that in (3.174) L stands for the period of the cosine function, while in (3.178) L is related to the value of the radius at which the function $J_0(2\pi \frac{r}{L})$ attains the first negative minimum. In fact by analogy L can be taken as the resolution parameter of the interface δH, in the sense that the $\cos 2\pi \frac{x}{L}$ has the first negative minimum at $0.5\ L$, while $J_0(2\pi \frac{r}{L})$ has the first negative minimum at $\sim 0.6L$ (see Fig. 3.1).

Remark 3.11 We note that the signal generated in the above example is quite weak. Just for comparison, let us compute the value of $g_Z(0)$ when the anomaly is contained in a sphere of radius R and with centered at depth $\bar{H}$. Assume one has

$$R = 0.8\bar{H};$$

then, according to (2.9), with a reversed sign (here the density contrast ρ is positive), we get

$$\begin{aligned} g_Z(0) &= \frac{G\frac{2}{3}\pi R^3 \bar{H}}{\bar{H}^3} = \\ &= \mu \frac{2}{3} 0.8^3 \bar{H} = \mu\, 0.34\, \bar{H}. \end{aligned} \tag{3.179}$$

Formula (3.78) with $\mu = 12.6$ mGal km^{-1} and various depths $\bar{H}$ gives

$\bar{H} =$	0.5	1	2	4	8	(km)
$g_Z(0) =$	2.1	4.3	8.6	17.2	34.4	(mGal).

As we see, this signal is much more intense than that in the previous example. This is quite understandable because in the "periodic" 2-layer case (linearized formula) there are excess and lack of masses alternating, not so far from the origin, while in the case of the sphere all masses are in excess, producing a more intense effect.

Let us observe that, following Remark 1.4, any bulky body produces a gravitational signal with a leading term of spherical form (the monopole term: see (1.79)), so that the above consideration should not be restricted to the case of the sphere.

Based on the above examples and reasonings, we conclude the section with a kind of rule of thumb that will be useful when we will apply it to the gravity interpretation in Part III of the book.

For a mass distribution with alternating density, $\pm\rho$, like an oscillating interface between two layers, the horizontal resolution, i.e., the minimum wavelength that can produce a signal above 0.1 mGal, can hardly be lower than the mean depth $\bar{H}$, depending on the value of ρ and the precise geometry of the anomaly. The signal of such an anomaly is attenuated with an exponential factor in $\bar{H}$,

A bulky body generally gives rise to a more intense signal with a higher horizontal resolution and an attenuation factor of the order of a power of $\bar{H}$.

3.9 Exercises

Exercise 1 Considering Eq. (3.127), prove that:

$$\int_{-\infty}^{+\infty} e^{-\pi x^2} dx = 1;$$

use the Gaussian identity (or Euler-Poisson integral):

$$\int_{-\infty}^{+\infty} e^{-x^2} dx = \sqrt{\pi}.$$

Solution To calculate the required integral we have just to replace x by $\frac{u}{\sqrt{\pi}}$, i.e., set $u = \sqrt{\pi}x$ and $dx = \frac{1}{\sqrt{\pi}}du$.

The integral therefore becomes

$$\frac{1}{\sqrt{\pi}} \int_{-\infty}^{+\infty} e^{-u^2} du,$$

which, given the Gaussian identity, is equal to 1.

Exercise 2 Let $\delta H(x) = e^{-\pi A^2 x^2}$. Compute $\delta \widehat{H}(p) = \mathscr{F}\{\delta H(x)\}$ and $\delta \widehat{g}_z(p)$.

Solution.

By the definition of FT,

$$\delta \widehat{H}(p) = \int_{-\infty}^{+\infty} e^{i2\pi px} \delta \widehat{H}(x) dx = \int_{-\infty}^{+\infty} e^{i2\pi px} e^{-\pi A^2 x^2} dx,$$

hence

$$\delta \widehat{H}(p) = \int_{-\infty}^{+\infty} e^{-\pi(-2ipx + A^2 x^2)} dx. \tag{3.180}$$

If we add and remove here $e^{-\pi\frac{p^2}{A^2}}$ we obtain

$$\delta\widehat{H}(p) = \int_{-\infty}^{+\infty} e^{-\pi\left(-2ipx+A^2x^2-\frac{p^2}{A^2}+\frac{p^2}{A^2}\right)}dx = \int_{-\infty}^{+\infty} e^{-\pi\left[(Ax-i\frac{p}{A})^2+\frac{p^2}{A^2}\right]}dx,$$

which can be rewritten as

$$\delta\widehat{H}(p) = e^{\frac{-\pi p^2}{A^2}}\int_{-\infty}^{+\infty} e^{-\pi(Ax-i\frac{p}{A})^2}dx. \tag{3.181}$$

By substituting in Eq. (3.181) $\xi = Ax - i\frac{p}{A}$ we have

$$\delta\widehat{H}(p) = e^{\frac{-\pi p^2}{A^2}}\int_{-\infty}^{+\infty} e^{-\pi\xi^2}\frac{1}{A}d\xi. \tag{3.182}$$

Here the argument is somehow heuristic but it can be rigorously justified, from Exercise 1, the integral in (3.182) is equal to $\frac{1}{A}$ and therefore $\delta\widehat{H}(p) = \frac{1}{A}e^{\frac{-\pi p^2}{A^2}}$, which is the answer to the first question.

Now $\delta\widehat{g}_z(p)$ can be computed by applying the convolution theorem or directly Eq. (3.126) as

$$\delta\widehat{g}_Z(p) \equiv \mathscr{F}\left\{\frac{\bar{H}}{\left(r^2+\bar{H}^2\right)^{\frac{3}{2}}}\right\}\mathscr{F}\left\{\delta H(x)\right\},$$

which gives

$$\delta\widehat{g}_Z(p) = \frac{2\pi}{A}e^{-2\pi|p|\bar{H}}e^{\frac{-\pi p^2}{A^2}}.$$

Exercise 3 Compute $\delta g_Z(0)$ for $\delta H(x) = e^{-\pi A^2x^2}$ by using its FT and recalling that $\delta g_Z(0) = \int_{-\infty}^{+\infty}\delta\widehat{g}_z(p)dp$.

Solution Considering the solution of Exercise 2, we have to compute the integral

$$\delta g(0) = \int_{-\infty}^{+\infty}\delta\widehat{g}_z(p)dp = \int_{-\infty}^{+\infty}\frac{2\pi}{A}e^{-2\pi|p|\bar{H}}e^{\frac{-\pi p^2}{A^2}}dp,$$

which can be rewritten as

$$\delta g(0) = \frac{2\pi}{A}\int_{-\infty}^{+\infty} e^{-\pi\left(2|p|\bar{H}+\frac{p^2}{A^2}\right)}dp.$$

By adding and removing in the exponent the term $A^2\bar{H}^2$ we have

$$\begin{aligned}\delta g(0) &= \frac{2\pi}{A}\int_{-\infty}^{+\infty} e^{-\pi\left(2|p|\bar{H}+\frac{p^2}{A^2}+A^2\bar{H}^2-A^2\bar{H}^2\right)}dp\\ &= \frac{2\pi}{A}\int_{-\infty}^{+\infty} e^{-\pi((A\bar{H}+\frac{p}{A})^2-A^2\bar{H}^2)}dp.\end{aligned}$$

Rearranging the terms yields

$$\delta g(0) = \frac{2\pi}{A}e^{\pi A^2\bar{H}^2}\int_{-\infty}^{+\infty} e^{-\pi(A\bar{H}+\frac{p}{A})^2}dp.$$

Substituting in the integral $\xi = A\bar{H} + \frac{p}{A}$ we obtain

$$\delta g(0) = \frac{2\pi}{A}e^{\pi A^2\bar{H}^2}\int_{-\infty}^{+\infty} e^{-\pi\xi^2}Ad\xi = 4\pi e^{\pi A^2\bar{H}^2}\int_{0}^{+\infty} e^{-H\xi^2}d\xi,$$

which gives

$$\delta g(0) = 2\pi e^{\pi A^2\bar{H}^2}\left[1+\operatorname{erf}\left(\sqrt{H}A^2\bar{H}\right)\right]\frac{\sqrt{\pi}}{2\sqrt{H}}.$$

Part II
The Preprocessing and Processing of Gravity Data: From Observations to a Gravity *map* on a Local Horizontal Plane

Chapter 4
The Gravity Field of the Earth

4.1 Outline of the Chapter

In Part I of the book we have studied the effects in terms of gravitation of an "anomalous body", namely a body with some density embedded in some larger body, e.g., a layer, of uniform density producing a constant gravitational field. Such effects are expressed in terms of Newton's integral and we have studied their properties with various mathematical tools.

However, the physical reality of the Earth is quite different, especially when we examine it at a global scale. At a first glance it appears as a planet of spherical form, with a radius of 6371 km and a mass $M = 5.98 \cdot 10^{24}$ kg, rigidly rotating around a fixed axis, with an angular rate $\omega = 7.292115 \cdot 10^{-5}$ rad s^{-1}, and producing a gravitational attraction of about 10 ms$^{-2} \equiv 10^3$ Gal.

At a closer look none of the above statements can be taken as true. So, we need an introductory knowledge of the physical reality of the Earth to model its gravitational and gravity field, at least for the features at large and medium scale, and subtract them from what we can observe on the Earth surface, or above it. In this way we are reduced to a model like the one described in Part I, and then we can study the problem of its inversion as we will do in Part III.

In this chapter, in particular, we will give the definition of gravity field of the Earth. We will introduce the global knowledge we have of it in terms of normal field and global models. Finally, we will define anomalous, or residual, quantities and their relationships.

Such items are usually treated in books on physical geodesy like [63, 86, 133] in much more details. Here we basically need a knowledge of concepts and results, so we send to the above literature for proofs not reported in this book.

F. Sansò, D. Sampietro, *Analysis of the Gravity Field*, Lecture Notes in Geosystems Mathematics and Computing, https://doi.org/10.1007/978-3-030-74353-6_4

4.2 The Gravity Field of the Earth. Physics and Geometry

What the Earth looks like at a first glance, we have recalled in Sect. 4.1; now we will examine to what extent the statements in Sect. 4.1 are not exact.

First of all, the Earth is not a sphere. If we use its mean radius as a constant all over the surface we commit an error which goes from $\sim -10\,\text{km}$ at the equator to $\sim +10\,\text{km}$ at the poles.

As we know, the Earth is much closer to an ellipsoid of revolution with (see [132])

$$a = 6378.1370\,\text{km} \qquad \text{equatorial radius}$$

$$b = 6356.7523\,\text{km} \qquad \text{polar radius}$$

$$e^2 = \frac{a^2 - b^2}{a^2} = 6.69438002290 \cdot 10^{-3} \quad \text{squared first eccentricity.}$$

This is indeed the effect of the rotation and of the centrifugal force.

One might object that it is also common knowledge that the Earth surface is in reality quite irregular, specially in the continental part, because the topographic surface can rise, in mountainous areas, up to $\sim$9 km, e.g., in the Himalayan region. This is indeed true, but what is extremely close to an ellipsoid is the surface of the oceans as well as its continuation under the continents, obtained by moving always in the horizontal direction. This surface is called the geoid and, as we will learn soon, is in fact an equipotential surface of the gravity field. While the relative error between geoid and sphere is in the range of 10^{-2}, the error between geoid and ellipsoid is in the range of 10^{-5}, a fact this of great importance when we need to linearize complicated nonlinear relations.

Second, the rotation axis of the Earth does not have a fixed direction, in an inertial system, because of the interaction with close planetary masses, primarily the Moon (the closest celestial body), the Sun (because of its large mass) and to a lesser extent the other planets.

Yet the motion of the rotation axis with respect to the fixed stars is small, very slow and predictable with high precision, basically with no impact on the gravity field. The same can be said of the magnitude of ω which evolves with a regular trend and shorter period variations related to friction in the motion of masses internal to the Earth and to variations of winds and oceanic currents. Relative variations of ω are in the range of 10^{-8} per day. Moreover, the spin axis has not a constant direction inside the Earth, but it exhibits what is called a polar motion, represented by the trajectory of the North pole with respect to the Earth body. This motion is slow and can amount to several meters per year; yet also this phenomenon is accurately monitored by the IERS (International Earth Rotation and Reference Systems Service) [10] and its perturbation effects on gravity can be computed and filtered away. On such matters a useful reference is [155].

Finally, the masses inside the Earth do not constitute a rigid body, but are generally in relative motion, starting from the continental crust that is structured in relatively rigid plates floating and shifting on a less viscous layer, the asthenosphere. This plate motion amounts to some centimeters per year and is also monitored on a global scale by the IERS. Its gravitational effect is somehow more important when the drift has locally a significant vertical component.

Masses are also in motion inside the Earth, both in the mantle and in the core. Such motions, though, are very slow and have gravity effects of very long wavelength, strongly attenuated by their distance from the first layer, the first 30 km of depth, which is the region where we want to study the inversion of the gravitational signal.

Finally, the masses external to the Earth, primarily the Moon and the Sun, do produce themselves a gravitational effect on a proof mass on the surface of the Earth or close to it (e.g., flying at a few kilometers above the ground or the ocean) and this effect is changing in time, due to the change in their relative position with respect to the Earth.

Furthermore, the Earth itself has a reaction of elastic deformation under this external attraction, known as Earth tide, which generates a gravitational signal, since the vertical amplitude of this tidal response is in the range of some decimeters. Its effect in terms of gravity is in the range of 0.1 mGal and therefore not negligible. The picture is further complicated along the coasts of the ocean, because the mean surface of the water has a tidal motion which can locally amount up to 10 m with respect to the continents, causing a so-called loading effect that induces a vertical elastic reaction in a stripe of, say, 100 km width. All these phenomena cause a vertical motion of masses and therefore they produce a gravimetric signal varying in time.

Fortunately, all tidal effects have a periodic character, with prevailing periods of 12 and 24 h; as such, they are detected, first of all by a worldwide network of permanent gravity stations [14], and the time periodic part of the gravity observations can be corrected for, as we shall see in the next chapter.

Of course such matters are discussed partly in Geodesy and more extensively in Geophysics. Here the literature is very large so we mention here only a few books [5, 42, 80, 153]

After this long digression, it is time to give the definition of the gravity field of the Earth. Let us consider a proof mass m in a terrestrial reference system, namely a system with origin at the barycenter of the Earth masses, the Z-axis along the rotation axis and co-rotating with the Earth at the same constant angular velocity ω.

The proof mass m feels a number of forces, in part due to the internal masses of the Earth, in part to the external masses, plus other apparent forces like the centrifugal force, due to the rotation of the reference system with respect to an inertial system, and the Coriolis force related to the possible motion of m with respect to the Earth.

We *define* the gravity field of the Earth $\boldsymbol{g}$ as the sum of the pure Newtonian gravitational force exerted by the internal masses $\boldsymbol{f}_N$, plus the centrifugal force $\boldsymbol{f}_C$ divided by the value of the proof mass $\boldsymbol{m}$, namely

$$g = \frac{1}{m}(f_N + f_C) \equiv g_N + g_C \tag{4.1}$$

where (see Sect. 1.2)

$$g_N(P) = -G \int_B \frac{\ell_{QP}}{\ell_{QP}^3} \rho(Q)\, dB_Q \tag{4.2}$$

and

$$g_C(P) = \omega^2 \left(x_P e_x + y_P e_y \right), \tag{4.3}$$

which is just the acceleration of a point moving with uniform angular velocity ω on a circle of radius $r = \sqrt{x^2 + y^2}$. We note that $\boldsymbol{g}$ is essentially the force per unit mass observed at a station rigidly attached to the Earth, eliminating the influence of external masses.

Indeed, the field $\boldsymbol{g}(P)$ is characterized by its modulus, $g(P) = |\boldsymbol{g}(P)|$, which is often called simply *gravity value* (*or modulus*) *at* P, and its direction $\boldsymbol{n}(P)$; in fact, what is mostly used is the unit vector

$$\boldsymbol{n}(P) = -\frac{\boldsymbol{g}(P)}{g(P)}, \tag{4.4}$$

which is also called the *direction of the vertical.*

Remark 4.1 We notice that due to mass density flows and tidal dislocations, $\boldsymbol{g}$ is a vector changing in time, even if we have eliminated from it the time varying attraction of the Sun and Moon. In particular, it is easy to see that the effects of the Earth rotation, which can amount to up to 3 Gal at the equator, undergo negligible time variations (in the range of parts of a μGal) for the phenomena described at the beginning of this section.

So, time variations in $\boldsymbol{g}$ are essentially induced via (4.2) by time variations in ρ due to tidal shifts.

In Chap. 5 we will learn how to eliminate these effects so that we are left with a field $\boldsymbol{g}$ that we shall consider as stationary in time.

The first fundamental property of the Earth gravity field $\boldsymbol{g}$ is that it is conservative, namely, it is the gradient of a potential function $W(P)$.

In fact, as shown in Sect. 1.3, we have

$$\begin{cases} g_N = \nabla u_N \\ u_N(P) = G \int_B \frac{\rho(Q)}{\ell_{PQ}} dB_Q. \end{cases} \tag{4.5}$$

At the same time,

$$\begin{cases} F_C = \nabla \frac{1}{2}\omega^2 \left(x_P^2 + y_P^2 \right) = \nabla u_C, \\ u_C = \frac{1}{2}\omega^2 \left(x_P^2 + y_P^2 \right) = \frac{1}{2}\omega^2 r^2, \end{cases} \tag{4.6}$$

so we can put

$$\begin{cases} g = \nabla W \\ W(P) = u_N(P) + u_C(P) \, . \end{cases} \tag{4.7}$$

A standard way to represent geometrically the field $\boldsymbol{g}$ is by the field lines, i.e., those lines that are tangent at every point to $\boldsymbol{g}$, namely to $\boldsymbol{n}$ given by (4.4), and by equipotential surfaces.

The field lines of $\boldsymbol{g}$ are called *lines of the vertical* or *plumb lines*, the equipotential surfaces are also called *horizontal surfaces*. It is common knowledge that since $\boldsymbol{g}$ is conservative, plumb lines and horizontal surfaces cross orthogonally to one another (see [133], or Exercise 2 in Chap. 1).

In particular, one of the equipotential surfaces, chosen so as to be close (within a couple of meters) to the mean surface of the oceans is called the *geoid* and has the special role to serve as a reference surface for a particular type of height coordinate, the orthometric height, defined as follows [132]: the orthometric height H_P of a point P close to the Earth surface is the arclength of plumb line between P and the geoid, counted positively outside and negatively inside the geoid. It has to be stressed that H_P should not be used too deep inside the masses, because there the geometry of plumb lines becomes more irregular.

Another common definition of coordinates is the couple (Λ, Φ) of astrogeodetic longitude and latitude. These are defined through the Cartesian components of $\boldsymbol{n}$ in a terrestrial system, by inverting the relations

$$\boldsymbol{n} \sim \begin{bmatrix} u_X \\ u_Y \\ u_Z \end{bmatrix} = \begin{bmatrix} \cos\Phi\cos\Lambda \\ \cos\Phi\sin\Lambda \\ \sin\Phi \end{bmatrix} . \tag{4.8}$$

We notice that (Λ, Φ, H) (or alternatively (Λ, Φ, W)) are curvilinear coordinates traditionally considered in geodesy [133] because they are related to classical geodetic observations, like astrogeodetic measurements to "fixed" stars and leveling. Much simper, and nowadays easily accessible by GNSS observations, are the so-called *ellipsoidal coordinates* (λ, φ, h) defined as follows. Let $\mathscr{E}$ be the Earth ellipsoid with geometric parameters (a, e), placed with its center at the barycenter, the symmetry axis along the terrestrial Z-axis and the couple of horizontal axes (x, y) conventionally chosen so that the (x, z)-plane goes through a given point on the Earth surface (e.g. Greenwich); in this case $\mathscr{E}$ is referred to the same Cartesian axes as the terrestrial system and it is rotating with it, with the same angular velocity. The couple (λ, φ) is related to the unit normal $\boldsymbol{v}$, to $\mathscr{E}$, by

$$\boldsymbol{v} \sim \begin{bmatrix} v_X \\ v_Y \\ v_Z \end{bmatrix} = \begin{bmatrix} \cos\varphi\cos\lambda \\ \cos\varphi\sin\lambda \\ \sin\varphi \end{bmatrix} , \tag{4.9}$$

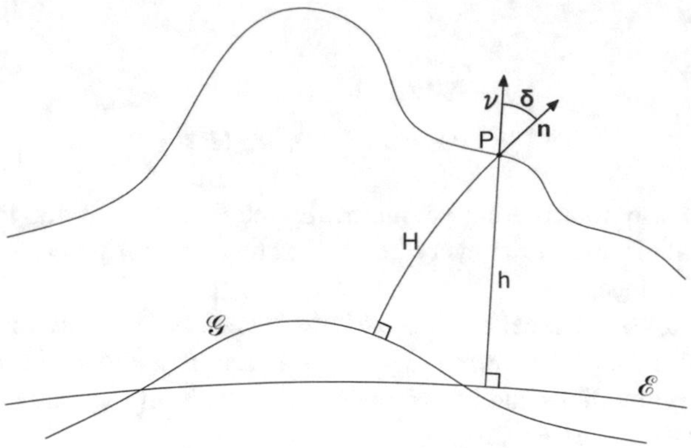

Fig. 4.1 Geometric relation between ellipsoidal (h) and orthometric heights (H) of a point P; δ represents the deflection of the vertical

while the ellipsoidal height h is the length of the normal to $\mathcal{E}$ through a point P, counted positively out of $\mathcal{E}$ and negatively inside $\mathcal{E}$ (see Fig. 4.1).

It is remarkable that the direction of $\boldsymbol{n}$, which reflects the mass distribution inside the Earth, is so well approximated by $\boldsymbol{v}$, namely

$$|\boldsymbol{n} - \boldsymbol{v}| \leq 3 \cdot 10^{-4}. \tag{4.10}$$

The difference

$$\boldsymbol{\delta} = \boldsymbol{n} - \boldsymbol{v}$$

is called in geodesy the *vector of the deflection of the vertical* [133].

Also the geometric relation between h and H is simple and illustrated in Fig. 4.1. The distance of the geoid $\mathcal{G}$ from the ellipsoid $\mathcal{E}$, measured along $\boldsymbol{v}$, is called the *geoid undulation* and denoted by N; then, thanks to the small inclination of the line of the vertical with respect to $\boldsymbol{v}$, we have with an excellent approximation the relation

$$h = H + N. \tag{4.11}$$

The relation (4.11) clarifies that knowing the geoid undulation one can transform the two main height systems into one another.

4.3 The Normal Gravity Field

To arrive at our goal of isolating, studying and interpreting the gravity field in a local area, we first need to identify components of the gravity field that are explained by global features of the mass distribution in the Earth. The first step in this case is related to the ellipsoidal shape of the Earth and to the fact that the main variations of the internal mass density can be described by an ellipsoidal layered model. One of the best known ellipsoidal layered model is the so called PREM (Preliminary Reference Earth Model [31]), roughly summarized in Table 4.1

The use of suitable ellipsoidal coordinates was already envisaged by the early study of Clairaut, who arrived at a famous equation [25] expressing a "hydrostatic" equilibrium distribution in ellipsoidal layers.

One has to arrive though, at the end of the nineteenth century, beginning of the twentieth century to find a proper definition and a closed solution for what is known nowadays as the normal gravity field [105–107, 141]. This is defined in terms of a normal potential $U(P)$ which is characterized by the following requirements:

(a) U has to be split into two potentials, one, $V(P)$, that is regular (i.e., tending to 0 at infinity) and harmonic outside the Earth ellipsoid, described in Sect. 4.2, and another, $u_C(P)$, that is a centrifugal potential identical to the one (see (4.6)) used for the actual gravity potential $W(P)$; so

$$U(P) = V(P) + u_C(P); \tag{4.12}$$

(b) the Earth ellipsoid $\mathscr{E}$ should be an equipotential surface for $U(P)$, as the geoid $\mathscr{G}$ is an equipotential of $W(P)$, so that

$$U(P)|_{\mathscr{E}} = U_0 = W_0 = W(P)|_{\mathscr{G}}, \tag{4.13}$$

(c) the value $U_0 = W_0$ has to be constrained to geometric parameters of $\mathscr{E}$, (a, e), and to physical parameters (M, ω) by the relation [133, §1.9]

$$U_0 = \frac{1}{3}\omega^2 a^2 + \frac{GM}{ae} \arctan \frac{e}{\sqrt{1-e^2}}. \tag{4.14}$$

We note that the above constraints imply that $V(P)$ must be a solution of the following boundary value problem for the Laplace equation:

$$\begin{cases} \Delta V = 0 & \text{outside } \mathscr{E}, \\ V|_E = U_0 - \frac{1}{2}\omega^2 \left(x^2 + y^2\right), \\ V(P) \to 0 \quad r_P \to \infty & \left(r_P = \sqrt{x^2 + y^2 + z^2}\right). \end{cases} \tag{4.15}$$

The problem was solved in closed form by Somigliana; a more elementary proof of the solution can be found in [133], Ch. 1, A.4. Here we are not so much

Table 4.1 The layered distribution of the internal mass density as function of depth according to the PREM model [31]

Radius [km]	Depth [km]	Density [g/cm^3]	Radius [km]	Depth [km]	Density [g/cm^3]
6371.0	0.0	1.02	5751.0	620.0	3.98
6369.0	2.0	1.02	5731.0	640.0	3.98
6368.0	3.0	2.60	5711.0	660.0	3.98
6366.0	5.0	2.60	5685.7	685.3	4.39
6364.0	7.0	2.60	5600.0	771.0	4.44
6362.0	9.0	2.60	5600.0	771.0	4.44
6361.0	10.0	2.60	5535.7	835.3	4.48
6359.0	12.0	2.60	5485.7	885.3	4.51
6357.0	14.0	2.60	5385.7	985.3	4.57
6356.0	15.0	2.90	5285.7	1085.3	4.62
6355.0	16.0	2.90	5185.7	1185.3	4.68
6353.0	18.0	2.90	5085.7	1285.3	4.74
6351.0	20.0	2.90	4985.7	1385.3	4.79
6349.0	22.0	2.90	4885.7	1485.3	4.85
6346.6	24.4	2.90	4835.7	1535.3	4.87
6346.0	25.0	3.38	4785.7	1585.3	4.90
6344.0	27.0	3.38	4735.7	1635.3	4.93
6342.0	29.0	3.38	4635.7	1735.3	4.98
6340.0	31.0	3.38	4535.7	1835.3	5.03
6338.0	33.0	3.37	4435.7	1935.3	5.08
6336.0	35.0	3.37	4335.7	2035.3	5.13
6326.0	45.0	3.37	4235.7	2135.3	5.18
6311.0	60.0	3.37	4135.7	2235.3	5.23
6291.0	80.0	3.37	4035.7	2335.3	5.28
6271.0	100.0	3.37	3985.7	2385.3	5.31
6251.0	120.0	3.37	3935.7	2435.3	5.33
6231.0	140.0	3.36	3835.7	2535.3	5.38
6211.0	160.0	3.36	3735.7	2635.3	5.43
6191.0	180.0	3.36	3635.7	2735.3	5.48
6171.0	200.0	3.36	3585.7	2785.3	5.51
6151.0	220.0	3.35	3485.7	2885.3	5.56
6151.0	220.0	3.43	3480.0	2891.0	5.56
6131.0	240.0	3.44	3480.0	2891.0	9.90
6111.0	260.0	3.45	3400.0	2971.0	10.0
6091.0	280.0	3.47	3200.0	3171.0	10.3
6071.0	300.0	3.48	3000.0	3371.0	10.6
6051.0	320.0	3.49	2800.0	3571.0	10.8
6031.0	340.0	3.50	2600.0	3771.0	11.0
6011.0	360.0	3.51	2400.0	3971.0	11.2
5991.0	380.0	3.53	2221.5	4149.5	11.4

(continued)

Table 4.1 (continued)

Radius [km]	Depth [km]	Density [g/cm^3]	Radius [km]	Depth [km]	Density [g/cm^3]
5971.0	400.0	3.54	2100.0	4271.0	11.5
5971.0	400.0	3.72	1900.0	4471.0	11.7
5951.0	420.0	3.74	1700.0	4671.0	11.8
5931.0	440.0	3.77	1500.0	4871.0	12.0
5911.0	460.0	3.79	1300.0	5071.0	12.1
5891.0	480.0	3.82	1221.5	5149.5	12.7
5871.0	500.0	3.84	1200.0	5171.0	12.7
5851.0	520.0	3.87	1000.0	5371.0	12.8
5831.0	540.0	3.90	800.0	5571.0	12.9
5811.0	560.0	3.92	600.0	5771.0	13.0
5791.0	580.0	3.95	400.0	5971.0	13.0
5771.0	600.0	3.97	200.0	6171.0	13.0
5771.0	600.0	3.97	0.0	6371.0	13.0

Table 4.2 Parameters of the normal gravity formula

γ_0	978.03267715
a_1	0.0052790414
a_2	0.0000232728
a_3	$1.262 \cdot 10^{-7}$
b_0	0.30877
b_1	0.00043
c_0	0.000072

interested in the closed form of $U(P)$, the use of which requires a non-elementary transformation from the above mentioned astrogeodetic coordinates to (λ, φ, h). We will rather give an approximate expression for the modulus of normal gravity which can be used in practice, with negligible error, in the range of h values that interest us, i.e., $0 < h < 10$ km.

Such an approximate formula for $\gamma(P) = |\nabla U(P)|$ is [151, §4.3]

$$\begin{aligned} \gamma(P) &= \gamma_0 \left(1 + a_1 \sin^2\varphi + a_2 \sin^4\varphi + a_3 \sin^6\varphi\right) + \\ &- \left(b_0 - b_1 \sin^2\varphi\right) h + c_0 h^2 \end{aligned} \tag{4.16}$$

where γ, γ_0 are given in Gal and h in km. The numerical values to be used in (4.16) are given in Table 4.2. Note that $\gamma(P)$ depends only on φ and h and not on λ, because of the cylindrical symmetry of problem (4.15). In case one needs the full normal vector $\boldsymbol{\gamma}(P)$, formula (4.16) can be generalized to

$$\begin{aligned} \boldsymbol{\gamma} \cdot \boldsymbol{v} &\equiv \gamma(\varphi, h), \\ \boldsymbol{\gamma} \cdot \boldsymbol{e}_\varphi &= -d_0 h \sin 2\varphi, \end{aligned} \tag{4.17}$$

where $\boldsymbol{e}_\varphi$ is the unit vector of the $\boldsymbol{\varphi}$ coordinate, pointing North, and

$$d_0 = 0.813 \cdot 10^{-3}; \tag{4.18}$$

note that $\boldsymbol{\gamma} \cdot \boldsymbol{e}_\varphi = 0$ when $h = 0$, as it should be. The reason for the first relation in (4.17) is just that $\boldsymbol{\gamma} \cdot \boldsymbol{v} = \gamma \cos \delta_0 \cong \gamma \left(1 - \frac{1}{2}\delta_0^2\right)$ and $|\delta_0| < 8 \cdot 10^{-6}$ (see [151, §4.2.3]).

The degree of approximation of the actual gravity field by the normal one can be summarized by three rough estimates of the order of magnitude of maximum errors, namely

$$\begin{cases} |W - U| \sim 2 \cdot 10^{-5} \, |U| \, , \\ |g - \gamma| \sim 10^{-4} \gamma, \\ |\boldsymbol{n} - \boldsymbol{v}| \sim 3 \cdot 10^{-4}. \end{cases} \tag{4.19}$$

The residual potential

$$T(P) = W(P) - U(P) \tag{4.20}$$

is called *anomalous gravity potential.*

A number of perturbative quantities related to the gravity field can be expressed as functionals of $T(P)$, that are generally nonlinear; yet the accuracies recalled in (4.19) are such that linearized functionals can be used. So we find the relations (see [133, §2.3])

$$\text{(geoid undulation)}\ N(P) = \frac{T(P)}{\gamma(P)}; \tag{4.21}$$

$$\text{(gravity disturbance)}\ \delta g(P) = g(P) - \gamma(P) = -\frac{\partial T(P)}{\partial h}; \tag{4.22}$$

$$\text{(deflection of the vertical)}\ \boldsymbol{\delta} = \boldsymbol{n} - \boldsymbol{v} = -\frac{\nabla_0 T(P)}{\gamma(P)}; \tag{4.23}$$

$$\left(\nabla_0 = \text{horizontal gradient} = \left(\nabla - \boldsymbol{v}\frac{\partial}{\partial h}\right)\right).$$

Remark 4.2 It is important to realize that $U(P)$ has been found by solving a boundary value problem and therefore it is not strictly associated to a mass distribution.

The formula expressing $U(P)$ does not loose a meaning even inside $\mathscr{E}$, where $U(P)$ then continues to be harmonic, almost down to the origin. Yet this harmonically continued potential looses its approximation properties, the deeper we go inside the Earth. So it is neither univocal nor advisable to define the anomalous potential $T(P)$ as in (4.20), apart maybe for the first very few kilometers. This phenomenon is due to the presence of a mass density $\rho \neq 0$, that changes the behaviour of W inside the masses (see (1.96) and (1.97)), while U is just downward continued in empty space.

4.4 Global Gravity Models

The anomalous potential $T(P)$ is still containing too many important global features of the Earth to be useful for the scope of a local interpretation. If we look at a map of $N(P)$ all over the Earth surface (Fig. 4.2), which is almost a representation of $T(P)$ because $\gamma(P)$ has a smooth and limited variation on $\mathscr{E}$ (see (4.16)), we immediately note that it still contains very long wavelength features plus more local variations related to the underlying topography. So we understand that, in order to localize the signal to be inverted, we must first subtract from T such effects.

This is accomplished by building so-called *global gravity models* (the interested reader can find a complete archive of existing global gravity field models at the International Centre for Global Earth Models—ICGEM—web page [8]), that are represented in the form of a truncated series, as

$$T_M(P) = \frac{GM}{R} \sum_{n=2}^{N} \sum_{m=-n}^{n} T_{nm} \left(\frac{R}{r}\right)^{n+1} Y_{nm}(\sigma) \tag{4.24}$$

where, M is the total mass of the Earth, and

$$T_{nm} \qquad n = 2, 3 \ldots N; m = -n, \ldots, 0, \ldots n$$

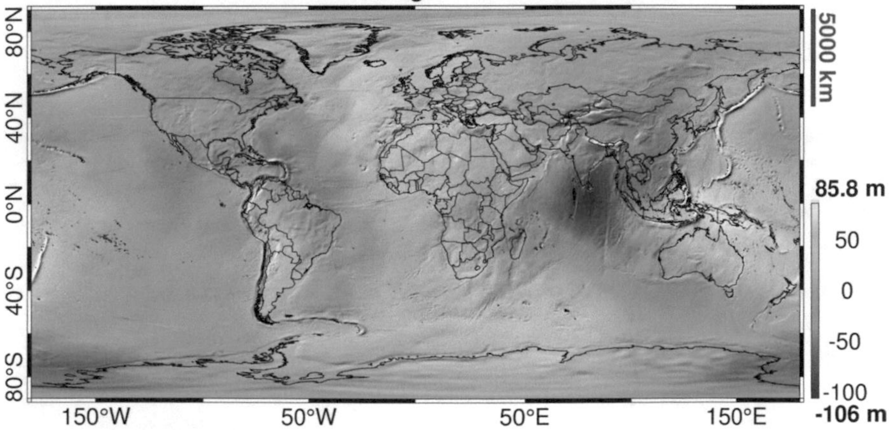

Fig. 4.2 EGM2008 [104] geoid model

are the potential spherical coefficients in a non-dimensional form (note that $\frac{GM}{R}$ has the dimension of a potential); (n, m) are called *degree and order of the coefficient* $T_{n,m}$; $Y_{nm}(\sigma)$ are so-called *spherical harmonics*;

$$\sigma \equiv (\vartheta, \lambda) \quad \text{(spherical colatitude and longitude, see Exercise 2)}$$

and

$$r = \left(x^2 + y^2 + z^2\right)^{-1/2}.$$

Note that in this section r represents the distance of P from the origin in 3D and not on the plane (2D) as in the rest of the text; this is because here the spherical geometry is necessary to represent the global gravity models, in contrast to the Euclidean geometry, which is more useful when describing the local gravity field. As for R, we can take any value close to the mean radius of the Earth $\bar{R} = 6371$ km; in fact, most of the authors prefer to take $R = a$, the equatorial radius of $\mathscr{E}$. Indeed, by changing R, the numerical values of T_{nm} have to be changed too.

The family $\{Y_{nm}(\sigma)\}$, when n and m are allowed to range from 0 up to infinity, shares many properties of the Fourier family $\{\sin n\vartheta, \cos n\vartheta\}$ (or $\{e^{2\pi i n\vartheta}\}$), transposing them from the circle $\{0 \le \vartheta < 2\pi\}$, where the Fourier sequence is L^2 orthogonal and complete, to the sphere where spherical harmonics form an L^2 orthogonal and complete system, too.

Thus, one has (see [133])

$$Y_{nm}(\sigma) = \bar{P}_{nm}(\vartheta) \begin{cases} \cos m\lambda, & \text{if } m \ge 0, \\ \sin |m| \lambda, & \text{if } m < 0, \end{cases} \tag{4.25}$$

where $\bar{P}_{nm}(\vartheta)$ are normalized associated Legendre functions of the first kind well-known in mathematical physics (see [2]). Combining the orthogonality properties of $\bar{P}_{nm}(\vartheta)$, namely

$$\frac{1}{4}\int_0^{\pi} \bar{P}_{nm}(\vartheta)\, \bar{P}_{\ell m}(\vartheta) \sin\vartheta\, d\vartheta = \delta_{n\ell} \tag{4.26}$$

and that of the Fourier basis, namely

$$\begin{aligned} &\frac{1}{\pi}\int_0^{2\pi} \cos m\lambda \cos j\lambda,\ d\lambda = (1 + \delta_{m0})\,\delta_{mj} \\ &\frac{1}{\pi}\int_0^{2\pi} \sin m\lambda \sin j\lambda,\ d\lambda = (1 - \delta_{m0})\,\delta_{mj} \\ &\frac{1}{\pi}\int_0^{2\pi} \cos m\lambda \sin j\lambda,\ d\lambda = 0 \end{aligned} \tag{4.27}$$

one obtains the L^2 orthogonality of $Y_{nm}(\sigma)$ on the sphere, namely

$$\frac{1}{4\pi}\int Y_{nm}(\sigma)\,Y_{\ell j}(\sigma)\,d\sigma = \delta_{n\ell}\delta_{mj}$$
$$(d\sigma = \sin\vartheta\, d\vartheta\, d\lambda)\,. \tag{4.28}$$

The completeness of the basis $\{Y_{nm}(\sigma)\}$ in L^2 is a deeper result for which we send to the literature (e.g., [133, §3.4]). Recall that the meaning of completeness is that, for any $g(\sigma) \in L^2$, one has

$$\begin{cases} g(\sigma) = \sum_{n=0}^{+\infty}\sum_{m=-n}^{n} g_{nm} Y_{nm}(\sigma), \\ g_{nm} = \frac{1}{4\pi}\int g(\sigma)\, Y_{nm}(\sigma)\, d\sigma, \end{cases} \tag{4.29}$$

where the series converges in the L^2 norm.

As we see, when r is kept constant, the expression (4.24) is a kind of series (4.30) truncated below and above. The elimination of degrees zero and one, namely $(n, m) = (0, 0), (1, -1), (1, 0), (1, 1)$, in the representation of T is related, first of all, to the fact that T_{00} is proportional to the mass generating T and this is zero because of requirement c) in Sect. 4.3 imposed on $U(P)$. On the other hand, the vector $(T_{1,-1}, T_{1,0}, T_{1,1})$ has components proportional to the Cartesian components of the barycenter, which are zero due to our choice to place there the origin of the reference terrestrial system. The maximum degree N, on the contrary, is related to the resolution of the representation (4.24). On the sphere the situation is not so neat, as on the circle where the maximum frequency readable with a spacing Δ of the data is regulated by the Nyquist theorem, yet one can establish an approximate rule of thumb stating that a representation up to degree 180 is equivalent to a grid of data of $1^\circ \times 1^\circ$; then degree 360 corresponds to a grid of size $0.5^\circ \times 0.5^\circ$, and so forth.

The factor $\left(\frac{R}{r}\right)^{n+1}$, which multiplies each spherical harmonic of degree n and any order, is such as to make the functions

$$s_{nm}(r, \sigma) = \left(\frac{R}{r}\right)^{n+1} Y_{nm}(\sigma) \tag{4.30}$$

harmonic everywhere in 3D space, outside the origin: $s_{nm}(r, \sigma)$ are called *solid spherical harmonics*.

It is interesting to observe that, since according to our rule of thumb n regulates the minimum wavelength

$$\Lambda \sim \frac{\pi R}{n} \tag{4.31}$$

of the harmonics of degree n, the factor $\left(\frac{R}{r}\right)^{n+1}$ damps the surface spherical harmonics, for $r > R$, i.e., outside the sphere of radius R, with a negative exponential factor $e^{-(n+1)\log\frac{r}{R}}$ and exponent proportional to Λ^{-1}, similarly to what happens in a Cartesian geometry where an oscillating plane Fourier function $e^{-i2\pi \boldsymbol{p}\cdot\boldsymbol{x}}$ is damped in the semispace $\{z \geq 0\}$ by the factor $e^{-2\pi p z}$ (see (3.14)).

More on this similarity is explained in Example 4.1 below. The calculation of $\{T_{nm}\}$ from observed data is a complex procedure combining several type of data:

- primarily gravity observations on the continents
- digital models of the topography
- tracking of satellites trajectories
- spatial gravimetry and gradiometry
- aerial and marine gravimetry
- satellite altimetry on the oceans, combined with circulation models.

As one can imagine, such data are inhomogeneous (for instance, the gravity coverage of Africa and South America is still rather poor), containing different biases and with a stochastic structure of the errors, often very roughly known.

The way in which a model is computed requires a longer reasoning, which is out of the scope of this book (see [133], especially Chaps. 3 and 6). At present the model with higher resolution and accuracy is determined with a maximum degree $N = 2159$, corresponding roughly to a wavelength of 10 km; this has been derived from a combination of satellite altrimetry, ground data, and satellite data of missions dedicated to the gravity field recovery [41, 104].

Two things are important to remember when using a global model to reduce the anomalous potential, or better its gravity disturbance δg_{M}, see (4.22), namely that we want to subtract the influence of mass anomalies located below 20/30 km because we are interested in inferring location and shape of the shallower anomalies from local data; that certainly the model δg_{M} at the maximum degree 2159 displays a notable correlation with the topography (see Fig. 4.3), meaning that a significant part of the gravity signal coming from the discontinuity between solid Earth and air is already captured by δg_{M}

This is the reason why for local geophysical applications we prefer to use a global model up to a maximum degree somewhere between 720 and 1440 (corresponding to wavelengths between 30 and 15 km), in order to avoid eliminating part of the signal that we want to identify.

This has two advantages. The first is to control the error intrinsic to the model; in fact, the so-called *commission error* (see [133, §3.8]) of δg is 1.5 mGal at $N = 720$, 2.4 mGal at $N = 1440$, and it would be 3.4 mGal at $N = 2159$.

Second, we notice that the use of (4.24) or of the corresponding formula for δg_{M}, which in spherical approximation reads

$$\begin{aligned}\delta g_{\mathrm{M}}(P) &= -\frac{\partial T_{\mathrm{M}}}{\partial h} \cong -\frac{\partial T_{\mathrm{M}}}{\partial r} = \\ &= \frac{GM}{R^2}\sum_{n=2}^{N}\sum_{m=-n}^{n}(n+1)\,T_{nm}\left(\frac{R}{r}\right)^{n+2} Y_{nm}(\sigma)\,,\end{aligned} \tag{4.32}$$

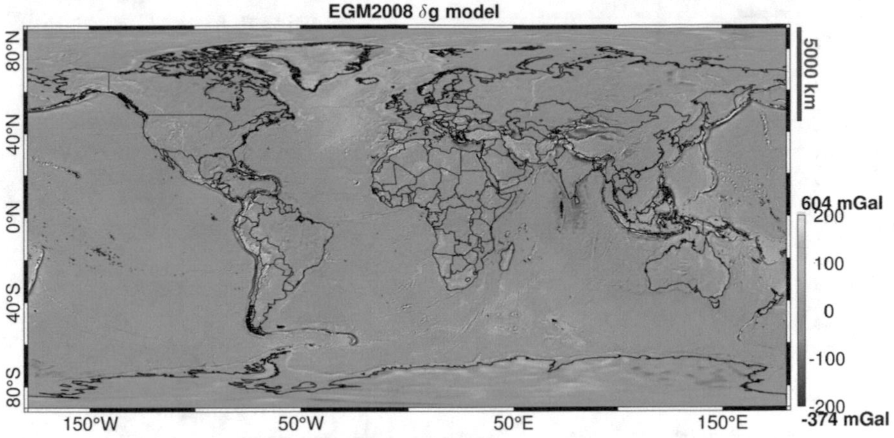

Fig. 4.3 The range of gravity disturbance at maximum degree 2159

requires the numerical calculation of $(N+1)^2$ spherical harmonics (i.e., $\sim 2 \cdot 10^6$ functions for $N = 1440$ and $4.6 \cdot 10^6$ functions for $N = 2159$). This, multiplied by the number of points where δg_M is needed, would soon become an impossible task, was it not for the fact that there are algorithms that, similar to the FFT, permit a very fast computation of regular data grids on the sphere, [64].

Remark 4.3 Let us return to our definition (4.24) and to the value of R to be used in it. The first remark is that in principle it seems that the problem should be irrelevant: in fact, if we have a certain set of coefficients $T_{nm}\,(R)$ associated to T_{M}, another set $T_{nm}(R')$

$$T_{nm}\left(R'\right) = \left(\frac{R'}{R}\right)^{n+1} T_{nm}\,(R) \tag{4.33}$$

followed by a change of R into R' in the factor $\left(\frac{R}{r}\right)^{n+1}$, gives exactly the same results. Moreover, since (4.24) is a finite sum, no convergence problem needs to be considered for its validity. In fact, it is clear that (4.24) could even be used with $r < R$, although the function obtained in this manner would loose its relationship to the actual gravity field of the Earth.

Nevertheless, it is important to note that if we want to consider (4.24) as a truncated series for a function $T_{\mathrm{M}}(P)$ that has to be regular at least down to the sphere S_R of radius R, we are expecting that $T_{nm} \to 0$ when $n \to \infty$; in fact a condition like

$$\sum_{n=2}^{+\infty} \sum_{m=-n}^{n} (T_{nm})^2 < +\infty \tag{4.34}$$

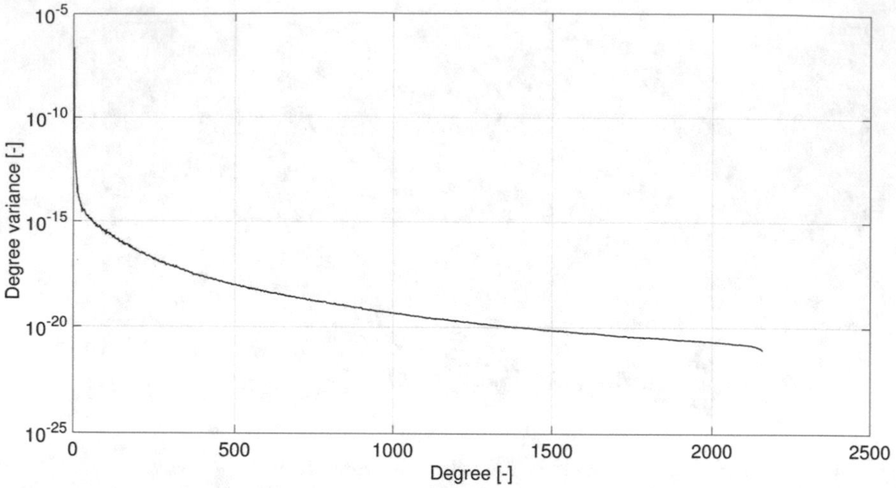

Fig. 4.4 The degree variances spectrum of the EGM2008 model

would guarantee that $T_M(P)$ is in $L^2(S_R)$. This is indeed a useful property if we want to improve our representation going to a higher N; in this case in fact we would like the added degrees to have a minor effect on T_M. The choice $R = a$ used in the model is reasonably satisfying this requirement.

One way in which the regularity of T_M is analyzed, is through its degree variances spectrum. The full power degree variances are defined by

$$c_n(T) = \sum_{m=-n}^{n} T_{mn}^2; \tag{4.35}$$

they measure the contribution of the full degree n to the construction of the complete L^2 norm of T as in (4.35). Such quantities must have a physical meaning as it is clearly seen by noting that the spectrum has a quite regular shape (Fig. 4.4).

If the degree variances quantify the spectral distributions of the power of the signal T_M, still another spectrum is important to qualify their accuracy. This is the error degree variances spectrum, which is defined as the total variance of the estimation errors ε_{nm} for each degree

$$\begin{cases} \varepsilon_{nm} = (T_M)_{nm} - T_{nm}, \\ \sigma_n^2 = \sum_{m=-n}^{n} E\left\{\varepsilon_{nm}^2\right\}. \end{cases} \tag{4.36}$$

The error degree variances of the EGM2008 model [104] are presented in Fig. 4.5.

One can observe that at very high degrees ($n > 1000$) the error degree variances tend to follow the degree variances $c_n(T)$; this is the effect of the regularization

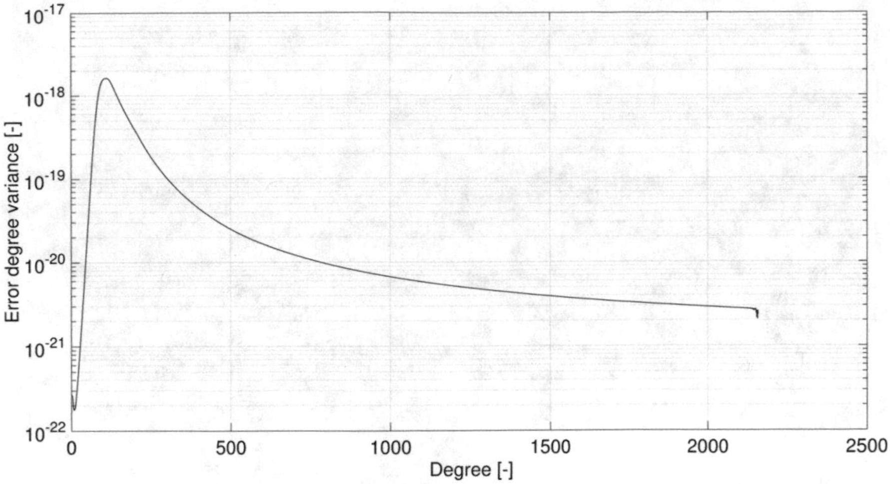

Fig. 4.5 Error degree variances of the EGM2008 model

(which will be studied in detail in Part III of the book), which however makes more questionable the reliability of the estimates. This is another reason why for our purposes it is preferable not to use actual global models up to their maximum degree.

Example 4.1 (The Moho Gravity Signal) This example has a twofold target: the first is to show how the Cartesian theory of the linearized gravity signal of the homogeneous layers (see Sect. 3.6) generalizes the spherical geometry; the second is to introduce, in a quantitatively realistic way, the effects in terms of gravity of the first important discontinuity in the Earth body, the Moho surface. This surface, first discovered by the analysis of seismological data [85], has a mean depth of ~30 km and is winding roughly between ~70 km and 10 km. Across such a surface not only the elastic parameters of the lithosphere undergo a sudden change, but also the mass density undergoes a jump of say 500 kgm^{-3}, in the mean. The effect of this is indeed a gravimetric signal that is represented in Fig. 4.6.

So in this example we study the analytical form of the gravity signal for a simplified spherical model, as presented in Fig. 4.7, as function of the undulation δR of the surface S_M with respect to the mean sphere S_{R_0}.

The body internal to S_M has density ρ_0, while the upper layer between S_M and S_R has density ρ_c. We put

$$\delta\rho = \rho_0 - \rho_c, \tag{4.37}$$

supposing a positive density downward variation. In this example we take $\rho_0 = 3270$, $\rho_c = 2670\,\text{kgm}^{-3}$ and $\delta\rho = 500\,\text{kgm}^{-3}$.

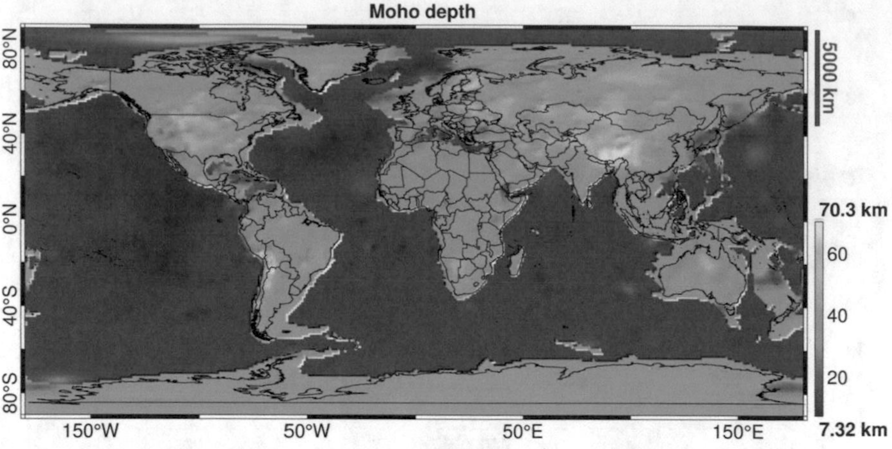

(a)

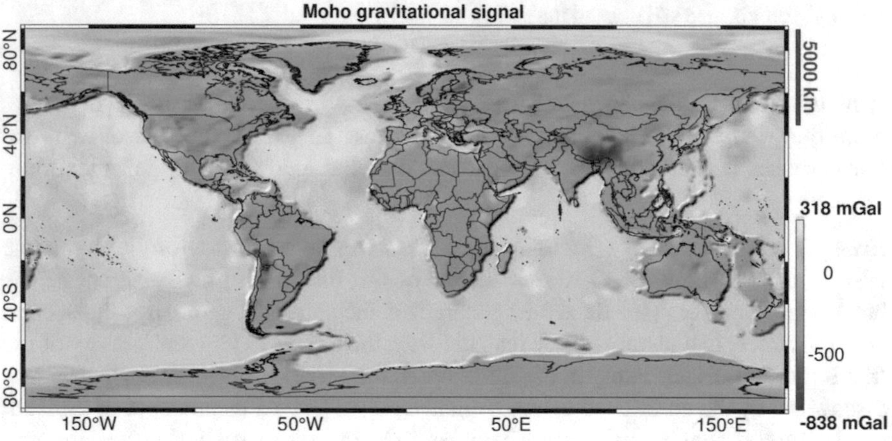

(b)

Fig. 4.6 (**a**) the Moho depth according to the model by Szwillus et al. [143]; (**b**) the gravimetric signal of the above Moho model

We also denote

$$\Delta R = R - R_0, \tag{4.38}$$

where R_0 is the mean radius of the Moho, which in this example is fixed at $R_0 = 6370 - 30\,\text{km}$. The Newtonian potential of the body in Fig. 4.7 is, for $r \geq R$

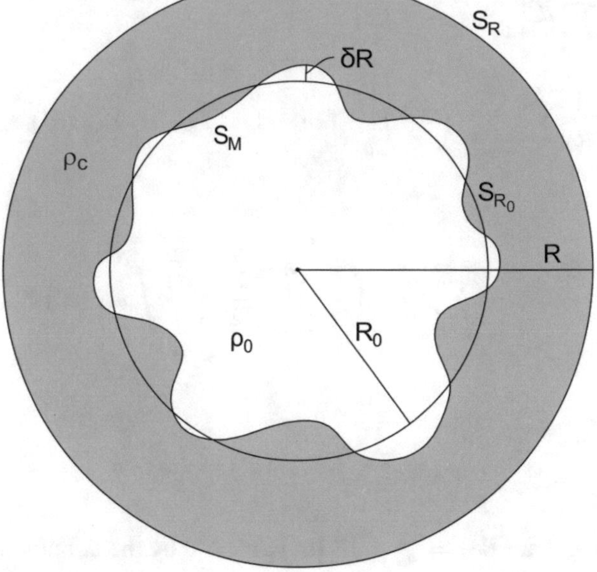

Fig. 4.7 A simplified Moho model: S_M the Moho surface, R_0 the mean Moho radius, δR the Moho undulation

$$
\begin{aligned}
u(r,\sigma) &= u_0 - G\delta\rho \int\int_{R_0}^{R_0+\delta R} \frac{r'^2 dr' d\sigma'}{\ell}, \\
\ell &= \left[r^2 + r'^2 - 2rr' \cos\psi_{\sigma\sigma'} \right]^{1/2}, \\
\psi &= \text{spherical angle between the radius } (r,\sigma) \text{ and } \left(r',\sigma'\right), \\
u_0 &= \frac{G(M_0+M_c)}{r}, \quad M_0 = \frac{4}{3}\pi R_0^3 \rho_0, \quad M_e = \frac{4}{3}\pi \left(R^3 - R_0^3\right)\rho_c.
\end{aligned}
\tag{4.39}
$$

From (4.39) it is clear that u_0 is just the normal potential for the configuration under study, so we can put, also linearizing the expression with respect to δR,

$$
\begin{aligned}
T(r,\sigma) &= -G\delta\rho \int\int_{R_0}^{R_0+\delta R} \frac{r'^2 dr' d\sigma'}{\ell} \cong -G\delta\rho \int \frac{R_0^2 \delta R}{\ell} d\sigma'. \\
&\left(\ell = \left[r^2 + R_0^2 - 2rR_0 \cos\psi_{\sigma\sigma'} \right]^{1/2} \right)
\end{aligned}
\tag{4.40}
$$

Now we use in (4.40) the following expansion of the Newton kernel (see [133])

$$\frac{1}{\ell} = \sum_{n=0}^{+\infty} \frac{R_0^n}{r^{n+1}} P_n(\cos\psi_{\sigma\sigma'}) = \\ = \frac{1}{R_0} \sum_{n=0}^{+\infty} \sum_{m=-n}^{n} \left(\frac{R_0}{r}\right)^{n+1} (2n+1)^{-1} Y_{nm}(\sigma) Y_{nm}(\sigma'), \tag{4.41}$$

which yields

$$T(r,\sigma) = -G\delta\rho R_0 \sum_{n=0}^{+\infty} \sum_{m=-n}^{n} \left(\frac{R_0}{r}\right)^{n+1} \frac{Y_{nm}(\sigma)}{2n+1} \int \delta R(\sigma') Y_{nm}(\sigma')\, d\sigma'. \tag{4.42}$$

Denoting

$$\delta R_{nm} = \frac{1}{4\pi} \int Y_{nm}(\sigma') \delta R(\sigma')\, d\sigma'$$

and considering that $\delta R_{00} = \frac{1}{4\pi} \int \delta R(\sigma')\, d\sigma' = 0$ by the definition of R_0, (4.42) gives the representation

$$T(r,\sigma) = -4\pi G\delta\rho R_0 \sum_{n=1}^{+\infty} \sum_{m=-n}^{n} \left(\frac{R_0}{r}\right)^{n+1} \delta R_{nm} \frac{Y_{nm}(\sigma)}{2n+1}. \tag{4.43}$$

We finally use $\delta g = -\frac{\partial T}{\partial r}$ in (4.43) and set $r = R$ because we want to have δg on the Earth surface, to get

$$\delta g = -4\pi G\delta\rho \sum_{n=1}^{+\infty} \sum_{m=-n}^{n} \left(\frac{R_0}{R}\right)^{n+2} \delta R_{nm} \frac{n+1}{2n+1} Y_{nm}(\sigma). \tag{4.44}$$

If we put

$$q = \frac{R_0}{R} = \frac{R_0}{R_0 + \Delta R} = 0.9953,$$

we can write (4.44) in terms of harmonic coefficients as

$$\delta g_{nm} = -4\pi G\delta\rho q^{n+2} \frac{n+1}{2n+1} \delta R_{nm}. \tag{4.45}$$

Note that with the rough but realistic estimate

$$c_n(\delta R) = \sum_{m=-n}^{n} \delta R_{nm}^2$$

one obtains from (4.45) the value

$$\sqrt{\sum_{n=1}^{+\infty} \delta g_{nm}^2} = 236\,\mathrm{mGal}$$

which, as we have seen from Fig. 4.6, seems realistic too.

Remark 4.4 (Isostasy) If we would interpret the Moho surface as the transition from crust to mantle, we could use (4.45) to derive the so-called isostatic compensation signal, according to one of the various hypotheses concerning its model. For instance, following the simplest Airy–Heiskanen model [63], we have that the load of topography h_t is compensated by the hydrostatic push on the root of a crust floating on the mantle, according to Fig. 4.8.

In this case the hydrostatic equilibrium condition reads

$$\rho_\mathrm{c} h_t = (\rho_0 - \rho_\mathrm{c}) h_r,$$

namely

$$h_\mathrm{r} = \frac{\rho_c}{\rho_0 - \rho_\mathrm{c}} h_t; \tag{4.46}$$

note that h_r should be δR of the Example 4.1. It is clear that a straightforward relation as (4.46) would imply that δR is positive under continental areas and negative under the sea. A look at Fig. 4.6 shows that this is not exactly the case; however, a certain tendency for $\delta R < 0$ in oceanic areas and viceversa, is clearly visible, implying that the simple Airy–Heiskanen mechanism, though not strictly real, has a validity in the mean over large areas. Indeed, the theory of isostatic

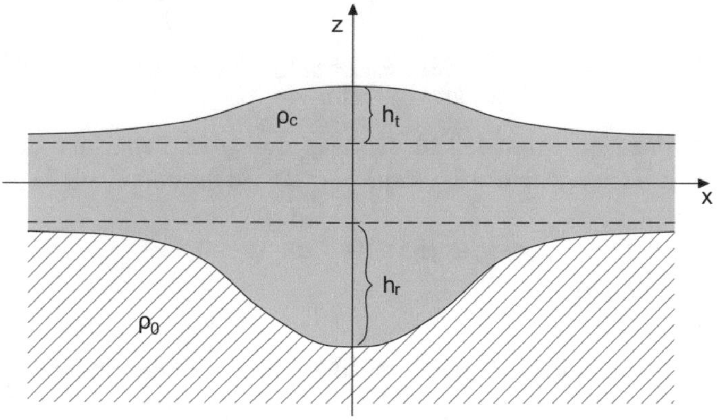

Fig. 4.8 The isostatic equilibrium model of Airy–Heiskanen: h_t = topographic height, h_r = depth of the root plunging into the mantle below the compensation depth

models has been evolving nowadays into a dynamic theory of the Earth interior. The interested reader can consult some literature (e.g. [158]).

4.5 Free Air Gravity Anomalies

Up to now we presented a number of reductions to be applied to observed gravity values, every time subtracting one part of the signal that we know can be attributed to physical components of $g(P)$ that have origin of global nature and below the 30 km, which is our deepest horizon.

Every step produces what we can call a *gravity anomaly.* This is our path to the residual gravity anomaly, which is the final local signal, cleaned by all other effects, so that we can hope that it is the result of some simple model that we have studied in Part I and then we are able to interpret, as we shall see in Part III.

For the moment, the steps undertaken are summarized in the following chain of formulas

$$g(P) \rightarrow g(P) - \gamma(P) = \delta g(P) \rightarrow \delta g(P) - \delta g_{\mathrm{M}}(P) = \delta g_{\mathrm{L}}(P) \tag{4.47}$$

meaning that from the gravity value $g(P)$, at the known point P in space, we have subtracted the normal gravity, to produce the first anomaly, that in accordance with geodetic tradition has been called *gravity disturbance;* subtracting a model, up to some wise maximum degree, from the gravity disturbance, we produce a new anomaly that we could call a *localized gravity disturbance.* More actions will be taken in Sect. 6.2 where we will study the topographic correction. Here however we must concentrate on a type of anomaly that is widely used because it was the only one that could be really computed, before the spatial era, with the advent of satellite based positioning systems, took over. In fact, the chain of (4.47) is meaningful if we know P, i.e., we know $(x, y, z)_P$ or $(\lambda, \varphi, h)_P$, so that *at the same point* we can compute γ.

Before continuing the reasoning, we must state a principle on the behaviour, in terms of variation, of the global gravity field.

Remark 4.5 Since the normal field captures the gross behaviour of the global gravity field, it is enough to use the formula (4.16) in the simplified form

$$\gamma = \gamma_0 + \gamma_0 5 \cdot 10^{-3} \sin^2 \varphi - 0.3h$$

to verify that:

- one can obtain a variation $|\delta g| = 1\,\mathrm{mGal}$ by moving vertically for $\delta h = 3.3\,\mathrm{m}$;
- one can obtain the same variation $|\delta g| = 1\,\mathrm{mGal}$ by moving horizontally for a distance $\delta L = R\delta\varphi = 1.2\,\mathrm{km}$ in a meridian direction.

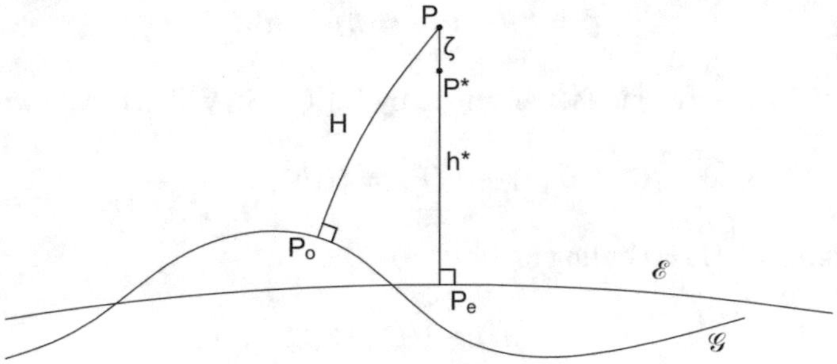

Fig. 4.9 The geometric meaning of H and h^*; $\mathscr{E}$ = ellipsoid, $\mathscr{G}$ = geoid

That is to say, there is a factor of ~360 between the scales of variation in the vertical and horizontal directions. This justifies the statement that with a *classical* position of P taken from a map, the horizontal position (λ, φ) is known to a sufficient accuracy (say 10/20 m) to compute γ, while the vertical position, traditionally given by an orthometric height, can never be confused with an ellipsoidal height, because the difference (i.e., the geoid N according to (4.11)) can amount up to 100 m.

This is the reason why already back in the nineteenth century, the introduction of a new kind of anomaly was deemed necessary: the free air gravity anomaly, universally denoted by Δg. Please note that here Δ has not the meaning of a Laplacian, but it is attached to g to form a unique symbol. This is defined following another definition, that of normal height h^*. We introduce h^*, slightly deviating from the classical definition, as the solution of the algebraic equation [133, pag. 48]

$$W(P) = W(\lambda, \varphi, h) = U(P^*) = U(\lambda, \varphi, h^*). \tag{4.48}$$

The geometrical meaning, illustrated in Fig. 4.9, is that the actual potential at P is the same as the normal potential at P^*, a point along the ellipsoidal normal passing through P.

The reason for introducing the concept of normal height is that it was understood that such a coordinate was much closer to observations than orthometric heights. In fact, a combination of gravimetry and leveling can provide values of $W(P)$ on the Earth surface (and then of h_P^*) without introducing any hypotheses on the internal mass density distribution (see [132], Chap. 6). So, assuming that $W(P)$ is known, we can claim that h_P^* is known too and we can define the free air gravity anomaly as

$$\Delta g = g(P) - \gamma(P^*). \tag{4.49}$$

Looking at Fig. 4.9 we can observe that the small segment $\zeta = \overline{P^*P}$, i.e.,

$$\zeta = h_P - h_{P*} = h_P - h_P^*, \tag{4.50}$$

called *height anomaly,* is in fact identified by Eq. (4.48), which can be written as

$$U(h) + T(P) = U(h^*). \tag{4.51}$$

Linearizing (4.51) according to

$$U(h^*) - U(h) \cong \gamma\zeta$$

(remember that $\boldsymbol{\gamma}$ is pointing downward), we get the equation

$$\zeta = \frac{T(P)}{\gamma}, \tag{4.52}$$

which is a straightforward generalization of (4.21).

The free air anomaly Δg should be considered the basis for further gravity elaboration, when $W(P)$ or h_P^* are known. This was the situation before GPS observations came into the geodetic arsenal, what made the ellipsoidal height h_P directly available.

Unfortunately, for more than one century geodesists and cartographers have decided to prefer orthometric heights, although such coordinates cannot be computed from observations, with a centrimetric accuracy, without introducing hypotheses on the density of topographic masses. Anyway, the abundance of the data set of orthometric heights has forced to study approximate formulas for the computation of Δg.

In this respect, if we define pseudo-free air anomalies by

$$\Delta\tilde{g} = g(P) - \gamma(H_P), \tag{4.53}$$

then the following relation can be derived (see the discussion in [133, §2.3]);

$$\Delta g = \left(1 + \frac{2H}{a}\right)\Delta\tilde{g} - 4\pi G\rho\frac{H^2}{a}. \tag{4.54}$$

It is easy to verify that below 1 km of altitude, (4.54) can be substituted by the simple relation

$$\Delta g \cong \Delta\tilde{g},$$

even in areas with intense anomalies, accepting errors well below the 0.1 mGal level. In mountainous areas, though, the use of (4.54) is recommended.

The relation between the free air anomaly and the anomalous potential is easy to find, we start with

$$\Delta g = g(P) - \gamma(P^*) = g(P) - \gamma(P) + \gamma(P) + \gamma(P^*) \cong \delta g + \frac{\partial \gamma}{\partial h}\zeta; \qquad (4.55)$$

then using (4.22) and (4.52) we get

$$\Delta g = -\frac{\partial T}{\partial h} + \frac{\partial \gamma}{\partial h}\frac{T}{\gamma}, \qquad (4.56)$$

known also as the *fundamental equation of physical geodesy* [133, §2.3].

Other gravity anomalies can be defined and could be used for further analysis. For instance, a model of the Moho could be subtracted too, as explained in Example 4.1: when this is based on some isostatic model, then we speak of *isostatic anomalies.*

Another kind of anomaly will be treated in Sect. 6.3, namely the *Bouguer anomaly,* which comes from an epoch when the data of Digital Terrain Models were scarce and inaccurate, and no global model was available. Nowadays it seems preferable to follow a different procedure, yet since there are so many data of Bouguer anomalies, the item will be anyway covered later in the book.

4.6 Exercises

Exercise 1 Given the points P_1, P_2, P_3 with ellipsoidal coordinates $\lambda_{P_1} = \lambda_{P_2} = \lambda_{P_3} = 0°$, $\varphi_{P_1} = \varphi_{P_2} = \varphi_{P_3} = 45°$, $h_{P_1} = 100\,\text{m}$, $h_{P_2} = 200\,\text{m}$, $h_{P_3} = 1000\,\text{m}$, compute in P, the values of γ, γ_φ, γ_λ and the angle θ between $\boldsymbol{\gamma}$ and $-\boldsymbol{\nu}$.

Solution $\gamma(P_1)$, $\gamma(P_2)$ and $\gamma(P_3)$ can be easily computed by recalling (4.16) and considering the parameters given in Table 4.2:

$$\begin{aligned}\gamma(P_1) = {} & 978.03267715(1 + 0.052790414 \sin^2\varphi + \\ & + 0.0000232728 \sin^4\varphi + 0.1262 \cdot 10^{-6} \sin^6\varphi) - \\ & - (0.30877 - 0.00043 \sin^2\varphi)h + 0.000072h^2.\end{aligned}$$

Therefore we obtain:

- $\gamma(P_1) = 980.589$ Gal;
- $\gamma(P_2) = 980.558$ Gal;
- $\gamma(P_3) = 980.311$ Gal.

As for the other quantities, we have from (4.17) that

$$\gamma_\varphi(P_1) = \boldsymbol{\gamma} \cdot \boldsymbol{e}_\varphi = -0.813 \cdot 10^{-3} h \sin^2(\varphi) = -4.0650 \cdot 10^{-4}\ \text{Gal};$$

and since for symmetry reasons $\gamma_\lambda = 0$, we can compute $\boldsymbol{\gamma} \cdot \boldsymbol{\nu}$ and as a consequence the angle between $\boldsymbol{\gamma}$ and $\boldsymbol{\nu}$ simply from $\sin(\theta) = \frac{\gamma_\varphi}{\gamma_\nu}$. Substituting the value for P_1we obtain $\theta = (-2.3752 \cdot 10^{-5})°$. Treating P_2 and P_3 similarly, we have:

- $\gamma_\varphi(P_2) = -8.13 \cdot 10^{-14}$ Gal; $\theta(P_2) = -(4.7505 \cdot 10^{-5})°$;
- $\gamma_\varphi(P_3) = -0.0041$ Gal; $\theta(P_3) = (-2.3759.10^{-4})°$.

The Matlab function to compute the above components of the normal gravity field is provide below.

```
% Matlab code for Exercise 1
function [gamma, gamma_phi, gamma_lambda, gamma_nu, theta] = eser-
cizio4_1(phi,la,h);
%Compute the modulus of normal gravity (gamma) and the component of
the gravity vector (gamma_phi, gamma_lambda, gamma_nu) of a point given
its ellispoidal geodetic
%coordinates (phi, lambda in degree, h in km). Outputs are in Gal.
%%define constant parameters
g0 = 978.03267715; %[Gal]
a1 = 0.0052790414;
a2 = 0.0000232728;
a3 = 1.262e-7;
b0 = 0.30877; %[1/m]
b1 = 0.00043; %[1/m]
c0 = 0.000072; %[1/m^2]
d0 = 8.13e-3;
%% compute normal gravity modulus
s2 = sind(phi).^2;
s4 = s2.^2;
s6 = s2.^4 .* s2.^2

gamma = g0 .* (1 + a1 .* s2 + a2 .* s4 + a3 .* s6) - (b0 + b1 .* s2).*h +
c0.*h.^2;
%% compute normal gravity components
gamma_phi = -d0 .* h .* s2; gamma_lambda = 0; gamma_nu = sqrt (gamma
.^2 - gamma_phi.^2);
%% compute angle between normal gravity and normal to the ellipsoid
theta = asind(gamma_phi/gamma_nu);
end
```

Exercise 2 Since GPS can give the ellipsoidal coordinates of a point, but the global model is expressed in terms of (geocentric) spherical coordinates, it is of interest to transform $(\lambda_e, \varphi_e, h)$ into $(\lambda_s, \varphi_s, r)$. This can be done by passing through the Cartesian geocentric coordinates of P. Given the ellipsoidal coordinates $(\lambda_e, \varphi_e, h)$, compute $(\lambda_s, \varphi_s, r)$ of the same point.

Solution We have first of all to compute the geocentric Cartesian coordinates (x_P, y_P, z_P) of the point P. This can be done by exploiting the following formulas:

$$N = \frac{a}{\sqrt{1 - e^2 \sin^2 \varphi_e}};$$

$$e = \sqrt{1 - (1 - f)^2};$$

$$e' = 1 - (1 - f)^2;$$

$$X = (N + h) \cos \lambda_e \cos \varphi_e;$$

$$Y = (N + h) \sin \lambda_e \cos \varphi_e;$$

$$Z = (N(1 - e') + h) \sin \varphi_e.$$

Here N is the radius of curvature in the prime vertical, a is the equatorial radius (or semi-major axis), $e = \sqrt{1 - \frac{b^2}{a^2}}$ is the first eccentricity, $e' = \sqrt{\frac{a^2}{b^2} - 1}$ is the second eccentricity, b is the polar radius (or semi-minor axis) and finally X, Y and Z are the geocentric Cartesian coordinates of P. Passing to the geocentric system, we can easily see that the radial coordinate is given by

$$r = \sqrt{X^2 + Y^2 + Z^2} = \sqrt{[(N + h) \cos \varphi_e]^2 + [(N(1 - e') + h \sin \varphi_e]^2}.$$

As for λ_s, we have:

$$\lambda_s = \tan^{-1} \frac{Y}{X} = \lambda_e.$$

The last coordinate, namely φ_s, is given by:

$$\varphi_s = \tan^{-1} \frac{Z}{\sqrt{X^2 + Y^2}} = \tan^{-1} \left[\frac{N(1 - e') + h}{N + h} \tan \varphi_e \right];$$

The Matlab code for the conversion is provided below.

```
% Matlab code for Exercise 2
function [X, Y, Z, lambda_s, phi_s, r] = esercizio4_2(lambda_e ,phi_e , h)
%Compute the Cartesian (X,Y,Z) and the geocentric (lambda_s,phi_s,r)
%coordinates of a point P of coordinates (phi_e, lambda_e, h_e). Angles are
%in degree, h and r in meters.

a = 6378137.0; % semi-major axis [m]
f= 1 / 298.257223563; % flattening
e2 = 1-(1-f)^2;

sinphi = sind(phi_e);
cosphi = cosd(phi_e);
N = a ./ sqrt(1 - e2 * sinphi.^2);
X = (N + h) .* cosphi .* cosd(lambda_e);
Y = (N + h) .* cosphi .* sind(lambda_e);
Z = (N .*(1 - e2) + h) .* sinphi;
r = sqrt(X.^2 + Y.^2 + Z.^2);
lambda_s = atan2d(Y, X);
phi_s = atand(Z./sqrt(X.^2 + Y.^2));
end
```

Exercise 3 Given a mountain chain (infinite in the y-direction) approximated by two rectangular prisms with side $2a = 20{,}000$ m and $2b = 10{,}000$ m respectively and height $h = 500$ m, compute the depth of the isostatic root and the gravitational signal on the top of the mountain at $x = 0$, knowing that the crust density is $\rho_c = 2670\,\text{kgm}^{-3}$, the mantle density is $\rho_m = 3330\,\text{kgm}^{-3}$ and supposing a compensation depth of 30 km.

Solution The isostatic root of the mountains can be found by applying the Airy–Heiskanen model, reported in (4.47):

$$h_r = \frac{\rho_c}{\rho_m - \rho_c} h_t = \frac{2670}{3330 - 2670} 1000 = 4045\,\text{m}.$$

As for the gravitational field at the top of the mountain, it can be computed by combining the effect of Bouguer plate with the effects of the several rectangular prisms, namely $\delta g(0) = R_1 + R_2 + R_3 + R_4 + B_1 + B_2$, see (2.65). Before applying the computation of δg we need to calculate the height of R_3 and R_4, namely h_3 and h_4. The former can be obtained applying again the isostatic formula, while the latter is given simply as $h_4 = h_r - h_3$.

$$h_3 = \frac{\rho_c}{\rho_m - \rho_c} h_t = \frac{2670}{(3330 - 2670)} 500 = 2022.5\,\text{m}.$$

and $h_4 = 2022.5$ m. The final δg is:

$$\delta g(0) = R_1 + R_2 + R_3 + R_4 + B_1 + B_2 =$$
$$= -54.2 - 53.3 + 10.8 + 5.2 + \text{const.} = -91.5 + \text{const.}$$

Note that the constant value strictly depends on the compensation depth, and on the depth of the base of the B_2 Bouguer plate.

Exercise 4 Given the point P ($\lambda_e = 0°, \varphi_e = 48°, h = 100$ m) and knowing that the corresponding geoid ondulation at the level of the ellipsoid is $N = 48$ m. Knowing that $g(P) = 980{,}589$ mGal, compute the gravity disturbance $\delta g(P)$ and the free air gravity anomaly $\Delta g(P)$.

Solution We start the exercise by computing the orthometric height of P. This can be done by recalling that $H_p = h_P - N(P)$, which gives $H_P = 100 - 48 = 52$ m.

As for the gravity disturbance δg, it can be computed as $g(p) - \gamma(P)$, where $\gamma(P)$ is computed accordingly to (4.16) thus obtaining $\delta g(P) = -0.01$ mGal.

The gravity anomaly can be computed simply by exploiting (4.53) and supposing that $h_P^* = H_P = 52$ m, thus obtaining $\Delta g(P) = -14.84$ mGal. Note that if we want to use the more precise definition of normal height we have [133]:

$$h_P^* - H_P = \left(\frac{g(P)}{\gamma} - 1 + 0.2012 \cdot 10^{-3} H_P\right) H_P = 5 \cdot 10^{-3} \text{ m},$$

which is negligible.

Chapter 5
Gravity Surveying and Preprocessing

5.1 Outline of the Chapter

After a first section where we clarify some type of observations that can be performed on the gravity field, we shortly review the measuring principles of existing most common gravimeters, especially with the purpose of specifying their typical use and their performance in terms of accuracy. Then we come to the observations of the most classical instrument, the gravity meter and its use in the relative gravimetry mode.

This is typically done in the form of repeated measurements in a gravity network. The observations display a clear time dependence, partly due to a drift of the instrument itself, and partly due to gravity changes in time, in particular tidal signals. Moreover, the constant that transforms a reading into a gravity value has also to be calibrated, because it varies from one measurement campaign to another. All these parameters describing the above phenomena enter into observation equations, that can be then compensated via a least squares solution of the gravity network. Notice that in this program we use empirical knowledge of the measuring process and a priori external information (e.g., tidal theory) to arrive at defining a field of δg or Δg (quasi-static) evaluated at observation points, without exploiting functional properties of the field; this is what we call *preprocessing*.

5.2 Observations of the Gravity Field

Observations on the gravity field can be performed either on the direction n or on the intensity of forces that affect one or more proof masses. The first type of measurement pertains more to Geodesy and is only slightly related with the subject of this book, so we shall not dwell upon it.

As for observations of the gravity intensity, we first of all define two families: kinematic (or dynamic) observations and static observations.

F. Sansò, D. Sampietro, *Analysis of the Gravity Field*, Lecture Notes in Geosystems Mathematics and Computing, https://doi.org/10.1007/978-3-030-74353-6_5

5.2.1 Kinematic Observations

These are observations performed by instruments mounted on moving platforms, such as cars, ships, aircraft or satellites. The point is that, to describe the relation between such observations and the gravity field, it is necessary to couple them with the dynamics of the support, which requires knowledge of special parts of mechanics, for which we send to the literature [71, 129].

This is in particular true for satellite observations, which in the last ten years, with dedicated missions like Champ, GRACE and GOCE [30, 114, 146], have allowed the reconstruction of the long/medium wavelengths of the gravity field, producing satellite-only models up to degree 250, i.e. with a ground resolution of $\sim$80 km. We consider the results of space gravimetry/gradiometry as an acquired datum which is incorporated into the already mentioned global model. Marine and airborne gravimetry/gradiometry are kind of measurements that are certainly employed for the type of interpretation that this book is concerned with. Even when the modality of the surveying decouples the attitude of the vehicle from the measurements, which is achieved by fixing the instruments to an inertially stabilized platform, the difficulty of performing such observations is related to the presence of mechanical vibrations that induce oscillating forces, which must be separated from the gravity variations along the trajectory. It is intuitively clear then that a low velocity kinematics is preferable to a high velocity one, because in this case the frequencies of the signal induced by gravity variations will be much more concentrated in a low frequency band and it will be easier to separate them from environmental vibrations [122].

Such a separation is carried out by a low-pass filtering, the simplest example of which is just taking a moving average over a certain time window. The effect of such a filter is that higher frequencies, even in the signal we want to achieve, are removed or damped. Lower frequencies are enhanced, with a more favourable signal to noise ratio.

It has to be mentioned too that in kinematic surveying, every proof mass will be submitted to a further apparent force derived from its motion with respect to the terrestrial reference system. Just to give an idea, such an Eötvös correction δg_{E}, for the vertical component of gravity can be expressed by the simple formula [151, §5.4.4]

$$\delta g_{\mathrm{E}} = 40 v \cos \varphi \sin \alpha + 1.2 \cdot 10^{-3} v^2, \tag{5.1}$$

where v is the horizontal velocity of the vehicle, given in km/h, α is the azimuth with respect to the North direction (also called heading in navigation literature), φ is the latitude, and δg_{E} is given in mGal.

As we see, for a ship moving at a velocity of 10 km/h, δg_{E} can amount to 400 mGal, while for an aircraft flying at 100 km/h it can be as large as 4014 mGal. This is just to underline that such an effect has to be taken carefully into account. As a matter of fact, to derive the gravitational force acting on a proof mass, one has also to know its position, velocity and acceleration with respect to the terrestrial reference frame.

This is provided by Inertial Navigation Units (IMU) and GNSS observations, the number and accuracy of which is continuously increasing, also due to constellation improvements [21]. All in all, the state of the art today could be summarized by the following Table 5.1. As a result, we can say that, for the purpose of local gravity interpretation, we can consider data from marine or airborne gravimetry, to provide the gravity intensity at given points in space, with a spacing equal to the resolution indicated above. The Eötvös kinematic corrections are supposed to be already applied to the above gravity data, so that they can undergo from that point on, the same preprocessing that we shall apply to static gravimetric data. Despite the improvements in the gravimetry technology and the development of stabilized platform systems, airborne gravimetry did not become fully operational until the advent of Global Positioning System (GPS) (1980s–1990s).

In the 1990s, the scientific community (i.e., the University of Calgary, the University FAF Munich, the National Survey ad Cadastre of Denmark, the Swiss Federal Institute of Technology, and others) in collaboration with industries and governments, started the first large-scale airborne gravity campaigns to assess the capabilities of this technique to measure the gravity field at a resolution of few km. In 1991–1992 one of the first wide-area surveys was conducted in Greenland [20, 40], and with various joint projects other surveys were conducted during the same years and in the following ones over the Antarctic region and in Switzerland. The new accuracies reachable after the advent of GPS were of few mGal at 5–6 km of spatial resolution.

Concurrently with the spreading of airborne gravimetry with gravity measured using a stabilized platform system, in the 1990s another type of system was deployed: the Strapdown Inertial Navigation System (SINS). It consists of a set of three orthogonal accelerometers and three gyroscopes. The objective of this system is the determination of the full gravity disturbance vector along the aircraft trajectory; when used, the technique is referred to as airborne vector gravimetry. The first tests and developments of this technique were conducted by the University of Calgary [138] and showed that comparable accuracies at the same spatial resolution of few kilometers could be achieved [159]. For geophysical applications, such as resource exploration, nowadays the classical technique adopted is the stabilized platform system. However, in recent years, vector gravimetry surveys begin to be employed by various geophysics companies. Strapdown gravimetry systems are attracting increasing interest of the scientific community due to the potential evolution of IMU [93].

Table 5.1 Typical accuracy and resolution of kinematic gravimetry with marine or airborne techniques

	Accuracy [mGal]	Resolution [km]
Marine	2–5	1–2
Airborne	1–3	2–10

5.2.2 Static Observations

The measurements we shall consider here are derived from only two types of instruments: gravity meters and gradiometers. The corresponding measuring techniques are called *gravimetry* and *gradiometry*.

Gravimetry can be performed by two kinds of instruments: absolute gravity meters or relative gravity meters.

5.2.3 Absolute Gravimetry

The purpose of the measurement is to obtain the gravity modulus $g(P)$. Apart from a historical period, dating back to G. Galilei and lasting three centuries, when the pendulum was used to measure g, nowadays such measurements are performed by observing the rise and fall of a proof mass in empty space.

Knowing the law of motion (see Fig. 5.1)

$$h(t) = h_0 + r_0 t - \frac{1}{2} g t^2 \tag{5.2}$$

and measuring heights versus time one can derive the coefficients of (5.2), and in particular g.

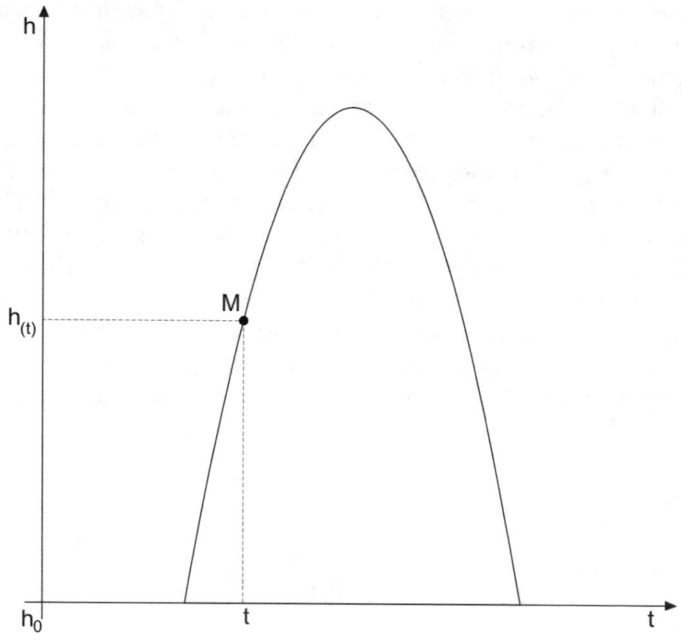

Fig. 5.1 Height trajectory of a proof mass *M* launched upward in a pipe where vacuum has been made

The apparatus of an absolute gravity meter is generally heavy and its main use is in a fixed, stable station. The accuracy that can be reached by an absolute gravimeter is in the range between one and few μGals; for this reason, too, it is important that instruments operate in very quite places, where noises, in terms of gravitation or vibrations, are very small. In any event, the main use of the worldwide network of absolute gravimeters is to provide a reference for all relative gravity measurements and to monitor temporal changes in the Earth gravity field.

In the context of this book, it is the first target that might be of interest, although this is not a critical issue in local gravity field interpretation.

Let us only mention here, that starting from the nineties (see e.g. [27]) a new technique of absolute gravity measurement, based on precision atom interferometry, was developed by physicists, which is capable of improving further the accuracy (see [131]).

5.2.4 Relative Gravimetry

The purpose of relative gravimetry measurements is to provide the value of the gravity difference between two points,

$$Dg = g(p) - g(Q), \tag{5.3}$$

or $D\Delta g$, $D\delta g$.

By measuring several differences in a gravity network, with an overdetermined number of observations with respect to the nodes, one can then determine gravity at all nodes, if this is assumed to be known at least at one station, or as an average over all the stations.

This is because in a relative gravity network there is always a rank deficiency of 1, exactly as for an electric network with measurements of potential differences, or like a surveying leveling network (Fig. 5.2).

The principle of the observation of a relative gravity meter is essentially that of measuring the elongation (in some cases the torsion) of a spring when moving the instrument between two points, P and Q. This can be done in several configurations, as it will be shown in the next Section. According to Hooke's law, the elongation $\delta\ell$ is just proportional to Dg

$$Dg = k\delta\ell \tag{5.4}$$

The data analysis of a gravity network at this point would be as simple as the compensation of a leveling network, was it not for the fact that the constant k cannot be calibrated in laboratory once and for all, but rather it exhibits a behaviour drifting in time, requiring so to say an a posteriori calibration procedure during compensation. For this reason, the observation equation of a spring gravimeter

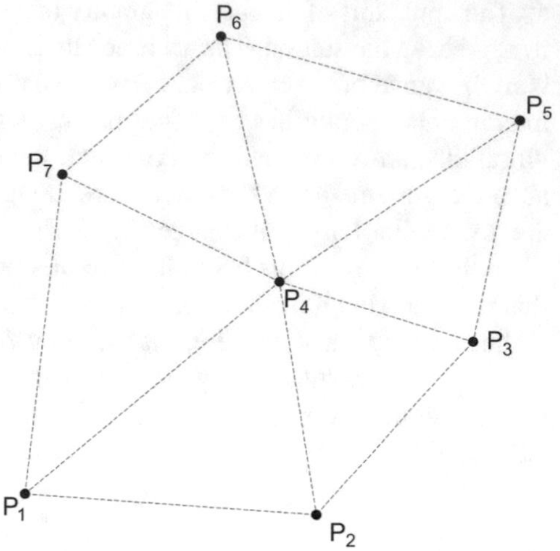

Fig. 5.2 Hypothetical scheme of a gravity network, with 7 nodes and 12 relative gravity measurements

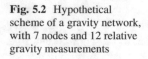

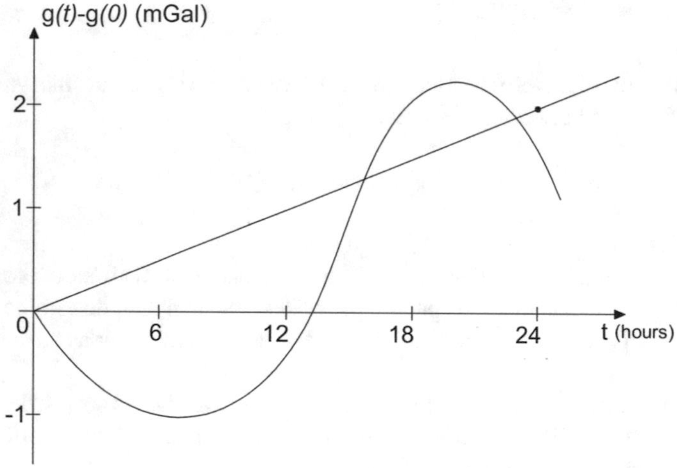

Fig. 5.3 Variation in time of the registration of a spring gravity meter at a fixed position on the Earth surface

needs to be carefully studied, together with measuring schemes that simplify the determination of the relevant parameters and of the static component, $g(P)$, of the gravity field. Just to provide a qualitative insight into the instrumental behaviour of a spring gravimeter, let us assume that we perform measurements over one-day period, keeping the instrument still at a fixed place. Neglecting other noises, we could register a result like that of Fig. 5.3, where $g(t) - g(0)$ is plotted.

The typical repeatability of a measurement of a spring gravimeter is in the range between 0.01 and 0.1 mGal.

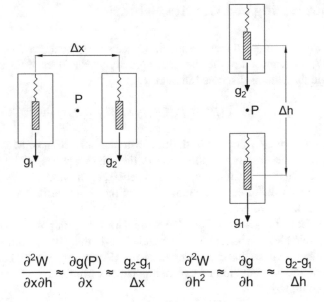

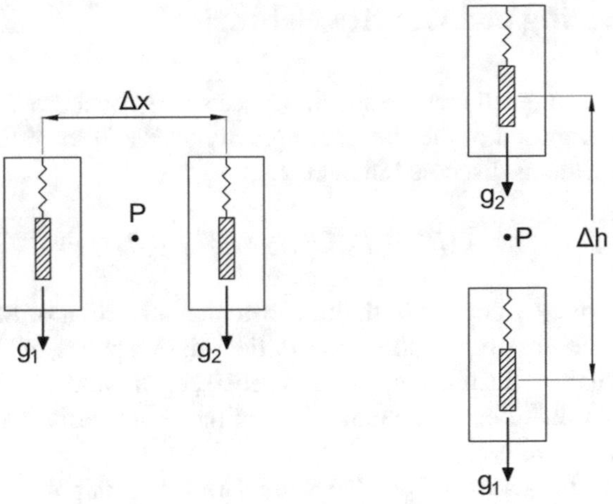

$$\frac{\partial^2 W}{\partial x \partial h} \approx \frac{\partial g(P)}{\partial x} \approx \frac{g_2 - g_1}{\Delta x} \qquad \frac{\partial^2 W}{\partial h^2} \approx \frac{\partial g}{\partial h} \approx \frac{g_2 - g_1}{\Delta h}$$

Fig. 5.4 Arrays of gravimeters used to measure second derivatives of the potential

5.2.5 *Gradiometry*

By simultaneously measuring gravity at close points one can achieve spatial derivatives of g, namely second derivatives of the gravity potential W. The principle in itself is elementary and we illustrate it in Fig. 5.4 for two components of the tensor of the second derivatives of W.

The unit of gravity gradient is usually the Eötvös: $1\text{E} = 10^{-9}\text{s}^2$. The name comes from L. Eötvös, the scientist who in 1900 invented a gradiometer that measures horizontal second derivatives by the principle of torsional balance, widely applied in the first half of the twentieth century. The accuracy of the gradiometers of the type shown in Fig. 5.4 is of the order of $1\ \text{mGalcm}^{-1} = 10^{-8}\ \text{s}^{-2} = 10\ \text{E}$. Other measurement principles are used too for the purpose of observing second derivatives of W.

We notice that various versions of airborne gradiometers are also available today, which can be used for local airborne surveying and are therefore interesting for the purpose of local gravity interpretation. We should also mention that a spatial version of the gradiometer has been flown on board the GOCE satellite of the European Space Agency (ESA). This mission has allowed the most accurate (at the moment) determination of the anomalous potential coefficients in the range of degrees between 100 and 250 [98].

5.3 Preprocessing of a Gravity Network

First of all, we have to recall that the specific force sensed by a proof mass, at point P and time t, is composed of the stationary gravity at P and the tidal attraction of the Moon and the Sun, as discussed in Sect. 4.2,

$$f(P,t) = g(P) + f_{\mathrm{M}}(P,t) \cdot \boldsymbol{\nu} + f_{\mathrm{S}}(P,t) \cdot \boldsymbol{\nu}. \tag{5.5}$$

The terms f_{M} and f_{S} contain both the Newtonian attraction of Moon and Sun, respectively, and the gravity variation due to the tidal response of the Earth. The scalar product with the ellipsoidal normal $\boldsymbol{\nu}$ is necessary to evaluate only the vertical components of the tidal effects. The magnitude of the above terms is in the range of a few 10^{-1}mGal.

All such effects (in fact including also those deriving from polar motion) can be computed by suitable models. Yet, local irregularities in the Earth structure can leave an unaccounted signal, so that exploiting our a priori knowledge that it could be well represented by the combination of two components with diurnal and semi-diurnal periods, we could write

$$\begin{aligned} & g(P,t) = g(P) + s(t), \\ & s(t) = a_1 \cos 2\pi \frac{t}{24} + b_1 \sin 2\pi \frac{t}{24} + a_2 \cos 2\pi \frac{t}{12} + b_2 \sin 2\pi \frac{t}{12}, \\ & (t \text{ in hours}), \end{aligned} \tag{5.6}$$

where we agree that $g(P,t)$ has already been corrected for modelled tidal forces and $s(t)$ is the unknown residual part, defined by means of 4 unknown parameters.

The reading of the instrument is related to the length of the spring $L(P,t)$ and needs generally a linear transformation to be reported to $f(P,t)$;

$$f(P,t) = B + kL(P,t); \tag{5.7}$$

yet, since the spring undergoes a drift in time, one has instead of (5.7)

$$f(P,t) = B + kL(P,t) + Dt. \tag{5.8}$$

Now, k varies from one measurement campaign to another, although remaining close to a calibrated value $\bar{k}$, so that we can write

$$\begin{cases} k = \bar{k}(1+\varepsilon), \\ \bar{k}L(P_i,t_i) = Y_i, \end{cases} \tag{5.9}$$

with Y_i considered as our new observations. Combining (5.6), (5.8), and (5.9) and linearizing, we can write

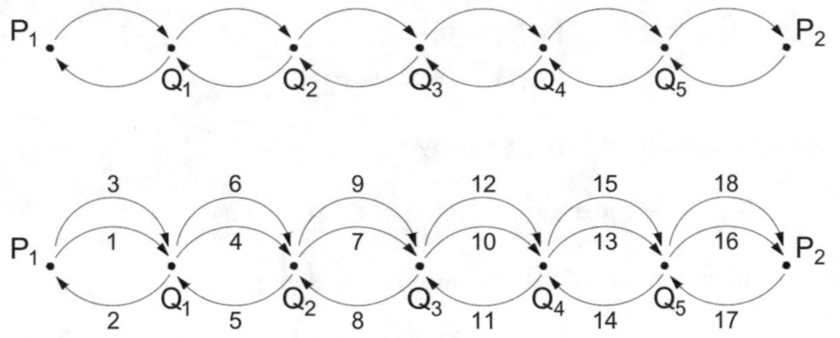

Fig. 5.5 Possible schemes of chain of relative gravity observations with multiple occupation of the stations

$$Y_i = -\varepsilon Y_i - B - Dt_i + g(P) + s(t_i), \tag{5.10}$$

where the always present measurement noise has been disregarded.

Subtracting two equations (5.10), for measurements at (P_i, t_i) and (P_j, t_j), we get the observation equation of relative gravimetry,

$$Y_i - Y_j = -\varepsilon(Y_i - Y_j) - D(t_i - t_j) + g(P_i) - g(P_j) + s(t_i) - s(t_j). \tag{5.11}$$

The unknown parameters in (5.11) are ε, D, a_1, b_1, a_2,b_2 and $g(P_i)$, $g(P_j)$. In principle we can use a least squares solution for the network with observations (5.11) if we fix one of the values $g(P_i)$ either because we know it, e.g., from absolute gravimetry, or because we give it a conventional value. All that if there is redundancy in the network.

It is for this reason, and in general to improve the accuracy of the estimates, that typical measurement schemes are followed, breaking each side of the network, e.g., $\overline{P_1 P_2}$ in Fig. 5.2, into a chain of observations, where each point is reoccupied by the instrument at different times.

For instance, one could perform the measurements according to one of the schemes of Fig. 5.5.

In fact, every time we return to the same point in the Eq. (5.11), $g(P_i) = g(P_j)$ cancel and an equation containing only the 6 instrumental parameters is obtained.

This increases the accuracy of the estimates: in a typical gravity network, the accuracy of the determined values of $g(P_i)$ is in the range of a few 10^{-2} mGals.

As for the statistical tool to be employed to estimate parameters and $g(P_i)$, as we said, simple least squares can be used. Recall that for a system of linear (or linearized) equations

$$\boldsymbol{Y}_0 = A\mathbf{p} + \boldsymbol{a} + \boldsymbol{v}, \tag{5.12}$$

with $\boldsymbol{Y}_0$ the vector of observations, $\mathbf{p}$ the vector of parameters, A a design matrix, $\boldsymbol{a}$ a constant vector, and $\boldsymbol{v}$ a vector of errors with stochastic structure:

$$\begin{cases} E\{\boldsymbol{\nu}\} = 0, \\ E\{\boldsymbol{\nu}\boldsymbol{\nu}^T\} = C_\nu = \sigma_0^2 C, \end{cases} \tag{5.13}$$

the least squares estimate of **p** is given by

$$\hat{\boldsymbol{p}} = (A^T C^{-1} A)^{-1} A^T C^{-1} (\boldsymbol{Y}_0 - \boldsymbol{a}) \tag{5.14}$$

and the covariance matrix of the estimation error is given by

$$C_{\mathbf{p}} = \sigma_0^2 (A^T C^{-1} A)^{-1}. \tag{5.15}$$

Finally, an unbiased estimate of σ_0^2 is given by

$$\hat{\sigma}_0^2 = \frac{\hat{\mathbf{u}}^T C^{-1} \hat{\mathbf{u}}}{m - n}, \tag{5.16}$$

with

$$\hat{\mathbf{u}} = \boldsymbol{Y}_0 - A\hat{\mathbf{p}} - \mathbf{a} \tag{5.17}$$

and, when A is of full rank, m is the number of the equations (rows of A) and n the number of parameters (columns of A). As we have already remarked, when the observation equations are of type (5.11) we need to fix one of the values of $g(P_i)$ to obtain a design matrix of full rank.

Note also that if we use only one instrument to survey the network, we can assume $C = I$, thus simplifying the computation.

Another, less rigorous, but as effective approach, could be to use first of all (5.11) only for measurements where the same point has been reoccupied, namely

$$\begin{cases} P_i = P_j, \quad t_i \neq t_j, \\ Y_i - Y_j = -\varepsilon(Y_i - Y_j) - D(t_i - t_j) + s(t_i) - s(t_j) + \nu. \end{cases} \tag{5.18}$$

The system (5.18) is solved via least squares to estimate the parameters ε, D, a_1, b_1, a_2, b_2. Such a system is never rank deficient, especially if the time span of the whole survey covers several hours. When the 6 parameters have been estimated, new "observations" can be defined as

$$Z_i = Y_i + \hat{\varepsilon} Y_i + \hat{D} t_i + \hat{s}(t_i); \tag{5.19}$$

note that, for each station P_i, only one Y_i has to be used, without repetition. Then new "observation equations" of the form

$$Z_i - Z_j = g(P_i) - g(P_j) + \nu \tag{5.20}$$

can again be solved for $g(P_i)$ by the least squares method, always after fixing one of them.

One final comment is that in principle the use of a so-called (physical) covariance function of the gravity field could improve the estimates, by applying a collocation solution with parameters. Nevertheless, on the one hand the improvement seems not to be worthwhile the effort, and on the other hand, such more sophisticated method will be treated in the framework of the next chapter.

Example 5.1 This example utilized data from a real gravity survey aiming at characterizing the gravimetric signal of the Grotta Gigante, a large cave in the Italian region of Friuli, close to the Slovenian border. The acquisitions were made by a traditional La Coste–Romberg model D relative gravimeter with optical reading. The gravitational field was measured at 75 stations distributed over the cave, over an area of about 200×200 m, according to the network shown in Fig. 5.6. In details, in order to solve the least squares system 30% of all measurements were made on a station that was already been occupied before. The raw gravity observations are corrected for the Earth tides and drift with the dbGrav software written and distributed by Dr. Sabine Schmidt of University of Kiel, Germany. In Fig. 5.7 the Earth tide and drift are plotted for the days spanning the data acquisition (15–19 March 2013). The drift is systematically in one direction, increasing the gravity value, with an average rate 0.012 ± 0.006 mGalh^{-1}. The loading and gravitational effect of the nearby Adriatic Sea has not been corrected, becausc a rcliablc testing of available ocean models for the area was not available at the time of the survey. The final free air anomaly for the area is shown in Fig. 5.8. The interested reader can find more details on this example in [19] and the reference therein.

5.4 Exercises

Exercise 1 Calculate the tidal acceleration a_T due to the Moon on a proof mass in free fall at the Earth's equator at lunar noon, i.e., with the Moon directly overhead.

Solution From Newton's law we have that the acceleration due to the Moon a_{M} acting on a point mass in free fall on the Earth surface is given by

$$a_{\mathrm{M}} = \frac{GM_{\mathrm{M}}}{(R_{\mathrm{M}} - a)^2}$$

where G is the universal gravitational constant, here considered equal to $6.67259 \cdot 10^{-11}\,\mathrm{m^3\,s^{-2}\,kg^{-1}}$, M_{M} is the Moon mass ($M_{\mathrm{M}} = 7.3547 \cdot 10^{22}$ kg), R_M is the distance from the centered of mass of the Earth to that of the Moon ($3.84 \cdot 10^8$ m), and a is the equatorial radius ($a = 6.378137 \cdot 10^6$ m). However, since also the Earth (i.e., its barycenter) is accelerated by the Moon attraction, with an acceleration approximatively given by $\frac{GM_{\mathrm{M}}}{R_{\mathrm{M}}^2}$, the total acceleration sensed by the gravimeter is

Fig. 5.6 Location of the gravity stations in relation to the urbanization above the cave. The yellow line is the outline of the cave

$$a_{\mathrm{M}} = GM_{\mathrm{M}} \left[\frac{1}{(R_{\mathrm{M}} - a)^2} - \frac{1}{R_{\mathrm{M}}^2} \right].$$

With the above values, the tidal acceleration is

$$a_{\mathrm{M}} = 3.441 - 3.328 = 0.113\,\mathrm{mGal}.$$

Exercise 2 A gravimeter is placed in a fixed position in a lab to derive its drift curve. Naturally the observed gravity variations will contain also the tidal effects. Observations together with the computed tidal effects and the time (expressed in hours from the first observation) are reported in Table 5.2. Apply a Least Squares Adjustment to estimate the instrumental drift.

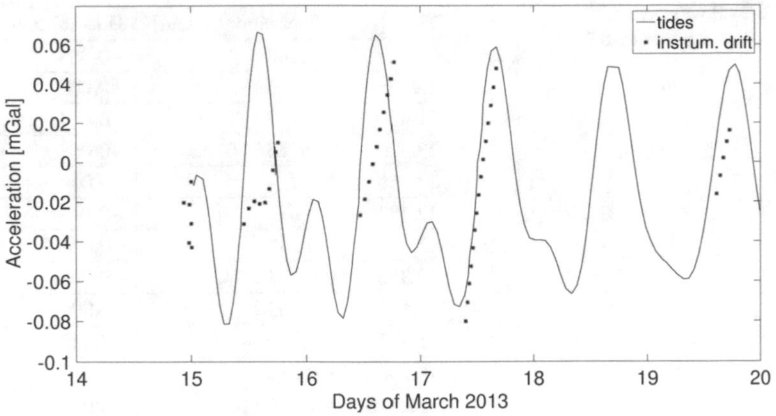

Fig. 5.7 Earth tides and instrumental drift for the Grotta Gigante survey

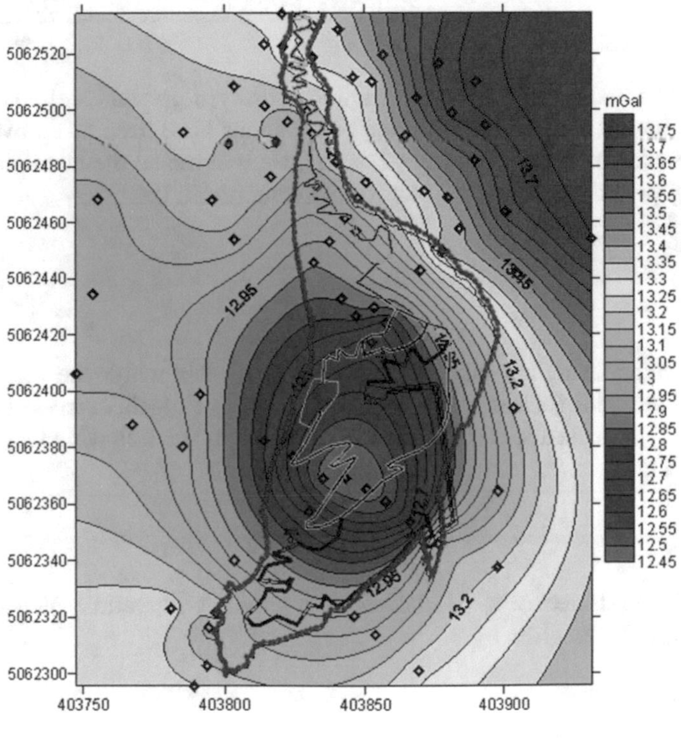

Fig. 5.8 Gravitational anomaly of the Grotta Gigante cave

Table 5.2 Raw gravity observations for Exercise 2

Time [h]	Observation [mGal]	Tide effect [mGal]
0	1.248	−0.063
1.783	1.206	−0.062
21.400	0.727	0.047
23.300	0.713	−0.008
23.766	0.724	−0.019
27.283	0.677	−0.012
47.300	0.332	0.021
49.350	0.272	−0.008
51.133	0.263	0.001
71.416	0.089	0.035
74.100	−0.059	0.014
117.116	−0.736	0.001
119.350	−0.701	0.025
122.266	−0.824	0.039
145.566	−1.149	0.034
164.083	−1.350	−0.065

Solution The first step is to remove from the observed gravitational field the tidal effect. As for the drift effect, since the observations have been taken over a quite large time interval (approximatively 1 week), we will model the instrumental drift d as a long-period (weekly) drift plus a daily component, i.e.,

$$d = c + d_{\mathrm{W}} t_{\mathrm{W}} + \sum_{i=1}^{7} d_i t_i;$$

where d_{W}, d_i and c are the unknown parameters collected within the vector $\underline{x}$, t_{W} is the time of the week (in hours) and t_i is the time of day i (again expressed in hours). The above system the above equations can be solved by the following Least Squares Adjustement:

$$\hat{\mathbf{x}} = \left(\mathbf{A}^T\mathbf{A}\right)^{-1}\mathbf{A}^T\mathbf{y};$$

where $\hat{\mathbf{x}}$ is the estimate of $\mathbf{x}$, $\mathbf{y}$ is the vector of gravity observations reduced for the tidal effect, and A is given by

$$A = \begin{bmatrix} 1 & 0 & 0 & 0 & 0 & \dots \\ 1 & 1.783 & 1.783 & 0 & 0 & \dots \\ 1 & 21.400 & 21.400 & 0 & 0 & \dots \\ 1 & 23.300 & 23.300 & 0 & 0 & \dots \\ 1 & 23.766 & 23.766 & 0 & 0 & \dots \\ 1 & 27.283 & 0 & 27.283-24 & 0 & \dots \\ 1 & 47.300 & 0 & 47.300-24 & 0 & \dots \\ 1 & 49.350 & 0 & 0 & 49.350-48 & \dots \\ 1 & 51.133 & 0 & 0 & 51.133-48 & \dots \\ 1 & 71.416 & 0 & 0 & 71.416-48 & \dots \\ 1 & 74.100 & 0 & 0 & 0 & \dots \\ 1 & 117.116 & 0 & 0 & 0 & \dots \\ 1 & 119.350 & 0 & 0 & 0 & \dots \\ 1 & 122.266 & 0 & 0 & 0 & \dots \\ 1 & 145.566 & 0 & 0 & 0 & \dots \\ 1 & 164.083 & 0 & 0 & 0 & \dots \end{bmatrix}.$$

The estimated values are

$$\hat{\mathbf{x}} = \begin{bmatrix} 1.2268 \\ -0.0171 \\ -0.0050 \\ -0.0047 \\ 0.0017 \\ -0.0114 \\ 0.0035 \\ 0.0067 \\ 0.0159 \end{bmatrix}.$$

Note that from the above Least Squares Adjustment it is possible also to estimate the residuals, i.e., $\boldsymbol{v} = \mathbf{y} - \mathbf{A}\mathbf{x}$, thus obtaining a piece of information on the accuracy of the analysed dataset. The estimated observation variance $\hat{\sigma}_o^2 = \frac{\mathbf{v}^T\mathbf{v}}{n-m}$ is 0:0068 mGal^2. Here n is the number of observations and m is the number of unknown in the least squares system. The Matlab code to solve the exercise is given below.

```
% Matlab code for Exercise 2
time = [ 0 1.783 21.400 23.300 23.766 27.283 47.300 49.350 51.133 71.416
74.100 117.116 119.350 122.266 145.566 164.083];

g = [1.248 1.206 0.727 0.713 0.724 0.677 0.332 0.272 0.263 0.089
-0.059 -0.736 -0.701 -0.824 -1.149 -1.35];

tide = [-0.063 -0.062 0.047 -0.008 -0.019 -0.012 0.021 -0.008 0.001
0.035 0.014 0.001 0.025 0.039 0.034 -0.065];

y = g' - tide';

A = zeros(16,9);

A(:,1) = 1;
A(:,2) = time';
for i = 3 : 9
        for j = 1 : 16
                if (time(j) > 24 * (i-3)) & (time(j) < 24 * (i-2))
                        A(j,i)=time(j)-24 * (i - 3);
                end
        end
end

xst=inv(A'*A)*A'*y;
ni = y - A *xst;
s2est = (ni' * ni) / (length(y) - length(xst));
```

Table 5.3 Relative corrected gravity observations for Exercise 3

Δg_{12}	−0.523 mGal
Δg_{23}	0.214 mGal
Δg_{31}	0.322 mGal
g_1	$9.8342124 \cdot 10^5$; mGal

Exercise 3 Consider the following gravimetric net (see Table 5.3), constituted by one loop only with three points P_1, P_2 and P_3 and the following observations already corrected for time dependent effects, where $\Delta g_{ij} = g_i - g_j$ and g_1 is the absolute gravity at the point P_1. Compensate the network to estimate the absolute gravity for the three points knowing that both relative gravimeter and the absolute one have an accuracy $\sigma = 0.1$ mGal.

Solution The network adjustment can be performed by means of a least squares solution. Considering the following observation vector **y**:

$$\mathbf{y} = \begin{bmatrix} -0.523 \\ 0.214 \\ 0.322 \\ 9.8342124 \cdot 10^5 \end{bmatrix},$$

the unknowns are:

$$\mathbf{x} = \begin{bmatrix} g_1 \\ g_2 \\ g_3 \end{bmatrix},$$

and the corresponding design matrix A is:

$$A = \begin{bmatrix} 1 & -1 & 0 \\ 0 & 1 & -1 \\ -1 & 0 & 1 \\ 1 & 0 & 0 \end{bmatrix}.$$

As for the cofactor matrix Q, since all the observation are independent we can use the identity matrix. The solution is given by

$$\hat{\mathbf{x}} = \left(\mathbf{A}^T\mathbf{A}\right)^{-1}\mathbf{A}^T\mathbf{y},$$

yielding

$$\hat{\mathbf{x}} = \begin{bmatrix} 98342.1240 \\ 98342.1767 \\ 98342.1558 \end{bmatrix}$$

and an the estimate of $\hat{\sigma}_0 = 0.053$ mGal.

The Matlab code to solve the exercise is given below.

```
% Matlab code for Exercise 3
Delta_g_12 = -0.523;
Delta_g_23 = +0.214;
Delta_g_31 = +0.322;
g_1 = 9.8342124*1e5;
s2 = 0.01;

y=[Delta_g_12 Delta_g_23 Delta_g_31 g_1]';

A =[1 -1 0; 0 1 -1; -1 0 1; 1 0 0];
Q = eye(4) .* 0.01;
Q(1,1) = 2 .* Q(1,1);
Q(2,2) = 2 .* Q(2,2);
Q(3,3) = 2 .* Q(3,3);
xst=inv(A'*inv(Q)*A)*A'*inv(Q)*y;
ni =y - A*xst;
s2est = (ni' * ni) / (length(y) - length(xst));
```

Chapter 6
Gravity Processing

6.1 Outline of the Chapter

According to the discussion in Chap. 5, one could say that, after having performed a gravity survey, one can apply the approach of Sect. 5.3 to reduce observations to a static picture of values $g(P_i)$, with P_i in a certain area A.

Then, following the discussion in Sect. 4.5, one can apply to our data set a normal gravity correction as well as a correction derived from some global model.

In this way, we arrive at a sample of what we have called a stationary localized field of gravity disturbances $\delta g_L(P_i)$ (or free air gravity anomalies $\Delta g_L(P_i)$).

On the other hand, it is well known that in $g(P_i)$ another signal is present, which explains in particular the shorter wavelength part of the spectrum of the data: this is the effect of topographic masses, i.e. those masses that are laying above the geoid.

Such an effect can be computed in a local area by means of one of the direct methods illustrated in Part I of the book, if we assume that the geometry of the topographic masses is known namely, if we have a Digital Model of the terrain and we assume that the mass density is more or less a known constant. When no more specific information is available, the mean value of the crust density, namely $\rho_c = 2670\ \mathrm{kg\,m^{-3}}$, can be used; otherwise, estimates for the local area can be derived from the values $\delta g_L(P_i)$ by the so-called *Nettleton method*, as it will be explained in this chapter.

Let $g_t(P_i)$ denote the gravitational attraction of the topographic masses at a point P_i on the Earth surface or above it, and for the moment assume that it can be computed without error. Then one is tempted as a first step to subtract it from $g(P_i)$, This is in fact what has been proposed long ago by Bouguer [17, 63, §3.3], especially when no global models of T were available.

This Bouguer theory, although no longer up to date, will be presented in this chapter, in view of its popularity among applied geophysicists.

The use of the global model, though, is important for reducing our data set to a local level, but it interferes with the concept of the terrain correction because the

F. Sansò, D. Sampietro, *Analysis of the Gravity Field*, Lecture Notes in Geosystems Mathematics and Computing, https://doi.org/10.1007/978-3-030-74353-6_6

model itself is capable of accounting for one part of the effect of the terrain, up to some medium degree. This leads to the need of identifying the model terrain correction, $g_{\mathrm{tM}}(P_i)$, so that to $\delta g_{\mathrm{L}}(P_i)$ we could apply only a *Residual Terrain Correction*, namely

$$\delta g_{\mathrm{t}}(P_i) = g_{\mathrm{t}}(P_i) - g_{\mathrm{tM}}(P_i). \tag{6.1}$$

In this way we can subtract from $\delta g_{\mathrm{L}}(P)$ only that part of the terrain correction that has not been accounted for by the model. This gives rise to the residual gravity disturbance $\delta g_{\mathrm{r}}(P)$ (or gravity anomaly $\Delta g_{\mathrm{r}}(P)$) defined by

$$\delta g_{\mathrm{r}}(P) = \delta g_{\mathrm{L}}(P) - \delta g_{\mathrm{t}}(P). \tag{6.2}$$

It has to be noticed that if the global model has a maximum degree of 720 or higher, it is clear that it should include the effect of the Moho discontinuity, too which, seen at the Earth surface, has hardly a resolution better than 30 km.

Therefore, the data set $\{\delta g_{\mathrm{r}}(P_i)\}$ will presumably contain only the local mass anomalies that have not yet been considered in our reduction process.

Eventually, we could expect a salt dome, an ophiolite body, etc., which create bulky bodies in our area, or a discontinuity surface of a more superficial layer of different density, for instance a sedimentary basin.

Anyway, in order to infer some information on this gravitational source, we often need to transform our original dataset $\delta g_{\mathrm{r}}(P_i)$ into another dataset of δg_{r} at the nodes of a regular grid $\{P_{jk}\}$, all at the same level. This can be done if we assume that all the mass anomalies lie below the level of the grid. This opens the way to applications of the FT, specially when the number of data is very large.

In the last section of the chapter, this matter is explained together with the necessary introduction to random fields linear prediction theory.

6.2 The Terrain Correction

In this section we first tackle the problem of computing the terrain correction $g_t(P)$, namely, the Newtonian attraction of topographic masses lying above the local area A', at points P on the Earth surface, or even at some altitude above it, when we have to correct aerogravimetric data. We start with the first of the two cases. As usual, we introduce some simplifying hypotheses. First of all, we assume that the density of topographic masses is constant and known. Second, we constrain the topographic body to be described in a spherical geometry by means of an area A' on the sphere of radius $R = 6371$ km, with a diameter of less than ~ 140 km, so that the maximum inscribed "spherical" square has a side of 100 km, and an upper surface S described by the altitude function

$$h = h(P_0) = h(\sigma) \quad \sigma = (\lambda, \varphi) \tag{6.3}$$

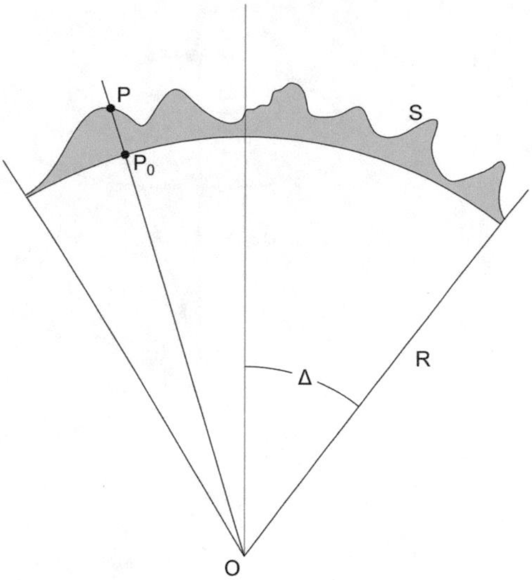

Fig. 6.1 The topographic masses in spherical geometry

The situation is presented in Fig. 6.1. With reference to it, we have

$$h_P \leq 9\,\text{km}, \quad \varepsilon_P = \frac{h_P}{R} \leq 10^{-3}, \tag{6.4}$$

$$\Delta \leq 1.1 \cdot 10^{-3}\ \text{rad}. \tag{6.5}$$

We maintain that by applying the mapping of Fig. 6.2, by the simple relations

$$r_{P_0'} = s_{P_0}, \alpha_{P_0'} = \alpha_{P_0}, z_P' = h_P, \tag{6.6}$$

the spherical formula for $g_t\,(P)$ goes into the corresponding Cartesian formula. Namely starting from

$$g_t\,(P) = -\mu \int_{A'} \int_0^{h_{Q_0}} \frac{(R + h_P) - (R + h) \cos \psi_{PQ}}{\ell_{PQ}^3} (R + h)^2 dh d\sigma_{Q_0}, \tag{6.7}$$

$\ell_{PQ} = \left[(R + h_P)^2 + (R + h)^2 - 2\,(R + h_P)\,(R + h) \cos \psi_{PQ} \right]^{1/2}$, ψ_{PQ} the spherical angular distance between the computation point P and the running point Q, this becomes approximately

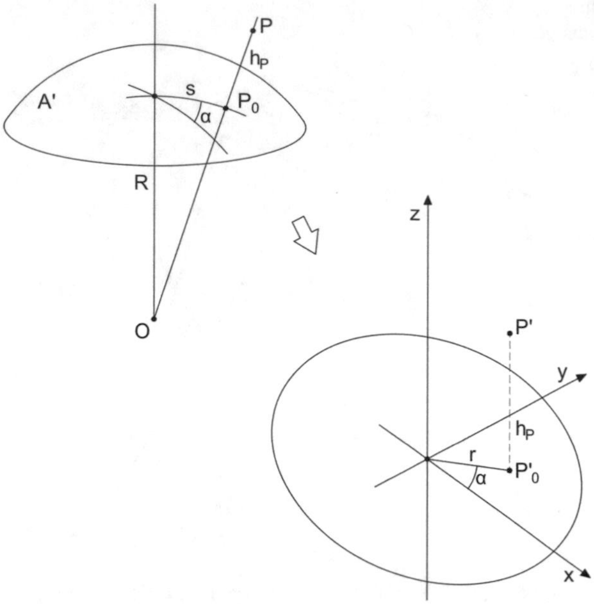

Fig. 6.2 The mapping *spherical to Cartesian*

$$g_t(P) \cong \mu \int_A \int_0^{h(\xi)} \frac{h_P - z}{\ell^3} dz d_2\xi, \tag{6.8}$$

$$\ell = \left[|\boldsymbol{x}_P - \boldsymbol{\xi}|^2 + (h_P - z)^2 \right]^{1/2},$$

where $\boldsymbol{\xi}$ is the position vector of the projection of the point Q on the horizontal plane.

Recall that, written in this way, (6.7) represents the component of $\boldsymbol{g}_t(P)$ in the direction $\overline{OP}$, while (6.8) gives the z-component of this vector.

A careful analysis of this approximation can be found for instance in [124], where it is proved that under our hypotheses, even in extreme conditions, the approximation of (6.7) by (6.8) implies an error of the order of 0.1 mGal.

In [124] it is also shown how to correct (6.8) by simple formulas, if needed, reducing the modelling error by at least one order of magnitude.

Just to give a general idea of why this procedure can work, we observe that it is based on the following approximations:

$$\begin{aligned}
\ell_{PQ}^2 &= \left[(R+h_P)^2 + (R+h)^2 - 2(R+h_P)(R+h)\cos\psi_{PQ} \right]^{1/2} \\
&\cong \left[(R+h_P)^2 + (R+h)^2 - 2(R+h_P)(R+h) + \left(R+h_p\right)(R+h)\,\psi_{PQ}^2 \right]^{1/2} \\
&\cong \left[\left(h_p - h\right)^2 + (1+\frac{h_P}{R})(1+\frac{h}{R})R^2\psi_{PQ}^2 \right]^{1/2} \\
&\cong \left[(h_P - h)^2 + r_{PQ}^2 \right]^{1/2},
\end{aligned} \tag{6.9}$$

which is justified because

$$\begin{aligned}
R^2\psi_{PQ}^2 &= s_{PQ}^2 = s_P^2 + s_Q^2 - 2s_{PQ}\cos\left(\alpha_p - \alpha_Q\right) \\
&= r_P^2 + r^2 - 2r_p\cos\left(\alpha_P - \alpha_Q\right) = r_{PQ}^2.
\end{aligned}$$

The other approximate relation necessary to understand (6.8) is

$$R^2 d\sigma \cong d_2 x, \tag{6.10}$$

because the projection of the spherical area on the plane is done by multiplying it by $\cos\psi$, which is practically 1, since $\frac{1}{2}\psi^2 < 10^{-6}$.

Given the above discussion and the quoted paper, we are left with the job of computing the integral (6.8). Recalling the theory if Sect. 2.2 and the notation therein, we can calculate the integral along dz to obtain

$$\begin{aligned}
g_t(P) &= g_t\left(\boldsymbol{x}, h_p\right) = -U_+(P) + U_-(P) = \\
&= -\mu \int_A \frac{d_2\xi}{\left[|\boldsymbol{x}-\boldsymbol{\xi}|^2 + \left(h_x - h_\xi\right)^2 \right]^{1/2}}. \\
&\quad + \mu \int_A \frac{d_2\xi}{\left[|\boldsymbol{x}-\boldsymbol{\xi}|^2 + h_x \right]^{1/2}}.
\end{aligned} \tag{6.11}$$

When A is just a rectangular region, the integral $U_-(P)$ is known and given by (2.23), if we put $H = h_x$. So we have to compute the integral $U_+(P)$, which is complicated because the surface of the topography $S \equiv \{H = h(\boldsymbol{x})\}$, can be very irregular.

From the point of view of existing data one has to acknowledge that after the SRTM satellite mission [69] a global Digital Terrain Model (DTM) is known for the whole Earth. This is complemented with a bathymetric data set for the oceanic part. The DTM from SRTM has a horizontal resolution of about 30 m with an accuracy of the height h, σ_h, better than 10 m; the bathymetric data set has a resolution of about 800 m, with an estimated accuracy $\sigma_d = 50$ m for the depth d. Such data can

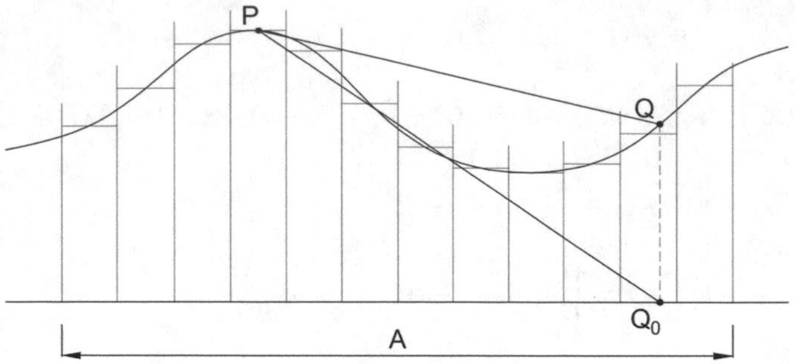

Fig. 6.3 The computation of $g_t\,(P)$ by the prism method

be found online in [37] and [72]. Better data exist locally in terms of both resolution and accuracy, although a lower resolution is often used to compute g_t.

A first obvious algorithm for the computation of $U_+\,(P)$ is just a discretization of this integral in terms of the top of a collection of prisms corresponding to the DTM available.

The situation is represented schematically in Fig. 6.3.

We note that the computation of the influence of an individual prism on $g_t\,(P)$ implies the contribution of the upper and the lower faces, see (2.23), yet the lower faces of all the prisms are put together to provide $U_-\,(P)$ by one calculation only. The upper faces, on the contrary, represent the discretized surface S on which the integral $U_+\,(P)$ is computed.

This approach is viable up to a certain number of prisms, depending on the computer power available. However this number can readily become overwhelming; for instance, an area $100 \times 100\ \text{km}^2$ with a DTM with a resolution of 20 m requires the computation of some $25 \cdot 10^6$ prism formulas, so it is desirable to try other approaches that can speed up the calculations.

The point is that if we could transform $U_+\,(P)$ into a convolution integral, then we could resort to the theory of FT studied in Chap.3, and then discretize by applying the DFT, introduced in Sect. 3.7.

This can be apparently achieved by writing

$$U_+\,(P) = \mu \int_A \frac{d_2\xi}{r_{x\xi}\left[1 + \left(\frac{h_x - h_\xi}{r_{x\xi}}\right)^2\right]^{1/2}}, \qquad (6.12)$$

$$\left(r_{x\xi} = |\boldsymbol{x} - \boldsymbol{\xi}|\right).$$

If we assume that the inclination of the terrain is always lower than 45°, the condition

$$\frac{\left|h_x - h_\xi\right|}{r_{x\xi}} \leq k < 1$$

is met and the square root in (6.12) can be expanded in a series, namely

$$\frac{1}{\left[1+\left(\frac{h_x-h_\xi}{r_{x\xi}}\right)^2\right]^{1/2}} = 1 - \frac{1}{2}\left(\frac{h_x - h_\xi}{r_{x\xi}}\right)^2 + \frac{3}{8}\left(\frac{h_x - h_\xi}{r_{x\xi}}\right)^4 + \ldots. \tag{6.13}$$

Therefore, (6.12) becomes

$$\begin{aligned} U_+ (P) = \mu \int_A \frac{d_2\xi}{r_{x\xi}} - \frac{1}{2}\mu \int_A \frac{\left(h_x - h_\xi\right)^2}{r_{x\xi}^3} d_2\varepsilon + \\ + \frac{3}{8}\mu \int_A \frac{\left(h_x - h_\xi\right)^4}{r_{x\xi}^5} d_2\xi + \ldots. \end{aligned} \tag{6.14}$$

A first comment on formula (6.14) is that each integral in it is weakly singular, because the area element $d_2\xi$, in a polar coordinate system centered at x, can be written as

$$d_2\xi = r dr d\vartheta.$$

However, none of these integrals, except for the first one, for which however we can derive an exact expression, has the form of a convolution. Yet by developing the binomials at the numerators of the integrands, convolutions appear, although together with an annoying drawback.

Let us explain the situation for the second term on the right:

$$\frac{\mu}{2}\int_A \frac{\left(h_x - h_\xi\right)^2}{r_{x\xi}} d_2\xi = \frac{\mu}{2} h_x^2 \int_A \frac{d_2\xi}{r_{x\xi}^3} - \mu h_x \int_A \frac{h_\xi}{r_{x\xi}^3} d_2\xi + \mu \int_A \frac{h_\xi^2}{r_{x\xi}^3} d_2\xi. \tag{6.15}$$

As we see, the three terms in (6.15) have the form of products of a known function by convolution integrals. In fact, after defining the characteristic function

$$\chi_A (\xi) = \begin{cases} 1, & \text{if } \xi \in A, \\ 0, & \text{if } \xi \notin A, \end{cases} \tag{6.16}$$

and extending the surface $\{Z = h_\xi\}$ with 0 outside A, namely

$$\bar{h}_\xi = \begin{cases} h_\xi, & \text{if } \xi \in A, \\ 0, & \text{if } \xi \notin A, \end{cases} \tag{6.17}$$

we can write

$$\int_A \frac{d_2\xi}{r_{x\xi}^3} = \left(\frac{1}{r^3}\right) * (\chi_A)\,, \tag{6.18}$$

$$\int_A \frac{h_\xi d_2\xi}{r_{x\xi}^3} = \left(\frac{1}{r^3}\right) * (\bar{h}_\xi)\,, \tag{6.19}$$

$$\int_A \frac{h_\xi^2 d_2\xi}{r_{x\xi}^3} = \left(\frac{1}{r^3}\right) * (\bar{h}_\xi^2), \tag{6.20}$$

The three integrals in (6.18), (6.19), (6.20) are now formally convolutions, yet none of them is particularly meaningful, because $g(\boldsymbol{\xi}) = \frac{1}{r^3}$ has a non-integrable singularity at the origin. The same is true for all the other terms in the series (6.14). To get access to the fast FT algorithm one needs to find a compromise between the slow, but exact discretization by prisms and the need to make calculations of convolutions.

This can be done as follows: let us define the window characteristic function by

$$\chi_\varepsilon(\boldsymbol{\xi}) = \chi_\varepsilon(|\xi_1|)\,\chi_\varepsilon(|\xi_2|)\,,$$

$$\chi_\varepsilon(t) = \begin{cases} 1, & \text{if } |t| < \varepsilon, \\ 0, & \text{if } |t| > \varepsilon, \end{cases} \tag{6.21}$$

which is equal to 1 in W_ε (see Fig. 6.4).

Now let us return for a moment to (6.12) and write it as

$$U_+(P) = U_p(P) + U_W(P)\,, \tag{6.22}$$

$$U_\mathrm{p}(P) = \mu \int_A \frac{\chi_\varepsilon(\boldsymbol{x} - \boldsymbol{\xi})}{[r_{x\xi}^2 + (h_x - h_\xi)^2]^{1/2}} d_2\xi, \tag{6.23}$$

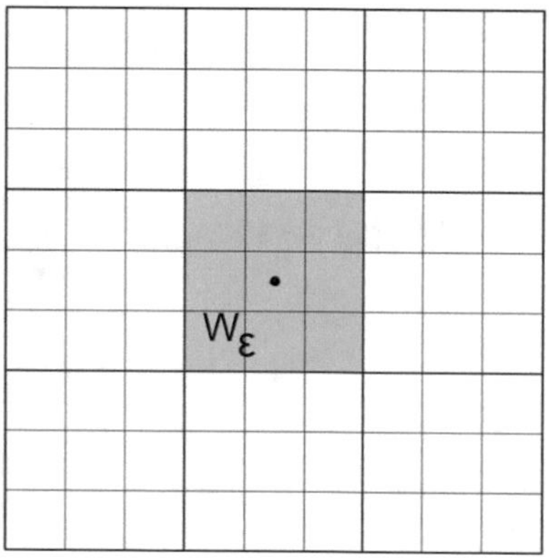

Fig. 6.4 The window W_ε and its disposition with respect to the grid of DTM

$$U_{\mathrm{W}}(P) = \mu \int_A \frac{1 - \chi_\varepsilon(\boldsymbol{x} - \boldsymbol{\xi})}{r_{x\xi}\left[1 + \frac{(h_x - h_\xi)^2}{r_{x\xi}^2}\right]^{1/2}} d_2\xi, \tag{6.24}$$

where the suffixes p an W stem from prism and window.

Let us consider first $U_{\mathrm{p}}(P)$ and imagine you want to compute it by the prism method at all centers of the cells of the DTM grid. Suppose that ε is chosen in such a way that W_ε is made only by 3×3 grid nodes. Indeed, for each prism with center at $\boldsymbol{x}$ you will need only to compute 9 values of the integral (6.23) because outside the window W_ε the integrand goes to zero. This means that, disregarding border effects, you need to compute formula (2.23) only $9N$ times, where N is the total number of prisms. This in contrast to the N^2 calculations needed if the job should be conducted over all the centers of the grid.

When N is of the order of 10^6, this becomes a quite substantial gain. Coming to U_{W}, we can apply to (6.24) the same expansion as in (6.14), through expanding the binomials; e.g., for that of the square term as in (6.15), we are left with the job of computing the integrals,

$$\begin{aligned} \int_A \frac{1 - \chi_\varepsilon(\boldsymbol{x} - \boldsymbol{\xi})}{r_{x\xi}^3} d_2\xi &= \int_{R_2} \frac{1 - \chi_\varepsilon(\boldsymbol{x} - \boldsymbol{\xi})}{r_{x\xi}^3} \chi_A(\xi)\, d_2\xi = \\ &= \left(\frac{1 - \chi_\varepsilon(\boldsymbol{\xi})}{r^3}\right) * (\chi_A), \end{aligned} \tag{6.25}$$

$$\int_{R_2} \frac{1 - \chi_\varepsilon(\boldsymbol{x} - \boldsymbol{\xi})}{r_{\boldsymbol{x}\boldsymbol{\xi}}^3} h_\xi d_2\xi = \left(\frac{1 - \chi_\varepsilon(\boldsymbol{\xi})}{r^3}\right) * (\bar{h}_\xi), \tag{6.26}$$

and

$$\int_{R_2} \frac{1 - \chi_\varepsilon(x - \xi)}{r_{x\xi}^3} h_\xi^2 d_2\xi = \left(\frac{1 - \chi_\varepsilon(\xi)}{r^3}\right) * (\bar{h}_\varepsilon^2). \tag{6.27}$$

As we see, the singularity of the integrands in (6.25), (6.26), (6.27) has been isolated and the integrals converge, so that the three convolutions can be safely computed.

The same reasoning applies to all the expansions of the binomials of the type (6.14) and we recognize that in general we have to compute convolutions of the type

$$\begin{aligned} J_{\ell k}(\boldsymbol{x}) &= \left(\frac{1 - \chi_\varepsilon(\boldsymbol{\xi})}{r^{2\ell+1}}\right) * (\bar{h}_\xi^k) \\ (k &= 0, 1, \ldots, 2\ell)\,. \end{aligned} \tag{6.28}$$

By applying the convolution theorem of Fourier Transform Theory and its converse, as in Proposition 3.2, we find

$$J_{\ell k}(\boldsymbol{x}) = \mathscr{F}^{-1}\left\{\mathscr{F}\left\{\frac{1-\chi_\varepsilon(\boldsymbol{\xi})}{r^{2\ell+1}}\right\}\cdot\mathscr{F}\left\{\bar{h}_\xi^k\right\}\right\}. \tag{6.29}$$

Since the computation of FT in the discretized form, DFT, can be very fast, specially when the dimensions of the grid are powers of 2 in both directions, as discussed in Sect. 3.7, the problem of efficiently computing U_W has been solved.

In literature [124] one can find examples with gains in computing time from a factor of 60 for a grid of 10^5 points, to a factor of 200 for a grid of $4 \cdot 10^5$ points. Clearly, the higher the number of points, the higher the efficiency of a method exploiting DFT computations.

Remark 6.1 In the above discussion we have presented the principles of the terrain effect theory, which is in reality just an application of formula (2.8), together with an algorithm to efficiently compute $g_t(P)$ on a grid of points $P_{\ell k}$ on the topographic surface S, corresponding to the grid of points where the DTM is given. In fact, several other algorithms have been presented and are implemented in software tools for the terrain correction computation, e.g., [9, 109, 123, 152, 154]. Yet we might be interested in computing a grid of g_t values at points placed at a constant altitude $\bar{h}$, above the topographic surface S, like in Fig. 6.5. This happens for instance when we have to apply $g_t(P)$ to offshore aerogravimetric measurements, which are typically taken on a flight at constant height.

Again we have to return to the formula

$$g_t(P) = -U_+(P) + U_-(P),$$

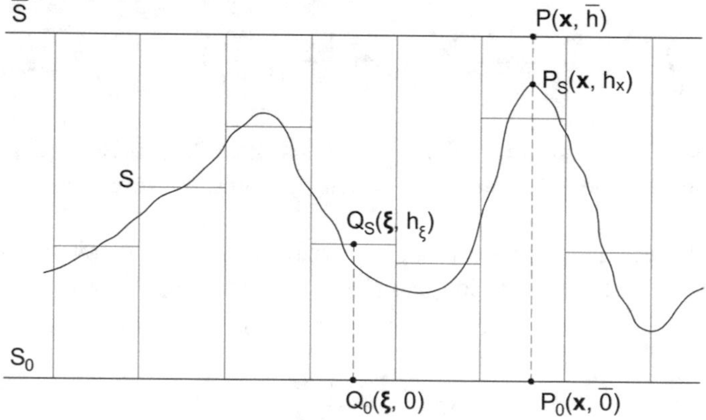

Fig. 6.5 Geometry and notation of terrain computation on a grid at height $\bar{h}$; notice that P is running on $\{z = \bar{h}\}$, while Q is running on $S \equiv \{z = h_{Q_0}\}$. P_0, Q_0 are both running on $S_0 \equiv \{z = 0\}$

where $U_-(P)$ is supplied by an exact expression. We need then to compute

$$U_+(P) = \mu \int_A \frac{d_2\xi}{[r_{x\xi}^2 + (\bar{h} - h_\xi)^2]^{1/2}}. \tag{6.30}$$

We notice that, since

$$h_\xi, \quad \forall \xi \in A,$$

the function $\ell_{PQ_s}^{-1} = [r_{x\xi}^2 + (\bar{h} - h_\varepsilon)2]^{1/2}$ is never singular, and we can therefore proceed as in (6.13), without any further concern. Namely, we write

$$\begin{aligned} \frac{1}{\ell} &= \frac{1}{[r_{x\xi}^2 + \bar{h}^2 - 2\bar{h}h_\xi + h_\xi^2]^{1/2}} = \\ &= \frac{1}{\bar{\ell}_{x\xi}} \frac{1}{\left[1 - 2\frac{\bar{h}}{\bar{\ell}_{x\xi}^2} + \frac{h_\xi^2}{\bar{\ell}_{x\xi}^2}\right]^{1/2}} \end{aligned} \tag{6.31}$$

where

$$\bar{\ell}_{x\xi} = [r_{x\xi}^2 + \bar{h}^2]^{1/2}; \tag{6.32}$$

we observe explicitly that $\bar{\ell}_{x\xi}$, which we shall denote simply by $\bar{\ell}$ in the next formulas, is in reality a function of $r_{x\xi} = |\boldsymbol{x} - \boldsymbol{\xi}|$. So, by the Taylor expansion of (6.31), we find

$$\frac{1}{\ell_{PQ}} = \frac{1}{\bar{\ell}} - \frac{1}{2}\frac{(-2\bar{h}h_\xi + h_\xi^2)^2}{\bar{\ell}^3} + \frac{3}{8}\frac{(-2\bar{h}h_\xi + h_\xi^2)^4}{\bar{\ell}^5} + \dots,. \tag{6.33}$$

Expanding the binomials and reordering in increasing powers of $\bar{\ell}^{-1}$, we see that the computation of (6.30) can be split into the computation of terms of the general form

$$I_{\ell k}(\boldsymbol{x}) = \mu \int_A \left(\frac{\bar{h}}{\bar{\ell}}\right)^{2\ell+1} \left(\frac{h_\xi}{\bar{h}}\right)^k d_2\boldsymbol{\xi}. \tag{6.34}$$

If we extend h_ξ by zero outside A as in (6.17), we see that (6.34) are all convolution integrals that can be quickly computed by FT. The only problem with (6.34) is that when $\bar{S}$ is close to the top of S, in the area around this top, the series (6.33) becomes slowly converging and the approximation by truncation can generate errors in the range of some mGal.

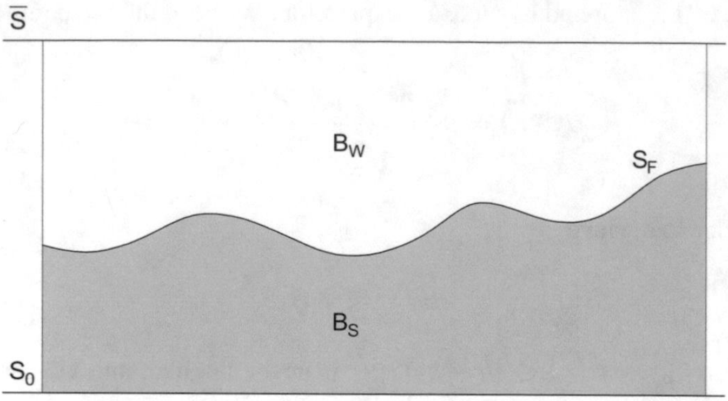

Fig. 6.6 Geometry and notation for the calculations of g_t for the sea; $\bar{S}$ sea surface, B_W water body, S_F sea floor, B_S solid matter below the sea floor (most probably sediments), S_0 reference surface

This drawback can be avoided by slicing horizontally the topographic body $B_t \equiv \{0 \le z \le h_x\}$; in this way the width of each slice can be reduced to a fraction of the distance between $\bar{S}$ and the top of the slice, so that convergence of (6.33) becomes faster. The effects of each slice can be added to find the total value of $g_t\,(P)$. More particulars can be found in [124]. Before closing this Remark, we observe that the above algorithm (and its SW implementation) is able to handle also the problem of computing $g_t\,(P)$ on the sea surface, due to density variations between the water body B_W and the body under the sea floor B_S (see Fig. 6.6).

The sea surface $\bar{S}$ is taken as (locally) flat, and the same for the reference surface S_0. If we let $g_t\,(\rho|B)$ denote the terrain correction at the level of $\bar{S}$ generated by the body B, we call $B_0 = B_W \cup B_S$ the prism encompassing the whole horizon of our calculation and finally we put

$$\rho = \rho\,(P) = \begin{cases} \rho_W, & \text{if } P \in B_W, \\ \rho_S, & \text{if } P \in B_S, \end{cases} \tag{6.35}$$

then the $g_t\,(P)$ $(P \in \bar{S})$ we need is exactly

$$g_t\,(P) = g_t\,(\rho\,|B_0)\,, \quad P \in \bar{S}. \tag{6.36}$$

On the other hand, thanks to the linearity of g_t in ρ, we can write

$$\begin{aligned} g_t\,(\rho\,|B_0) &= g_t\,(\rho_W\,|B_W) + g_t\,(\rho_S\,|B_S) = \\ &= g_t\,(\rho_W\,|B_0) + g_t\,(\rho_S - \rho_W\,|B_S)\,. \end{aligned} \tag{6.37}$$

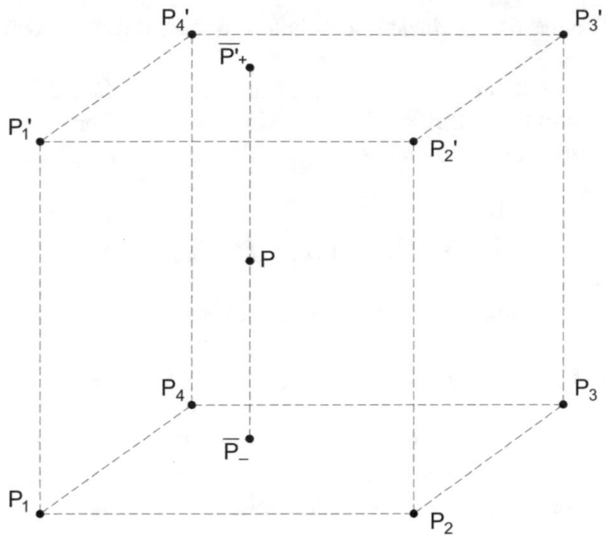

Fig. 6.7 Geometry of the interpolation between two grids in free air

Now $g_t\left(\rho_W\,|B_0\right)$ is given just by the prism formula, applied to the big prism B_0 with density ρ_W; $g_t\left(\rho_S-\rho_W\,|B_S\right)$ is exactly given by the algorithm discussed in this Remark, applied to the body B_S, only with the density $\rho_S-\rho_W$.

Up to here we have learned how we can compute a terrain correction $g_t\ (P)$, either when P belongs to a grid on the terrain surface S, or when P belongs to a grid on a flat surface $\bar{S}$ in the air, above the topographic masses. However, if such corrections have to be applied to observed values of $g(P)$, or of $\delta g\ (P) = g\ (P) - \gamma\ (P)$, we need to move the computation to a general point on S in the first case, or to a general point close to $\bar{S}$ in the second case. We treat first the latter case, noticing that the need of computing g_t at a point close to $\bar{S}$ comes, when we treat aerogravimetric observations, from the fact that the aircraft has generally a trajectory wandering about a mean height, with possible variations of a dozen of meters; moreover, the point P along the trajectory to which we attribute the observation of g is generally not a node of the grid.

The solution is to compute two grids of g_t values, at altitudes $\bar{h}$ and $\overline{h'}$, corresponding to the minimum and maximum altitude of the flight. The interpolation of g_t at a point P between the two planes $\overline{S}$ and $\overline{S'}$ (see Fig. 6.7) can be performed in two steps. First we use the values of g_t at (P_1, P_2, P_3, P_4) to produce by a bilinear function an estimate of g_t at $\overline{P}_-$; we then proceed similarly with points (P_1', P_2', P_3', P_4'), to produce an estimate at $\bar{P}_+'$. The choice of a bilinear function is because it has 4 coefficients univocally determined by the 4 corner values and because it is sufficiently well approximating the actual value of $g_t(\bar{P}_-)$ when the horizontal shift between P_i and $\bar{P}_-$, P_i' and $\bar{P}_+'$ is typically below 100 m. Finally, the value at P comes from a linear interpolation in the z-direction, between $g_t(\bar{P}i_+)$ and $g_t(\bar{P}'_-)$.

On the contrary, if the nodes of the grid are on the surface S, the easiest is to assume that the surface S itself can be approximated by a bilinear function, and that the bilinear interpolation of the values of g_t at the nodes can correctly interpolate g_t (P). Namely, if we call

$$B(\boldsymbol{x}_P;\ f(P_1), f(P_2), f(P_3), f(P_4))$$

the bilinear interpolation of the corner values of the function f at $\boldsymbol{x}_P$, we just put

$$g_t(\boldsymbol{x}_P) = B(\boldsymbol{x}_P; g_t(P_1), g_t(P_2), g_t(P_3), g_t(P_4)), \tag{6.38}$$

if P belongs to the cell with four corners $(P_1,\ P_2,\ P_3,\ P_4)$. Since the points of measurement of g usually have associated a height h_P, when this value is significantly different from the interpolated height

$$\tilde{h}_P = B(\boldsymbol{x}_P; h_{P_1}, h_{P_2}, h_{P_3}, h_{P_4}),$$

one must apply a more complicated formula that involves the vertical derivatives of g_t (P_t) at the grid points $g_t'(P_i)$. In this case one should keep in mind that such derivatives are discontinuous across the top of the prism, so external derivatives have to be used if $h_P > \tilde{H}_P$, while internal derivatives have to be used when $H_P > \tilde{h}_P$ (cf. [89]). The final expression of g_t (P) is then

$$\begin{aligned} g_t(P) = B(\boldsymbol{x}_P; g_t(P_1), g_t(P_2), g_t(P_3), g_t(P_4)) + \\ + B(\boldsymbol{x}_P; {g'}_t(P_1), {g'}_t(P_2), {g'}_t(P_3), {g'}_t(P_4))(h_P - \tilde{h}_P), \end{aligned} \tag{6.39}$$

where appropriate derivatives have to be inserted. In the following we will briefly present an example of TC computation for a real aerogravimetric survey.

Example 6.1 (Terrain Correction) The data for this example are taken from a real airborne acquisition performed in the framework of the CarbonNet project [23]. The dataset is made of 404384 real airborne observations acquired in 2011 by Sander Geophysics Ltd. to provide a better understanding of the onshore, nearshore and immediate offshore geology of the Gippsland Basin, a sedimentary basin situated in south-eastern Australia, about 200 km east of the city of Melbourne. In details, more than 10,000 km length of data were acquired along flight lines oriented northeast and southwest, with 1 km line spacing. Morevoer, a 9 km wide strip along the coast was flown at 500 m line spacing. Tie lines were flown on the northwest southeast direction at 10 km line spacing. The survey was carried out at a speed of about 50 ms^{-1} at an altitude of about 165 m above the ocean offshore and following the topography onshore with a maximum altitude of 369 m.

As for the terrain model we used a DTM with spatial resolution of 250 m, based on AusGeo model [160], and covering the region between 37.3° S and 39.4° S, and

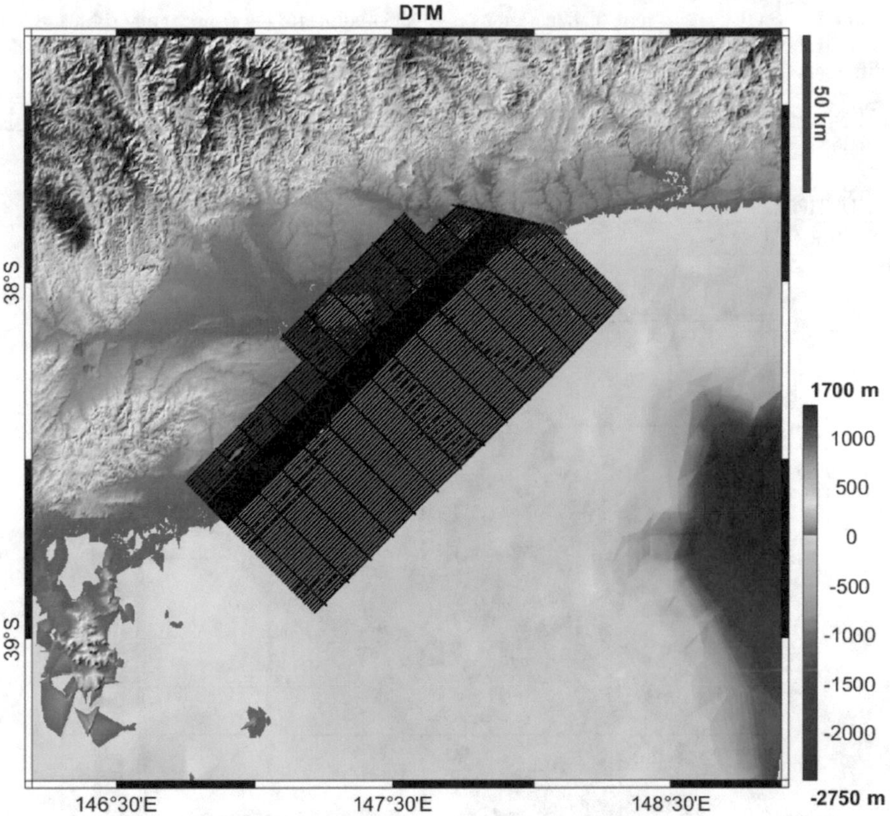

Fig. 6.8 Digital terrain model used for the TC computation. Black dots represent observation points downsampled by a factor of 50

146.2° E and 148.9° E for a total number of 819 by 1093 grid cells. The height of the DTM is ranging between 1700 m of the Mount Howitt and −2754 m in correspondence of the beginning of the Bass Canyon in the sea floor, with an average height of only 20 m and a standard deviation of 503 m. The DTM used as well as the survey tracks are shown in Fig. 6.8.

The computations were carried out made with different software packages on a single node of a supercomputer equipped with two 8-core Intel Haswell 2.40 GHz processors (for a total of 16 cores) with 128 GB RAM. In particular, a solution based on the Nagy equation for the rectangular prism (2.24), [89], is compared in terms of accuracy and computation time, with the above presented strategy based on DFT, and with the multipole approximation introduced in Exercise 3 §2. The effect of the spherical approximation, numerically evaluated, is also computed. Results are reported in Table 6.1, where it can be seen that the use of the DFT in the considered example is able to speed up the computation, compared with pure use of rectangular prisms, by a factor of 15, without loosing accuracy. A similar speed-up, but with worse results in terms of accuracy, is obtained by applying multipole approximation.

Table 6.1 Statistics and computational time on 404,384 points for the different algorithms tested. PRISM shows statistics on the computed signal. For the other rows the statistics are referred to the difference between each result and the terrain effect computed with the PRISM algorithm

Profile name	Time [s]	Mean [mGal]	Std [mGal]	Max [mGal]	Min [mGal]
PRISM	**$1.5 \cdot 10^4$**	**−0.67**	**4.44**	**20.21**	**−8.46**
DFT	1112	−0.043	0.016	0.001	−0.082
MULTIPOLE	2621	1.200	0.310	3.100	0.270
SPHERICAL	–	0.062	0.021	0.110	0.015

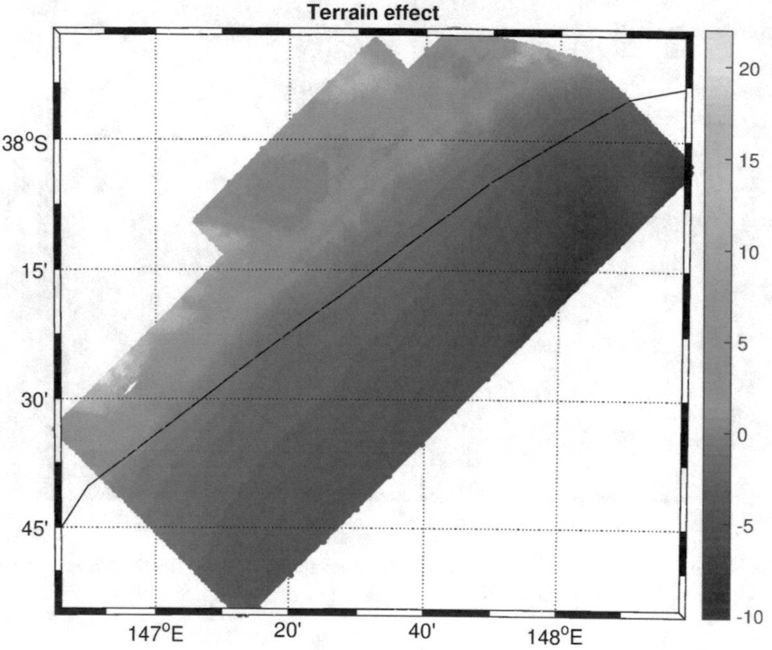

Fig. 6.9 Computed terrain effect. Unit mGal

Finally, it can also be observed that in this specific case, where areas smaller than 300 km × 300 km are considered, the planar approximation is sufficient, as the effect of the spherical correction is of the order of 0.02 mGal, in terms of standard deviation. The gravitational effect of the terrain computed at the observation points along the aircraft tracks is showed in Fig. 6.9.

6.3 Bouguer and Nettleton

The Bouguer theory culminates in an approximate formula for the fast computation of the terrain correction $g_t(P)$, in a Cartesian framework, relying on hypotheses similar to those used in Sect. 6.2; in particular, we assume that the density of topographic masses is a constant ρ. Contrary to our previous theory, however, we assume that the base of the topographic body is the whole plane $\mathbb{R}^2$; in other words, we obtain the Bouguer correction as the limit of (6.11) when $A \to \mathbb{R}^2$. At the same time, in computing $U_+(P)$ from (6.14), we limit the approximation to the second-order term.

So we start from

$$\begin{aligned} g_t &= U_-(P) - U_+(P) \cong \\ &\cong \mu \int_A \left[\frac{1}{(r_{x\xi} + h_x^2)^{1/2}} - \frac{1}{r_{x\xi}} \right] d_2\xi + \\ &+ \frac{1}{2}\mu \int_A \frac{(h_x - h_\xi)^2}{r_{x\xi}^3} d_2\xi, \end{aligned} \tag{6.40}$$

and let A go to $\mathbb{R}^2$.

In particular, in the first term to the right of (6.40), when A is for instance a disc of radius A and centered $\boldsymbol{x}$, we recognize the value of the attraction of a cylinder of height h_x at the centered of the top face.

It is clear that when $R \to \infty$ the integral tends to the attraction of a Bouguer plate of width h_x, namely

$$\begin{aligned} &\lim_{R\to\infty} \mu \int_A \left[\frac{1}{(r_{x\xi}^2 + h_x^2)^{1/2}} - \frac{1}{r_{x\xi}} \right] d_2\xi = \\ &= -2\pi\mu h_x = g_{B_0}(\boldsymbol{x}). \end{aligned} \tag{6.41}$$

The Bouguer plate correction (6.41) is sometimes called the simple Bouguer correction in literature; and indeed its computation is really very simple. So, with the above result, we obtain from (6.40), taking the limit for $R \to \infty$ as in (6.41),

$$\begin{aligned} g_B(P) &= -2\pi\mu h_x + \frac{1}{2}\mu \int_{R^2} \frac{(h_x - h_\xi)^2}{r_{x\xi}^3} d_2\xi = \\ &= g_{B_0}(\boldsymbol{x}) + \delta g_B(\boldsymbol{x}). \end{aligned} \tag{6.42}$$

It turns out that $\delta g_B(\boldsymbol{x})$ is typically one to two orders of magnitude smaller than $g_{B_0}(\boldsymbol{x})$; indeed, in a flat area $\delta g_B(\boldsymbol{x})$ becomes very small, as it depends on the square of the mean inclination of the terrain, $\frac{|h_x - h_\xi|}{r_{x\xi}}$.

A doubt that might rise in applying the Bouguer formula (6.42) is that, by extending A to the whole $\mathbb{R}^2$, we might perturb too much the attraction of topography, by introducing a lot of masses to form the Bouguer plate. However, this is not the case, as we will understand after discussing the following example.

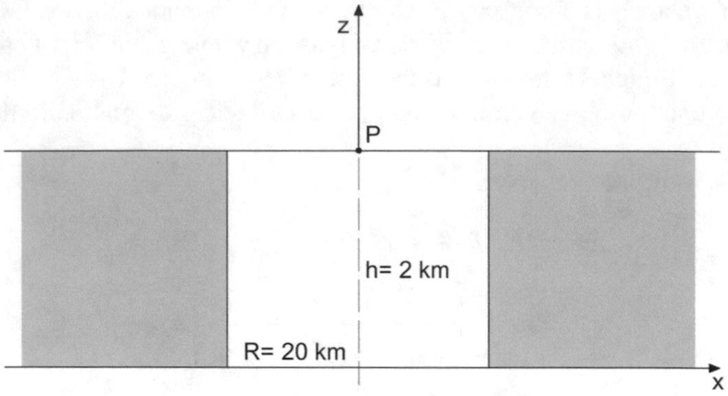

Fig. 6.10 The masses added to a cylinder to form a complete Bouguer plate

Example 6.2 Assume our terrain is a cylinder of radius $R = 20$ km and height $h = 2$ km (Fig. 6.10). We want to see in this case, by computing the attraction at the center of the topography by the exact cylinder formula or by the Bouguer correction, what is the level of the error committed.

For the cylinder one has

$$g_t(P) = 2\pi\mu(\sqrt{R^2 + h^2} - R - h),$$

while the Bouguer correction formula gives

$$\begin{aligned} g_B(P) &= g_{B_0}(P) + \delta g_B(P) = \\ &= -2\pi\mu h + \frac{1}{2}\mu \int_0^{2\pi} \int_R^{+\infty} \frac{h^2}{r^3} r\, dr d\alpha = \\ &= -2\pi\mu h + \frac{1}{2} 2\pi\mu \frac{h^2}{R}. \end{aligned}$$

Considering that, when $\rho = 2670\ \text{kgm}^{-3}$, one has $2\pi\mu = 0.1119\ \text{mGal m}^{-1}$, with the actual values one finds

$$\begin{aligned} g_t(P) - g_B(P) &= 2\pi\mu(\sqrt{R^2 + h^2} - R - \frac{1}{2}\frac{h^2}{R}) = \\ &= -0.028\ \text{mGal} \end{aligned}$$

Just to better understand the example, it is useful too to display separately the numerical value of each term, namely

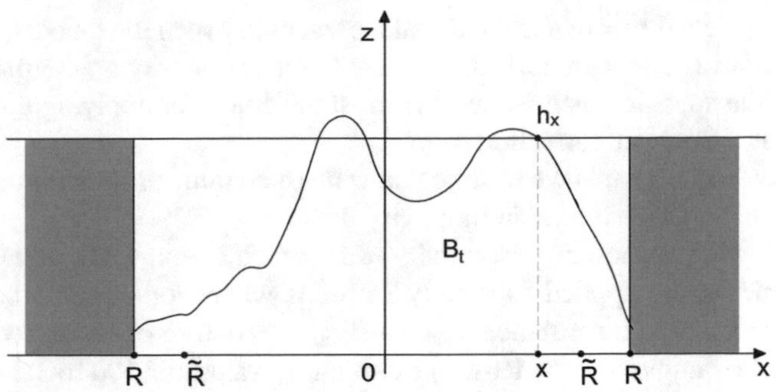

Fig. 6.11 The topographic body B_t extending up to a distance R from the origin; $\tilde{R}$ the radius inside which the Bouguer formula gives a reasonable approximation, due to the border discontinuity of h_x

$$g_t(P) = -212.638 \text{ mGal},$$
$$g_{B_0}(P) = -223, 800 \text{ mGal},$$
$$\delta g_B(P) = 11.19 \text{ mGal}.$$

Despite the elementary nature of the Example 6.2, we can draw from it some general conclusions.

First of all, we see that even when the topographic body B_t is much more complicated than a simple cylinder, like in Fig. 6.11, but the base of B_t is contained in a disc C_R of radius R, the large quantity of masses added to B_t outside C_R to arrive at a Bouguer plate configuration does not ruin the approximation of g_t, because the Bouguer correction δg_B compensates this effect, leaving a small error.

A second warning, coming from the Example 6.2, is that we have computed g_B at the center of the cylinder; if, on the contrary, we had tried to go close to the border, we would have seen that the term δg_B tends to infinity; this is because such a term combined with a Bouguer plate provides an approximation to g_t only if the expansion (6.13) holds, i.e., if the topographic surface has an inclination less than 45°. This is not the case for the cylinder, that has a vertical lateral wall.

The example suggests that, when computing a Bouguer correction, it is wise to consider it only in a central region, away from the border at a distance at least 5 $\sim$ 10 times the maximum jump of the topography on the border itself. The situation is illustrated in Fig. 6.11.

Up to here we have assumed that the density of topographic masses is constant, and known. Yet geologically, in a surface layer, ρ is known to attain possibly a large variety of values: from 2200 kgm^{-3} to 2700 kgm^{-3} for various kinds of soils, for instance loams, clays, sands, etc.; from 2500 kgm^{-3} to 3000 kgm^{-3} for sedimentary rocks like chalk, limestone, slate shales etc, or volcanic rocks like basalt, granite, quartzite, marble etc. (cfr. [68]).

It is the particular mix of such materials, present between the topographic surface and the geoid in a certain area where we want to make our gravity interpretation, that determines the specific mean value of ρ most adequate for applying a topographic correction or a Bouguer correction formula.

So in, say, a $1^\circ \times 1^\circ$ area we assume that ρ has a certain constant (in reality mean) value that we want to estimate from gravity data.

The following reasoning, essentially due to Nettleton [94], but adapted to our notation, can be applied especially in areas where topographic heights have significant variations, for instance $h_{\text{max}} - h_{\text{min}} = 1000$ m or more, so that, as we have seen in Example 6.2, the term g_{B_0} can attain values of 100 mGal or more. In such a situation one can write for a gravity disturbance (but a similar equation holds for a gravity anomaly)

$$\delta g = \delta g_{\text{M}_0} + g_{\text{t}} + \delta g_{\text{r}}; \tag{6.43}$$

in this formula δg_{M_0} is a low-degree gravity model, e.g., up to degree 180, that in a $1^\circ \times 1^\circ$ window appears almost a constant, g_{t} is the terrain correction and δg_r the residual gravity that is supposed to present an oscillating pattern in our window. The Bouguer approximation for (6.43) is

$$\delta g \cong \delta g_{\text{M}_0} + g_{\text{B}_0} + \delta g_{\text{B}} + \delta g_{\text{r}}; \tag{6.44}$$

here δg_{B}, δg_{r} can be in the range of milligals, up to a few dozens; $g_{\text{B}_0} = 2\pi\mu h$ is as irregular as h with values in the range of hundreds milligals, when h reaches 1 km, while δg_{M_0} can be as large as g_{B_0}, but is very smooth, almost constant.

It follows that, by subtracting from (6.44) its mean value in the area, denoted by an overbar, one can write

$$\delta g - \delta\bar{g} \cong 2\pi\mu(h - \bar{h}) + \text{residuals}\,; \tag{6.45}$$

in this formula the residuals are significantly smaller than $2\pi\mu(h - \bar{h})$ and are oscillating in the area.

The idea is then that an estimate of $\mu = G\rho$, can be derived from (6.45) by applying a simple least squares formula, namely

$$\widehat{\mu} = \frac{1}{2\pi}\frac{M\{(\delta g - \delta\bar{g})(h - \bar{h})\}}{\sigma^2(h)}, \tag{6.46}$$

where M is an area averaging operator, i.e.,

$$M\{f\} = \frac{1}{A}\int_A f(P)dS_P, \tag{6.47}$$

and the variance $\sigma^2(h)$ is

$$\sigma^2(h) = M\{(h - \bar{h})^2\}. \tag{6.48}$$

Actually the Nettleton estimator is written in (6.46), (6.47), (6.48) in a continuous form, but it is enough to introduce the corresponding discrete operator

$$M|f| = \frac{1}{N}\sum_{i=1}^{N} f(P_i) \tag{6.49}$$

to get the corresponding discrete formulas to be employed when only sample values $\{\delta g(P_i); 1 = 1, 2 \ldots N\}$, $\{h(P_i); i = 1, 2 \ldots N\}$ are given in the area. An approach of this kind is meaningful only if the dispersion of $\delta g - \delta\bar{g}$ around (6.45), namely

$$\sigma^2(\text{residuals}) \cong M\{[\delta g - \delta\bar{g} - 2\pi\widehat{\mu}(h - \bar{h})]^2\}, \tag{6.50}$$

is significantly smaller than the variance of δg, i.e.,

$$\sigma(\delta g) = M\{(\delta g - \delta\bar{g})^2\}. \tag{6.51}$$

One of the reasons why this might not hold can be that, in reality, the linear regression model (6.45) is too poor to represent the specific geological situation of the area. Imagine, for instance that, on the basis of prior geological information, we know that the data window A can be split into two regions A_1 and A_2, where we believe that the mean densities of the topographic masses are different (see Fig. 6.12a). In this case we can imagine that the plot of $\delta g - \delta g_{M_0}$ against h can contain simultaneously two regression lines with different slope corresponding to ρ_1 and ρ_2. The situation is represented schematically in Fig. 6.12a,b.

Sometimes the decomposition into geological provinces is not certain, maybe not even their number. In such cases, it might be useful to analyze the plot $(\delta g, h)$ trying to find clusters of points, indicating two or more regression lines, as it happens in the next Example 6.3; in such a case it is useful to apply a pull-back mapping from the plot $(\delta g, h)$ to the geographic area, to verify whether the clusters correspond to geographic areas. We observe that such a procedure is completely analogous to what is done in classifying remotely sensed data.

Example 6.3 (Gravity-Based Classification) We consider a window of about 100 km × 100 km in Italy, with a topographic surface represented in Fig. 6.13; as we see, the heights vary from 0 to about 2000 m. The corresponding map of free air anomaly is displayed in Fig. 6.14.

The range of Δg is from −27 mGal to 319 mGal. The plot of Δg versus h is presented in Fig. 6.15; a linear feature of Δg as function of h is indeed quite evident and, as a first exercise, one can compute the slope of the regression, resulting in an estimate of $\bar{\rho} \cong 2464$ kg m^{-3}. With this value a simple Bouguer correction can be computed and subtracted from the data. The resulting map of Bouguer anomalies,

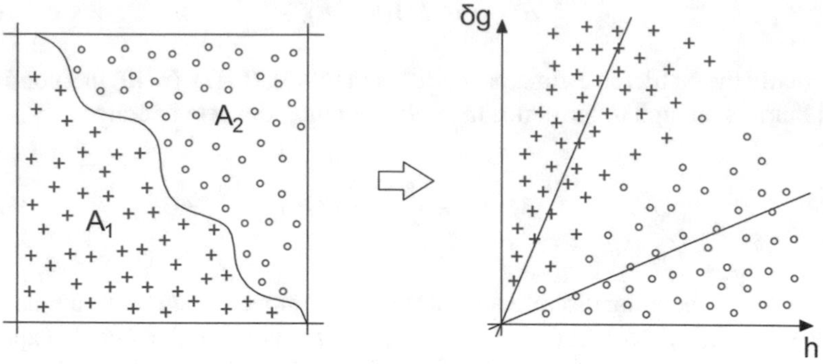

Fig. 6.12 (**a**) the region A displays two geological provinces A_1, A_2 with different densities ρ_1, ρ_2 and the relative measurement points (+ in A_1) (∘ in A_2); (**b**) the plot of δg vs. h shows a clustering of + and ∘ measurements around different regression lines, with slopes $\mu_1 = 2\pi G\rho_1$, $\mu_2 = 2\pi G\rho_2$

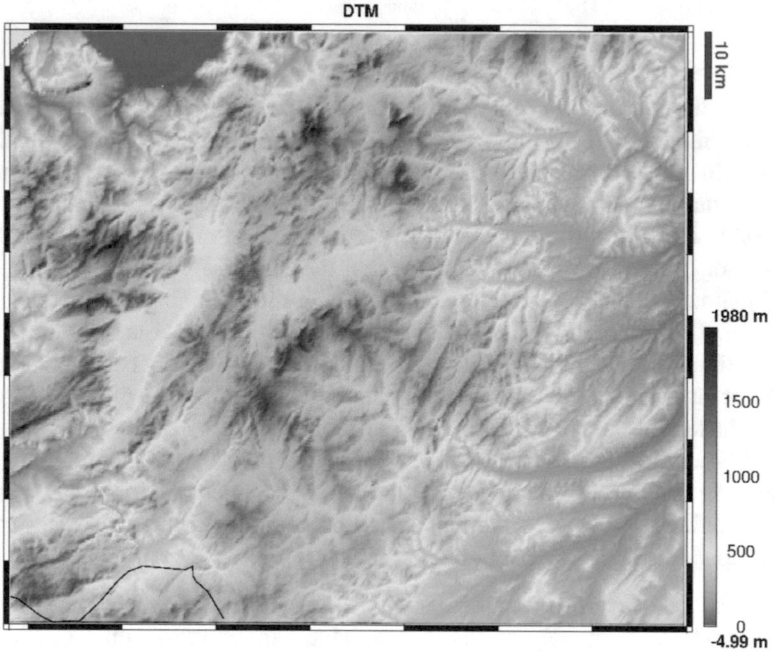

Fig. 6.13 Digital terrain model of Example 6.3

i.e., $\Delta g - g_{B_0}$, is shown in Fig. 6.16. As we see by comparing Figs. 6.16 and 6.13, there is still a correlation between the two maps.

A more careful inspection of Fig. 6.14 suggests that there might be at least two lines that correspond to two clusters associated at different densities. The corresponding estimates of ρ give

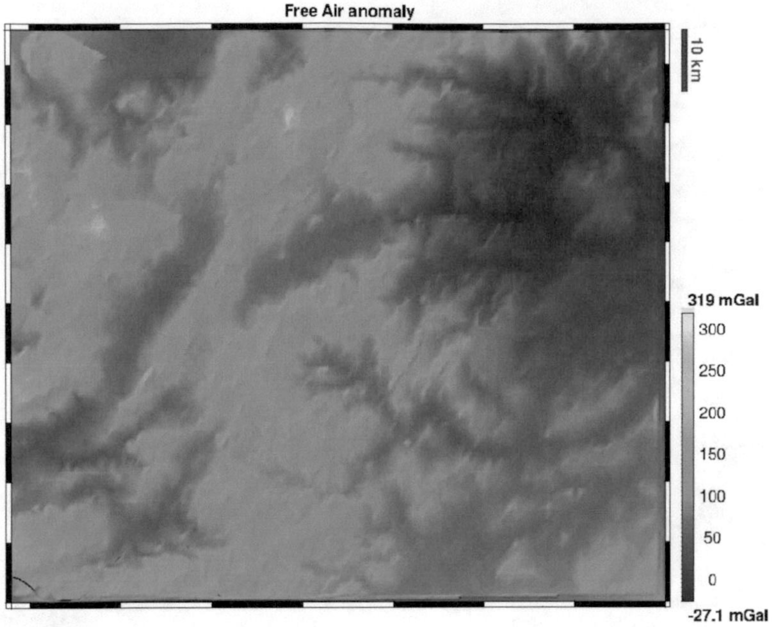

Fig. 6.14 Free air anomaly of Example 6.3

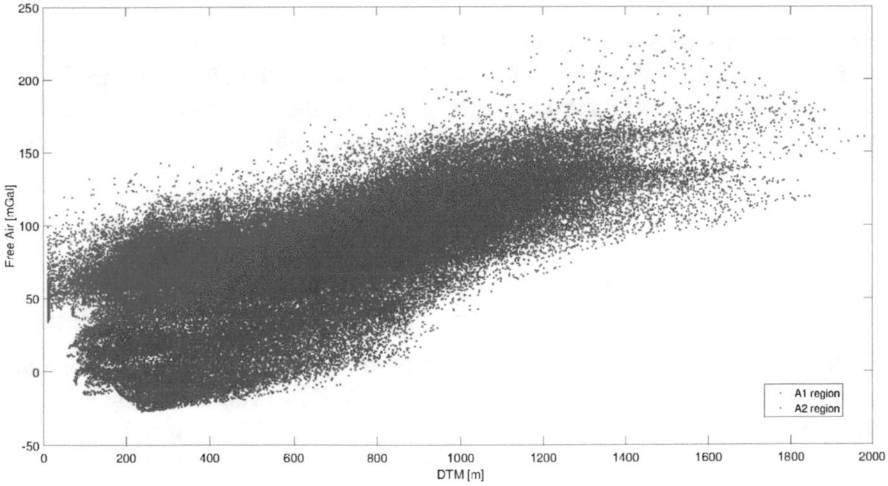

Fig. 6.15 Δg versus h for Example 6.3

$$\rho_1 = 2628 \text{ kg m}^{-3},$$

$$\rho_2 = 2181 \text{ kg m}^{-3},$$

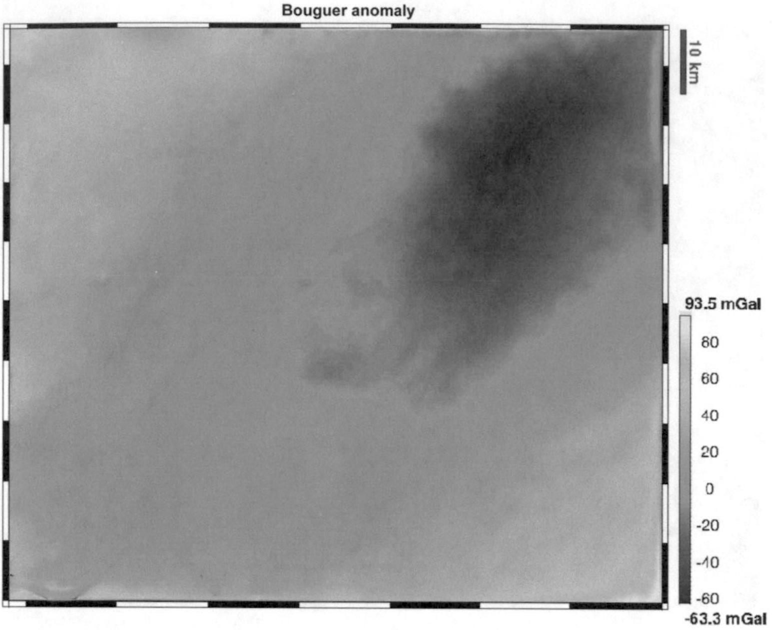

Fig. 6.16 Bouguer anomaly of Example 6.3

and the pull-back map is shown in Fig. 6.17.

The geographic coherence of this map suggests that the clustering has in fact caught a significant geological classification. The area A_1 could correspond to denser rocks of the Appenninc Platform; in the area A_2 there could be a prevailing presence of more recent sediments.

6.4 The Residual Terrain Correction (RTC)

Let us remind once more that the purpose of gravity processing is to isolate the gravimetric signal of an anomalous body B, embedded in a volume that, to fix ideas, has an horizontal extension of about 100 km × 100 km ($\sim 1° \times 1°$), and is bounded above by the topographic surface $S \equiv \{z = h(\boldsymbol{x})\}$, on which gravity values are available, and below by the Moho surface, which is "smoothly" wandering at a depth of $\sim$ 30 km (see Fig. 6.18).

At the times of Bouguer, and successively, until the age of satellite geodesy took over in the late sixties [73], very little was known on the global gravity potential, apart from the formulas that derived form the hypothesis that the Earth has ellipsoidal shape, namely those related to the normal potential. Therefore, the analysis and interpretation of the gravity field was based on the following simplifications.

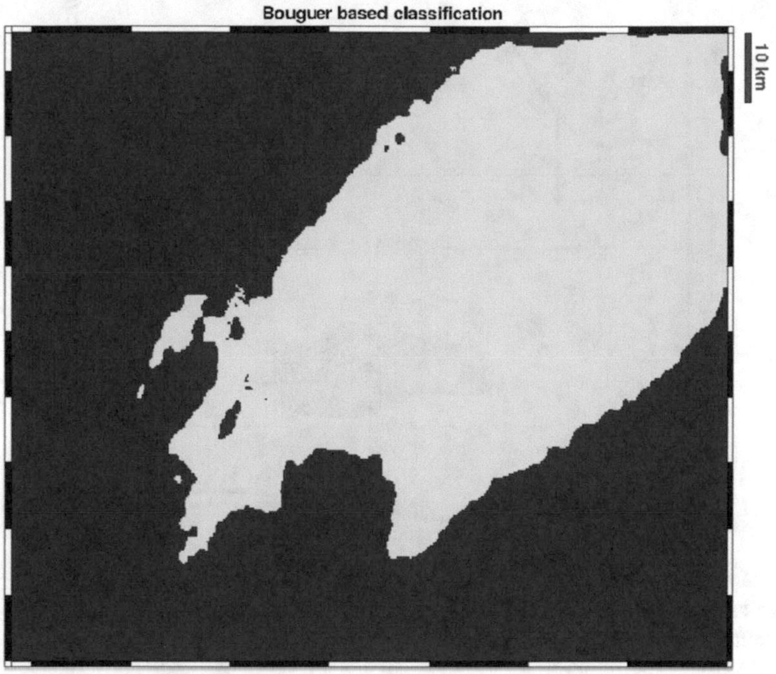

Fig. 6.17 Region classification of Example 6.3, based on the Bouguer anomaly: blue and yellow correspond to the provinces A_1 and A_2

After the computation of free air gravity anomalies, what is left in Δg is coming from three sources: mass inhomogeneities below the Moho, including the discontinuity of the Moho surface: anomalous masses in the crust, which are in fact the objects of our theory, topographic masses between the geoid and the actual topographic surface. The topographic signal was considered as computable by the algorithms presented in Sect. 6.3.

As a matter of fact, the classical Bouguer procedure tries to account also for the altitude of the observation point, by reducing at the same time $\Delta g(P)$ to $\Delta g(P_0)$, with $P_0$the projection of P on the geoid. This is not essential in the present discussion. Further on, the influence of the masses below the Moho was deemed to be of such a long wavelength (on account of the smoothing effect of propagating a gravitational signal from lower to higher altitudes), that in a window of $\sim$ 100 km $\times$ 100 km it would appear at most as a linear function (bias and tilt) in terms of geographic coordinates. The effect of Moho discontinuity was computed on the basis of some isostatic model. What was left was considered as our target residual gravity anomaly.

This old point of view has been revolutionized by the present-day capacity of estimating global models. As discussed in Sect. 4.4, such a model can arrive nowadays to a maximum degree $N = 2159$, although with the present computing

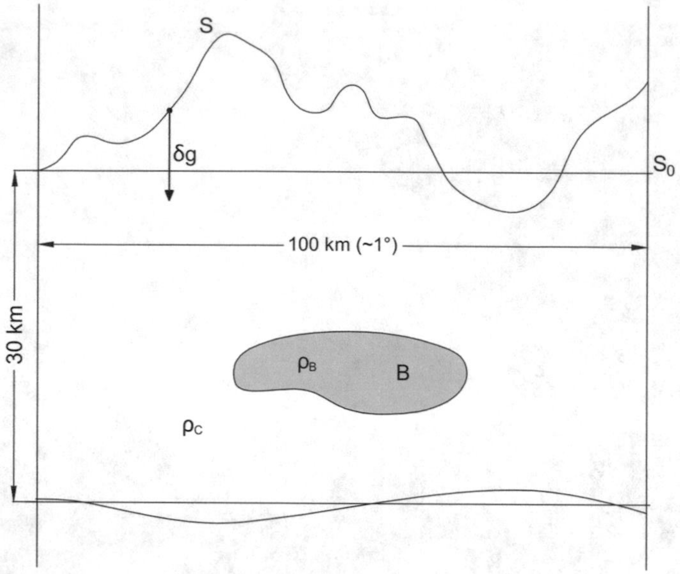

Fig. 6.18 Attraction generated by an anomalous body B with density $\rho_B > \rho_C$, and dimensions of the volume of interest

capabilities and available data, a doubling (or more) of this figure can be easily predicted in next years.

So, according to our rule of thumb (4.32), we could say that the gravimetric signal is explained down to a wavelength of $\sim$ 10 km. This is too much for our purposes and it is for this reason that we suggested to use a global model T_M up to a maximum degree of 720, which so to say accounts for a spectral part of the signal with wavelengths longer than $\sim$ 30 km. For instance, the signal of the Moho surface as perceived at the Earth surface is totally contained in the model.

The crucial point here however is to understand how much of the topographic signal is captured by T_M, when we compute a localized gravity disturbance, δg_L, from the corresponding free air

$$\delta g_L(P) = \delta g(P) - \delta g_M(P),$$

with P on the Earth surface, S. To explain this we have to return at least to a "spherical" model, to better understand how T_M is generated; it is only in the final algorithmic phase that we can come back to a Cartesian approximation and apply the formulas studied in Sect. 6.3. Here we follow basically the ideas of the nice OSU Report [39]. To make the presentation more transparent, we need a precise notation that we introduce, for the sequel of the section, with the help of Fig. 6.19.

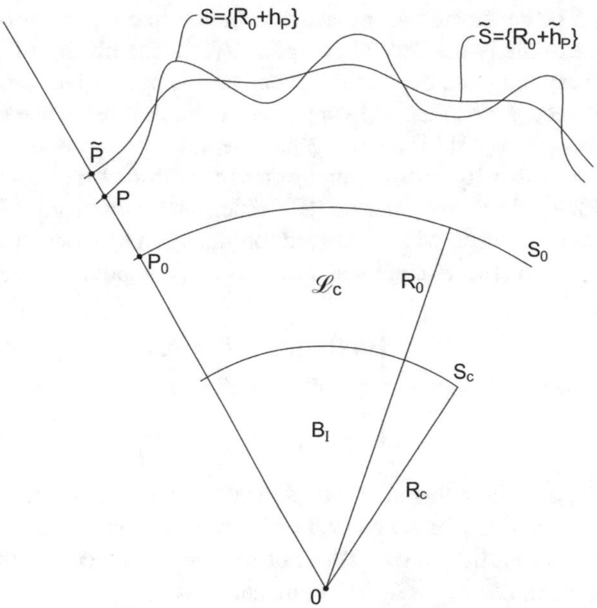

Fig. 6.19 Geometry of the different layers of density. S_c base of the crust also (improperly) identified with the Moho and modelled by a sphere B_I; S_0 ellipsoid also modelled by a spherical surface of radius R_0, S actual topographic surface, $\tilde{S}$ "modeled surface"; $\mathscr{L}_t$ and $\tilde{\mathscr{L}}_t$ are respectively the layers between S_0 and S and between S_0 and $\tilde{S}$

We define the sets

$$
\begin{aligned}
B_I &= \{0 \le r_P \le R_c\}, \\
\mathscr{L}_c &= \{R_c \le r_P \le R_0\}, \\
\mathscr{L}_t &= \{R_0 < r_P \le R_0 + h_P\}, \\
\widetilde{\mathscr{L}}_t &= \{R_0 < r_P \le R_0 + \tilde{h}_P\};
\end{aligned}
$$

B_I is the internal body, modelled as a sphere of radius R_c; $\mathscr{L}_c$ is the layer between S_c and S_0, a sphere of radius R_0 representing the ellipsoid in the actual approximation; $\mathscr{L}_t$ the topographic layer between S_0 and S_t, $\widetilde{\mathscr{L}}_t$ the "model" topographic layer between S_0 and $\widetilde{S}_t$.

What precisely $\tilde{S}_t$ is, will be explained in the sequel. For the moment it appears as a smoothed version of S_t. The use of a spherical approximation in the present reasoning is justified because we will not build on the full density ρ, which generates the whole gravity potential $W(P)$, but rather on the anomalous density $\delta\rho$ that is obtained form ρ by subtracting a normal density model ρ_M

$$\delta\rho = \rho - \rho_M. \tag{6.52}$$

This ρ_M can be considered as a mass distribution in the interior of the ellipsoid that generates externally the normal potential $U(P)$. The history of the determination of ρ_M is very long and beyond the scope of this book, since we will soon see that we do not need ρ_M explicitly; in any event a classical reference is [31] and one can consult Chap. 5 in [86] for a more general review of the problem.

The anomalous density distribution $\delta\rho$ generates in fact the anomalous gravity potential $T(P)$, a field of the order of 10^{-5} times the total potential $W(P)$, so that, as we said, we can proceed with linearized formulas in a spherical approximation.

Referring to Fig. 6.19, we can break $\delta\rho$ into three components, namely

$$\delta\rho = \begin{cases} \delta\rho_{\mathrm{I}}(P), & \text{if } P \in B_I, \\ \delta\rho_{\mathrm{c}}(P), & \text{if } P \in \mathscr{L}_{\mathrm{C}}, \\ \rho_{\mathrm{t}}, & \text{if } P \in \mathscr{L}_{\mathrm{t}}; \end{cases} \tag{6.53}$$

in particular, in $\mathscr{L}_{\mathrm{t}}$, where there the model density vanishes, ρ_{t} will be assumed for simplicity to be a constant, as we have done in previous sections. To the partitioning of $\delta\rho$ corresponds a partitioning of T and of $\delta g = -\frac{\partial T}{\partial r}$, as seen on the topographic surface and upward, i.e., $r_P > R_0 + h_P$, namely

$$\begin{aligned} T(P) &= G\int_{B_{\mathrm{I}}} \frac{\delta\rho_I(Q)}{\ell_{PQ}} dB_Q + G\int_{\mathscr{L}_{\mathrm{c}}} \frac{\delta\rho c(Q)}{\ell_{PQ}} dB_Q + \mu\int_{\mathscr{L}_{\mathrm{t}}} \frac{1}{\ell_{PQ}} dB_Q = \\ &= T_{\mathrm{I}}(P) + T_{\mathrm{c}}(P) + T_{\mathrm{t}}(P). \end{aligned} \tag{6.54}$$

with $P \in S$, or $r_P > R_P + h_P$, and $\mu = G\rho_{\mathrm{t}}$.

Furthermore,

$$\begin{aligned} \delta g(P) &= -\frac{\partial T_{\mathrm{I}}}{\partial r} - \frac{\partial T_{\mathrm{c}}}{\partial r} - \frac{\partial T_{\mathrm{t}}}{\partial r} = \\ &= \delta g_{\mathrm{I}}(P) + \delta g_{\mathrm{c}}(P) + \delta g_{\mathrm{t}}(P). \end{aligned} \tag{6.55}$$

Now we have to understand how the model T_{M} is constructed from $\delta g(P)$ data, given on S. We go along with N. K. Pavlis (see Chap. 6 in [133]) but try to follow a simplified (by spherical approximation) reasoning and use only the principal terms, when we come to the issue of the so-called downward continuation.

We claim that T_{M}, similarly to T, can be split into three parts

$$T_{\mathrm{M}} = T_{\mathrm{M_I}} + T_{\mathrm{M_c}} + T_{\mathrm{M_t}}, \tag{6.56}$$

to which correspond, too, three model gravity disturbances.

To explain each of the three terms, let us introduce the truncation operator, $\mathscr{T}$, acting on function of σ according to the following rule

$$\begin{cases} g(\sigma) = \sum_{n=m}^{+\infty} \sum_{m=-n}^{n} F_{nm} Y_{nm}(\sigma), \\ \mathscr{T} f(\sigma) = \sum_{n=0}^{N} \sum_{m=-n}^{n} f_{nm} Y_{nm}(\sigma); \end{cases} \tag{6.57}$$

so $\mathscr{T} f$ is a smoothed version of f with a maximum degree N, which for the sake of definiteness we have fixed to $N = 720$. Now, since T_{I} is generated by the (bounded) density $\delta\rho_{\mathrm{I}}$, it can be expressed for $r_P \geq R_{\mathrm{c}}$ (see Fig. 6.19) as

$$T_{\mathrm{I}}(P) = \frac{GM_{\mathrm{I}}}{R_{\mathrm{c}}} \sum_{n=0}^{+\infty} \sum_{m=-n}^{n} T_{nm}^{\mathrm{I}} \left(\frac{R_{\mathrm{c}}}{r}\right)^{n+1} Y_{nm}(\sigma) \tag{6.58}$$

with M_{I} the total mass contained in B_{I}, and we define

$$T_{M_{\mathrm{I}}}(P) = \mathscr{T}\, T_{\mathrm{I}}(P) = \frac{GM_{\mathrm{I}}}{R_{\mathrm{c}}} \sum_{n=0}^{\mathrm{N}} \sum_{m=-n}^{n} T_{nm}^{\mathrm{I}} \left(\frac{R_{\mathrm{c}}}{r}\right)^{n+1} Y_{nm}(\sigma). \tag{6.59}$$

Accordingly, we have

$$\delta g_{M_{\mathrm{I}}}(P) = -\frac{\partial T_{M_{\mathrm{I}}}}{\partial r} = \frac{GM_{\mathrm{I}}}{R_{\mathrm{c}}^2} \sum_{n=0}^{\mathrm{N}} \sum_{m=-n}^{\mathrm{n}} T_{nm}^{\mathrm{I}} (n+1) \left(\frac{R_{\mathrm{c}}}{r}\right)^{n+1} Y_{nm}(\sigma). \tag{6.60}$$

Note that the factor $\frac{GM_{\mathrm{I}}}{R_{\mathrm{c}}^2}$ in front of the sum in (6.60) is used to make the coefficients T_{nm}^{I} non-dimensional, exactly as we did in the definition of the total model δg_{M} in (4.32).

Now one can also take into account that the ratio between the total mass of the Earth ($M \sim 6 \cdot 10^{24}$kg) and that of the internal body B_{I} is roughly $7 \cdot 10^{-6}$ in relative terms.

Moreover, the relative difference between $R_0$and R_{c} is about $4.7 \cdot 10^{-3}$, so that for the rough computation we are making here the two factors $\frac{GM_{\mathrm{I}}}{R_{\mathrm{c}}^2}$ and $\frac{GM}{R_0^2}$ could be considered as equal.

Accordingly, we can compute, in $\{r \geq R_{\mathrm{c}}\}$, the difference

$$\delta g_{\mathrm{I}}(P) - \delta g_{M_{\mathrm{I}}}(P) = \frac{GM}{R_0^2} \sum_{n=N+1}^{+\infty} \sum_{m=-n}^{n} T_{nm}^{\mathrm{I}} \left(\frac{R_{\mathrm{c}}}{r}\right)^{n+1} (n+1) Y_{nm}(\sigma). \tag{6.61}$$

If we specialize this relation to $P \in S_0$ ($r = R_0$) and compute the $L^2(\sigma)$ norm, or, what amounts to the same (cf. (4.37)), the sum of the full degree variances of $\delta g_{\mathrm{I}} - \delta g_{M_{\mathrm{I}}}$, we get

$$||\delta g_{\rm I} - \delta g_{\rm M_I}||^2_{L^2_\delta} = \sum_{n=N+1}^{+\infty} c_n(\delta g_{\rm I} - \delta g_{\rm M_I}) =$$
$$= \left(\frac{GM}{R_0^2}\right) \sum_{n=N+1}^{+\infty} \left[\sum_{m=-n}^{n} (T^{\rm I}_{nm})^2\right] \left(\frac{R_{\rm c}}{R_{\rm O}}\right)^{2n+1} (n+1)^2. \tag{6.62}$$

Of course we do not know exactly the values of

$$c_n(T_{\rm I}) = \sum_{m=-n}^{n} (T^{\rm I}_{nm})^2;$$

however we can make a reasonable, conservative hypothesis. By analyzing the full range of the T_{nm} available from EGM08, referred to the sphere $S_{\rm O}$, we can formulate a law of decay of $c_n(T)$ of the form (see [133], §3.8)

$$c_n(T) = \frac{3.9 \cdot 10^{-8} (0.999443)^n}{(n-1)(n-2)(n+4)(n+17)}. \tag{6.63}$$

Other authors find slightly different expressions (see [64]), depending on the weight one puts on the very high degrees ($N > 1800$), but that would not change our conclusions. Now in such T_{nm} one can see the high variability of ρ close to $S_{\rm O}$, including the topography causing a jump of ~ 2670 kgm^{-3} through S, as we shall see later on. The conclusion is that, if we choose

$$c_n(T_{\rm I})|_{S_{\rm c}} \cong c_n(T)|_{S_{\rm O}}, \quad n > 720,$$

we are likely to be quite pessimistic. Therefore, substituting the expression (6.63) into (6.62) and simplifying, we find

$$||\delta g_{\rm I} - \delta g_{\rm M_I}||^2_{L^2_\sigma} \le \left(\frac{GM}{R_0^2}\right)^2 3.9 \cdot 10^{-8} \sum_{n=N+1}^{+\infty} \frac{p^n}{n^2} \tag{6.64}$$

with

$$p = 0.999443 \left(\frac{R_{\rm c}}{R_{\rm O}}\right)^2 = 0.990053.$$

An approximate calculation of the series (6.64) gives

$$||\delta g_{\rm I} - \delta g_{\rm M_I}||_{L^2_\sigma} \le 0.067 \text{ mGal}. \tag{6.65}$$

On account of our hypotheses, we can consider such a figure negligible. In other words, we are claiming that since from $S_{\rm c}$ to $S_{\rm O}$ we have to raise the harmonic function T_{I} by 30 km, this upward continuation operation smoothes the high frequency of T_{I} to the extent that T_{I} can be completely represented by $T_{M_{\mathrm{I}}}$ on S_0.

Continuing our reasoning we could claim that the part of the anomalous potential generated in $\mathscr{L}_{\mathrm{c}}$ is, for $r_P \geq R_0$,

$$T_{\mathrm{c}}(P) = \frac{GM_{\mathrm{c}}}{R_0} \sum_{n=0}^{+\infty} \sum_{m=-n}^{n} T_{nm} \left(\frac{R_0}{r}\right)^{n+1} Y_{nm}(\sigma) \tag{6.66}$$

and the corresponding model for the gravity anomaly δg_{M_c} becomes

$$\delta g_{\mathrm{M_c}}(P) = \mathscr{T}\,\delta g_{\mathrm{M}}(P) = \frac{\mathrm{GM_c}}{R_0^2} \sum_{n=0}^{N} \sum_{m=-n}^{n} T_{nm}^{\mathrm{c}}(n+1)\left(\frac{R_0}{r}\right)^{n+1} Y_{nm}(\sigma). \tag{6.67}$$

Using the expansion (4.41) of Newton's kernel in the definition of $T_{\mathrm{c}}(P)$, from (6.54), we see that

$$T_{nm}^{\mathrm{c}} = \frac{1}{2n+1}\left[\frac{1}{M_c}\int_{\mathscr{L}_{\mathrm{c}}} \delta\rho_{\mathrm{c}}(Q)\left(\frac{r_Q}{R_0}\right)^n Y_{nm}(\sigma_Q)dB_0\right]. \tag{6.68}$$

If we further use the representation of $\delta\rho_{\mathrm{c}}(Q)$ in terms of spherical harmonics,

$$\delta\rho_{\mathrm{c}}(Q) = \sum_{n=0}^{+\infty} \sum_{m=-n}^{\mathrm{n}} \delta\rho_{nm}^{c}(r_Q)Y_{nm}(\sigma_Q), \tag{6.69}$$

certainly valid on any sphere in $\mathscr{L}_{\mathrm{c}}$ because $\delta\rho_{\mathrm{c}}$ is a measurable bounded function in such a set, and exploite the orthogonality relation (4.29), we obtain

$$T_{nm}^{\mathrm{c}} = \frac{4\pi R_0^3}{2n+1}\left[\frac{1}{M_c}\int_{R_c}^{R_0} \delta\rho_{nm}^{\mathrm{c}}(r)\left(\frac{r}{R_0}\right)^{n+2} d\left(\frac{r}{R_0}\right)\right]. \tag{6.70}$$

Note that dimensionally the term in square parenthesis in (6.70) is the inverse of a volume, so that $\{T_{nm}^{\mathrm{c}}\}$ turn out to be non-dimensional numbers, as foreseen.

According to our discussion in Sect. 4.4 we know that $T_{\mathrm{M_c}}$ is basically generated by the components of the anomalous density in $\mathscr{L}_{\mathrm{c}}$ with a "wavelength" down to $\sim$ 30 km, while shorter-wavelength components do not enter into our model. Besides, we observe that the mass of the crust M_{c} is about $7 \cdot 10^{-6}$ times the total mass of the Earth.

Summarizing this discussion, we could say that the difference $\delta g_{\mathrm{c}} - \delta g_{\mathrm{M_c}}$ is exactly the sought-for residual gravity disturbance δg_{r},

$$\delta g_{\mathrm{r}}(P) = \delta g_{\mathrm{c}}(P) - \delta g_{\mathrm{M_c}}(P), \tag{6.71}$$

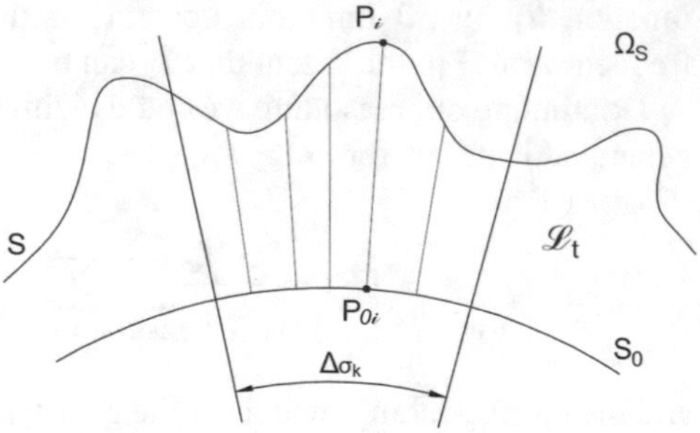

Fig. 6.20 Downward continuation of a discrete sample from S to S_0 and subsequent averaging on geophysically regular squares $\Delta\sigma_k$. $\mathscr{L}_\mathrm{t}$ is the actual topographic layer and Ω_S is the space outside S

namely, the gravitation signal generated in the layer $\mathscr{L}_\mathrm{c}$ by mass anomalies with dimensions (wavelengths in terms of harmonic representation) up to $\sim$ 30 km.

We come now to the third term, T_t, in (6.54). In Sect. 6.3 we have studied how to compute T_t, or better δg_t, on a local basis, in Euclidean approximation. However, such an approach cannot be extended to the whole Earth, primarily because of the dimension of the problem. If δg_t has to be computed from a global digital terrain model given with a resolution of 100 m, one has to pave the Earth surface with some $5 \cdot 10^{10}$ squares, each with an area 0.01 km^2. When the effect has to be computed, for instance, on a geographic grid of 1,296,000 nodes (corresponding to $3' \times 3'$ squares), each at a different altitude, one is facing a very hard numerical problem. What is done in practice is a so-called downward continuation of the effect from S to S_0.

Strictly speaking, this is not a sound mathematical operation because continuing a harmonic function from a higher S to a lower S_0 is generally impossible and, when possible, it is an exponentially blowing up, unstable operation. However, there are ways to control the instability, by using some prior information, so we shall agree that, when speaking of downward continuation (DC), we mean a regularized DC.

A thorough discussion on this problem can be found in [133, 134].

So $\delta g_\mathrm{t}(P_i)$ values are downward continued to $\delta g_\mathrm{t}^\mathrm{d}(P_{oi})$ and then area-averaged on blocks $\Delta\sigma_n$ (see Fig. 6.20) to generate a regular distribution of values on S_0.

The smoothing effect of block averaging is then counteracted at the level of spherical harmonic analysis by so-called de-smoothing factors (cfr [133], Chap. 6). We will ignore this particular item and we proceed with the continuous picture of the problem. We first try to understand what is $\delta g_\mathrm{t}^\mathrm{d}(P)$ on S_0. We do that by the mass coating method, which is basically equivalent to a first-order Taylor formula for δg_t as function of h. Although higher-order Taylor expansions are used especially in mountainous areas, the first-order term gives the largest contribution, and we shall content ourselves with analysing this. Let us observe that one can write

$$\delta T_\mathrm{t}(P) = \mu \int_{\mathscr{L}_\mathrm{t}} \frac{1}{\ell_{PQ}} dB_Q = \mu \int\int_{R_0}^{R_0+h_Q} \frac{1}{\ell_{PQ}} r_Q^2 \, dr_Q d\sigma_Q. \tag{6.72}$$

The following approximation is certainly valid to the first order in h_Q (with respect to R_0)

$$\int_{R_0}^{R_0+h_Q} \frac{1}{\ell_{PQ}} r_Q^2 \, dr_Q \cong \frac{1}{\ell_{PQ_0}} R_Q^2 h_{Q_0}, \tag{6.73}$$

with Q_0 the projection of Q on S_0. By using such an approximation we can write

$$\delta T_t(P) \cong \delta T_t^d(P) = \mu \int \frac{R_0^2 h_{Q_0}}{\ell_{PQ_0}} d\sigma_{Q_0} \quad (P \in S). \tag{6.74}$$

We observe that (6.74) amounts to substituting the Newton integral through the topographic masses with a single layer integral on S_0; the elementary mass of the single layer, i.e., $\mu \left(R_0^2 d\sigma\right) h$, corresponds to squeezing the mass column above Q_0 onto S_0 generating a surface density μh_{Q_0}. This is why the method is also known as *mass coating*.

The important point is that, if we accept the approximation (6.74), we have achieved the downward continuation of T_t down to S_0, because δT_t^d is clearly harmonic outside this surface. This is even better perceived if we use the expansion (4.41) for $\frac{1}{\ell_{PQ_0}}$ and we represent h by its series in spherical harmonics

$$h_P \equiv h(P) = \sum_{n=0}^{+\infty} \sum_{m=-n}^{n} h_{nm} Y_{nm}(\sigma_P); \tag{6.75}$$

the result is

$$T_t^d(P) = 4\pi \mu R_0 \sum_{n=0}^{+\infty} \sum_{m=-n}^{n} h_{nm} \left(\frac{R_0}{r_P}\right)^{n+1} \frac{Y_{nm}(\sigma_P)}{2n+1}, \tag{6.76}$$

where the harmonicity of T_t^d for $r_P \geq R_0$ is quite clear. If we further compute $-\frac{\partial}{\partial r} T_t^d(P)$, we get

$$\delta g_t^d(P) = 4\pi \mu \sum_{n=0}^{+\infty} \sum_{m=-n}^{n} h_{nm} \frac{n+1}{2n+1} \left(\frac{R_0}{r_P}\right)^{n+2} Y_{nm}(\sigma_P). \tag{6.77}$$

Let us note explicitly that we expect $\delta g_t^d(P)$ to be a good approximation of $\delta g_t(P)$ only when P is on S or outside it, but the approximation breaks down whenever P goes inside the masses, as it is well illustrated by Example 1.11.

What enters into the model δg_M is just $\delta g_t^d(P)$, given by (6.77), truncated at the maximum degree N:

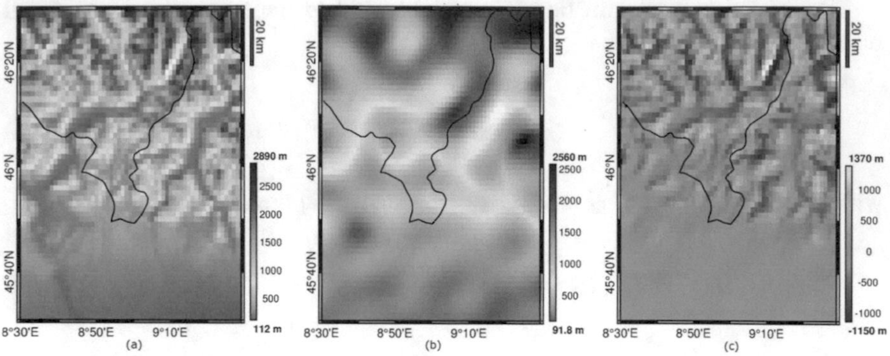

Fig. 6.21 (**a**) the surface S; (**b**) the surface $\tilde{S}$, with a truncation at degree $N = 720$; (**c**) the height differences, in a $1^\circ \times 1^\circ$ area in the alpine region

$$\delta g_{\mathrm{Mt}}(P) = \mathscr{I}\delta g_{\mathrm{t}}^{\mathrm{d}} = 4\pi\mu \sum_{n=0}^{N}\sum_{m=-n}^{n} h_{nm}\frac{n+1}{2n+1}\left(\frac{R_0}{r_P}\right)^{n+2} Y_{nm}(\sigma_P). \tag{6.78}$$

We shall use the following remark; let us introduce a surface $\tilde{S}$ defined by

$$\tilde{h}_P = \tilde{h}(P) = \sum_{n=0}^{N}\sum_{m=-n}^{n} h_{nm} Y_{nm}(\sigma_P). \tag{6.79}$$

This is exactly the surface $\tilde{S}$ that we have represented in Fig. 6.19 and it is indeed a “smoothed version” of S, as Fig. 6.21 shows for a mountainous region of the Alps.

Now looking at the form of $\delta g_{\mathrm{t}}^{\mathrm{d}}$ in (6.77) we realise that if instead of h_P we used $\tilde{h}_p$ given by (6.79) we would find exactly (6.78), namely we could say that $\delta g_{\mathrm{Mt}}(P)$ is just the downward continued version of the terrain effect implied by the approximate surface $\tilde{S}$.

At this point it is useful to introduce a new symbol: we call $\delta T_{\mathrm{t}}(P, \mathscr{L})$ the Newton integral (6.72) and $\delta g_{\mathrm{t}} = -\partial_r \delta T_{\mathrm{t}}(P, \mathscr{L}) = \delta g_{\mathrm{t}}(P, \mathscr{L})$ the corresponding gravity anomaly, which are clearly functionals of the layer $\mathscr{L}$ between S_0 and some surface $S \equiv \{r = R_0 + h_P\}$, the upper boundary of $\mathscr{L}$. As such δg_{t} can be considered, for each fixed P, a functional of $\mathscr{L}$ or of $\{h_P\}$, as preferred. Let us stress that when we use $\delta g_{\mathrm{t}}(P, \mathscr{L})$ we mean precisely a numerical algorithm that is capable of computing the value of δg_{t} at P. As we have seen, this is more easily done when P is outside the masses of $\mathscr{L}$ or at most on S. With the above specification when we take $\mathscr{L} = \mathscr{L}_{\mathrm{t}}$ and S is actually the topographic surface of the Earth, we have

$$\delta g_{\mathrm{t}}(P) = \delta g_{\mathrm{t}}(P, \mathscr{L}_{\mathrm{t}}) \tag{6.80}$$

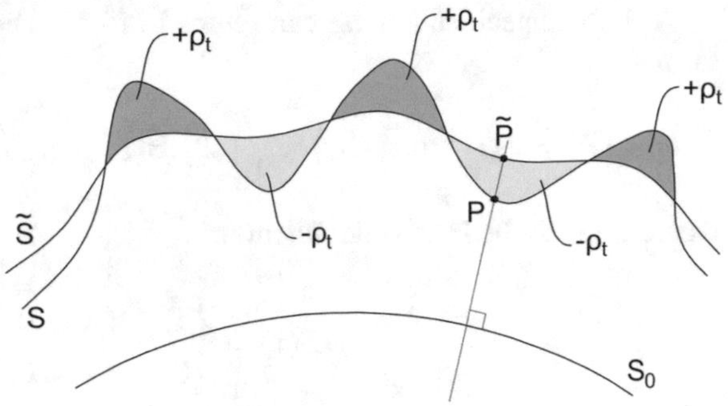

Fig. 6.22 The body $\mathscr{L}_S \div \mathscr{L}_{\tilde{S}}$, and its density with alternating signs. Note that there are points P such that $h_P > h_{\tilde{P}}$

with P on S or outside it, namely in Ω_S (see Fig. 6.20). When we take $\mathscr{L} = \widetilde{\mathscr{L}}_t$ and $\tilde{S}$ is just the approximate surface, we have

$$\delta \tilde{g}_t(P) = \delta g_t\left(P, \widetilde{\mathscr{L}}_t\right). \tag{6.81}$$

Then, if we squeeze the masses of $\widetilde{\mathscr{L}}_t$ on S_0 we get exactly

$$\delta \tilde{g}_{Mt}(P) = \delta g_t^d\left(P, \widetilde{\mathscr{L}}_t\right). \tag{6.82}$$

We stress once more that, at least approximately,

$$\delta \tilde{g}_{Mt}(P) = \delta \tilde{g}_t(P) \tag{6.83}$$

when $P \in \Omega_{\tilde{S}}$, i.e., it lies outside $\tilde{S}$, while the approximation does not hold when P is inside $\widetilde{\mathscr{L}}_t$, because δg_{Mt} is harmonic there, while $\delta \tilde{g}_t$ is not. Nevertheless, when $h_P < h_{\tilde{P}}$, as it can happen (see Fig. 6.22), but $\Delta h = h_P - h_{\tilde{P}}$ is small, the downward continuation $\delta g_{Mt}(P)$ can be expressed, in a linear approximation, as

$$\delta g_{Mt}(P) \cong \delta g_t\left(\tilde{P},\ \widetilde{\mathscr{L}}_t\right) + \delta g'_{t+}\left(\tilde{P},\ \widetilde{\mathscr{L}}_t\right)\Delta h, \quad P \in \widetilde{\mathscr{L}}_t; \tag{6.84}$$

in the RHS of (6.84) the prime stands for the vertical derivative and the index + means that such a derivative is computed immediately above $\tilde{S}$, i.e., in free air.

On the other hand, when $h_P < h_{\tilde{P}}$, we have, at the first order in $\tilde{h}$,

$$\delta g_t\left(P | \widetilde{\mathscr{L}}_t\right) = \delta g_t\left(\tilde{P} \middle| \widetilde{\mathscr{L}}_t\right) + \delta g'_{t-}\left(\tilde{P} \middle| \widetilde{\mathscr{L}}_t\right)\Delta h, \tag{6.85}$$

where now one appends to the vertical derivative δg_t the index – because it has to be computed on $\tilde{S}$, but inside the masses. The relation (6.85) can be written also as

$$\delta g_t\left(\tilde{P} \middle| \widetilde{\mathscr{L}}_t\right) = \delta g_t\left(P | \widetilde{\mathscr{L}}_t\right) - \delta g'_{t-}\left(\tilde{P} \middle| \widetilde{\mathscr{L}}_t\right)\Delta h. \tag{6.86}$$

With this specification we can return to (6.83) and write $\delta g_{\mathrm{Mt}}(P)$ in a unified formula, as

$$\delta g_{\mathrm{Mt}}(P) = \delta g_{\mathrm{t}}\left(P \mid \widetilde{\mathscr{L}}_{\mathrm{t}}\right) + \left[\delta g'_{\mathrm{t}+}\left(\tilde{P} \middle| \widetilde{\mathscr{L}}_{\mathrm{t}}\right) - \delta g'_{\mathrm{t}-}\left(\tilde{P} \middle| \widetilde{\mathscr{L}}_{\mathrm{t}}\right)\right] \Delta h \cdot H(-\Delta h), \tag{6.87}$$

where $H(x)$ is the Heaviside function:

$$H(x) = \begin{cases} 1, & \text{if } x > 0, \\ 0, & \text{if } x < 0. \end{cases} \tag{6.88}$$

By using (6.87) we can finally write

$$\begin{aligned} \delta g_{\mathrm{t}}(P) - \delta g_{\mathrm{Mt}}(P) &= \delta g_{\mathrm{t}}(P \mid \mathscr{L}_{\mathrm{t}}) - \delta g_{\mathrm{t}}\left(P \mid \widetilde{\mathscr{L}}_{\mathrm{t}}\right) + \\ &- \left[\delta g'_{\mathrm{t}+}\left(\tilde{P} \middle| \widetilde{\mathscr{L}}_{\mathrm{t}}\right) - \delta g'_{\mathrm{t}-}\left(\tilde{P} \middle| \widetilde{\mathscr{L}}_{\mathrm{t}}\right)\right] \Delta h \cdot H(-\Delta h). \end{aligned} \tag{6.89}$$

The first two terms in (6.89) representing the attraction of the topographic masses at the same point P, can be put together as the anomalous gravity generated by he body $\mathscr{L}_{\mathrm{t}} \div \mathscr{L}_{\mathrm{t}}$ with an alternating density, as shown in Fig. 6.22

$$\delta g_{\mathrm{t}}(P \mid \mathscr{L}_{\mathrm{t}}) - \delta g_{\mathrm{t}}\left(P \mid \widetilde{\mathscr{L}}_{\mathrm{t}}\right) = \delta g_{\mathrm{t}}\left(P \mid \mathscr{L}_{\mathrm{t}} \div \widetilde{\mathscr{L}}_{\mathrm{t}}\right) = \delta g_{\mathrm{tr}}(P). \tag{6.90}$$

This is the residual topographic correction as computed by many software tools with a large diffusion; whence the index r. As for the third term in (6.89), we need to compute the jump of δg_{t} across the surface $\tilde{S}$.

This could be calculated, e.g., by some prism formula, at the cost of considerable numerical work. On the contrary, a simple alternative is to compute it from the simple Bouguer correction, i.e., using only the main term linear in h (cf. (6.42)). The situation is represented in Fig. 6.23. Where $h_P > h_{\tilde{P}}$, i.e., P is outside $\tilde{S}$, we have

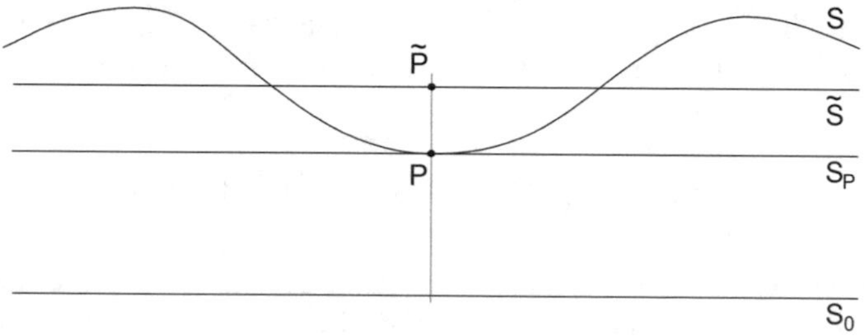

Fig. 6.23 The Bouguer layer used to compute the jump of $\delta g\prime$ across $\tilde{S}$

$$g_{\text{B0+}}(P) = -2\pi\mu h_{\tilde{P}}; \tag{6.91}$$

since $g_{\text{B0+}}$ is constant, with respect to h_P, one has

$$g'_{\text{B0+}}(P) = 0. \tag{6.92}$$

On the contrary, when $h_P < h_{\tilde{P}}$ as in the figure, one has

$$g_{\text{B0-}} = -2\pi\mu h_P + 2\pi\mu\left(h_{\tilde{P}} - h_P\right) \tag{6.93}$$

considering that the attraction of the two layers (S_0, S_P), $(S_P, \tilde{S})$ is expressed in terms of the vertical component with respect to the Z-axis directed upward.

Therefore, one has

$$g'_{\text{B0-}} = -4\pi\mu$$

and we can put

$$\delta g'_{\text{t+}}\left(\tilde{P}\middle|\widetilde{\mathscr{L}}_{\text{t}}\right) - \delta g'_{\text{t-}}\left(\tilde{P}\middle|\widetilde{\mathscr{L}}_{\text{t}}\right) \cong g'_{\text{B0+}} - g'_{\text{B0-}} = 4\pi\mu. \tag{6.94}$$

Combining (6.94) with (6.90), (6.89) we get

$$\delta g_{\text{RTC}}(P) = \delta g_{\text{t}}(P) - \delta g_{\text{Mt}}(P) = \delta g_{\text{tr}}(P) + 4\pi\mu\left(h_{\tilde{P}} - h_P\right)H\left(h_{\tilde{P}} - h_P\right). \tag{6.95}$$

This expression is known as *Residual Terrain Correction*, $\delta g_{\text{RTC}}(P)$.

We note that, to compute δg_{RTC}, we could use the algorithm of Sect. 6.3 for the computation of $\delta g_{\text{tr}} = \delta g_{\text{t}}(P|\mathscr{L}_{\text{t}}) - \delta g_{\text{t}}\left(P|\widetilde{\mathscr{L}}_{\text{t}}\right)$ with the only warning that in general $\delta g_{\text{t}}\left(P|\widetilde{\mathscr{L}}_{\text{t}}\right)$ has to be computed at a point P which may lay outside or inside $\tilde{S}$.

To conclude our reasoning, we can now return to the definition of δg_{L} and notice that

$$\begin{aligned}\delta g_{\text{L}}(P) = \delta g(P) - \delta g_{\text{M}}(P) = \delta g_{\text{I}}(P) - \delta g_{\text{M}_\text{I}}(P) + \\ + \delta g_{\text{c}}(P) - \delta g_{\text{M}_\text{c}}(P) + \delta g_{\text{t}}(P) - \delta g_{\text{M}_\text{t}}(P).\end{aligned} \tag{6.96}$$

The first contribution from B_{I} was shown above to be negligible, that from $\mathscr{L}_{\text{c}}$ is in fact our target function $\delta g_{\text{r}}(P) = \delta g_{\text{c}}(P) - \delta g_{\text{M}c}(P)$, that from topographic masses is just δg_{RTC} and we have learned how to compute it. So we can rewrite (6.96) in the form

$$\delta g_{\text{r}}(P) = \delta g_{\text{L}}(P) - \delta g_{\text{RTC}}(P), \tag{6.97}$$

where we know how to compute the terms on the right, and the function on the left is exactly what we wanted to obtain. Of course, in practice formula (6.97) will provide us with $\delta g_{\rm r}$ computed at points where gravity measurements have been performed, namely at points irregularly distributed in both horizontal and vertical coordinates. How to transform this data set to a regular grid of $\delta g_{\rm r}$ values at constant height, will be discussed in the next section. We conclude this section by a remark.

Remark 6.2 The reader probably noticed that in the computation of the various components $T_{\rm I}$, $T_{\rm c}$, $T_{\rm t}^{\rm d}$, when coming to a spherical harmonic analysis, we have started the series from 0 instead of 2, as usual for the anomalous potential. The reason is that, on a logical ground, the coefficients $T_{0,0}$, $T_{1,-1}$, $T_{1,0}$, $T_{1,1}$ have to be set to zero, but not those of each individual component of T.

Yet, due to the overwhelming importance of $T_{\rm I}$, in terms of internal masses, one could see that each other component could move the barycenter at most of a few centimeters.

In any way, the argument is irrelevant for the matter treated in this section, because long-wavelength contributions for T are filtered out by subtracting U and $T_{\rm M}$, and we will not be bothered by them.

6.5 Gridding Gravity Disturbances on a Plane at Constant Height: Deterministic Approaches

Let us first summarize what is the data set $\{\delta g_{\rm r}(P_i)\}$, computed from (6.97) at observation points $\{P_i\}$. First of all, at the level of notation we shall drop the index r and recall that, according to our conventions,

$$\delta g\,(P) = -\delta g_Z\,(P) = -\frac{\partial T}{\partial z}, \tag{6.98}$$

where $T(P) \equiv T_{\rm r}(P)$ is the anomalous potential generated in $\{h_p \geq 0\}$ by mass distribution contained in the layer between $S_{\rm c}$ and S_0.

Both $T(P)$ and $\delta g(P)$ are harmonic functions in $\mathbb{R}^3_+ \equiv \{h_p \geq 0\}$; our purpose is to obtain a set of values δg at the nodes of a regular grid on a plane at constant altitude $\bar{h}$, because such a set simplifies the further analysis of the inverse problem, namely the reconstruction of $\rho(Q)$, specially if we use Fourier methods, opening the way to the application of the FFT algorithm.

We observe immediately that $\bar{h}$ can be any non-negative value. As a matter of fact, the best would be to take $\bar{h} = 0$, because then the resulting δg data set is as close as possible to the masses. Yet this implies a downward continuation of the data from S to S_0 that, as we will better see later on, tends to amplify the residual errors in the data. So for the present section we consider only two alternatives: a) $\bar{h}$ is the mean height of S, and b) $\bar{h}$ is the maximum of h_P over S.

(a) We make this choice, particularly when the oscillation of h_P around $\bar{h}$ is small; then, knowing that δg has a little, if not exactly zero, correlation with $\delta h_P = h_P - \bar{h}$, we impose

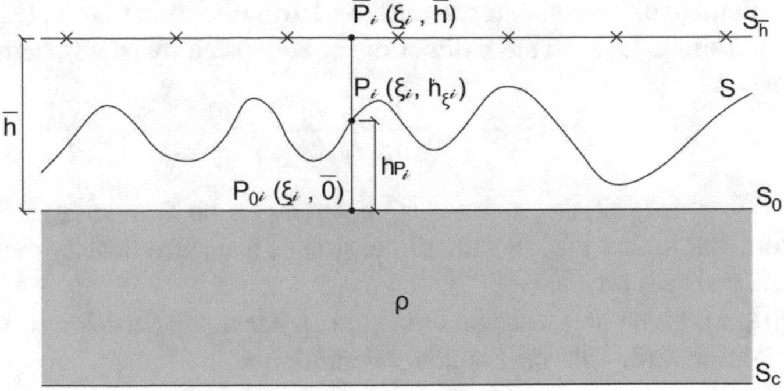

Fig. 6.24 Geometry of $\delta g_r(P)$ observation points, generated only by the density ρ, and the gridding plane $S_{\bar{h}}$

$$\delta g\,(P_i) \cong \delta g\left(\bar{P}_i\right). \tag{6.99}$$

With (6.99) we can assume to have a data set irregularly distributed on $S_{\bar{h}}$. After that, any reasonable gridder, like a spline interpolator or a moving average with weights dependent on the distance between the nodes (prediction point) and the data points in a neighbouring window can be used [157]. What is important is that the sides of the grid have a length larger than the mean (horizontal) distance between data points $\{\bar{P}_i\}$; otherwise we simply pretend to multiply a (non-existing) information. In other words, the number of the nodes of the grid should be comparable or smaller than the number of data (Fig. 6.24).

In case there are holes, area-wise, in the data distribution, we advise the reader that it is better to use the technique presented in the next section.

(b) When the conditions on the oscillation of h_P are not satisfied, we can resort to the choice $\bar{h} \geq \max h_P$.

We observe that in this case $S_{\bar{h}}$ is completely embedded into the harmonicity domain $\Omega = \{z \geq h_{\xi}\}$. So the task now is basically to solve the Dirichlet problem

$$\begin{cases} \Delta\delta g = 0 & \text{in } \Omega, \\ \delta g\left(\xi, h_{\xi}\right) = \delta g_0\,(\xi) & \text{on } S, \end{cases} \tag{6.100}$$

in a discretized form; in fact $\frac{\partial u}{\partial z}$ is hamonic where u is.

This is one of the oldest problems in the theory of partial differential equations and a number of techniques are available in both the general mathematical literature and the geophysical one. We mention here the boundary elements method [36] which, by applying the third Green's identity, transforms the problem to that of solving a singular integral equation. Yet, to the knowledge of the authors there is little experience of the application of this approach to the problem at hand.

As an alternative, one can represent the harmonic function $\delta g(\xi, z)$ as the derivative of a single layer in the z-direction, as suggested by (6.98), namely

$$\delta g\left(\boldsymbol{\xi}, z\right) = -\frac{\partial}{\partial z}\int_S \frac{\nu\left(\boldsymbol{\eta}\right)}{\ell_{PQ}} dS_Q, \tag{6.101}$$

where $P \equiv (\boldsymbol{\xi}, z)$ is in Ω, i.e., above S, $Q \equiv (\boldsymbol{\eta}, h_{\boldsymbol{\eta}})$ is on S and $\nu(\boldsymbol{\eta})$ is the surface density. Note that in fact $\nu(\boldsymbol{\eta})$ has the dimension of a surface density multiplied by the universal constant G.

That $\delta g(\boldsymbol{\xi}, z)$, given any reasonable $\nu(\boldsymbol{\eta})$, is a harmonic function in $\Omega \equiv \{z > h_{\boldsymbol{\xi}}\}$, is straightforward. That the boundary condition

$$\delta g\left(\boldsymbol{\xi},\ h_{\boldsymbol{\xi}}\right) = \delta g_0\left(\boldsymbol{\xi}\right) \tag{6.102}$$

is satisfied depends on the choice of $\nu(\boldsymbol{\eta})$. To obtain the relation (6.102) we have to take the limit of (6.101) for z tending to $h_{\boldsymbol{\xi}}$. This is given by a well-known relation in potential theory, also used in geodesy in Molodensky's formulation of the geodetic boundary value problem (cf. [63], §1.3), namely

$$\delta g_0\left(\boldsymbol{\xi}\right) = \lim_{\substack{Z \to h_{\boldsymbol{\xi}} \\ (Z > h_{\boldsymbol{\xi}})}} \delta g\left(\boldsymbol{\xi}, z\right) = 2\pi\nu\left(\boldsymbol{\xi}\right)\boldsymbol{e}_Z \cdot \boldsymbol{n}\left(\boldsymbol{\xi}\right) - \int_S \nu\left(\boldsymbol{\eta}\right)\frac{\partial}{\partial Z}\frac{1}{\ell_{PQ}} dS_Q \tag{6.103}$$

Since the normal vector to S at P is given by

$$\boldsymbol{n}\left(\boldsymbol{\xi}\right) = \frac{\boldsymbol{e}_Z - \nabla_{\boldsymbol{\xi}} h_{\boldsymbol{\xi}}}{\left[1 + \left|\nabla_{\boldsymbol{\xi}} h_{\boldsymbol{\xi}}\right|^2\right]^{1/2}} = \cos I_{\boldsymbol{\xi}}\left(\boldsymbol{e}_Z - \nabla_{\boldsymbol{\xi}} h_{\boldsymbol{\xi}}\right), \tag{6.104}$$

where $I_{\boldsymbol{\xi}}$ is just the inclination of the surface S with respect to the direction of the vertical $\boldsymbol{e}_Z$, the first term in the RHS of (6.104) can be written as $2\pi\nu(\boldsymbol{\xi})\cos I_{\boldsymbol{\xi}}$. To get any reasonable solution of our Dirichlet problem we have to assume that

$$\cos I_{\boldsymbol{\xi}} \geq c_0 > 0, \tag{6.105}$$

i.e., that S has nowhere a vertical tangent plane. We note as well that $\cos I_{\eta} dS_Q = d_2\eta$.

Moreover, since

$$-\frac{\partial}{\partial z}\frac{1}{\ell_{PQ}}\bigg|_S = -\,\boldsymbol{e}_Z \cdot \nabla_P \frac{1}{\ell_{PQ}}\bigg|_S = \frac{h_{\boldsymbol{\xi}} - h_{\eta}}{\ell_{PQ}^3}, \tag{6.106}$$

we can finally write (6.105) as

$$\delta g_0\left(\boldsymbol{\xi}\right) = 2\pi\nu\left(\boldsymbol{\eta}\right)\cos I_{\boldsymbol{\xi}} + \int_{\mathbb{R}^2} \frac{\nu\left(\boldsymbol{\eta}\right)}{\cos I_{\eta}}\frac{h_{\boldsymbol{\xi}} - h_{\eta}}{\ell_{PQ}^3} d_2\eta. \tag{6.107}$$

If we put

$$\nu\left(\boldsymbol{\eta}\right)=\cos I_{\eta}\,\omega\left(\boldsymbol{\eta}\right),$$

that, given that the condition (6.105) is a one-to-one relation between $\nu(\boldsymbol{\eta})$ and the new unknown $\omega(\boldsymbol{\eta})$, we finally have

$$\delta g_0\left(\boldsymbol{\xi}\right)=2\pi\omega\left(\boldsymbol{\xi}\right)\cos^2 I_{\xi}+\int_{\mathbb{R}^2}\omega\left(\boldsymbol{\xi}\right)\frac{h_{\xi}-h_{\eta}}{\ell_{PQ}^3}d_2\eta. \tag{6.108}$$

A bit of caution has to be applied to interpret (6.108). In fact, the integral kernel

$$k\left(\xi,\boldsymbol{\eta}\right)=\frac{h_{\xi}-h_{\eta}}{\ell_{PQ}^3}, \tag{6.109}$$

when h_{ξ} has a finite gradient, i.e. the normal $\boldsymbol{n}_{\xi}$ to S is well defined, has a singularity of the order of $\frac{1}{\ell_{PQ}^2}$ when $\boldsymbol{\eta}$ approaches $\boldsymbol{\xi}$, and such a singularity is not integrable with respect to $d_2\eta$; therefore, (6.108) is a singular integral equation [83]. According to the specific theory, (6.108) has a meaning only if the integral can be considered in the sense of Cauchy principal part, i.e., if the integral along a circle C_{ε} of radius ε and centered at $\boldsymbol{\xi}$ satisfies

$$\lim_{S\to\infty}\int_{C_{\varepsilon}(\boldsymbol{\xi})}k\left(\boldsymbol{\xi},\boldsymbol{\eta}\right)d\alpha=0,$$

where α is the angular coordinate on C_{ε}. This is in fact the case, for instance, if h_{ξ} is continuous up to the second derivatives, so that one has

$$h_{\eta}-h_{\xi}=\left(\boldsymbol{\eta}-\boldsymbol{\xi}\right)\cdot\nabla h_{\xi}+O\left(|\eta-\xi|^2\right). \tag{6.110}$$

So, assuming that both (6.105) and (6.109) are satisfied, we are in a position to solve numerically (6.108), e.g., by discretization. Once $\omega(\boldsymbol{\xi})$ (and hence $\nu(\boldsymbol{\xi})$) is found, (6.101) can be used to compute $\delta g(\boldsymbol{\xi},z)$ anywhere in $\{z\geq h_{\xi}\}$ and therefore also for $z=\bar{h}$.

Such an approach is indeed classical in mathematics, but one cannot say that it is easy to implement.

The idea of the method can be considered as a particular case of what is called in geophysics the method of equivalent sources (see [11] and the references therein).

We will not continue along this line; instead, we will present an algorithm, used in geodesy [133], that is much easier to handle although it has a mathematical justification only in a perturbative sense. Since the following arguments refer to the solution of the Dirichlet problem in general terms, we shall abandon the notation δg, δg_0 and we shall rather use the notation $u(\boldsymbol{\xi},z)$ for a general harmonic function.

Before going on with this matter, we present, as an example, the solution of the Dirichlet problem when S is itself a plane, that we can take as S_0. This allows us to introduce the Poisson kernel and the Poisson integral operator.

Example 6.4 (Poisson Kernel) We assume that $S \equiv \{z = 0\}$ and we are given a function $u_0(\boldsymbol{\xi})$ on S such that $u_0 \in L^2\left(\mathbb{R}^2\right)$ and $u_0 \in L^1\left(\mathbb{R}^2\right)$, too.

We want to find a $u(\boldsymbol{\xi}, z)$ harmonic in $\Omega = \{z > 0\} \equiv \mathbb{R}^3_+$, that is at least bounded in every domain $\{z \geq h > 0\}$, for every positive and constant h, and such that

$$\lim_{z \to 0} u\left(\boldsymbol{\xi}, z\right) = u_0\left(\boldsymbol{\xi}\right). \tag{6.111}$$

We claim that the solution of the problem is given by the integral formula

$$u\left(\boldsymbol{\xi}, z\right) = \frac{1}{2\pi} \int_{\mathbb{R}^2} \frac{z}{\ell_{PQ}^3} u_0\left(\boldsymbol{\eta}\right) d_2\eta, \tag{6.112}$$

with $P \equiv (\boldsymbol{\xi}, z)$, $Q \equiv (\boldsymbol{\eta}, 0)$ and, as usual

$$\ell_{PQ} = \left(|\boldsymbol{\xi} - \boldsymbol{\eta}|^2 + z^2\right)^{1/2}. \tag{6.113}$$

The kernel

$$P\left(\boldsymbol{\xi},\ z\right) = \frac{1}{2\pi} \frac{z}{\left(|\boldsymbol{\xi}|^2 + z^2\right)^{3/2}} \tag{6.114}$$

is called the *Poisson kernel*, and with the notation (6.114) we write (6.112) as

$$u\left(\boldsymbol{x}, z\right) = \int_{\mathbb{R}^2} P\left(\boldsymbol{\xi} - \boldsymbol{\eta},\ z\right) u_0\left(\boldsymbol{\eta}\right) d_2\eta. \tag{6.115}$$

We note that (6.112) could be proved at once by exploiting (6.101) and (6.107), which implies $\delta g_0(\boldsymbol{\xi}) = \nu(\boldsymbol{\xi})$ because $\cos I_{\mathrm{fv}} = 0$ and $h_{\boldsymbol{\xi}} = 0$ in this case.

Nevertheless, we prove (6.112) anew because en route we take the opportunity to provide a Fourier representation of this a formula.

We start by assuming that $u\left(\boldsymbol{\xi}, z\right) \in L^2\left(\mathbb{R}^2\right)$ for each fixed z, a property that we will verify a posteriori to be satisfied. Therefore we can define

$$\hat{u}\left(\boldsymbol{p}, z\right) = \int_{R^2} e^{i2\pi \boldsymbol{p}\cdot\boldsymbol{\xi}} u\left(\boldsymbol{\xi}, z\right) d_2\xi. \tag{6.116}$$

Now we note that the Laplace equation for $u(\boldsymbol{\xi}, z)$ can be written as

$$\frac{\partial^2}{\partial z^2} u\left(\boldsymbol{\xi}, z\right) + \Delta_{\boldsymbol{\xi}} u\left(\boldsymbol{\xi}, z\right) = 0$$

and use the identity

$$\int e^{i2\pi \boldsymbol{p}\cdot\boldsymbol{\xi}} \Delta_{\xi} u\left(\boldsymbol{\xi}, z\right) d_2\xi = \left(-4\pi^2 p^2\right) \int e^{i2\pi \boldsymbol{p}\cdot\boldsymbol{\xi}} u\left(\boldsymbol{\xi}, z\right) d_2\xi, \tag{6.117}$$

which is a straightforward generalization of (3.18). This implies that $\hat{u}(\boldsymbol{p}, z)$ must satisfy

$$\frac{\partial^2}{\partial z^2} \hat{u}\left(\boldsymbol{p}, z\right) - 4\pi^2 p^2 \hat{u}\left(\boldsymbol{p}, z\right) = 0. \tag{6.118}$$

The general solution of (6.118) is

$$\hat{u}\left(\boldsymbol{p}, z\right) = A\left(\boldsymbol{p}\right) e^{-2\pi p z} + B\left(\boldsymbol{p}\right) e^{2\pi p z}, \tag{6.119}$$

but if we are to get a bounded solution for $z \to \infty$, we must assume that

$$B\left(\boldsymbol{p}\right) \equiv 0. \tag{6.120}$$

Now we can impose the boundary condition by using the Fourier transform of $u_0(\boldsymbol{\xi})$, i.e., write

$$\hat{u}_0\left(\boldsymbol{p}\right) = \hat{u}_0\left(\boldsymbol{p}, 0\right) = A\left(\boldsymbol{p}\right). \tag{6.121}$$

So the solution found is

$$\hat{u}\left(\boldsymbol{p}, z\right) = \hat{u}_0\left(\boldsymbol{p}\right) e^{-2\pi p z}, \tag{6.122}$$

Recalling Proposition 3.3, we see that

$$P\left(\boldsymbol{\xi}, z\right) = \mathscr{F}^* \left\{ e^{-2\pi p z} \right\}$$

and so, by Proposition 3.2, we obtain (6.115).

That $u(\boldsymbol{\xi}, z)$ is, for fixed z, an $L^2\left(\mathbb{R}^2\right)$ function is immediate. In fact, by Parseval's identity and (6.121),

$$\left\| u\left(\boldsymbol{\xi}, z\right) \right\|^2_{L^2(R^2)} = \int_{\mathbb{R}^2} e^{-4\pi p z} \left| \hat{u}_0\left(P\right) \right|^2 d_2 p < +\infty. \tag{6.123}$$

As a matter of fact, since $0 < e^{--4\pi p z} < 1$ for all $p \neq 0$, (6.123) says that

$$\left\| u\left(\boldsymbol{\xi}, z\right) \right\|_{L^2(\mathbb{R}^2)} < \left\| u_0\left(\boldsymbol{\xi}\right) \right\|_{L^2(\mathbb{R}^2)}, \tag{6.124}$$

i.e., the upward continuation operator $\mathscr{U}_h$ between the plane $\{z = 0\}$ and $\{z = h\}$ is non-expansive in $L^2\left(\mathbb{R}^2\right)$.

Furthermore, one can notice that

$$|P(\boldsymbol{\xi}-\boldsymbol{\eta},z)| \leq \frac{2\pi}{z^2}$$

so that, from (6.115),

$$|u(\boldsymbol{\xi},z)| \leq \frac{2\pi}{z^2}\int_{R^2}|u_0(\boldsymbol{\eta})|\,d_2\eta; \tag{6.125}$$

since $u_0 \in L^1$ by hypothesis, (6.125) says that

$$\lim_{z\to\infty} u(\boldsymbol{\xi},z) = 0 \tag{6.126}$$

uniformly on $\mathbb{R}^2$; therefore the requirement that $u(\boldsymbol{\xi},z)$ be bounded in every set $\Omega_h \in \{z \geq h > 0\}$ (h constant) is satisfied.

We conclude the example by observing that the Poisson kernel is always positive and satisfies the identity

$$\int_{\mathbb{R}^2} P(\boldsymbol{\xi},z)\,d_2\xi \equiv 1 \quad \forall z. \tag{6.127}$$

Since for every $\boldsymbol{\xi}$, $|\boldsymbol{\xi}| > 0$,

$$\lim_{z\to 0}\frac{1}{2\pi}\frac{z}{\left(|\xi|^2+z^2\right)^{3/2}} \equiv 0, \tag{6.128}$$

we conclude that

$$\lim_{z\to 0} P(\xi,z) = \delta(\boldsymbol{\xi}), \tag{6.129}$$

which incidentally was implicit in (6.111) combined with (6.112).

The Fourier transformed version of (6.129) is particularly evident, since it reads

$$\lim_{z\to 0}\hat{P}(p,z) = \lim_{z\to 0} e^{-2\pi pz} \equiv 1.$$

Indeed, to give a precise meaning to (6.129), one has to use the definition of a distribution shortly recalled in Remark 3.4.

One says that the family of kernels $\{P(\boldsymbol{\xi},z);\, z > 0\}$ provides an approximation to the identity operator.

With what we learned from the discussion of the above example, we consider now the situation when $S \equiv \{z = h_\xi\}$ is above $S_0 \equiv \{z = 0\}$. We even assume that

$$h_\xi \geq h_0 > 0, \tag{6.130}$$

for some positive constant h_0.

We define the upward continuation operator $\mathscr{U}_S$ from $L^2(S_0) \equiv L^2(\mathbb{R}^2)$ to $L^2(S)$ as

$$u(\xi, h_\xi) \equiv \mathscr{U}_S u_0 = \int_{R^2} P(\xi - \eta, h_\xi)\, u_0(\eta)\, d_2 y = \\ = \frac{1}{2\pi} \int_{R^2} \frac{h_\xi}{\left(|\xi - \eta|^2 + h_\xi^2\right)^{3/2}} u_0(\eta)\, d_2 y. \tag{6.131}$$

Since, when $h_\xi \leq \bar{h} < +\infty$, we have

$$0 < P(\xi - \eta, h_\xi) \leq \frac{1}{2\pi} \frac{\bar{h}}{\left(|\xi - \eta|^2 + h_0^2\right)^{3/2}} = \frac{\bar{h}}{h_0} P(\xi - \eta, h_0), \tag{6.132}$$

we also have

$$|u(\xi, h_\xi)| \leq \frac{\bar{h}}{h_0} \int_{\mathbb{R}^2} P(\xi - \eta, h_0)\, |h_0(\eta)|\, d_2 y \tag{6.133}$$

and so

$$\|u(\xi, h_\xi)\|_{L^2(\mathbb{R}^2)} \leq \left(\frac{\bar{h}}{h_0}\right) \|h_0\|_{L^2(\mathbb{R}^2)} \tag{6.134}$$

proving that $\mathscr{U}_S$ is indeed a bounded operator from L^2 into L^2.

A much finer analysis, though based on a perturbative argument in the parameter $\frac{\bar{h}-h_0}{h_0}$, can show that if we work with functions having only a finite spectrum, i.e.,

$$\hat{u}_0(\boldsymbol{p}) = 0 \quad \text{for} \quad |\boldsymbol{p}| > |\bar{\boldsymbol{p}}|, \tag{6.135}$$

we find that the operator $I - \mathscr{U}_S$, restricted to such a subspace, is a proper contraction, namely

$$\|I - \mathscr{U}_S\| < 1. \tag{6.136}$$

A "proof" of (6.136) in a spherical approximation setting can be found in [133].

The condition (6.135) is less restrictive than one way think, because in practice using a grid of known values $u_0(P_i)$, with a spacing Δ, we erase any information on the Fourier transform of u_0 beyond the wave number (recall (3.171))

$$\bar{p} = \frac{1}{2\Delta}$$

and we can typically assume that $\hat{u}_0(\boldsymbol{p})$ is zero beyond this limit.

Table 6.2 The Change of Boundary iterative algorithm

$$
\begin{array}{lcl}
S & & S_0 \\
u_0(\xi) & \xrightarrow{\;\;PB\;\;} & u_0(\xi) \\[2em]
u_1 = \mathcal{U}_S u_0 & & \\
\downarrow & & \\
r_1 = u_0 - u_1 & \xrightarrow{\;\;PB\;\;} & r_1 \qquad r_1 = (I - \mathcal{U}_S)u_0 \\[2em]
u_2 = \mathcal{U}_S r_1 & & \\
\downarrow & & \\
r_2 = r_1 - u_2 & \xrightarrow{\;\;PB\;\;} & r_2 \qquad r_2 = (I - \mathcal{U}_S)r_1 = (I - \mathcal{U}_S)^2 u_0
\end{array}
$$

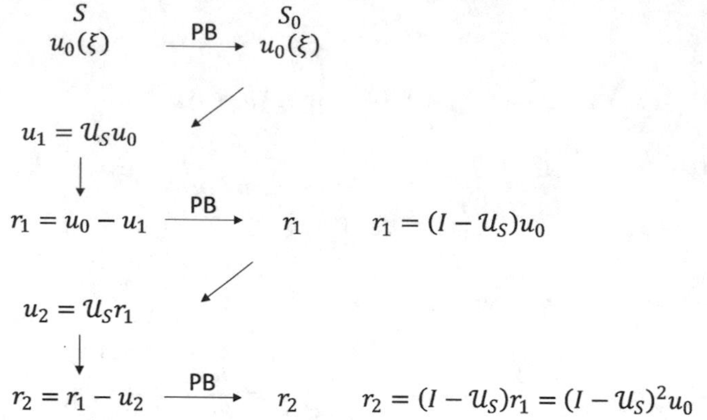

So, accepting the conjecture (6.136), we can establish an iterative algorithm, called by one of the authors *Change of Boundary*, with the hope that it will converge.

If we take a function defined on S, $u(\xi, h_\xi) = u_0(\xi)$ as defined directly on S_0, we construct an operator, called a pullback operation (PB). Then we compute a sequence $\{u_n\}$ according to the scheme presented in Table 6.2.

If (6.136) is correct, then

$$
\|r_n(\xi)\|_{L^2(\mathbb{R}^2)} \leq \|I - \mathcal{U}_S\|^n \|u_0\|_{L^2(\mathbb{R}^2)}, \tag{6.137}
$$

and $r_n(\xi)$ tends to 0 in $L^2(\mathbb{R}^2)$ when $n \to \infty$.

Since adding n terms of the first column in the Table 6.2, we have on S

$$
u_0(\xi) - u_1(\xi) - u_2(\xi) \cdots - u_n(\xi) = r_n(\xi) \tag{6.138}
$$

the fact that $r_n(\xi) \to 0$ implies

$$
u_0(\xi) = \sum_{n=1}^{+\infty} u_n(\xi). \tag{6.139}
$$

Since the convergence of this series on the boundary S implies the convergence of the series

$$
u(\xi, z) = \sum_{n=1}^{+\infty} u_n(\xi, z) \tag{6.140}
$$

everywhere in $\Omega \equiv \{z > h_\xi\}$, we have found our solution of the Dirichlet problem in Ω, and, in particular, we can compute $u(\xi, z)$ on $\bar{S} \equiv \{z = \bar{h}\}$.

Note that each $u_n(\boldsymbol{\xi}, z)$ is derived as a Poisson integral of a function defined on S_0, so its computation on $\bar{S}$ is easily accomplished by a Fourier algorithm, duly discretized, as seen in Example 6.4.

Numerical experiments with the algorithm seem to confirm the above theory, as we will show in the next Example 6.5.

The algorithm is generally convergent on S and above, however there is no guarantee that it should converge, for instance, on S_0, too; on the contrary, we expect that since going from S to S_0 we have to continue downward our harmonic field, the instability of such an operation can blow up errors, making the approximation not usable.

Example 6.5 (Upward 1 Downward Continuation) In this numerical example we will move the real airborne dataset already introduced in Example 6.1 from the actual position (with altitude defined by the acquisition height of the airborne survey) to a constant altitude, here fixed at 500 m, and then down to the zero level.

Before applying the upward/downward continuation algorithm, the actual gravity signal is properly reduced by removing the low frequencies from a GGM, and the high frequencies from a RTC. This operation allows to decrease the signal amplitude, thereby reducing the intrinsic error (due to the discretization of the Poisson kernel) when moving up and down the gravity signal.

Once reduced, the along-track signal can be interpolated (disregarding the effect of vertical shifts) on a regular grid, e.g., according to the Least Squares Collocation procedure described in the next section. The heights of the observations are interpolated on the same grid by means of a kriging procedure [97]. This operation would allow to "move" the observations from a set of sparse points along the aircraft trajectory to a grid, regular in the X and Y directions, but with varying height. Note that this simplification is allowed only if the height differences between observation points is small (of about 200 m) and if the signal has been properly reduced. Within this test, the area has been modelled by a grid of 801×801 knots in the X and Y directions, with a pixel size of about 200 m.

The grid of the reduced field, which shows a signal amplitude of about 5 mGal (in terms of standard deviation), is presented in Fig. 6.25. while in Fig. 6.26 the actual acquisition altitude of the dataset is shown. At this point, the change of boundary approach can be applied to continue upward the reduced gridded signal from the actual height to a constant altitude of 500 m. The effect of this upward continuation is shown in Fig. 6.27. It can be seen that the largest effect, which as expected is located in the South-West corner of the survey where the largest signal and the lowest altitude are present, reaches 4.8 mGal. However, the standard deviation of this upward continuation is of the order of just 0.3 mGal. The accuracy of the upward continuation, based on the change of boundary approach, can be roughly evaluated by continuing the signal downward back from the constant altitude of 500 m to the initial surface. Note that this operation does not require the use of the change of boundary since it is performed from an initial constant altitude. In any case, the standard deviation of the errors between the original signal and the solution obtained, after a single cycle of upward-downward continuation, is smaller than $3 \cdot 10^{-3}$ mGal (considering only the region where observations are available).

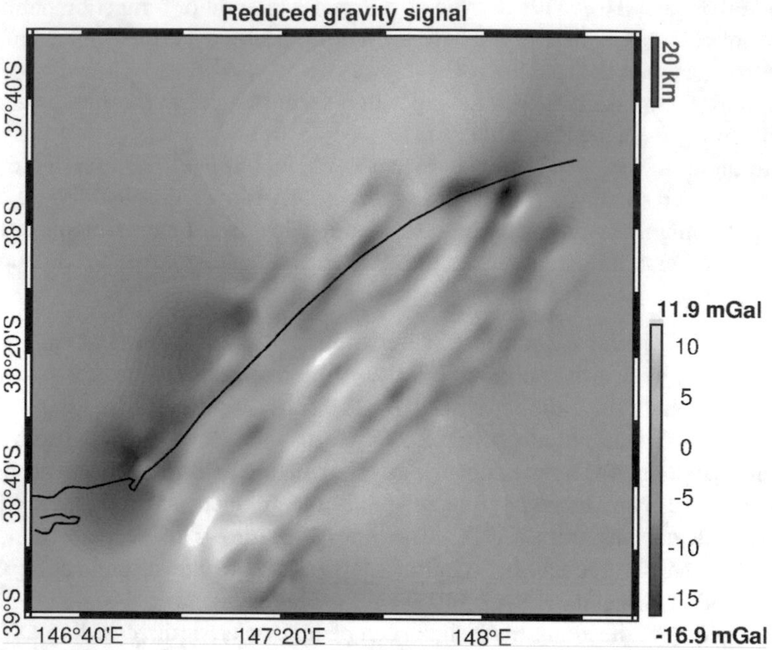

Fig. 6.25 The grid of the reduced gravity signal of Example 6.5

Once the signal is available on a given constant surface, we can apply the downward continuation algorithm to move it to any other height of interest. In particular, we moved here the signal down to the zero level. In this case, the effect of the downward continuation is of the order of 1.4 mGal (standard deviation), with features similar to those presented in Fig. 6.27, therefore it is not negligible. As for the overall accuracy, again it can be evaluated by moving the "zero level" grid upward to the starting point and comparing the results with the initial values. For the current test this shows a standard deviation of 10^{-3} mGal with a maximum value of about 0.5 mGal in the border area. Considering the computational time, when moving from/to a non-constant altitude, it required up to 4 hours, while when only constant altitudes are processed, the computational time considerably decreases to about 1 minute, as a result of exploiting FFT (all the test have been performed on a desktop computer equipped with an Intel Core i7-4790K CPU 4.00 GHz with 32 GB RAM).

Before concluding this numerical example, a last remark is in order: the downward continuation of the data was performed here without using any regularization algorithms. This is basically due to the very low noise level of the initial signal, especially in the high frequencies, which was filtered out through the gridding procedure, presented in the next section, and to the resolution of the computational grid, which is comparable with the relative small variation in altitude of the different surfaces. For more details on the results presented in this example the interested reader can refer to [79].

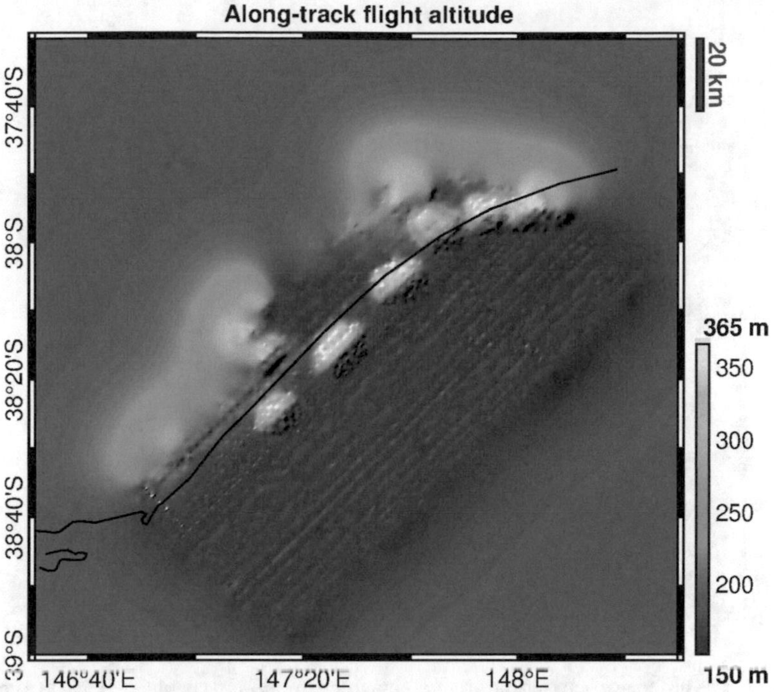

Fig. 6.26 The grid of the along-track flight altitude of Example 6.5

6.6 Gridding the Gravity Disturbances on the S_0 Plane, by a Stochastic Approach

The theory of generalized random fields (RF) is shortly reviewed in Appendix B.

From a Bayesian standpoint, its application is legitimated by the idea that an unknown RF δg has to be described as a (functional) random variable whose posterior probability distribution is conditioned by the observations $\delta g(P_i)$, allowing the determination of the best linear predictor of δg (or of some of its functionals) at any point P of its domain of definition.

As shown in Appendix B, the best linear prediction essentially depends on the covariance of the RF; if we assume $\delta g(P)$ to have zero average and finite variance, the covariance is defined as

$$C\left(P, Q\right) = E\left\{\delta g\left(P\right)\delta g\left(Q\right)\right\}. \tag{6.141}$$

In our case the domain of definition of $\delta g(P)$ is $\Omega_0 \equiv \{z \geq 0\}$, where $\delta g(P)$ has to be harmonic by assumption.

This harmonic character of $\delta g(P)$ has to be reflected by the covariance function. In fact

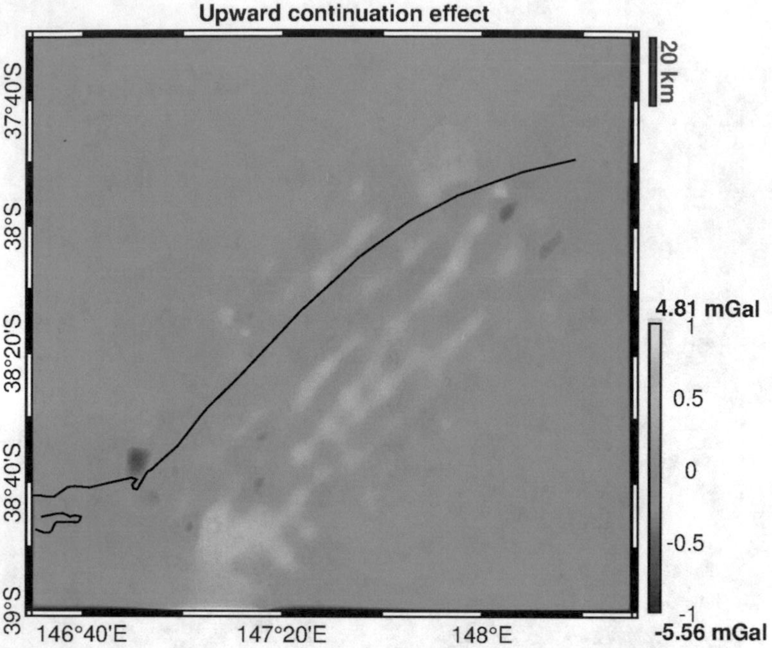

Fig. 6.27 The difference between the original signal and the upward continued signal to a constant height of 500 m of Example 6.5

$$\begin{aligned} \Delta_P C\,(P, Q) &= \Delta_P E\,\{\delta g\,(P)\,\delta g\,(Q)\} = \\ &= E\,\{\Delta_P \delta g\,(P)\,\delta g\,(Q)\} = 0; \end{aligned} \tag{6.142}$$

due to the symmetry of $C(P,Q)$, (6.142) implies too

$$\Delta_Q C\,(P, Q) = 0. \tag{6.143}$$

Further on, since $\delta g(P)$ is harmonic in Ω_0, it must be connected with its boundary values $\delta g_0(P_0) \equiv \delta g_0(\boldsymbol{\eta})$ via relation (6.115); so we can write

$$\delta g\,(\boldsymbol{\xi}, z) = \int_{\mathbb{R}^2} P\,(\boldsymbol{\xi} - \boldsymbol{\eta}, z)\,\delta g_0\,(\boldsymbol{\eta})\,d_2\eta. \tag{6.144}$$

If $\delta g(\boldsymbol{\xi}, z)$ has to have a zero mean, we have to require that this property holds for $\delta g_0(\boldsymbol{\eta})$ too.

If we devote by

$$C_0\left(\boldsymbol{\eta}, \boldsymbol{\eta}'\right) = E\left\{\delta g_0\,(\boldsymbol{\eta})\,\delta g_0\left(\boldsymbol{\eta}'\right)\right\} = C\,(P_0, Q_0) \tag{6.145}$$

the covariance of δg_0 on the S_0 plane, propagating the functional relation (6.114), we get, with obvious notation

$$\begin{aligned} C(P,Q) &= C\left(\boldsymbol{\xi}, z;\ \boldsymbol{\xi}', z'\right) = \\ &= \int\int P(\boldsymbol{\xi}-\boldsymbol{\eta}, z)\, P\left(\boldsymbol{\xi}'-\boldsymbol{\eta}', z'\right) C_0\left(\boldsymbol{\eta}, \boldsymbol{\eta}'\right) d_2\eta' d_2\eta. \end{aligned} \tag{6.146}$$

We want to show that if we assume that the boundary covariance C_0 is that of a homogeneous, isotropic RF, i.e., with a little abuse of notation,

$$C_0\left(\boldsymbol{\eta}, \boldsymbol{\eta}'\right) = C_0\left(\boldsymbol{\eta}-\boldsymbol{\eta}'\right) = C_0\left(\left|\boldsymbol{\eta}-\boldsymbol{\eta}'\right|\right), \tag{6.147}$$

then the spatial covariance $C(P, Q)$ takes a special form.

In fact, if $C_0(r)$ is a regular, bounded $L^1\left(\mathbb{R}^2\right)$ function, then in view of the Wiener–Khinchin theorem (see Appendix B), there is a power spectral density $S_0(p)$ such that

$$C_0(r) = \int e^{-2\pi i \boldsymbol{p}\cdot\boldsymbol{r}} S_0(p)\, d_2p. \tag{6.148}$$

Inserting this expression into (6.146) and rearranging the integrals, we get

$$\begin{aligned} C\left(\boldsymbol{\xi}, z\,;\boldsymbol{\xi}', z'\right) &= \int S_0(p) \int P(\boldsymbol{\xi}-\boldsymbol{\eta}, z)\, e^{-12\pi \boldsymbol{p}\cdot\boldsymbol{\eta}} s_2\eta \times \\ &\quad \times \int P\left(\boldsymbol{\xi}'-\boldsymbol{\eta}'.z'\right) e^{i2\pi \boldsymbol{p}\cdot\boldsymbol{\eta}'} d_2\eta' d_2p. \end{aligned} \tag{6.149}$$

If we put

$$\boldsymbol{\eta} = \boldsymbol{\xi} + \boldsymbol{t}, \quad \boldsymbol{\eta}' = \boldsymbol{\xi}' + \boldsymbol{t}'$$

in (6.149), then observing that $P(\boldsymbol{\xi}, z) = P(|\boldsymbol{\xi}|, z) = P(-\boldsymbol{\xi}, z)$, and recalling Proposition 3.3, we finally get

$$\begin{aligned} C\left(\boldsymbol{\xi}, z\,;\boldsymbol{\xi}', z'\right) &= \int S_0(p) e^{-i2\pi \boldsymbol{p}\cdot\boldsymbol{\xi}} e^{i2\pi \boldsymbol{p}\cdot\boldsymbol{\xi}'} d_2p \times \\ &\quad \times \int P(\boldsymbol{t}, z)\, e^{-i2\pi \boldsymbol{p}\cdot\boldsymbol{t}} \int P\left(\boldsymbol{t}', z'\right) e^{i2\pi \boldsymbol{p}\cdot\boldsymbol{t}'} d_2t' d_2t = \\ &= \int S_0(p) e^{-2\pi p(z+z')} e^{-i2\pi \boldsymbol{p}\cdot(\boldsymbol{\xi}-\boldsymbol{\xi}')} d_2p. \end{aligned} \tag{6.150}$$

Since $S_0(p)e^{-2\pi p(z+z')}$ is a function of p only, (6.150) shows at once that $C(\boldsymbol{\xi}, z\,; \boldsymbol{\xi}', z')$ depends on $\boldsymbol{\xi}, \boldsymbol{\xi}'$ through $|\boldsymbol{\xi} - \boldsymbol{\xi}'|$ only. Moreover, it clearly depends on z, z' through $z + z'$ only. So, again abusing notation, we can write

$$C(\boldsymbol{\xi}, z; \boldsymbol{\xi}', z') = C(|\boldsymbol{\xi} - \boldsymbol{\xi}'|, z + z'); \tag{6.151}$$

in particular, the field δg seen at an altitude $z = h$ is again a homogeneous, isotropic random field with covariance

$$C(|\xi - \xi'|, 2h) = \int S_0(P) e^{-2\pi p(2h)} e^{-2\pi i p \cdot (\xi - \xi')} d_2 p. \tag{6.152}$$

As we see, moving up by a certain height has the effect of damping the boundary spectrum of $\delta g_0(\boldsymbol{\xi})$ at high frequencies by the exponential factor $e^{-4\pi ph}$, i.e., smoothing the boundary covariance $C_0(|\boldsymbol{\xi} - \boldsymbol{\xi}'|)$.

As always in linear prediction theory of RF functionals, a crucial issue is the determination of the covariance function from empirical data.

In fact, limiting ourselves to the prediction of $\delta g(P)$ at any point $P \in \Omega$, our aim is to compute (see Appendix B) the Wiener–Kolmogorov or collocation estimator

$$\delta \hat{g}(P) = \sum_{k,\ell=1}^{N} C(P, P_k)\{C(P_k, P_\ell) + \sigma_\nu^2 \delta_{k\ell}\}^{(-1)} \delta g_{\mathrm{obs}}(P_\ell), \tag{6.153}$$

where $\delta g_{\mathrm{obs}}(P_\ell)$ are the values of observations at P_ℓ ($\ell = 1, \ldots N$), σ_ν^2 is the white noise variance representing the residual of the data with respect to our prediction model, $\{a_{ik}\}^{(-1)}$ is just the (i, k) entry of the inverse of the matrix $A \equiv \{a_{ik}\}$.

It is important to understand that the meaning of the noise ν, of which σ_ν^2 is the variance, is not only that of the measurement noise, but it includes also the errors done in reducing the original observations by a RTC with a constant density, as well as the component of δg that is not "isotropic and homogeneous", i.e., that cannot be explained by a covariance $C_0(|\boldsymbol{\xi} - \boldsymbol{\xi}'|)$. This point has been clarified several times in both Geodesy, where the relevant theory is known as *collocation theory* (see [126, 133]), as well as in Geostatistics, where σ_ν^2 is known as the *nugget effect* (see [26, 156]).

As we see also in Appendix B, the predictor (6.153) is endowed with an estimate of the prediction error, namely

$$\begin{aligned} E^2(P) &= E\{[\delta\hat{g}(P) - \delta g(P)]^2\} = \\ &= C(P, P) - \sum_{k,\ell=1}^{N} C(P, P_k)\{C(P_k, P_\ell) + \sigma_\nu^2 \delta_{k\ell}\}^{(-1)} C(P_\ell, P). \end{aligned} \tag{6.154}$$

This is particularly important, not only because we can have an idea of the prediction error to be expected, but also because such error can become significantly large, a circumstance that we want to be aware of, especially if the height of the prediction point h_p is significantly lower than the heights of the observation points.

In fact, we observe that the shape of the covariance (6.152), with an integrable $S_0(p)$, includes the harmonicity of $\delta\hat{g}$ in the whole Ω_0, so that (6.153) can be used to predict at points at S_0 even if the observation points are higher; the prediction, however, can have an error variance as large as the signal if we have observations at great altitude, suggesting that it is better to create a grid of predicted values at a height $h \neq 0$.

In any event, to work with (6.153), (6.154) we need to know the covariance function $C(P,Q)$. Before addressing the estimation of $C(P, Q)$ from data, we emphasize that this job can be performed by a somewhat coarse method, because fortunately we have a theorem showing that errors in the estimate of C reflect only into second-order effects in (6.153) (see [135]).

In particular, we know a priori that if we estimate $C(r)$ at regular values $r_k = k\Delta$ $(k = 1, 2, \dots k)$, the empirical values $\hat{C}(r_k)$ become quite uncertain at large values of r_k, so that typically we use only the first $10 \sim 20$ values of $\hat{C}(r_k)$ to get an estimate of the theoretical covariance. We present below two estimators of $C(r)$, one in the geometry domain, the other in the frequency domain.

Before doing that we need to give a warning.

Remark 6.3 When data are already on a surface not very far from a plane, thanks to the little correlation of δg with heights, we can roughly stake the observations as if they were on a plane at mid altitude and proceed to the estimation of the covariance, by using only their planar coordinates. Otherwise we have first to perform the first step as described above, then to predict the vertical derivatives (see the next section) of δg at P_k, and reduce with them the values at mid altitude and repeat the estimation.

6.6.1 Covariance Estimation by the Empirical Covariance

We define an empirical covariance $\hat{C}(r)$, which is estimated at points

$$r_k = k\Delta, \quad k = 0, 1, \dots, K. \tag{6.155}$$

First, we order the observation points $\{\xi_k\}$ in a sequence $k = 1, \dots, K$ and then define

$$\begin{aligned}
r_{j\ell} &= \left|\xi_j - \xi_\ell\right|, \quad j < \ell, \\
I_k &= \left[r_k - \frac{\Delta}{2}, \quad r_k + \frac{\Delta}{2}\right], \quad j = 1, 2, \dots, K, \\
I_0 &= \left(0, \frac{\Delta}{2}\right], \\
N_k &= \mathcal{N}\left\{r_{j\ell} \in I_k\right\}, \quad k \neq 0, \\
N_0 &= \mathcal{N}\left\{r_{j\ell} \in I_0\right\}.
\end{aligned} \tag{6.156}$$

where $\mathcal{N}\{A\}$ denotes the number of points in A. Note that O does not belong to I_0, so if $r_{j\ell} \in I_0$, it means that $j \neq \ell$. Observe also that the number of couples (j, ℓ) belonging to I_k or to I_0 is counted only once because $j < \ell$.

The estimated or empirical covariance function at r_k, $\hat{C}\,(r_k)$, is given by

$$\hat{C}\,(r_k) = \frac{1}{N_k} \sum_{r_{j\ell} \in I_k} \delta g(\boldsymbol{\xi}_j) \delta g(\boldsymbol{\xi}_\ell), \quad k = 0, 1, \ldots, K. \tag{6.157}$$

The condition for (6.157) to be meaningful is that $N_k > 0$, $\forall k = 0, 1, \ldots, K$ and at the same time $C(r)$ is well approximated by a linear function in I_k. Then, splitting the observations $\delta g\left(\boldsymbol{\xi}_j\right)$ into signal s (with covariance $C(r)$) and noise $\boldsymbol{\nu}$, namely

$$\delta g(\boldsymbol{\xi}_j) = s(\boldsymbol{\xi}_j) + \boldsymbol{\nu}_j, \tag{6.158}$$

and assuming that s and $\boldsymbol{\nu}$ are not correlated, we have, noting that $j \neq \ell$,

$$\begin{aligned} E\{\delta g(\boldsymbol{\xi}_j)\delta g(\boldsymbol{\xi}_\ell)\} &= E\{s(\boldsymbol{\xi}_j)s(\boldsymbol{\xi}_\ell)\} + E\{\boldsymbol{\nu}_j \boldsymbol{\nu}_\ell\} = C(r_{j\ell}) = \\ &\cong C(r_k) + C'(r_k)(r_{j\ell} - r_k). \end{aligned}$$

If the distances $r_{j\ell}$ are evenly distributed around r_k, we can assume that

$$E\{\hat{C}(r_k)\} \equiv C(r_k) + \frac{C'(r_k)}{N_k} \sum_{r_{j\ell} \in I_k} (r_{j\ell} - r_k) \cong C(r_k).$$

In any case, the error term satisfies

$$\left| \frac{C'(r_k)}{N_k} \sum_{r_{j\ell} \in I_k} (r_{j\ell} - r_k) \right| \leq \frac{C'(r_k)}{2} \Delta,$$

and this, when Δ is suitably small, should guarantee that our empirical value $\hat{C}(r_k)$ is a reasonable estimate of $C(r_k)$.

An estimate of σ_ν^2 can then be derived from the relation

$$E\left\{\frac{1}{N}\sum_{j=1}^{N} \delta g(\boldsymbol{\xi}_j)^2\right\} = E\left\{\frac{1}{N}\sum s(\boldsymbol{\xi}_j)^2 + \frac{1}{N}\sum \nu_j^2\right\} = C(0) + \sigma_\nu^2, \tag{6.159}$$

which shows that $\frac{1}{N}\sum_{j=1}^{N} \delta g(\boldsymbol{\xi}_j)^2 - \hat{C}(0)$ is a good estimator of σ_ν^2.

Now $C(r)$ can be chosen to interpolate $\hat{C}(r_k)$, which has to be done by fixing a certain parametric model for $C(r)$ and then adjusting the parameters to the empirical values, e.g., by simple least squares. The reason why we have to use a parametric model is that, as we know from Appendix B, $C(r)$ has to be a positive definite function in 2D, and such a property has either to be forced through a family of functions that is known to possess it, or by exploiting the condition that the spectrum $S(p)$ has to be positive.

The above short description shows that there are subjective elements in such a classical procedure. But even more important is the following remark.

Remark 6.4 Note that once $C(r)$ has been correctly estimated on a certain plane, to perform predictions of δg at different altitudes we have to propagate $C(r)$ by (6.146), which might be not very easy for a $C(r)$ with a general analytic expression.

For this reason, we add here to the standard list of covariance families in 2D (see Appendix B), two other families that are easily seen to provide valid models for harmonic covariance functions.

The first family is

$$C(r; z+z') = \frac{A}{2\pi} \frac{He^{-2\pi H(z+z')}}{(r^2+(H+z+z')^2)^{3/2}} \quad (r = |\boldsymbol{\xi} - \boldsymbol{\xi}'|). \tag{6.160}$$

In fact we see that

$$C_0(r) = C(r, 0) = \frac{A}{2\pi} \frac{H}{(r^2+H^2)^{3/2}}, \tag{6.161}$$

which is indeed a positive definite function on $\mathbb{R}^2$, since its Fourier transform, namely $e^{-2\pi pH}$, is a non-negative function.

Moreover, if we fix $\boldsymbol{\xi}'$ and z', it is obvious that $C(r, z+z')$ given by (6.160) is harmonic in $(\boldsymbol{\xi}, z)$ in the half space $\Omega_0 \equiv \{z \geq 0\}$. Since C_0 is also symmetric in $(\boldsymbol{\xi}, \boldsymbol{\xi}')$, we have that the same holds in $(\boldsymbol{\xi}, z)$, $(\boldsymbol{\xi}', z')$.

Thus, (6,160) is a valid family of harmonic covariance functions. It is interesting to remark that the function (6.160) can be extended harmonically even below the plane $S_0 \equiv \{z = 0\}$; in fact, it is harmonic down to any $S_{-\bar{h}} \equiv \{z = -\bar{h}\}$, for $\bar{h} < \frac{H}{2}$.

The parameters A and H can be estimated to adjust C_0 to an empirical covariance function. However, the function (6.160) is a smooth, bell-shaped function on $\mathbb{R}^2$, going to zero like r^{-3} at infinity, as such, it is not apt to model oscillating empirical covariance functions. A different family that could be used in this case is

$$C(r; z+z') = A2\pi J_0(2\pi Hr)e^{-2\pi H(z+z')}. \tag{6.162}$$

Note that here H is not a height, but on the contrary, it is a constant with the dimension of the inverse of a length. As before, we can verify that

$$C_0(r) = C(r, 0) = 2\pi AJ_0(2\pi Hr), \tag{6.163}$$

is positive definite on account of the Fourier relation

$$\begin{aligned} &A\int_{R^2} 2\pi J_0(2\pi H r)e^{i\cdot 2\pi p\cdot r}d_2r = \\ &= \frac{A}{H}\int 2\pi H J_0(2\pi H r)2\pi J_0(2\pi pr)tdr = \\ &= \frac{A}{H}\delta(p-H), \end{aligned} \tag{6.164}$$

as already observed in (3.176).

Now $\delta(p-H)$ is not a genuine function, but only a distribution; yet it is a positive distribution, because for any $\varphi(\boldsymbol{p})$ in $\mathscr{D}$, $\varphi(\boldsymbol{p}) \geq 0$, we have

$$\langle\delta(p-H), \varphi(\boldsymbol{p})\rangle = \int_0^{2\pi}\int_0^{+\infty} p\delta(p-H)\varphi(p,\vartheta)dpd\vartheta = H\int_0^{2\pi}\varphi(H,\vartheta)d\vartheta \geq 0.$$

This is strictly analogous to what happens with the $\mathbb{R}^1$ Fourier transform of the function $f(t) = \cos\omega t$, $(\omega > 0)$. The conclusion is that (6.163) is a positive definite function and therefore an admissible covariance function on $\mathbb{R}^2$.

Moreover, recalling that $J_0(t)$ satisfies the Bessel differential equation

$$J_0''(t) + \frac{1}{t}J_0' + J_0(t) = 0, \tag{6.165}$$

it is easy to see that (6.162) is harmonic on Ω_0. In fact, expressing the Laplacian in cylindrical coordinates (see (1.56)) and noting that C does not depend on the anomaly λ, we obtain, using (6.165),

$$\begin{aligned} &\frac{\partial^2}{\partial r^2}C(r\,;z+z') + \frac{2}{r}C(r\,;z+z') + \frac{\partial^2}{\partial z^2}C(r,z+z') = 0 \\ &= 2\pi A\left[\left(J_0''4\pi^2H^2 + \frac{4\pi^2H^2}{2\pi H r}J_0'\right)e^{-2\pi H(z+z')} + 4\pi^2H^2J_0e^{-2\pi H(z+z')}\right] = \\ &= 2\pi A\left[-4\pi^2H^2J_0e^{-2\pi H(z+z')} + 4\pi^2H^2J_0e^{-2\pi H(z+z')}\right] = 0. \end{aligned}$$

Therefore, the function (6.161) is harmonic in Ω_0 with respect to $(\boldsymbol{\xi}, z)$ and, by symmetry, with respect to $(\boldsymbol{\xi}', z')$ too.

We note that (6.162) cannot be continued harmonically below S_0.

The parameters A, H can be used to adapt (6.162) to the empirical covariance function. Let us observe that in this case, we can use for instance H to interpolate the first zero r_1 of the empirical covariance, which has to coincide with the first zero of J_0, $(J_0(2.4) = 0)$, namely we can impose

$$2\pi H r_1 = 2.4\,;$$

it can happen that the second zero r_2 of the empirical covariance used in the model gives a value $J_0(2\pi Hr_2) \neq 0$, because $2\pi Hr_2 \neq 5.5$, $(J_0(5.5) = 0)$. This is indeed because the model (6.161) has only two parameters to adjust it to the empirical covariance. This problem could be alleviated by recalling that a linear combination of functions of the form (6.161), with positive coefficients, is still a positive definite function; thus introducing more (constrained) parameters, it is possible that one could solve the mentioned problem of interpolating the first two empirical zeros.

We conclude by noting that both families (6.160) and (6.162) have a slow decrease for $r \to \infty$; this however is not a major problem in the process of interpolating the empirical covariance, because, as we said, the values of $\hat{C}(r)$ for large r become more and more uncertain, so that only values close to $r = 0$ have to be used in the adjustment.

6.6.2 Covariance Estimation by the Empirical Spectrum

First of all, we have to determine the empirical spectrum. This is done in two steps: the former consists in making a coarse (non-optimal) gridding of the data $\delta g(\xi)$, so that we can easily compute its Fourier transform by the DFT algorithm. This is done in view of the already mentioned robustness of the estimator (6.153) against variations of the covariance model and therefore against the spectrum model too. Once we have a (first) grid of d $\delta\hat{g}(p)$ values in the frequency domain, we can apply the spectral estimator

$$\hat{S}(\bar{p}) \cong \frac{N}{\bar{N}} \sum_{i=1}^{\bar{N}} |\delta\widehat{g}(\boldsymbol{p}_i)|^2, \tag{6.166}$$

where $\boldsymbol{p}_i$ are the grid points falling in the circular crown

$$\bar{p} - \frac{\Delta}{2} \leq p \leq \bar{p} + \frac{\Delta}{2},$$

$\bar{N}$ is the number of $\{\boldsymbol{p}_i\}$ and N is the total number of grid points.

Therefore (6.166) gives an empirical value of $S(p)$ at $\bar{p}$. Repeating the procedure on a sequence of $p_k = k\Delta$ values yields the empirical spectral function $\hat{S}(k\Delta)$.

We notice for later use that $\hat{S}(p_k) > 0$ by definition.

A justification of (6.166) is found in the discrete estimator of the spectral density discussed in Appendix B, where it is shown, in that case for a 1D process, that

$$E\{|\delta\hat{g}(\boldsymbol{p}_i)|^2\} = \frac{1}{N} S(\boldsymbol{p}_i); \tag{6.167}$$

the formula holds in 2D as well.

So (6.62) is nothing but an application of (6.167) to all the points of the grid in the frequency domain that share approximately the same value of the argument $|\boldsymbol{p}|$, where $S(|\boldsymbol{p}|)$ is always the same. Averaging on the circular crown has the effect of reducing the estimation error of the many $|\delta\hat{g}(\boldsymbol{p}_i)|^2$ that have the same mean.

The idea now is to take the finite interval where the spectrum $\hat{S}(p)$ shows a significant amplitude (remember that $\hat{S}(p)$ cannot go beyond the Nyquist frequency $\frac{1}{2\Delta}$), partition it into subintervals I_k,

$$I = \bigcup_{k=1}^{\mathrm{K}} I_k, \tag{6.168}$$

and approximate the spectrum $S(p)$ on each I_k by a constant.

To be more precise, let us notice that in general our grid will be at altitude $\bar{h} > 0$, so we know that the relation between the spectrum $S(p)$ at altitude $\bar{h}$, which is the FT of the covariance (6.150), and the spectrum $S_0(p)$ at $h = 0$, has the form

$$S(p) = e^{-2\pi p 2\bar{h}} S_0(p). \tag{6.169}$$

This is because we have eliminated the influence of all the masses above $\{z = 0\}$ and we know that δg is therefore harmonic in Ω_0.

Accordingly, what we approximate by a piecewise constant function is $S_0(p)$. Namely, using the characteristic functions

$$\chi_k(p) = \begin{cases} 1, & \text{if } p \in I_k, \\ 0, & \text{if } p \notin I_k, \end{cases}$$

we put

$$S_0(p) = \sum_{k=1}^{K} S_{0k}\, \chi_k(p). \tag{6.170}$$

Note that $S_0(p)$ and $S(p)$ are indeed non-negative functions when $S_{0k} \geq 0$; so they provide legitimate spectral estimators under the above condition.

We estimate S_{0k} by imposing a least squares adjustment of $S(p)$, given by (6.168), to the data $\{\widehat{S(p_\ell)}\}$.

The criterion is to minimize the function

$$F\left(S_{00}, \ldots, S_{0K}\right) = \sum_{\ell=1}^{L} \left[\widehat{S(p_\ell)} - e^{-4\pi \bar{h} p_\ell} \sum_{k=0}^{K} S_{0k}\, \chi_k(p_\ell) \right]^2. \tag{6.171}$$

We notice that since each $p_\ell \in I_k$ for a specific k and therefore $\chi_j(p_\ell) = 0$ for $j \neq k$, we can rewrite (6.171) as

$$F\left(S_{00},\ldots,S_{0k}\right)=\sum_{j=1}^{K}\sum_{(P_\ell\in I_j)}[\hat{S}(p_\ell)-e^{-4\pi\bar{h}p_\ell}S_{0j}]^2. \tag{6.172}$$

The minimization is now elementary and comes from the solution of the normal equations

$$-\frac{1}{2}\frac{\partial F}{\partial S_{0k}}=\sum_{p_\ell\in I_k}\hat{S}(p_\ell)e^{-4\pi\bar{h}p_\ell}-\left(\sum_{p_\ell\in I_k}e^{-8\pi\bar{h}p_\ell}\right)S_{0k}=0,$$

namely

$$S_{0k}=\frac{\sum\limits_{p_\ell\in I_k}\hat{S}(p_\ell)e^{-4\pi\bar{h}p_\ell}}{\sum\limits_{p_\ell\in I_k}e^{-8\pi\bar{h}p_\ell}}. \tag{6.173}$$

An important remark is that S_{0k} are always non-negative, as required.

When $\bar{h}=0$ the exponentials in (6.172) become identically 1 and the formula gives S_{0k} as the simple average of $\hat{S}(p_\ell)$ in I_k. We observe that an approximation of the type (6.170) could be applied directly to $S(p)$ even when $\bar{h}>0$, but in this case we would loose a piece of information and we would not be able to predict $\delta g(\boldsymbol{P})$ at any point P below the plane $\{z=\bar{h}\}$.

In this case the covariance at $z=0$, $C_0(r)$, has a simple analytic form that we determine.

We first note that (6.170) can be put in an alternative form.

In fact, if $I_k=(a_{k-1},a_k]$ we can write

$$\chi_k(p)=\Gamma_k(p)-\Gamma_{k-1}(p), \tag{6.174}$$

where

$$\Gamma_k(p)=\begin{cases}1, & \text{if } 0\le p\le a_k,\\ 0, & \text{if } p>a_k.\end{cases} \tag{6.175}$$

Using these functions we can write

$$\begin{aligned}S_0(p)&=S_{01}\chi_1(p)+S_{02}\chi_2(p)+\cdots S_{0k}\chi_j(p)=\\&=S_{01}\Gamma_1(p)+S_{02}[\Gamma_2(p)-\Gamma_1(p)]\cdots+S_{0k}[\Gamma_k(p)-\Gamma_{k-1}(p)]=\\&=(S_{01}-S_{02})\Gamma_1(p)+(S_{02}-S_{03})\Gamma_2(p)+\cdots+S_{0k}\Gamma_k(p)= \qquad (6.176)\\&=\sum_{k=1}^{K}b_k\Gamma_k(p). \qquad (6.177)\end{aligned}$$

With the help of this representation, we have

$$\begin{aligned} C_0(r) &= \sum_{k=1}^{\mathrm{K}} b_k 2\pi \int_0^{\mathrm{K}\Delta} p J_0(2\pi pr)\Gamma_k(p)dp = \\ &= \sum_{k=1}^{K} b_k 2\pi \int_0^{a_k} p J_0(2\pi pr)dp. \end{aligned} \tag{6.178}$$

Recalling the relation

$$x J_0(x) = D_x \left[x J_1(x)\right],$$

already used in the Example 6.4, it is easy to see that (6.177) gives

$$C_0(r) = \sum_{k=1}^{K} b_k a_k \frac{J_1(2\pi a_k r)}{r}. \tag{6.179}$$

The relation (6.179) is easy to handle when $C_0(r)$ only is required. If, on the contrary, one would need to use the covariance $C(r, z+z')$ for arbitrary values of r, z, z', it is easier to resort to a numerical computation by discretizing the inverse FT of $S_0(p)e^{2\pi p(z+z')}$ (see Sect. 3.7).

Remark 6.5 An important characteristic of the stochastic theory is that, beside the predicted value (6.154), it provides also a variance of the estimation error, namely

$$\begin{aligned} \mathscr{E}^2(\boldsymbol{\xi}, z) &= E\{[\delta\hat{g}(\boldsymbol{\xi}, z) - \delta g(\boldsymbol{\xi}, z)]^2\} = \\ &= C(0, 2z) - \sum_{j,k}^{N} C(|\boldsymbol{\xi} - \boldsymbol{\xi}_j|,\ z + z_j) \\ &\{C(\boldsymbol{\xi}_j - \boldsymbol{\xi}_k|,\ z_j + z_k) + \sigma_\nu^2 \delta_{jk}\}^{(-1)} C(|\xi - \xi_k|,\ z + z_k), \end{aligned} \tag{6.180}$$

where

$(\boldsymbol{\xi}, z)$ is the prediction point,

$(\boldsymbol{\xi}_j, z_j)$ is the observation point,

σ_ν^2 is the noise variance.

However, attention should be paid to the fact that (6.180) is the variance of the estimation error of that part of $\delta g_{\mathrm{obs}}(P_i)$ that is in fact represented by the estimated covariance. If the noise ν, with variance σ_ν^2, is due to the non-modelled part of δg more than to the measurement noise, due to factors independent of the non-modelled

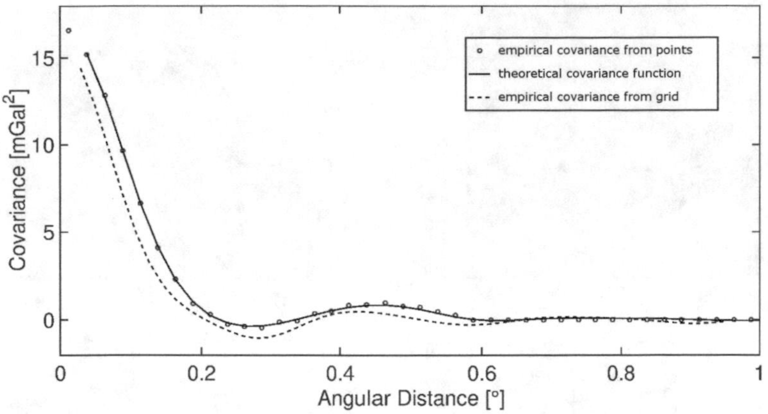

Fig. 6.28 Estimated covariances by means of the empirical covariances method and by means of the empirical spectrum

mass distribution, a more realistic and conservative estimate of the prediction error variance is given by

$$\sigma^2(\text{Pred}) = \mathscr{E}^2(\boldsymbol{\xi}, z) + \sigma_\nu^2. \tag{6.181}$$

Some number between (6.180) and (6.181) is probably the correct figure.

Example 6.6 (Data Gridding) We consider again the real dataset used in Example 6.1. Similarly to what has been done in the Downward Continuation example, the signal has been reduced by removing the effect of a so-called reference field, made by the harmonic synthesis of a GGM and a proper RTC. Here the former was computed using the GECO model [53], while the RTC was reckoned as the difference between a full-resolution and a smoothed digital elevation model, the latter obtained by applying a moving average window of about 20 km to the full-resolution one. The covariance of the reduced signal was computed by means of the two methods described in this section, i.e., from the reduced observations empirical covariance and from the empirical spectrum and are shown in Fig. 6.28.

It can be seen that the two covariances have a quite similar behaviour, with the first zero at around 0.21° (corresponding roughly to 23 km), and a variance of about 17 mGal^2. The main differences between the two covariances are due to the fact that the two methods use slightly different inputs, in fact in order to compute the empirical spectrum the whole survey was rotated so as to make the investigated area as close as possible to a rectangle with sides parallel to the X and Y axes. After that the signal has been (bi-linearly) interpolated on a regular grid (reducing the area of the survey to that of the largest rectangle fitting within the surveyed area) so as to make the use of the DFT possible.

Once the signal covariance has been modelled, it is possible to apply the Least Squares Collocation algorithm in order to grid the observations. For this purpose, the

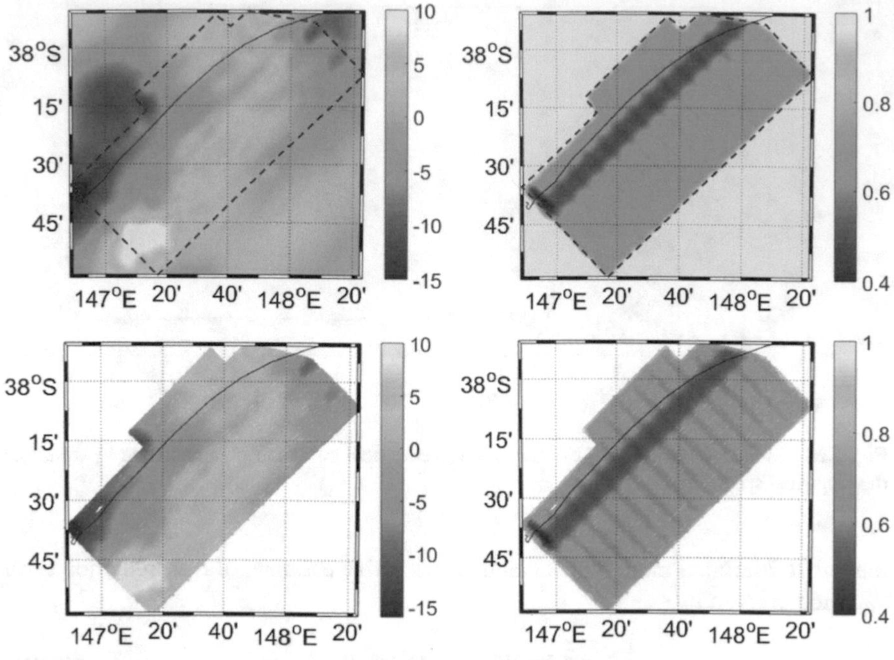

Fig. 6.29 Estimated corrections to the reference signal (left panels) and corresponding accuracies (rights panels) for gridded (top) and along-track points (bottom). Red dotted line represents the contour of the flight area; unit (mGal).

dataset was down-sampled by a factor 10, reducing the number of the observations to about 40,000 points. This is justified by the fact that the initial dataset was sampled at about 25 m and it was corrupted by a noise with approximate correlation length of about 2 km. As a consequence, the contribution to the final estimate of points closer than 250 m can be safely assumed to be limited. The final results in terms of along-track and gridded reduced gravity disturbances, and their predicted errors are shown in Fig. 6.29. It can be seen that the predicted error (referred to the data at flight altitude or to the grid at mean flight level) is of the order of 0.6 mGal, decreasing to only 0.4 mGal when the distance between any two consecutive flight tracks is of 500 m.

6.7 From Grids of Gravity to Grids of Different Functionals

The motivation for this short section lies is that, if we look at the map of gridded values of δg, when this is the residual gravity disturbance that is free of the long wavelength gravity signal and of the topographic masses too, we see a kind of blurred image of the underlying anomalous masses at short wavelength.

This is particularly evident if we use the linearized formula (3.125). There we see that an interface between two layers with depth $z = -(\bar{H} - \delta H)$, in the Fourier domain at height $z = 0$, gives an image which is the FT of δH multiplied by $e^{-2\pi p\bar{H}}$, which leaves almost unaltered the wavelengths of the order of $\bar{H}$ or longer, and damps the wavelengths shorter than $\bar{H}$. One says that $e^{-2\pi p\bar{H}}$ is a low-pass filter and its effect in the geometry domain, namely on $\delta g(\boldsymbol{\xi})$ ($\boldsymbol{\xi} \in S_0$), is to smooth, i.e., to blur, the image of $\delta H(\boldsymbol{\xi})$. Now we borrow from the theory of image analysis [137] a tool used to identify edges in images, namely discontinuities in the field of grey densities that have been previously smoothed to reduce the noise.

The tool is to compute functionals of the smoothed field δg, like

$$|\nabla \delta g(\boldsymbol{\xi})| = v_1(\boldsymbol{\xi}), \tag{6.182}$$

$$-\Delta \delta g(\boldsymbol{\xi}) = v_2(\boldsymbol{\xi})\,; \tag{6.183}$$

here the differential operators have to be interpreted as acting on $\mathbb{R}^2$, namely, in Cartesian coordinates,

$$\nabla = \boldsymbol{e}_1 \frac{\partial}{\partial \xi_1} + \boldsymbol{e}_2 \frac{\partial}{\partial \xi_2}, \tag{6.184}$$

$$\Delta = \frac{\partial^2}{\partial \xi_1^2} + \frac{\partial^2}{\partial \xi_2^2}. \tag{6.185}$$

To justify this choice, we present first of all an elementary example, based on the discussion of Remark 2.5.

Example 6.7 Let us consider a profile of δg along the x-axis on S_0, $\delta g(\boldsymbol{x})$, where the underlying anomalous mass is a semi-slab with the discontinuity just below the origin. The slab is infinite in the y-direction and in the negative x-direction.

Recalling formulas (2.68), (2.63), we know that the corresponding signal in the z-direction is given by

$$\begin{cases} \delta g_Z = -\pi\mu\delta H + \mu F(x), \\ F(x) = x \log \frac{x^2+H_-^2}{x^2+H_+^2} + 2H_- \arctan \frac{x}{H_-} - 2H_+ \arctan \frac{x}{H_+}; \end{cases}$$

here $\mu = G\rho$, as always.

Given the way in which $F(x)$ has been obtained (see (2.62), (2.63) it is obvious that

$$F'(x) = \log \frac{x^2 + H^2}{x^2 + H_+^2}$$

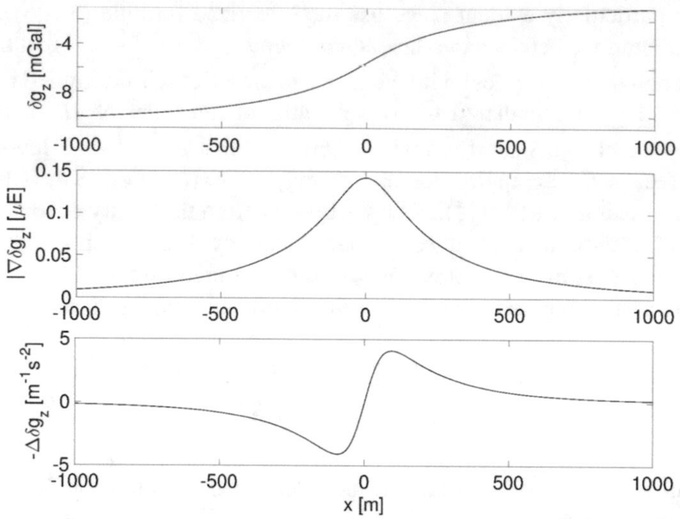

Fig. 6.30 The three functions δg_Z, v_1 and v_2 when $H_+ = 100$ m, $H_- = 500$ m and $\delta\rho = 500$ kgm^{-3}

and

$$F''(x) = \frac{2x}{x^2 + H_-^2} - \frac{2x}{x^2 + H_+^2}.$$

Since δH is constant and F depends only on x, it is obvious that

$$v_1(x) = |\nabla \delta g_Z| = \mu F'(x)$$

and

$$v_2(x) = -\Delta \delta g_Z = -\mu F''(x).$$

The three functions δg_Z, $|\nabla \delta g_Z|$ and $-\Delta \delta g_Z$ are represented in Fig. 6.30. As it appears quite clearly, it is much easier to identify the discontinuity at $z = 0$ in $v_1(x)$ and $v_2(x)$, rather than in δg_Z itself.

Already the above Example 6.7 clarifies that in general discontinuities will have to be sought by identifying large values of $v_1(\boldsymbol{\xi})$ and zeroes of $v_2(\boldsymbol{\xi})$. The problem of computing v_1, v_2 can be solved in two ways, depending also on the approach followed in the gridding plane: namely one can use the grid to calculate v_1, v_2 by discretization as we show in the point a) below or, when the covariance $C_0(r)$ of the gridded values is available, one can use the functional relation that will be shown in point (b) below.

(a) *Calculation by discretization.* The discretization of the derivative implicit in (6.183), (6.184) can be done in different ways, the most straightforward of which is given by the following formulas.

Let us put

$$f_{jk} = \delta g_Z \left(\boldsymbol{\xi}_{\mathrm{jk}}\right), \tag{6.186}$$

where $\boldsymbol{\xi}_{jk}$ is the node (j, k) of the grid, namely

$$\boldsymbol{\xi}_{\mathrm{jk}} = (j\Delta)\boldsymbol{e}_x + (k\Delta)\boldsymbol{e}_y. \tag{6.187}$$

$$j = 0, 1, \dots J, \; k = 0, 1, \dots, K.$$

If $J \times K$ are the dimensions of the grid, then

$$v_1(\boldsymbol{\xi}_{\mathrm{jk}}) = \frac{1}{2\Delta}\{[f_{j+1,k} - f_{j-1,k}]^2 + [f_{j,k+1} - f_{j,k-1}]^2\}^{1/2} \tag{6.188}$$

and

$$v_2(\boldsymbol{\xi}_{\mathrm{jk}}) = -\frac{1}{\Delta^2}\{f_{j+1,k} + f_{j,k+1} + f_{j-1,k} + f_{j,k-1} - 4f_{j,k}\}. \tag{6.189}$$

Indeed, (6.188), (6.189) can only be computed at points (j, k) with $j = 1, \dots, J-1$, $k = 1, \dots, K-1$, leaving out the external contour of the grid.

One important remark is that formulas like (6.188), (6.189) should always be applied when we are sure that $\delta g_Z(\boldsymbol{\xi})$ has been suitably smoothed, so avoiding to compute derivatives of noisy data that have the unpleasant effect of amplifying the noise influence.

Remark 6.6 The above formulas can also be conveniently computed by FT algorithms.

We provide here the continuous forms since, it is clear how they can be discretized by the DFT algorithm.

We have, recalling (3.18),

$$\begin{aligned} v_1(\xi) &= |\mathscr{F}^*\{\mathscr{F}\{\nabla\delta g_{\mathbf{Z}}\}\}| = |\mathscr{F}^*\{-2\pi \boldsymbol{i}\boldsymbol{p}\delta\widehat{g}(\boldsymbol{p})\}| \\ &= 2\pi|\mathscr{F}^*\{\boldsymbol{p}\delta\widehat{g}_Z(\boldsymbol{p})\}| \end{aligned} \tag{6.190}$$

and

$$\begin{aligned} v_2(\boldsymbol{\xi}) &= \mathscr{F}^*\{\mathscr{F}\{-\Delta\delta g_Z\}\} = \\ &= 4\pi^2\mathscr{F}^*\{p^2\delta\widehat{g}_Z\}. \end{aligned} \tag{6.191}$$

For large grids, the application of (6.190), (6.191) is convenient from the numerical point of view, because it reduces the computation time. Note that since the harmonic upward continuation operator is just a multiplication by $e^{-2\pi pz}$, (6.191) can also be interpreted as

$$v_2(\boldsymbol{\xi}) = \left.\frac{\partial^2}{\partial z^2}\delta g(\boldsymbol{\xi}, z)\right|_{z=0} . \tag{6.192}$$

The form (6.192) is also present in geophysical literature, sometimes with the other transform

$$\begin{aligned} v_1'(\boldsymbol{\xi}) &= \left|\frac{\partial}{\partial z}\delta g_Z\ (\boldsymbol{\xi}, z)|_{z=0}\right| = \\ &= 2\pi|\mathscr{F}^*\{p\delta\widehat{g}_Z(p)\}|. \end{aligned} \tag{6.193}$$

Since in general

$$|\mathscr{F}^*\{p\delta\widehat{g}_Z(p)\}| \neq |\mathscr{F}^*\{\boldsymbol{p}\delta\widehat{g}_Z(p)\}|,$$

the two fields v_1' and v_1 bear different information on δg_Z.

(b) *Direct prediction of* v_1, v_2. Let us return to the stochastic predictor (6.153) and notice that when we apply that formula to a given data set $\{\delta g_{\text{obs}}(P_\ell)\}$ it is always convenient to compute first the numerical vector

$$\begin{cases} \boldsymbol{\Lambda} = \{\Lambda_k; k = 1, \dots, N\}, \\ \Lambda_k = \sum_{\ell=1}^N \{C(P_k, P_\ell) + \sigma_\nu^2\delta_{k\ell}\}^{(-1)}\delta g_{\text{obs}}(P_\ell), \end{cases} \tag{6.194}$$

and subsequently the function

$$\begin{cases} \delta\widehat{g}(\boldsymbol{p}) = \boldsymbol{\Lambda}^{\mathrm{T}}\boldsymbol{C}(P), \\ \boldsymbol{C}(P) = \{C(P, P_k)\,; k = 1, \dots, N\}, \end{cases} \tag{6.195}$$

which can be further evaluated at points P coinciding with the grid knots. The form (6.195) is useful for predicting also linear functionals of $\delta\widehat{g}(P)$; for instance, we can put

$$\begin{cases} \nabla\delta\hat{g}(P) = \boldsymbol{\Lambda}^{\mathrm{T}}\nabla_P\boldsymbol{C}(P), \\ \nabla_P\boldsymbol{C}(P) = \{\nabla_P C(P, P_\ell)\}; \end{cases} \tag{6.196}$$

this allow to predict

$$v_1 = |\nabla\delta\hat{g}(P)|$$

at any node of the grid, or, if required, at any point P where $C(P, Q)$ is defined. Similarly, we have

$$-\Delta\delta\hat{g}(P) = \boldsymbol{\Lambda}^{\mathrm{T}}(-\Delta_P \boldsymbol{C}(P)). \tag{6.197}$$

The above formulas can obviously be generalized to any continuous linear functional $\mathscr{L}$ by

$$\begin{aligned} \mathscr{L}(\delta\hat{g}) &= \boldsymbol{\Lambda}^{\mathrm{T}}\mathscr{L}_P[C(P)] \equiv \\ &\equiv \sum_{k=1}^{N} \Lambda_k \mathscr{L}_P C(P, P_k); \end{aligned} \tag{6.198}$$

in this formula $\mathscr{L}_P$ means that the linear functional $\mathscr{L}$ has to be applied to $C(P, P_k)$, as function of P, considering P_ℓ constant. Indeed, similarly to (6.180), one can as well compute a prediction error for $\mathscr{L}(\delta\hat{g})$: by adopting the classical notation

$$C(\mathscr{L}_P, \mathscr{L}_Q) = \mathscr{L}_Q\{\mathscr{L}_P C(P, Q)\},$$

we have

$$\begin{aligned} &\mathscr{E}^2(\mathscr{L}_P(\delta\hat{g})) = C(\mathscr{L}_P, \mathscr{L}_P)+ \\ &-\sum_{\ell,k=1}^{N} C(\mathscr{L}_P, P_\ell)\{C(P_\ell, P_k) + \sigma_\nu^2\delta_{\ell k}\}^{(-1)} C(P_k, \mathscr{L}_P). \end{aligned} \tag{6.199}$$

Remark 6.7 Since we have given a model for the covariance $C(P, Q)$ in the plane S_0, which can be automatically estimated, from the empirical spectrum, namely (see (6.14) for the symbols),

$$\begin{aligned} C_0(P, C) &= C_0\left(r_{PQ}\right) = \\ &= \sum\nolimits_{k=1}^{K} b_k a_k \frac{J_1(2\pi a_k r_{PQ})}{r_{PQ}}, \\ r_{PQ} &= \left|\boldsymbol{\xi}_P - \boldsymbol{\xi}_Q\right|, \end{aligned} \tag{6.200}$$

it might be useful to give an easy rule to compute (6.196), (6.197), which however can be applied only when both P and Q belong to S_0.

In this case in fact we have

$$\begin{aligned} &\nabla_P C_0(P, Q) = \nabla_{\xi_P} C_0\left(r_{PQ}\right) = \\ &= \sum_{k=1}^{K} b_k a_k \frac{\partial}{\partial r} \left.\frac{J_1(2\pi a_k r)}{r}\right|_{r=r_{PQ}} \cdot \nabla_{\xi_P} |\boldsymbol{\xi}_P - \boldsymbol{\xi}_Q| \end{aligned} \tag{6.201}$$

On the other hand, we know that (see Example 6.7)

$$J_1'(x) = J_0(x) - \frac{1}{x} J_1(x),$$

so that

$$\begin{aligned} \frac{\partial}{\partial r} \frac{J_1(2\pi a_k r)}{r} &= \frac{2\pi a_k}{r} J_1'(2\pi a_k r) - \frac{J_1(2\pi a_k r)}{r^2} \\ &= \frac{1}{r} \left(2\pi a_k a J_0\left(2\pi a_k r\right) - \frac{2}{r} J_1\left(2\pi a_k r\right) \right). \end{aligned} \tag{6.202}$$

Moreover,

$$\nabla_{\xi_P} |\xi_P - \xi_Q| = \frac{\xi_P - \xi_Q}{r_{PQ}},$$

so that (6.201) becomes

$$\nabla_{\xi_P} C_0(r_{PQ}) = \sum_{k=1}^{K} a_k b_k \left(2\pi a_k J_0(2\pi a_k r_{PQ}) - \frac{2}{r_{PQ}} J_1(2\pi a_k r_{PQ}) \right) \frac{\xi_P - \xi_Q}{r_{PQ}^2}; \tag{6.203}$$

from (6.203) and (6.196) one can compute v_1.

Furthermore one can write

$$\begin{aligned} \Delta_P \frac{J_1(2\pi a_q r_{PQ})}{r_{PQ}} &= \left(\frac{\partial^2}{\partial r^2} + \frac{1}{r} \frac{\partial}{\partial r} \right) \frac{J_1(2\pi a_k r)}{r} \bigg|_{r = r_{PQ}} \\ &= \frac{(2\pi a_k)^2}{r_{PQ}} J_1''(2\pi a_k r_{PQ}) - \frac{(2\pi a_k)}{r_{PQ}^2} J_1'(2\pi a_k r_{PQ}) \\ &\quad + \frac{1}{r_{PQ}^3} J_1(2\pi a_k r_{PQ}). \end{aligned} \tag{6.204}$$

We can use again the relation

$$J_1'(x) = J_0(x) - \frac{1}{x} J_1(x) \tag{6.205}$$

together with its derivative

$$J_1''(x) = J_0'(x) - \frac{1}{x} J_1'(x) + \frac{1}{x^2} J_1(x),$$

with the other known relation (see [1], Chap. 9)

$$J_0'(x) = -J_1(x),$$

to obtain

$$J_1''(x) = \left(\frac{2}{x^2} - 1\right) J_1(x) - \frac{1}{x} J_0(x). \tag{6.206}$$

Inserting the expression (6.206) and (6.205) we obtain the final formula

$$\begin{aligned} \Delta_P \frac{J_1(2\pi a_k r_{PQ})}{r_{PQ}} &= -\frac{4\pi a_k}{r^2} \quad J_0(2\pi a_k r) + \\ &+ \frac{1}{r}\left[\frac{4}{r^2} - (2\pi a_k)^2\right] J_1(2\pi a_k r)|_{r=r_{PQ}}, \end{aligned} \tag{6.207}$$

so that $C_0(PQ)$ is easily computed and (6.197) too.

6.8 Exercises

Exercise 1 Given a set of three observations of gravity anomalies $\delta g_1 = 3.2$ mGal, $\delta g_2 = 5.4$ mGal and $\delta g_3 = 2.1$ mGal on the same plane at points $P_1 = (0, 0)$, $P_2 = (0, 30)$ and $P_3 = (30, 0)$, respectively, affected by a white noise with standard deviation of $\sigma_\nu = 1$ mGal and assuming the signal is homogeneous and isotropic and characterized by a covariance function $C(d) = 2e^{-d/100}$, where d is the distance, compute the predicted value and its accuracy at $Q = (15, 15)$.

Solution We simply have to apply (6.153) and (6.154). First of all we have to compute the distance d_{ij} between point P_i and P_j for all the possible couples:

$$\begin{bmatrix} d_{11} & d_{12} & d_{13} \\ d_{21} & d_{22} & d_{23} \\ d_{31} & d_{32} & d_{33} \end{bmatrix} = \begin{bmatrix} 0 & 30 & 30 \\ 30 & 0 & 42.4 \\ 30 & 42.4 & 0 \end{bmatrix}.$$

Now knowing the covariance function we can compute the signal covariance matrix $\{C(P_k, P_i)\}$:

$$\{C(P_k, P_i)\} = \begin{bmatrix} 2 & 1.48 & 1.48 \\ 1.48 & 2 & 1.31 \\ 1.48 & 1.31 & 2 \end{bmatrix}.$$

The covariance of the noise will be a diagonal matrix with σ_ν on the main diagonal. In a similar way, but considering the distances between points P_1, P_2 and P_3 and Q we can compute $C(P_i, Q)$:

$$C(P_i, Q) = [1.62\ 1.62\ 1.62].$$

We have now all the elements required to compute the estimate at the point Q applying (6.153), which yields $\delta g_Q = 3.62$ mGal and the estimate of the prediction error $\mathscr{E}_Q = 0.62$ mGal. The numerical solution can be obtained by using the following Matlab function.

```
function [xst,est]=LSC(xp,yp,zp,ni,xq,yq,A,s)
% Matlab code for Exercise 1
% The function computes the LSC interpolation at point (xq,yq) starting
% from the data at point xp, yp, with value zp.
% The LSC suppose a covariance function of the form C(d)=A*exp(-d/s),
% where d is the distance, and white noise with variance ni.
% The function can be used to solve the exercise by putting
% xp = [0 0 30];
% yp = [0 30 0];
% zp = [3.2 5.4 2.1];
% xq =15;
% yq =15;
% A = 2;
% s = 100;

% vectorialize the inputs
xp = xp(:);
yp = yp(:);
zp = zp(:);
xq = xq(:);
yq = yq(:);
% compute distances between observation points
[X,Y]=meshgrid(xp(:),yp(:));
distxy=sqrt((X-X').^2+(Y-Y').^2);
% compute signal covariance matrix
Css = A.*exp(-distxy./s); % compute observation error matrix
Cnini = ni.*diag([ones(size(xp))]); % invert covariance matrix
iCss=inv(Css+Cnini);
% compute prediction
iCss_s=iCss*zp;
for i=1:length(xq)
        disto=sqrt((xp-xq(i)).^2+(yp-yq(i)).^2);
        Csso = A.*exp(-disto'./s);
        xst(i)=Csso*iCss_s;
        est(i)=sqrt(abs(Css(1,1)-Csso*iCss*Csso'));
end
end
```

Exercise 2 Write a code to move upward a grid of gravity anomalies by applying the Poisson Kernel (6.114).

Solution The Matlab function is reported in the following box.

```
function [cont_field]=Poisson(x,y,dg,z1,z2,fourier)
% function to compute the upward continuation of a gravity field grid from
% height z1 to height z2 by means of the Poisson Integral.
% input are:
% x,y coordinates of the grid in meters;
% dg gravity singal to be continued in mGal;
% z1 starting height;
% z2 ending height;
% flag to use fourier solution or dicretized integral solution;

% vectorialize the inputs
x = x(:);
y = y(:);
% define the spacing in Y direction
dx = abs(x(1)-x(2));
dy = abs(y(1)-y(2));
% Build the grid
[X,Y] = meshgrid(x,y);
% Compute the height difference
deltaZ = z2-z1;
% initialize the output signal
cont_field = zeros(size(dg));
if fourier==0 %upward continuation using discrete integral
    for i = 1:length(x)
        for j = 1:length(y)
            l = ((X-X(j,i)).^2+(Y-Y(j,i)).^2+(deltaZ).^2).^(3/2);
            cont_field(j,i) = sum(sum(dg.*deltaZ./l));
        end
    end
    cont_field = cont_field.*dx.*dy./2./pi;
elseif fourier==1 %upward continuation using Fourier convolution
    kernel = deltaZ./((X-X(1,1)).^2+(Y-Y(1,1)).^2+(deltaZ).^2).^(3/2);
    kernel = CompleteKernel( kernel );
    cont_field = conv2(dg,kernel,'same')./2./pi.*dx.*dy;
end
end


function [ CompleteKernel ] = CompleteKernel( I_Matrix ) % compute the
complete kernel given the kernel in the fouth quadrant
 FourthQuadrant = I_Matrix(2:end, 2:end); R = I_Matrix(1, 2:end); L =
I_Matrix(2:end, 1); FirstQuadrant = flipud(FourthQuadrant); ThirdQuadrant
= fliplr(FourthQuadrant); SecondQuadrant = fliplr(FirstQuadrant); Com-
pleteKernel = [ SecondQuadrant flipud(L) FirstQuadrant; ... [ fliplr(R);
ThirdQuadrant] I_Matrix]; end
```

•

Part III
Inverse Theory and Applications

Chapter 7
Elementary Inverse Theory

7.1 Outline of the Chapter

In this chapter, as well as in the next one, we move from where we left at the end of Part II, namely we shall assume to have on a plane, that we agree to take at zero height, a grid of gravity anomalies $\{\delta g_Z(P_{ik})\}$ which, thanks to all the processing procedures, reflects only the effects of anomalous masses, present in a layer of maximum depth $\sim$ 30 km, below the grid itself.

An important point to understand is that we assume the grid area to be wide enough to capture all the essential features of the anomaly of interest, while it is not influenced by other mass anomalies, as discussed in Sect. 6.3.

Accordingly, we shall assume that outside the data window δg_Z decays smoothly to zero in a "natural way". Then, as customary in the book, we shall consider, at our convenience, as data a continuous data set on the whole plane, of which the grid constitutes a finite sample.

At first we will discuss what information on this anomalous mass distribution can be exactly derived from the data: such are for instance the global amount of the anomalous masses, M, and the position of their barycenter or, in the case of a two-layers configuration, the mean depth of the layers, interface.

Equally important are exact inequalities that can be derived from Newton's law, particularly upper bounds (UB) for the minimum depth of anomalous bodies; this in fact tells us that if we want to dig to reach the anomalous mass, our excavation or drilling will have to be less or at most equal to UB.

Finally, connecting to the theory developed in Chap. 2, we shall see how to determine geometrical parameters of elementary bodies (sphere, ellipsoid, cylinder, prism) or some combination of them assumed as the source of the anomaly analysed.

F. Sansò, D. Sampietro, *Analysis of the Gravity Field*, Lecture Notes in Geosystems Mathematics and Computing, https://doi.org/10.1007/978-3-030-74353-6_7

7.2 Determination of the Anomalous Mass M

We want to determine the mass anomaly M that generates a given gravity anomaly $\delta g_Z(\boldsymbol{x})$ on the plane S_0; a classical reference on this and other items treated in the next paragraphs is Grant and West [59].

We refer to Fig. 7.1 for notation. Note that we do not make any assumption on the mass distribution in B, so that our results will have a general validity.

We start again from Newton's integral

$$\delta g_Z(\boldsymbol{x}) = G \int_B \frac{\zeta}{\ell_{PQ}^3} \rho(Q) dB_Q, \tag{7.1}$$

$$\ell_{PQ} = \sqrt{r^2 + \zeta^2}, \quad r^2 = |\boldsymbol{x} - \boldsymbol{\xi}|^2, \tag{7.2}$$

assuming that $\delta g_Z(\boldsymbol{x})$ is given on the horizontal plane S_0.

As a first step, we compute the integral

$$I = \int_{S_0} \delta g_Z(P) dS_P. \tag{7.3}$$

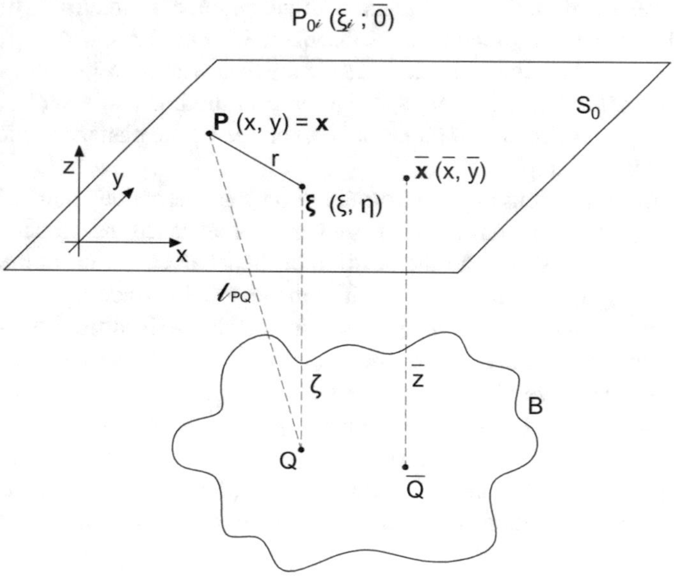

Fig. 7.1 The anomalous body B; $\overline{Q}$ is the barycenter of B, Q the running point in B, P the data point on $S_0 \equiv \{z = 0\}$

Substituting (7.1) in (7.3) we get

$$I = G \int_B \rho(Q) \int_{S_0} \frac{\zeta}{(r^2 + \zeta^2)^{3/2}} dS dB_Q. \tag{7.4}$$

Now switching to polar coordinates on S_0 centered at P, so that $dS = r\,dr\,d\alpha$, we get

$$I = G \int_B \rho(Q) 2\pi \int_0^{+\infty} \frac{\zeta r dr}{(r^2 + \zeta^2)^{3/2}} dB_Q. \tag{7.5}$$

Assuming that $\zeta < 0$ and substituting

$$r^2 = \zeta^2 t,$$

it is easy to verify that

$$\int_0^{+\infty} \frac{\zeta\, r dr}{\left(r^2 + \zeta^2\right)^{3/2}} = -1,$$

so that (7.5) yields

$$I = -2\pi G \int \rho(Q) dB_q = -2\pi G M. \tag{7.6}$$

Therefore, the anomalous mass M is given by

$$M = -\frac{1}{2\pi G} \int_{S_0} \delta g_Z(P) dS_p. \tag{7.7}$$

It is not difficult to recognize that (7.6) is just another way of expressing the Gauss theorem applied to the gravity field, namely (1.66), when the total surface S enclosing the mass M consists of the horizontal plane and a hemisphere in the lower half-space $\{z < 0\}$, with a radius $R \to \infty$.

Remark 7.1 Since many times it is convenient to use the Fourier Transform of the data $\delta \widehat{g}_z(\boldsymbol{p})$ to express the sought quantities, in the present case it is obvious that, in view of definition (3.78), one has

$$M = -\frac{1}{2\pi G} \delta \widehat{g}_Z(0). \tag{7.8}$$

7.3 Determination of the Barycenter $\bar{Q}$, Horizontal Coordinates

Always referring to Fig. 7.1, we would like to determine $\bar{\mathbf{x}}$; to this aim we compute the integral

$$J = \int_{S_0} \boldsymbol{x}_P \delta g_Z(P) dS_P. \tag{7.9}$$

Substituting here the expression (7.1) and reordering the integrals, we get

$$\begin{aligned} J &= G \int_B \rho(Q)\zeta \int_{S_0} \frac{\boldsymbol{x}_P}{\ell_{PQ}^3} dS_P dB_Q = \\ &= G \int_B \rho(Q)\zeta \left[\int_{S_0} \frac{\boldsymbol{x} - \boldsymbol{\xi}}{\ell_{PQ}^3} dS_P + \boldsymbol{\xi} \int_{S_0} \frac{dS_P}{\ell_{PQ}^3} \right] dB_Q. \end{aligned} \tag{7.10}$$

Considering the two integrals in square brackets we observe that, keeping $\boldsymbol{\xi}$ as a fixed vector, $\frac{\boldsymbol{x}-\boldsymbol{\xi}}{\ell_{PQ}^3}$ is an odd function of $(\boldsymbol{x} - \boldsymbol{\xi})$ so that its integral on S_0 vanishes. As for the second integral, fixing the origin of a polar coordinate system on S_0 at $\boldsymbol{\xi}$, we find, as we have seen in Sect. 7.2,

$$\zeta \int_{S_0} \frac{dS}{\ell_{PQ}^3} = 2\pi \int_0^{+\infty} \frac{\zeta r dr}{(r^2 + \zeta^2)^{3/2}} = -2\pi.$$

Therefore, returning to (7.10),

$$J = -2\pi G \int_B \rho(Q) \boldsymbol{\xi}_Q dB_Q. \tag{7.11}$$

On the other hand, by definition of barycenter,

$$\bar{\boldsymbol{x}} = \frac{1}{M} \int_B \boldsymbol{\xi}_Q \rho(Q) dB_Q,$$

so that (7.11) becomes

$$\bar{\boldsymbol{x}} = -\frac{1}{2\pi GM} \int_{S_0} \boldsymbol{x}_P \delta g_Z(P) dS_P. \tag{7.12}$$

Finally, recalling (7.7), we see that (7.12) can also be written as

$$\bar{\boldsymbol{x}} = \frac{\int_{S_0} \boldsymbol{x}_P \delta g_Z(P) dS_p}{\int_{S_0} \delta g_Z(P) dS_p}. \tag{7.13}$$

Remark 7.2 Willing to express the above formulas by means of the Fourier Transform $\delta \widehat{g}_z(\boldsymbol{P})$, we first note that from the definition (3.78) it follows that

$$\nabla_P \delta \widehat{g}_Z(0) = 2\pi i \int_{S_0} \boldsymbol{x}_P \delta g_Z(P) dS_P. \tag{7.14}$$

Therefore, (7.12) becomes

$$\bar{\boldsymbol{x}} = -\frac{1}{4\pi^2 i GM} \nabla_P \delta g_Z(0). \tag{7.15}$$

Finally, by using (7.8), we have

$$\bar{\boldsymbol{x}} = \frac{1}{2\pi i} \frac{\nabla_P \delta \widehat{g}_Z(0)}{\delta \widehat{g}_Z(0)}. \tag{7.16}$$

7.4 Determination of the Depth of the Barycenter $\bar{Q}$

We want to determine the depth $\bar{Z}$ of $\bar{Q}$.

In this case it is convenient to go directly through the Fourier representation of $\delta g_Z(\boldsymbol{x})$.

Recalling (3.103), (3.104), (3.105), and since $\zeta < 0$, we know that

$$\mathscr{F}\left\{\frac{\zeta}{(r^2+\zeta^2)^{3/2}}\right\} = -2\pi e^{-2\pi(p|\zeta| - i\boldsymbol{p}\cdot\boldsymbol{\xi})} \quad (\zeta < 0), \tag{7.17}$$

where $\boldsymbol{x}$ is transformed into the frequency vector $\boldsymbol{p}$. If we call ϑ the angular anomaly of $\boldsymbol{\xi}$ in the plane S_0, and α the angular anomaly of $\boldsymbol{p}$, we write (7.11) in the form

$$\mathscr{F}\left\{\frac{\zeta}{\left(r^2+\zeta^2\right)^{3/2}}\right\} = -2\pi e^{-2\pi p(|\zeta| - is\cos(\vartheta-\alpha))}, \tag{7.18}$$

where $s = |\boldsymbol{\xi}|$. Accordingly, the Fourier Transform of (7.1) becomes

$$\delta \hat{g}_Z(\boldsymbol{p}) = -2\pi G \int_B e^{-2\pi p(|\zeta| - is\cos(\vartheta-\alpha))} \rho(Q) dB_Q. \tag{7.19}$$

We now consider $\delta\widehat{g}_Z$ in polar coordinates on the plane, namely

$$\delta\widehat{g}_Z(\boldsymbol{p}) = \delta\widehat{g}_Z(p, \alpha),$$

and proceed by computing the radial derivatives $\frac{\partial}{\partial p}$ at the origin, in direction α. So we put

$$\frac{\partial}{\partial p}\delta\widehat{g}_Z(\boldsymbol{p})\bigg|_{p=0} = \frac{\partial}{\partial p}\delta\widehat{g}_Z(0, \alpha).$$

From (7.19) we have

$$\frac{\partial}{\partial p}\delta\widehat{g}_Z(0, \alpha) = 4\pi^2 G \int_B (|\zeta| - is\cos(\vartheta - \alpha))\,\rho(Q)dB_Q. \tag{7.20}$$

As we see, in general such a function depends on α, meaning that the shape of $\delta\widehat{g}_Z(\boldsymbol{p})$ close to $\boldsymbol{p} = 0$ is that of a cusp, with different radial derivatives in different directions.

Now taking the angular average of (7.20) and taking into account that

$$\frac{1}{2\pi}\int_0^{2\pi} \cos(\vartheta - \alpha)\,d\alpha = 0,$$

we find

$$\begin{aligned} \frac{1}{2\pi}\int_0^{2\pi} \frac{\partial}{\partial p}\delta\widehat{g}_Z(0, \alpha)d\alpha &= 4\pi^2 G \int_B |\zeta|\,\rho(Q)dB_Q = \\ &= -4\pi^2 G \int \zeta\rho(Q)dB_Q = -4\pi^2 GM\bar{Z}. \end{aligned} \tag{7.21}$$

This formula solves our problem, i.e.,

$$\bar{Z} = -\frac{1}{8\pi^3 GM}\int_0^{2\pi} \frac{\partial}{\partial p}\delta\hat{g}_Z(0, \alpha)d\alpha. \tag{7.22}$$

Once more, recalling (7.8), one can write as well

$$\bar{Z} = \frac{1}{4\pi}\frac{\int_0^{2\pi^2} \frac{\partial}{\partial p}\delta\hat{g}_Z(0, \alpha)d\alpha}{\delta\hat{g}_Z(0)}. \tag{7.23}$$

Notice that, contrary to the derivative in the numerator, the denominator $\delta\hat{g}_Z(\boldsymbol{p})\big|_{p=0}$, does not depend on α.

7.5 Determination of the Mean Depth of the Interface Between Two Layers

We apply a reasoning quite similar to that in Sect. 7.3, to obtain an estimate of the mean depth of the interface between two layers.

We assume to be in the same situation as that described in Sect. 3.6 (see Fig. 3.4) and we base our derivation on the linearized theory expressed in formula (3.125), namely

$$\delta\widehat{g}_Z(\boldsymbol{p}) = \delta\widehat{g}_Z(p,\alpha) = 2\pi\mu e^{-2\pi p\bar{H}}\delta\widehat{H}(p,\alpha), \tag{7.24}$$

with $\mu = G\rho$ and $\rho = \rho_1 - \rho_2$, i.e., the density contrast of the two layers.

Before continuing, we propose a formula already seen in Sect. 7.3. Namely, let

$$\begin{aligned}\delta\widehat{H}(p,\alpha) &= \int e^{i2\pi\boldsymbol{p}\cdot\boldsymbol{\xi}}\delta H(\boldsymbol{\xi})d_2\xi = \\ &= \int e^{i2\pi ps\cos(\vartheta-\alpha)}\delta H(\boldsymbol{\xi})d_2\xi.\end{aligned} \tag{7.25}$$

Then we have

$$\int_0^{2\pi}\frac{\partial}{\partial p}\delta\widehat{H}(0,\alpha)d\alpha = i2\pi\int\delta H(\boldsymbol{\xi})s\left(\int_0^{2\pi}\cos(\vartheta-\alpha)\,d\alpha\right)d_2\xi = 0. \tag{7.26}$$

Such a conclusion holds if

$$\int|\delta H(\boldsymbol{\xi})|\,|\boldsymbol{\xi}|d_2\xi < +\infty, \tag{7.27}$$

what we definitely assume to be true; so we can say that relation (7.26) is a general relation for the Fourier Transform of a function satisfying the condition (7.27). After this statement we return to (7.24) and compute

$$\begin{aligned}&\frac{1}{2\pi}\int_0^{2\pi}\frac{\partial}{\partial p}\delta\widehat{g}_Z(0,\alpha)d\alpha = -2\pi\bar{H}\mu\int_0^{2\pi}\delta\widehat{H}(0,\alpha)d\alpha = \\ &= -4\pi^2\bar{H}\mu\delta\widehat{H}(0),\end{aligned} \tag{7.28}$$

because $\delta\widehat{H}(0)$ does not depend on α.

On the other hand,

$$\delta\widehat{H}(0,\alpha) = \delta\widehat{H}(0) = \int\delta H(\boldsymbol{\xi})d_2\boldsymbol{\xi}.$$

Hence, returning to (7.24) we have

$$\delta\widehat{g}_Z(0) = 2\pi\mu\delta\widehat{H}(0),$$

which in conjunction with (7.28) yields

$$\frac{1}{2\pi}\int_0^{2\pi}\frac{\partial}{\partial p}\delta\widehat{g}_Z(0,\alpha)d\alpha = -2\pi\bar{H}\delta\widehat{g}_Z(0).$$

This solves the problem of expressing $\bar{H}$ in terms of the known function $\delta g_Z(\boldsymbol{\xi})$, namely

$$\bar{H} = -\frac{1}{4\pi^2}\frac{\int_0^{2\pi}\frac{\partial}{\partial p}\delta\widehat{g}_Z(0,\alpha)d\alpha}{\delta\widehat{g}_Z(0)}. \tag{7.29}$$

Although for the present configuration of masses it would be improper to speak of barycenter, we note the formal similarity of (7.29) with (7.23); so similar formulas are giving $\bar{Z}$ or $\bar{H}$ depending whether we assume that the underlying distribution is that of a bulky body or that of two layers. Note that in (7.29) we expect $\bar{H}$ to be positive, while in (7.22) we expect that $\bar{Z} < 0$.

Remark 7.3 Since often profiles of gravity anomalies are used in exploration geophysics, we deem it useful to give the analogue of formula (7.29) for two layers in a 2D configuration.

We start from formula (3.131), namely

$$\delta\widehat{g}_Z(p) = 2\pi\mu e^{-2\pi|p|\bar{H}}\delta\widehat{H}(p), \tag{7.30}$$

where $|p|$ is just the modulus of the scalar variable p.

We first recall that

$$\frac{\partial}{\partial p}\delta\widehat{H}(p) = \int_{-\infty}^{+\infty} 2\pi i\xi e^{i2\pi p\xi}\delta H(\xi)d\xi \tag{7.31}$$

so that, if we assume that

$$\int_{-\infty}^{+\infty}|\xi|\delta H(\xi)d\xi < +\infty,$$

(7.31) provides a continuous function of p, in particular at $p = 0$.

Then we proceed by differentiating (7.30), recalling that

$$\frac{\partial}{\partial p}|p| = \text{sign}(p) = \begin{cases} +1, & \text{if } p > 0, \\ -1, & \text{if } p < 0; \end{cases}$$

we obtain

$$\begin{aligned}\delta\widehat{g'}_Z(p) &= -4\pi^2\mu\overline{H}\text{sign}(p)e^{-2\pi|p|\overline{H}}\delta\widehat{H}(p)+\\ &+ 2\pi\mu e^{-2\pi|p|\bar{H}}\delta\widehat{H'}(p).\end{aligned} \tag{7.32}$$

As we see, this derivative has a jump at the origin, due to the first term on the right-hand side, while the second is continuous, thanks to the previous remark. Therefore, if we compute the jump of $\delta\widehat{g'}_z$ across 0, we get

$$\delta\widehat{g'}_Z(0_+) - \delta\widehat{g'}_Z(0_-) = -8\pi^2\bar{H}\delta\widehat{H}(0). \tag{7.33}$$

Since

$$\delta\widehat{g}_Z(0) = 2\pi\mu\delta\widehat{H}(0),$$

(7.33) can be written in the form

$$\delta\widehat{g'}_Z(0_+) - \delta\widehat{g'}_Z(0_-) = -4\pi\bar{H}\delta\widehat{g}_Z(0). \tag{7.34}$$

which solves the problem of computing $\bar{H}$ from $\delta\widehat{g}_Z(p)$.

7.6 Upper Bounds for the Minimum Depth of a Mass Anomaly

We have already mentioned that in exploration geophysics it is sometimes important to have an upper bound (UB) for the depth (d) of the top of a mass anomaly (see Fig. 7.2).

Formally,

$$-d = \sup_{Q\subset B} z_Q \tag{7.35}$$

and UB is such that

$$0 < d \le \text{UB} \tag{7.36}$$

First of all we observes that the exact relations in Sects. 7.1, 7.2, 7.3 hold irrespectively of the behaviour of the density function $\rho(Q)$. In this section though, we shall assume that $\rho(Q)$ has always the same sign in B. To fix that ideas, we shall assume

$$\rho(Q) \ge 0,$$

so that we expect that $\delta g_Z < 0$ on S.

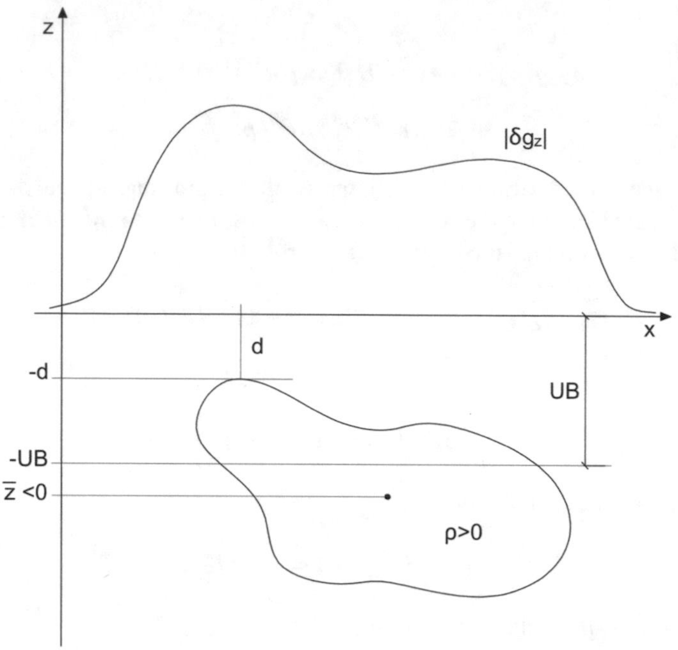

Fig. 7.2 The configuration of a positive mass anomaly and the definition of d, UB

Please notice that the hypothesis of positivity of ρ is a real assumption and not a statement that we can deduce from the sign of δg_Z as the following counterexample shows. Imagine you have a small uniform sphere of mass $-m$, buried below S_0, surrounded by a buried uniform spherical shell, with the same center, bearing a mass $M > m$. Then indeed on S_0 we see the same signal δg_Z as that of a mass $M - m > 0$, buried at the center of the spheres, although the distribution had a variable sign.

A second remark is that, to be useful, an upper bound UB must be smaller than $|\bar{Z}|$, because by the definition of the barycenter one always has

$$d < |\bar{Z}| .$$

In order to avoid reasoning on signs, we start by writing the Newton law in the form

$$|\delta g_Z(P)| = G \int_B \frac{|\zeta|}{\left(r^2 + \zeta^2\right)^{3/2}} \rho(Q) dB_Q, \tag{7.37}$$

where all relevant quantities are positive. We stress that in this case we have to assume that δg_Z has a constant sign on the plane.

We start noting that

$$|\zeta| < \left(r^2 + \zeta^2\right)^{1/2} \quad \forall P, Q \in S,$$

so that

$$\frac{|\zeta|}{\left(r^2 + \zeta^2\right)^{3/2}} \leq \frac{|\zeta|}{|\zeta|^3} = \frac{1}{\zeta^2}.$$

Since for any point in B

$$\frac{1}{|\zeta|} < \frac{1}{d},$$

implies that (7.37)

$$|\delta g_Z(P)| \leq G \int_B \frac{1}{d^2} \rho(Q) d B_Q = \frac{MG}{d^2}; \tag{7.38}$$

here, given our hypotheses, $M > 0$.

The relation (7.38) gives

$$d \leq \sqrt{\frac{MG}{|\delta g_Z(P)|}},$$

for any $P \in S_0$.

We note that the information given by UB is more valuable the smaller this bound is; since (7.39) has to hold for any P, we can conveniently choose

$$\mathrm{UB} = \sqrt{\frac{MG}{|\delta g_Z(P)|_{\max}}}. \tag{7.39}$$

In this formula, the term MG can be computed according to (7.7), which in this case reads

$$MG = \frac{1}{2\pi} \int_{S_0} |\delta g_Z| dS. \tag{7.40}$$

We provide a guide (Exercise 3) to show that, for a homogeneous cylinder, (7.39) can improve on the plane estimate $\mathrm{UB} = -\bar{Z}$. We note too that the example of a sphere would not work because any homogenous sphere of given mass and center has the same δg_Z on the plane S_0, therefore we cannot improve on the evaluation $\mathrm{UB} = -\bar{Z}$.

Remark 7.4 Several other rules can be found to give values of UB based on $|\delta g_z|$ and on some of its functionals, like $|\nabla \delta g_Z|$ or $|\Delta \delta g_Z|$, derived as discussed in Sect. 6.7.

For instance, from (7.27), written without modulus, we first deduce that

$$\nabla_P \delta g_Z = G \int_B \zeta \nabla_P \frac{1}{(r^2 + \zeta^2)^{3/2}} \rho(Q) dB =$$
$$= -3G \int_B \frac{\zeta r}{(r^2 + \zeta^2)^{5/2}} \frac{\boldsymbol{x}_P - \boldsymbol{x}_Q}{r} \rho(Q) dB,$$

so that, taking the modulus of vectors and considering that $\frac{|\boldsymbol{x}_P - \boldsymbol{x}_Q|}{r} = 1$, we get

$$(\rho \geq 0), \quad |\nabla_P \delta g_Z| \leq 3G \int_B \frac{|\zeta| \, r}{\left(r^2 + \zeta^2\right)^{5/2}} \rho(Q) dB. \tag{7.41}$$

On the other hand, the following inequalities are obvious, using the notation $t = \frac{r}{|\zeta|}$:

$$\frac{r}{(r^2 + \zeta^2)} = \frac{1}{|\zeta|} \frac{t}{t^2 + 1} \leq \frac{1}{2\,|\zeta|} \leq \frac{1}{2d}.$$

Hence, returning to (7.41), we get

$$\begin{aligned} |\nabla \delta g_Z| &\leq 1.5 \frac{G}{d} \int_B \frac{|\zeta|}{(r^2 + \zeta^2)^{3/2}} \rho(\Omega) dB = \\ &= 1.5 \frac{|\delta g_Z|}{d}. \end{aligned} \tag{7.42}$$

We reverse this inequality to get

$$d < 1.5 \frac{|\delta g_Z(P)|}{|\nabla \delta g_Z(P)|}, \tag{7.43}$$

and since this is true for any P in the plane S_0, one can even minimize the right-hand side. However, since we expect the anomaly (δg_Z) to go to zero outside the area of major influence, and since the evaluation of δg_Z is more uncertain on its tails, also due to the presence of other anomalies, a reasonable compromise is to look for the point $\bar{P}'$ where $|\nabla \delta g_Z|$ attains its maximum, i.e.

$$\left| \nabla \delta g_Z(\bar{P}') \right| = |\nabla \delta g_Z(P)|_{\max} ;$$

then one can take

$$d \leq 1.5 \frac{|\delta g_Z(\bar{P}')|}{|\nabla \delta g_Z(\bar{P}')|} = \text{UB}. \tag{7.44}$$

The above reasoning can be refined, with a more involved procedure (see [16]), to get

$$d \leq \frac{48\sqrt{5}}{125} \frac{|\delta g_Z|_{\max}}{|\nabla \delta g_Z|_{\max}} = \text{UB} \tag{7.45}$$

where now the numerator and the denominator are computed at different points. Indeed, in general $|\delta g_Z|_{\max} \geq |\delta g_Z(\bar{P}')|$, but on the other hand $\frac{48\sqrt{5}}{125} \cong 0.86$, so that one cannot immediately say when (7.45) is better than (7.44).

A similar reasoning can be applied to a 2D mass distribution, $\rho = \rho(\xi, \zeta)$, namely to Newton's law

$$\begin{aligned} \delta g_Z &= G \int_V \int_{-\infty}^{+\infty} \frac{\zeta}{[(x-\xi)^2 + \zeta^2 + \eta^2]^{3/2}} \rho(\xi, \zeta) d\eta dV = \\ &= 2G \int_V \frac{\zeta}{(x-\xi)^2 + \zeta^2} \rho(\xi, \zeta) dV, \end{aligned} \tag{7.46}$$

where V denotes the intersection of the body, elongated in the Y-direction, with the vertical plane (x, y).

Starting from $\delta g_Z'(x)$ and repeating the above reasoning, one gets

$$d \leq \frac{|\delta g_Z(x)|}{|\delta g_Z'(x)|}, \quad \forall x. \tag{7.47}$$

Here also we can find a refined estimation [16], namely

$$d \leq \frac{3\sqrt{3}}{8} \frac{|\delta g_Z|_{\max}}{|\delta g_Z'|_{\max}}. \tag{7.48}$$

Another inequality involves $\Delta_P \delta g_Z$. In this case we first compute

$$\Delta_P \delta g_Z(P) = G \int_B \Delta_P \frac{\zeta}{(r^2 + \zeta^2)^{3/2}} \rho(Q) dB_Q.$$

Namely, observing that here Δ_P is the 2D Laplacian on S_0, we can take advantage of the polar form of the operator in the plane, without any effect of the angular part because $\frac{\zeta}{(r^2+\zeta^2)^{3/2}}$ does not depend on it:

$$\Delta_P = \frac{\partial^2}{\partial r^2} + \frac{1}{r}\frac{\partial}{\partial r} + \frac{1}{r^2}\frac{\partial}{\partial \alpha^2};$$

We get then

$$\Delta\delta g_Z(P) = -3G\int_B \frac{\zeta^2 - 3r^2}{(r^2+\zeta^2)^{7/2}}\zeta\rho(Q)dB_Q. \tag{7.49}$$

Passing to the modulus we have

$$|\Delta\delta g_Z(P)| \leq 3G\int_B \frac{|\zeta^2 - 3r^2|}{(r^2+\zeta^2)^{7/2}}|\zeta|\,\rho(Q)dB_Q. \tag{7.50}$$

Now, setting as before $t = \frac{r}{|\zeta|}$, we establish the inequality

$$\frac{|2\zeta^2 - 3t^2|}{(r^2+\zeta^2)^2} = \frac{1}{\zeta^2}\frac{|2-3t^2|}{(t^2+1)^2} \leq \frac{2}{d^2}. \tag{7.51}$$

The last step in (7.51) is due to the shape of

$$f(t) = \frac{|2-3t^2|}{(t^2+1)^2}$$

in $t > 0$, where the inequality

$$0 < f(t) \leq f(0) = 2$$

is easily verified.

Using (7.51) in (7.50) we find

$$|\Delta\delta g_Z(P)| \leq \frac{6G}{d^2}\int_B \frac{|\zeta|}{(r^2+\zeta^2)^{3/2}}\rho(Q)dB = \frac{6}{d^2}\,|\delta g_Z(P)|\,, \tag{7.52}$$

or

$$d \leq 2.45\sqrt{\frac{|\delta g_Z(P)|}{|\Delta\delta g_Z(P)|}}; \tag{7.53}$$

this inequality has to hold for all $P \in S_0$, and it can be optimized, e.g., by choosing the maximum of $|\Delta\delta g_Z|$.

There is no analogue in [16], where only directional second derivatives instead of the Laplacian, are studied.

Finally, if we go to the 2D case, expressed by formula (7.46), and repeat the same reasoning, we get

$$d \leq \sqrt{2}\sqrt{\frac{|\delta g_Z(x)|}{|\delta g_Z''(x)|}}, \quad \forall x. \tag{7.54}$$

This is, as a matter of fact, the same relation found in [16].

A comparison of the various inequalities is performed in the Exercise 4 at the end of the chapter.

7.7 Inversion by Simple Bodies

Sometimes it is useful to try to interpret an isolated gravity anomaly in terms of a body with a given simple shape, like for example a sphere, a prism or a cylinder.

We already know from Sects. 7.1, 7.2, 7.3 how to determine the mass and barycenter of such bodies; however, given the elementary nature of the bodies, we pursue here an approach that can give us answers in a very simplified manner. So we proceed examining the three examples.

Example 7.1 (The Sphere) First of all, we notice that we can reasonably try to interpret an anomaly as produced by a spherical body, only if the pattern of gravity on S_0 is that of a "hill" with cylindrical symmetry around its top (or the negative of that, depending on the sign of the density ρ). To be comfortable in the further computations, we can choose our coordinate system with the origin at the maximum (or minimum) of $\delta g_Z(\boldsymbol{x})$ (see Fig. 7.3).

Educated by the result in Sect. 7.2 we know that the barycenter of the sphere, namely its geometrical center, has to be buried at some depth $z = -H$ on the Z-axis. We know from Chap. 2 that in this case $\delta g_Z(\boldsymbol{x})$ is the same as that of a point mass placed at $z = -H$, $x = 0$, $y = 0$, and more precisely (see (2.9)) that

$$\delta g_Z(\boldsymbol{x}) = \frac{G(-M)H}{(r^2 + H^2)^{3/2}} \qquad (r = |\boldsymbol{x}|). \tag{7.55}$$

Indeed, when we have a number of observations $\delta g_Z(\boldsymbol{x}_k)$ $(k = 1, \dots, N)$, the usual approach to estimate the source is an application of standard least squares to (7.55) to obtain the unknown parameters $(H, -M)$. Yet we give here a quick recipe to estimate them, so that, in case, such coarse values can be used as a starting guess to linearize (7.55) and then apply the least squares method.

In fact, assume you can determine the distance from the maximum (here the origin) at which the gravity anomaly is halved, $r_{0.5}$. Then one can write

$$\delta g_Z(r_{0.5}) = \frac{-GM \cdot H}{(r_{0,5}^2 + H^2)^{3/2}} = \frac{1}{2}\delta g_Z(0) = \frac{1}{2} \cdot \frac{-GM}{H^2},$$

or

$$H^3 = 0.5(r_{0.5}^2 + H^2)^{3/2}.$$

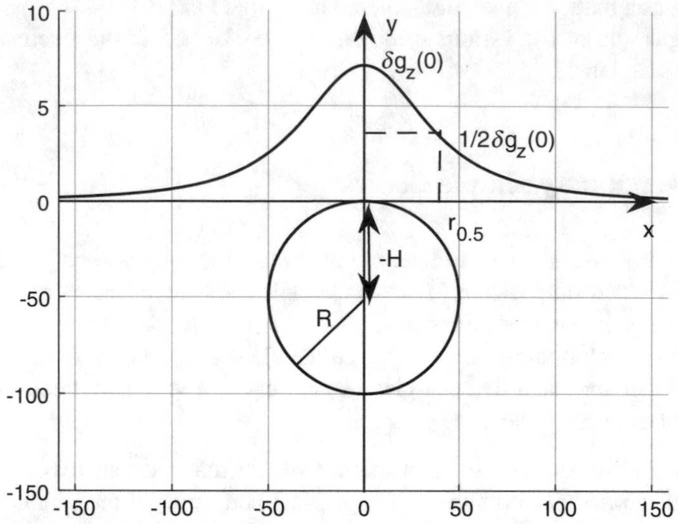

Fig. 7.3 A sphere (with negative density) and its gravity anomaly

The solution to this equation is

$$H = 1.3048 r_{0.5}. \tag{7.56}$$

The mass $(-M)$ is then given by

$$-M = \frac{1}{G}\delta g_Z(0) H^2. \tag{7.57}$$

Indeed, $(-M)$ is the only parameter estimable from δg_Z; if we want to have an idea about R, and consequently about the depth $d = H - R$, we have to make an assumption fixing the value of the density $-\rho$. When we do that, we receive

$$R = \left[\frac{3}{4\pi}\frac{\delta g_Z(0)}{G}\frac{H^2}{(-\rho)}\right]^{1/3}. \tag{7.58}$$

Example 7.2 (The Prism) In Fig. 7.3 we present the gravity anomaly $-\delta g_Z(\boldsymbol{x})$ of a prism with a rectangular basis, with sides $(2a, 2b)$ and width 2Δ, with the upper face buried at a depth H. As we see, we are again in the presence of a "hill" with isolines, especially close to the top, that present a certain eccentricity when $a \neq b$.

The first operation we want to do is to go from the axes (x, y) to the axes (x', y') in the plane S_0 (see Fig. 7.4), namely we want to shift the origin at $(x_{0,} y_0)$ (the horizontal coordinates of the maximum) and then rotate the axes to bring them parallel to the sides of the prism. To do that, we first identify (x_0, y_0) and then

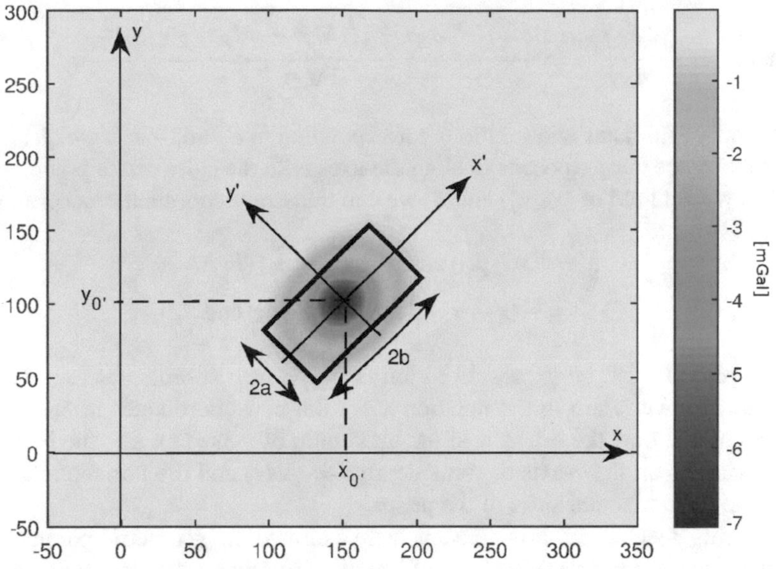

Fig. 7.4 The gravity anomaly of a prism with sides ($2a$, $2b$, $2c$), in a general position in the plane S_0

compute the "dispersion matrix". The unit vectors $\boldsymbol{e}'_x$ and $\boldsymbol{e}'_y$ in the direction of the new axes are in fact eigenvectors of the matrix.

$$D = \begin{bmatrix} M_{X^2} & M_{XY} \\ M_{XY} & M_{Y^2} \end{bmatrix}, \tag{7.59}$$

where

$$\begin{cases} M_{X^2} = \int (x - x_0)^2 \, |\delta g_Z| \, dS_0, \\ M_{Y^2} = \int (y - y_0)^2 \, |\delta g_Z| \, dS_0, \\ M_{XY} = \int (x - x_0)(y - y_0) \, |\delta g_Z| \, dS_0. \end{cases} \tag{7.60}$$

Indeed, in practice the integrals in (7.60) can be replaced by discrete sums.

The eigenvalues of (7.59) are

$$\lambda_{1,2} = \frac{1}{2}\left(M_{X^2} + M_{Y^2} \pm \sqrt{(M_{X^2} - M_{Y^2})^2 + 4M_{XY}^2} \right). \tag{7.61}$$

We just choose one of the two, e.g., λ_1, and then the angle ϑ between $\boldsymbol{e}'_X$ and $\boldsymbol{e}_X$ is given by

$$\tan \vartheta = \frac{\lambda_1 - M_{X^2}}{M_{XY}} = \frac{M_{Y^2} - M_{X^2} + \sqrt{(M_{Y^2} - M_{X^2})^2 + 4M_{XY}^2}}{2M_{XY}}. \tag{7.62}$$

Naturally (7.62) has two solutions corresponding to $\boldsymbol{e}'_X$ and $-\boldsymbol{e}'_X$, which are both acceptable since our purpose is to align the axes with the sides of the prism.

Once we get hold of (x_0, y_0) and ϑ, we can transform coordinates according to

$$x' = (x - x_0) \cos \vartheta + (y - y_0) \sin \vartheta,$$
$$y' = -(x - x_0) \sin \vartheta + (y - y_0) \cos \vartheta,$$

and so $-\delta g_Z(\boldsymbol{x})$ can be expressed in terms of the new coordinates as $-\delta g_Z(\boldsymbol{x}')$. From now on we return to the notation $\boldsymbol{x}$ for the new coordinates in S_0, with the convention that now the origin is at the maximum of $-\delta g_Z(\boldsymbol{x})$, i.e., the barycenter of the prism is on the z-axis at some depth $z = -\bar{H}$, and the horizontal axes are parallel to the horizontal sides of the prism.

Once this is achieved, it is convenient to estimate the geometric parameters of the prism from the Fourier transform of its anomaly. Without repeating too much of the previous arguments, we can use (7.6) and (7.23) to get hold of

$$M = \rho(2a2b2\Delta) = \rho V \tag{7.63}$$

and

$$\bar{Z} = -\bar{H}. \tag{7.64}$$

On the other hand, recalling (3.115), we see that

$$\delta g_Z(x) = -G\rho e^{-2\pi p \bar{H}} \frac{\mathrm{sh}(2\pi p \Delta)}{p\Delta} \mathrm{sinc}(2\pi pa) \mathrm{sinc}(2\pi qb) V. \tag{7.65}$$

Accordingly, since

$$\mathrm{sinc}(0) = 1; \quad \mathrm{sinc}\,(x) > 0, \quad 0 < x < \pi; \quad \mathrm{sinc}(\pi) = 0,$$

i.e., the sinc(x) function is positive from the origin up to π, where it gets to zero, we can analyze $\delta g_Z(x, y)$ along the x- and y-axes, and we should find a line parallel to the y-axis at $x = \frac{a}{2}$ and a horizontal line at $y = \frac{b}{2}$, where δg_Z goes to zero. When such lines are identified, we find a and b as well. Therefore (7.63), if we assume that we know ρ, provides Δ. Thus, all the parameters of the prism are found.

A *numerical example*. In the following we report a small numerical example for the inversion of a set of prismatic bodies of known density contrast. The example is based on the work presented in [125]. The study area is located in the Eastern region of the Mediterranean Sea and extends from 31° E to 34° E and from 33° N to 36° N. It is mainly characterized by the presence of Cyprus and of Eratosthenes

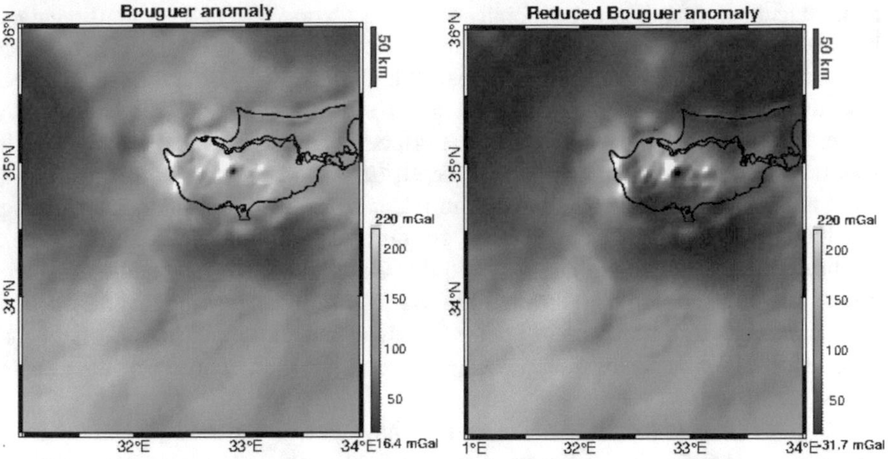

Fig. 7.5 The Bouguer anomaly synthesized from the GECO global gravity model and reduced for the ETOPO1 topographic layer (left). The Bouguer anomaly corrected for the effect of the subduction plate beneath the Cyprus arc (right)

Seamount. Before applying the gravity inversion procedure, it is necessary to remove from the gravitational field the effects of topography and bathymetry. The gravity field data have been synthesized from the GECO model [53]. In detail, the spherical harmonics of the GECO model have been synthesized in terms of gravity disturbances, on a grid at a constant altitude of 3500 m with a spatial resolution of 1 arc-min, which coincides to about 1:5 km in the West-East direction and 1:8 km in the South-North direction, respectively. Within the data reduction phase, a stripe 2° wide with respect to the actual extension of the study area has been added, in order to mitigate expected errors due to border effects. The altitude of 3500 m has been selected in such a way as to be as close as possible to the Earth surface, but always outside the topographic masses.

To remove the gravitational effects of the topographic layers, namely the bathymetry and topography, the ETOPO1 model [4] has been used. In detail, a complete Bouguer anomaly (see Fig. 7.5, left) was computed by estimating the gravitational effects of topography and bathymetry with the GTE package [124], using a reference density value of 2670 kg/m^3.

Although the sedimentary layer, which is known to have large effects in this specific area, has not yet been considered, it can be noticed that the Bouguer anomaly map already reflects the main geological characteristics of the region.

In particular, the highly positive regions (yellow area of Fig. 7.5, left) delineate well the oceanic crust of the Herodotus Basin, crossed by the Mediterranean Ridge (an accretionary wedge caused by the African plate that subducts under the Eurasian and Anatolian plates), as well as of the Cyprus arc. Moreover, the gravitational effects of the structures lying beneath the Eratosthenes Sea-mount and the boundaries of the Levantine Basin can be identified too. The maximum value of the signal is found in correspondence of the Cyprus arc and on the boundary between the Levantine and the Herodotus basins. The former signal is due to the presence of a crustal plate subducting beneath Cyprus [36], while the latter can be attributed to the presence of localized ophiolite structures [38, 121].

In order to reduce the subduction plate effects, a simple model of the plate itself was built. Both the geometry up to a depth of 40 km and the density contrast, fixed at 380 kg/m^3, of such a model are taken from [33], and the corresponding gravitational effect has been obtained by means of a point mass approximation, i.e., by discretizing the plate with a set of about 6000 point masses, and adding up their effect. The reduced signal is shown in Fig. 7.5 (right). It can be seen that the applied reduction leads to a much more homogeneous signal, in which the effects of the subduction plate basically disappeared, thus proving the adequacy of the simple model considered.

As for the others localized maxima, they have been removed by exploiting the stochastic characteristics of the gravitational field itself with a Kriging procedure [156]. Basically, the effects of these local anomalies are disentangled by considering an appropriate mask in correspondence with the anomalies, then by estimating an empirical and theoretical variogram from the data located outside the mask, and finally by interpolating on the whole study area, applying a Kriging technique using only the data outside the mask. This procedure gives a new field in which the aforementioned localized maxima anomalies are removed. Analysing the difference between the new field and the original one (see Fig. 7.6, left), some information on the causative bodies can be derived.

It is possible, for instance, to estimate the parameters (depth and size in the x-, y-, and z-directions) of a prism centered in correspondence of the maximum value of the anomalous field. A constant density contrast of 650 kgm^{-3} (corresponding to the difference between the densities of the ophiolites and sediments, 2850 kgm^{-3} and 2200 kgm^{-3}, respectively) was considered here. Once the best prism parameters have been estimated, one can reduce the observed signal for the effect of the prism itself and iterate the procedure. In Fig. 7.6 (right), the inversion-based estimated thickness, positions and size are presented. This simple modelling allows for reducing the signal standard deviation from 10 mGal to about 2 mGal. A more detailed modelling of these anomalies, at this point, seems therefore not to be required, since the residual effects are smaller than the global gravity observation error. In any case, results are highly-consistent with previous studies [38], with local thickness of the ophiolites up to 10 km.

Example 7.3 (The Vertical Cylinder) We know that the modulus of the gravitation signal of a vertical cylinder on S_0 has a maximum corresponding to its axis, it is bell-shaped, and has a circular symmetry on the plane.

If we place the origin of coordinates at the maximum of $-\delta g_Z(\boldsymbol{x})$, we can simplify further computations.

In this case we go directly to the Fourier transform of δg_Z that, recalling (3.119) reads

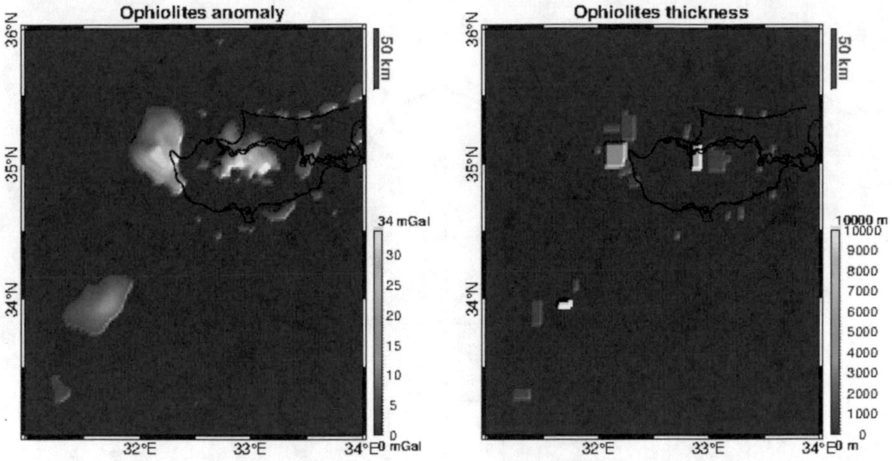

Fig. 7.6 The isolated signal due to the presence of ophiolites (left) and the thickness and position of the main ophiolite structures as estimated in the inversion (right)

$$-\delta\widehat{g}_Z(p) = \mu e^{-2\pi p\bar{H}} \frac{Sh(2\pi p\Delta)}{p} \frac{J_1(2\pi pR)}{p} R. \tag{7.66}$$

So, first of all we realize that $\delta\widehat{g}_z(p)$ also has a circular symmetry in the p plane, and it has to go to zero with $J_1(2\pi pR)$. All the other functions in (7.66) do not attain the zero value (Fig. 7.7).

Since the first zero of J_1 is (Standard Math tables)

$$J_1(x) = 0, \; x = 3.8,$$

we deduce that if p_0 is the first value for which

$$\delta g_Z(p_0) = 0,$$

then

$$R = \frac{3.8}{2\pi p_0}. \tag{7.67}$$

Next, when we want to estimate Δ, we see that this cannot be achieved unless we make some hypothesis on the density ρ.

In fact, recall that we are able from Sect. 7.1 to estimate M and we know that

$$M = \pi R^2 \cdot 2\Delta \cdot \rho, \tag{7.68}$$

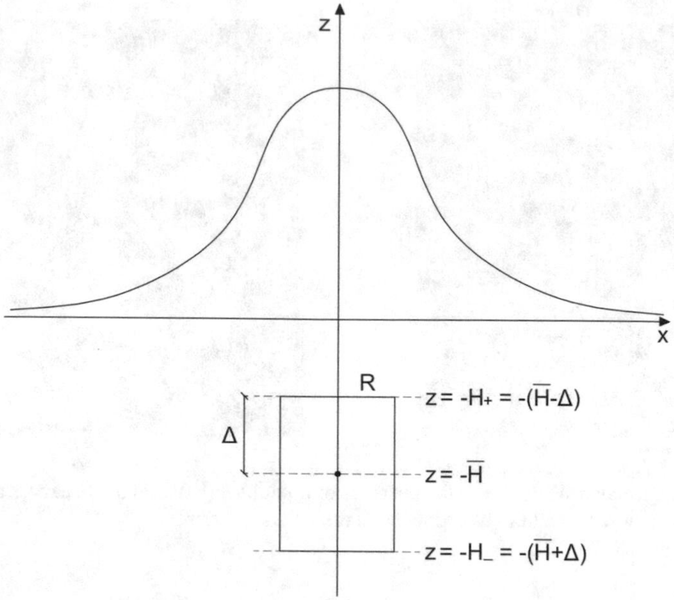

Fig. 7.7 The buried cylinder configuration

so, once we assign a value to ρ, from the estimates of M and R, we get via (7.68) the estimate of 2Δ.

7.8 Parker's Ideal Body

In some papers in the 70s, R. L. Parker [101, 102] stated and settled the following problem: given the gravity anomaly $\delta g_Z(\boldsymbol{x})$ at some points $\{\boldsymbol{x}_1, \ldots, \boldsymbol{x}_N\}$, in the plane S_0, find the body with the smallest constant density ρ_0, in the lower half-space $\mathbb{R}^3 \equiv \{z < 0\}$, capable of producing the given values $\delta g_Z(\boldsymbol{x}_1), \ldots, \delta g_Z(\boldsymbol{x}_N)$. Incidentally, Parker observed that the theory developed to solve the above problem, was able as well to answer, in the affirmative, the question: given a certain value of density ρ_0, is there a deepest body that with the above density is capable to reproduce the values $\delta g_Z(\boldsymbol{x}_1), \ldots, \delta g_Z(\boldsymbol{x}_N)$?

We review Parker's theory, but in the subsequent example we will limit ourselves to the 2D case, when only the values $\delta g_Z(\boldsymbol{x}_1)$ and $\delta g_Z(\boldsymbol{x}_2)$ are given, because this case can be completely worked out analytically.

Before getting started we need a few definitions. Let us denote, for the sake of simplicity,

$$N(\boldsymbol{x}, \boldsymbol{y}) = G\frac{|\zeta|}{\ell_{\boldsymbol{xy}}^3},$$
$$\boldsymbol{x} \in S_0; \quad \boldsymbol{y} = (\boldsymbol{\xi}, \zeta) \in \mathbb{R}^3; \quad \boldsymbol{\xi} \in S_0 \quad (\zeta < 0) \tag{7.69}$$
$$\ell_{\boldsymbol{xy}} = (r^2 + \zeta^2)^{\frac{1}{2}}, \quad r = |\boldsymbol{x} - \boldsymbol{\xi}|$$

and note that, although $\zeta < 0$,

$$N(\boldsymbol{x}, \boldsymbol{y}) > 0. \tag{7.70}$$

Furthermore we put

$$\boldsymbol{\alpha} = [\alpha_1 \ldots \alpha_N]^T \quad (\alpha_k \in \mathbb{R}), \tag{7.71}$$

$$F(\boldsymbol{y}, \boldsymbol{\alpha}) = \sum_{k=1}^{N} \alpha_k N(\boldsymbol{x}_k, \boldsymbol{y}) \tag{7.72}$$

and

$$B_+(\boldsymbol{\alpha}) = \{y; \zeta < 0, F(\boldsymbol{y}, \boldsymbol{\alpha}) > 0\}. \tag{7.73}$$

Then the following theorem holds:

Theorem 7.1 *Assume there are constants $\boldsymbol{\alpha}$ and ρ_0 such that*

$$|\delta g_Z(\boldsymbol{x}_k)| = \rho_0 \int_{B_+(\alpha)} N(\boldsymbol{x}_k, \boldsymbol{y})\, dB. \tag{7.74}$$

Then $B_+(\boldsymbol{\alpha})$ is an "ideal body" and ρ_0 is the least upper bound of the positive distributions $\rho(\boldsymbol{y})$ that reproduce the data; i.e., denoting

$$\mathscr{S} \equiv \{\rho(\boldsymbol{y}) \geq 0, \boldsymbol{y} \in \mathbb{R}^3; \int_{R^3} N(\boldsymbol{x}_k, \boldsymbol{y})\rho(\boldsymbol{y}) d_3\boldsymbol{y} = |\delta g_Z(\boldsymbol{x}_k)|\} \tag{7.75}$$

and, for $\rho \in \mathscr{S}$,

$$\bar{\rho}(\rho) = \sup_{\mathbb{R}^3_-} \rho(\boldsymbol{y}) \tag{7.76}$$

one has

$$\rho_0 = \inf_{\rho \in \mathscr{S}} \bar{\rho}(\rho). \tag{7.77}$$

Moreover, ρ_0 and $B_+(\boldsymbol{\alpha})$ are unique.

Proof We preliminarily notice that $\mathbb{R}^3_-$, which is an open half-space, can be split according to

$$\mathbb{R}^3_- \equiv B_+(\boldsymbol{\alpha}) \cup B_0(\boldsymbol{x}) \cup B_-(\boldsymbol{\alpha}) \tag{7.78}$$

where $B_+(\boldsymbol{\alpha})$ is given by (7.73),

$$B_-(\boldsymbol{\alpha}) = \{y; \zeta < 0, F(\boldsymbol{y}, \boldsymbol{\alpha}) < 0\} \tag{7.79}$$

and

$$B_0(\boldsymbol{\alpha}) = \{y; \zeta < 0, F(\boldsymbol{y}, \boldsymbol{\alpha}) = 0\}. \tag{7.80}$$

In particular, we observe that $F(\boldsymbol{y}, \boldsymbol{\alpha})$ is harmonic in $\mathbb{R}^3_-$ and therefore very smooth there. So it cannot be zero on a set of positive measure because the surface of equation $F(y, \boldsymbol{\alpha}) \equiv 0$ in $\mathbb{R}^3_-$ is an equipotential surface of $F(\boldsymbol{y}, \boldsymbol{\alpha})$ and, as recalled in Parker [102], it cannot have a positive measure. In other words, letting $m(B)$ denote the Lebesgue measure of a set B, we must have

$$m(B_0(\boldsymbol{\alpha})) \equiv 0, \quad \forall \boldsymbol{\alpha}. \tag{7.81}$$

As a consequence, any integral of a measurable function on $\mathbb{R}^3_-$ can be limited to the set $B_+(\boldsymbol{\alpha}) \cup B_-(\boldsymbol{\alpha})$, without changing its value.

Therefore, recalling (7.74) and (7.75), we have, for every $\rho \in \mathscr{S}$,

$$\begin{aligned} |\delta g_Z(\boldsymbol{x}_k)| &= \int_{B(\boldsymbol{\alpha})} \rho_0 N(\boldsymbol{x}_k, \boldsymbol{y}) dB = \\ &\equiv \int_{B_+(\boldsymbol{\alpha})} \rho(\boldsymbol{y}) N(\boldsymbol{x}_k, y) dB + \int_{B_-(\boldsymbol{\alpha})} \rho(\boldsymbol{y}) N(\boldsymbol{x}_k, \boldsymbol{y}) dB. \end{aligned} \tag{7.82}$$

Multiplying (7.82) by α_k and summing over k, we obtain

$$\int_{B_+(\boldsymbol{\alpha})} \rho_0 F(\boldsymbol{y}, \boldsymbol{\alpha}) dB = \int_{B_+(\boldsymbol{\alpha})} \rho(\boldsymbol{y}) F(\boldsymbol{y}, \boldsymbol{\alpha}) dB + \int_{B_-(\boldsymbol{\alpha})} \rho(\boldsymbol{y}) F(\boldsymbol{y}, \boldsymbol{\alpha}) dB,$$

or, with an obvious notation,

$$I_+ + I_- = \int_{B+} [\rho(y) - \rho_0] F(\boldsymbol{y}, \boldsymbol{\alpha}) dB + \int_{B_-} \rho(y) F(\boldsymbol{y}, \boldsymbol{\alpha}) dB = 0. \tag{7.83}$$

The relation (7.83) has to hold for every $\rho \in \mathscr{S}$. Assume that there is a $\tilde{\rho} \in \mathscr{S}$ such that

$$\bar{\rho}(\tilde{\rho}) < \rho_0; \tag{7.84}$$

then indeed for this $\tilde{\rho}$ one has

$$I_+ < 0 \text{ and } I_- \leq 0.$$

This would imply

$$I_+ + I_- < 0,$$

contrary to (7.83), and (7.77) is proved by contradiction.

Now, if we assume that (7.74) holds, it is clear that there cannot be another $\rho_0' < \rho_0$ such that the same relations are true with the same $B_+(\boldsymbol{\alpha})$. Then, the only non-uniqueness can concern the ideal body $B_+(\boldsymbol{\alpha})$. So assume we have another $\tilde{\rho}(\boldsymbol{y})$ such that

$$\bar{\rho}(\tilde{\rho}) = \rho_0$$

but the corresponding "ideal body" $\tilde{B}$ does not coincide with $B_+(\boldsymbol{\alpha})$.

This can happen only in two ways: either $\tilde{\rho} < \rho_0$ in a subset of $B_+(\boldsymbol{\alpha})$ of positive measure, or $\tilde{\rho} > 0$ in a set of positive measure contained in $B_-(\boldsymbol{\alpha})$.

In the first case $I_+ < 0$ in (7.83) and $I_- \leq 0$; in the second case we might have $I_+ = 0$, but certainly $I_- < 0$. In either cases $I_+ + I_- < 0$ and (7.83) is contradicted.

Naturally, if both cases mentioned above are verified, then we have $I_+ < 0$ and $I_- < 0$ so the same conclusion holds. □

Remark 7.5 We note incidentally that the fact that $B_+(\boldsymbol{\alpha})$, and therefore $B_-(\boldsymbol{\alpha})$, are unique does not mean that $\boldsymbol{\alpha}$ is unique too. In fact, it is obvious that, if $(\rho_0, \boldsymbol{\alpha})$ satisfies (7.74), then so do $(\rho_0, \lambda\boldsymbol{\alpha})$ for any $\lambda > 0$, because the shapes of $B_+(\lambda\boldsymbol{\alpha})$, $B_-(\lambda\boldsymbol{\alpha})$ are the same as those of $B_+(\boldsymbol{\alpha})$, $B_-(\boldsymbol{\alpha})$.

This is not the only type of non-uniqueness of the $\boldsymbol{\alpha}$ vector, as proved by counterexamples in [102].

Theorem 7.1 admits an important corollary that, among other reasons, justifies why Parker's theory belongs to this chapter.

Corollary 7.1 *Let us define the conditional ideal body $B_+(\boldsymbol{\alpha}, d)$ as*

$$B_+(\boldsymbol{\alpha}, d) \equiv \{\boldsymbol{y};\ \zeta < -d,\ F(y, \boldsymbol{\alpha}) > 0\} \tag{7.85}$$

and similarly $B_-(\boldsymbol{\alpha}, d)$, $B_0(\boldsymbol{\alpha}, d)$, so that

$$\mathbb{R}^3_{-d} \equiv \{\boldsymbol{y};\ \zeta < -d\} = B_+(\boldsymbol{\alpha}, d) \cup B_0(\boldsymbol{\alpha}, d) \cup B_-(\alpha, d), \tag{7.86}$$

such that

$$|\delta g_Z(\boldsymbol{x}_k)| = \rho_0 \int_{B_+(\boldsymbol{\alpha},d)} N(\boldsymbol{x}_k, \boldsymbol{y}) dB, \tag{7.87}$$

then no other $\tilde{\rho}(\mathbf{y})$ reproducing the data and positive in $\mathbb{R}^3_{-d}$ can satisfy $\bar{\rho}(\tilde{\rho}) < \rho_0$; in other words, ρ_0 is the least upper bound of the distributions with support in $\mathbb{R}^3_-$ that reproduce the data.

The proof of the above corollary is identical to that of Theorem 7.1.

We notice that this approach does not solve the existence problem: Theorem 7.1 claims that "if" there are constants $(\rho_0, \boldsymbol{\alpha})$ that, through (7.74), allow the reproduction of data, "then" $B_+(\boldsymbol{\alpha})$ is an ideal body and ρ_0 is the least upper bound of any positive distribution generating the data.

The existence problem, in a general form, is a difficult one and requires more advanced functional tools.

Here we restrict ourselves to studying the case of two observations only and, in addition, we shall assume that the anomalous body has a 2D configuration. In this way we will be able to give an explicit solution to the problem.

The analysis of 2D bodies proceeds exactly as in the 3D case, with the only difference that in 2D the Newton kernel (7.69) becomes

$$\begin{aligned} &N(\mathbf{x}, \mathbf{y}) = G\frac{|\zeta|}{(x-\xi)^2+\zeta^2}, \\ &\mathbf{x} = x\mathbf{e}_X \text{ on the } X\text{-axis, } \mathbf{y} = \xi\mathbf{e}_X + \zeta\mathbf{e}_Z \text{ in the } (X, Z)\text{-plane.} \end{aligned} \tag{7.88}$$

Example 7.4 Assume two values of gravity anomalies $\delta g_1, \delta g_2$ are given, generated by a 2D body and observed along the X-axis at two points x_1, x_2; to make it simple, we assume that

$$x_2 = 0, \quad x_1 = 1.$$

This does not reduce the generality, because we can choose origin and the unit of distance, as we like. Furthermore, we take the labels 1, 2 in such a way that

$$\delta g_2 > \delta g_1 > 0;$$

again to simplify the computation and to make the results more readable, we adapt the unit of gravity so that

$$\delta g_2 = 1.$$

Our aim is to determine whether there are always constants $(\rho_0, \alpha_1, \alpha_2)$ such that Theorem 7.1 holds, to find the shape of the corresponding ideal body and the least upper bound ρ_0.

We begin by forming the function

$$
\begin{aligned}
F(\mathbf{y}, \boldsymbol{\alpha}) &= \alpha_1 N(x_1;\ \xi, \zeta) + \alpha_2 N(x_2;\ \xi, \zeta) = \\
&= G\left[\frac{\alpha_1 |\zeta|}{(1-\xi)^2 + \zeta^2} + \frac{\alpha_2 |\zeta|}{\xi^2 + \zeta}\right] = \\
&= G|\zeta|\left[\frac{\alpha_1}{(1-\xi)^2 + \zeta^2} + \frac{\alpha_2}{\xi^2 + \zeta^2}\right].
\end{aligned}
\tag{7.89}
$$

Since the sign of $F(y, \boldsymbol{\alpha})$, which we are interested in, is invariant under multiplication by a positive factor (see Remark 7.5), we can decide to take $\alpha_2 = 1$ in (7.89). Since we want to find the domain, in $\mathbb{R}^3_-$, where $F(y, \boldsymbol{\alpha})$ is positive, bounded below by the surface $\{F(y, \boldsymbol{\alpha}) = 0\}$, we expect that α_1 should have a negative sign; so we put

$$
\alpha_1 = -\beta^2,
$$

and study the equation

$$
F(\mathbf{y}, \beta^2) = 0.
$$

This, after a reordering, reads

$$
\xi^2 + \frac{2}{\beta^2 - 1}\xi + \zeta^2 = \frac{1}{\beta^2 - 1};
\tag{7.90}
$$

we will put

$$
\gamma = \frac{1}{\beta^2 - 1}.
\tag{7.91}
$$

for the sake of brevity.

As (7.90) has to hold for $\zeta \le 0$, it is the equation of a semi-circle $\mathscr{C}$ with center $C \equiv (-\gamma, 0)$ and radius

$$
R = \beta\gamma
\tag{7.92}
$$

(see Fig. 7.8).

To be more precise we see that, if we also exploite the relation $\gamma + \gamma^2 = \beta^2\gamma^2$, the function $F(\mathbf{y}, \beta^2)$ can be written as

$$
\begin{aligned}
F(\mathbf{y}, \beta^2) &= G|\zeta| \frac{(1-\beta^2)\xi^2 - 2\xi + (1-\beta^2)\zeta^2 + 1}{(\xi^2 + \zeta^2)[(\xi - 1)^2 + \zeta^2]} = \\
&= G|\zeta|(1-\beta^2)\frac{(\xi - \gamma)^2 + \zeta^2 - \beta^2\gamma^2}{(\xi^2 + \zeta^2)[(\xi - 1)^2 + \zeta^2]}.
\end{aligned}
\tag{7.93}
$$

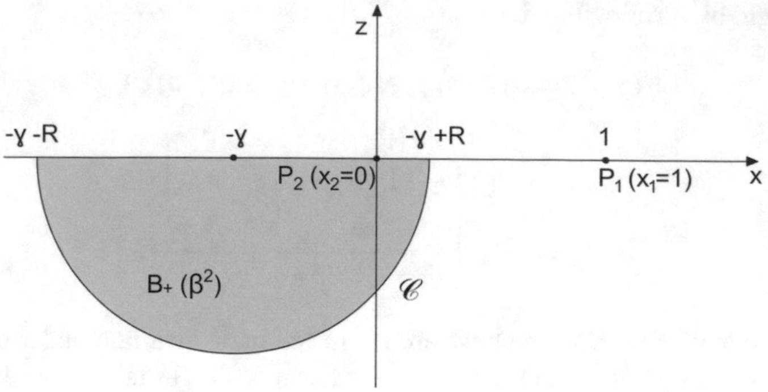

Fig. 7.8 The ideal body for two gravity values in 2D

Now it is clear from (7.93) that if $\beta > 1$, $F(\mathbf{y}, \beta^2) > 0$ in the bounded region between $\mathscr{C}$ and the $X-$axis, while when $\beta < 1$, $F(\mathbf{y}, \beta^2)$ is positive only in the unbounded region outside $\mathscr{C}$, in $\mathbb{R}^3$. Since an ideal body, with constant density $\rho_0 > 0$, can only be bounded, in this case, if it has to generate a finite gravity anomaly, we need to consider for $B_+(\beta^2)$ only values $\beta > 1$. With this specification we notice that $\gamma > 0$ and that the point P_2 always belongs to the top surface of the ideal body $B_+(\beta^2)$, while P_1 is outside it.

We want to express δg_2 and δg_1 as functions of β^2 and ρ_0. This is easily done, by taking advantage of the result of Exercise 5 in Chap. 2. We only need to take into account that in Exercise 5 the origin was placed at the center of $\mathscr{C}$, whence in that formula one has to set $x = \gamma$ for the point P_2 and $x = 1 + \gamma$ for the point P_1 (see Fig. 7.3). The result, after some simplifications, is

$$\delta g_2 = G\rho_0 \left[\log \frac{\beta + 1}{\beta - 1} + 2\frac{\beta}{\beta^2 - 1} \right] \tag{7.94}$$

$$\delta g_1 = G\rho_0 \left[-\log \frac{\beta + 1}{\beta - 1} + 2\frac{\beta}{\beta^2 - 1} \right] \tag{7.95}$$

Recalling that we have set $\delta g_2 = 1$, we can divide (7.95) by (7.94), obtaining

$$\begin{aligned} \delta g_1 = \frac{\delta g_1}{\delta g_2} &= \frac{-\log \frac{\beta+1}{\beta-1} + \frac{2\beta}{\beta^2-1}}{\log \frac{\beta+1}{\beta-1} + \frac{2\beta}{\beta^2-1}} = \\ &= \frac{-(\beta^2 - 1)\log \frac{\beta+1}{\beta-1} + 2\beta}{(\beta^2 - 1)\log \frac{\beta+1}{\beta-1} + 2\beta}. \end{aligned} \tag{7.96}$$

To this equation we can adjoin Eq. (7.94), rewritten in the form,

$$\rho_0 = G^{-1}\left[\log\frac{\beta+1}{\beta-1}+\frac{2\beta}{\beta^2-1}\right]^{-1} = \\ = \frac{1}{G}\frac{\beta^2-1}{(\beta^2-1)\log\frac{\beta+1}{\beta-1}+2\beta}. \tag{7.97}$$

Now we can consider (7.96) and (7.97) as the parametric form of the curve

$$\rho_0 = \rho_0(\delta g_1)$$

(or, if you like, $\rho\left(\frac{\delta g_1}{\delta g_2}\right)$) in the $(\rho_0, \delta g_1)$-plane (see Fig. 7.10). We first of all notice that (7.96), implies that

$$0 < \delta g_1 < \delta g_2 = 1,$$

as required.

The second inequality is trivial, while the first requires some study of the numerator to show that it is always positive in the range $1 < \beta < +\infty$, as it has to be, being the integral of a positive function.

At any rate, it is easy to see, from the second form of (7.96), that

$$\lim_{\beta\to 1} \delta g_1 = 1 (= \delta g_2),$$

while by the first form one shows that, when $\beta \to \infty$, both the numerator and the denominator tend to zero, but the first faster than the second, i.e.,

$$\lim_{\beta\to+\infty} \delta g_1 = 0.$$

Since $\delta g_1(\beta)$ is continuous in the open interval $(1, \infty)$, we deduce that all the values between 0 and $1 = \delta g_2$ are swept by δg_1.

Similarly, the second expression in (7.97) implies that

$$\lim_{\beta\to 1} \rho_0 = 0$$

whereas the first one implies that

$$\lim_{\beta\to+\infty} \rho_0 = +\infty$$

The example is continued by examining ideal bodies constrained to lay below the depth $\zeta = -d$. In this case, since the function $F(y, \boldsymbol{\alpha})$ is the same as before, we need only to intersect the semicircle of Fig. 7.8 with $\mathbb{R}^2_{-d}$, to get the ideal body $B_+(\boldsymbol{\alpha}, d)$. This has therefore a shape of the cap of circle as shown in Fig. 7.9. The solution of

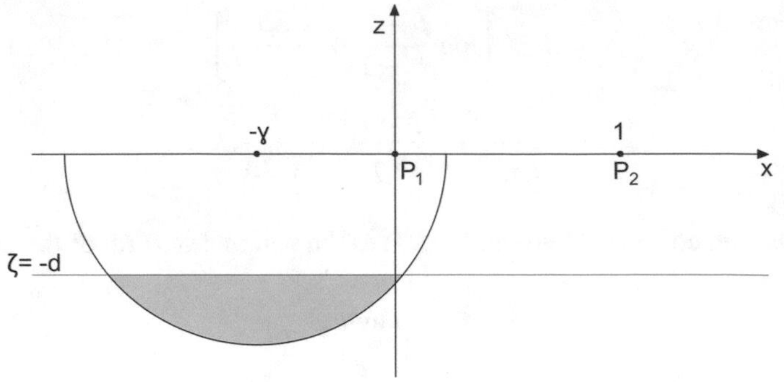

Fig. 7.9 The 2D ideal body with two observations $0 < \delta g_1 < 1 = \delta g_2$, constrained to $\mathbb{R}^2_{-d}$

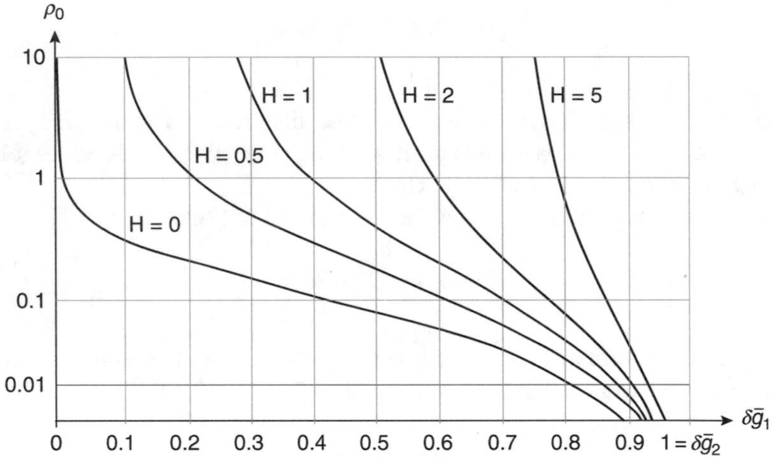

Fig. 7.10 $(\rho_0, \delta g_1)$ for the ideal body in 2D with observations of $0 < \delta g_1 < \delta g_2 = 1$. The horizontal scale is linear, the vertical scale is logarithmic. Notice that the unit length is $\overline{P_1 P_2}$, so $H = 1$ means a depth equal to the separation between P_1 and P_2

the problem of determining δg_1 and ρ_0 as functions of β^2 is still available in terms of elementary functions, however since it is much less compact than that of Exercise 5 in Chap. 2, we do not reproduce it here. We rather represent the result in terms of the plots of $(\rho_0, \delta g_1)$ in Fig. 7.10, for some of the depths. As one immediately sees, the range swept by δg_1 does not longer reach the value $\delta g_1 = 0$, but it has a lower bound corresponding to a vertical asymptote. This is an intrinsic feature relating to a constraint on $\left|\frac{\partial}{\partial x}\delta g_Z(x)\right|$, for a body deeper than d, that can be derived with techniques similar to those used in Sect. 7.5.

In any event, as it seen from Fig. 7.10, if we want to reproduce a certain value $\delta \bar{g}_1$ with bodies constrained to increasing depths, we need also an increasing density ρ_0, as suggested by physical intuition.

The situation with 3D ideal bodies is essentially the same and the qualitative content of Fig. 7.10 is perfectly preserved, as shown in [102].

As observed by the author of the theory, there is still another way in which the information content of the family of curves of Fig. 7.8, or the analogous one in a 3D context, can be used; this is expressed in Theorem 7.2.

Theorem 7.2 *Assume there is a conditional ideal body $B_+(\boldsymbol{\alpha}, d)$, with density ρ_0, that reproduces the data. Then no other body $\tilde{B}$ with the same constant density ρ_0, that reproduces the data $\delta g_Z(\boldsymbol{x}_1), \delta g_Z(\boldsymbol{x}_2)$, can have a larger minimum depth. Or, in other words, if we define*

$$\begin{aligned}\widetilde{\mathscr{S}}_0 \equiv \{\tilde{\rho}(P) = \rho_0; \quad P \in \tilde{B} \subset \mathbb{R}^3, \quad \tilde{\rho}(P) = 0, \quad P \in \tilde{B}^{\mathrm{c}}, \\ \rho_0 \int_{\tilde{B}} N(\boldsymbol{x}_k, \boldsymbol{y}) dB = \delta g_Z(\boldsymbol{x}_k), \quad k = 1, \ldots, N\};\end{aligned} \tag{7.98}$$

and put, for each $\tilde{\rho} \in \widetilde{\mathscr{S}}_0$,

$$d(\tilde{\rho}) = \inf_{P \in \tilde{B}} |\zeta_P|, \tag{7.99}$$

then one has

$$d_0 = \inf_{\tilde{\rho} \in \widetilde{\mathscr{S}}_0} d(\tilde{\rho}). \tag{7.100}$$

Proof The result is just a by-product of the uniqueness asserted by Theorem 7.1. In fact, assume there is body $\widetilde{B}$ such that $d(\tilde{\rho}) > d_0$. Then $\widetilde{B}$ is an ideal body, with density ρ_0, lying entirely below the level $-d(\tilde{\rho}) < -d_0$. But this cannot be true, because, by Theorem 7.1, $B_+(\alpha, d_0)$ is an ideal body constrained to $\mathbb{R}^3_{-d_0}$, with minimum density ρ_0, so any other body with the same minimum density must have the same shape. We reached a contradiction. □

In practice, if we reduce to the case of two gravity values generated by a 2D body, if we assume to know ρ_0 and δg_1 (or, if you like, $\frac{\delta g_1}{\delta g_2}$), we have to find on the plot of Fig. 7.10 the corresponding point and read out the value of H for the curve passing through it.

If the point $(\rho_0, \delta g_1)$ falls outside the area swept by the curves, it means that the hypotheses made are not consistent.

Remark 7.6 To conclude this chapter we note that, although the theory of this section seems to be too complicated to be applicable to a realistic case, with bodies in 3D and a large number of observed gravity anomalies, this actually is not the case.

In fact, one can organize the data into couples $\delta g_Z(\boldsymbol{x}_j)$, $\delta g_Z(\boldsymbol{x}_k)$ and find for each of them the corresponding greatest minimum depth, d_{jk}.

If all the data are generated by a body B with density ρ_0 and minimum depth $d(B)$, one must have

$$d(B) \leq d_{jk}$$

for all (j, k); namely

$$d_0 = \min_{(j,k)} d_{jk} \tag{7.101}$$

is the sought-for upper bound of $d(B)$.

7.9 Exercises

Exercise 1 Prove the validity of (7.6) in the case of a simple body such as a buried homogeneous sphere of mass M.

Solution
Considering a homogeneous buried sphere of mass M, centered at $x = 0$, $y = 0$, and at a depth H with respect to the observation plane S_0, we have, from (2.9):

$$\delta g_Z(x, y) = -\frac{GMH}{(x^2 + y^2 + H^2)^{\frac{3}{2}}}. \tag{7.102}$$

Substituting this expression in (7.6) we obtain:

$$M = \frac{1}{2\pi G} \iint_{S_o} \frac{GMH}{(x^2 + y^2 + H^2)} dxdy. \tag{7.103}$$

Moving now from a Cartesian to a polar coordinates system in the plane, (7.103) simplifies to:

$$M = MH \int_0^{+\infty} \frac{rdr}{(r^2 + H^2)^{\frac{3}{2}}}. \tag{7.104}$$

Putting in (7.104) $r = Ht$ we obtain:

$$M = M \int_0^{+\infty} \frac{tdt}{(t^2 + 1)^{\frac{3}{2}}}. \tag{7.105}$$

The above integral can be calculated by putting $u = t^2 + 1$:

$$M = \frac{M}{2} \int_1^{+\infty} \frac{du}{u^{\frac{3}{2}}} = M,$$

as it was meant to prove.

Exercise 2 Prove that the barycenter of the sphere in Exercise 1 has horizontal coordinates (0, 0) and $\bar{Z} = -H$.

Solution
Recalling (7.12) and (2.9) we have that the planar coordinates of the barycenter $\bar{\mathrm{x}}$ are given by

$$\bar{\mathbf{x}} = \frac{1}{2\pi MG} \iint_S \frac{GMH\mathbf{x}}{(x^2 + H^2)^{\frac{3}{2}}} dS, \tag{7.106}$$

where x is the magnitude of the $\mathbf{x}$ vector, spanning S. Note that the integrand is an odd function with respect to $\mathbf{x}$ since it is a product of an odd function, $\mathbf{x}$, by an even function, namely $\frac{1}{(x^2+H^2)^{\frac{3}{2}}}$. As a consequence, the result of (7.106) is $\bar{\mathbf{x}} = (0, 0)$.

Furthermore, we have

$$\delta\widehat{g}_Z(p) = \mathscr{F}\{\delta g_Z\} = -GH\mathscr{F}\left\{\frac{H}{(r^2 + H^2)^{\frac{3}{2}}}\right\} = -GM2\pi e^{-2\pi pH}.$$

Therefore:

$$\frac{1}{2\pi} \int \frac{\partial}{\partial p} \delta\widehat{g}_Z(0, \alpha) d\alpha = GMH4\pi^2 e^{-2\pi pH}|_{p=0} = GM4\pi^2 H.$$

However, since $\delta\widehat{g}_Z(0) = -2\pi GM$, we have

$$\bar{Z} = \frac{1}{4\pi^2} \frac{GM8\pi^3 H}{(-2\pi GM)} = -H,$$

as claimed.

Exercise 3 Given a cylinder of density $\rho > 0$ and radius R, with top at $-H_+$, bottom at $-H_-$, with the axis coinciding with the Z-axis, and supposing $R \ll H_+ < H_-$, show that (7.38) gives a better estimate of the depth of the top surface d than the barycenter.

Solution
The $|\delta g_Z(P)|$ generated by the cylinder has its maximum at $P = 0$, given by

$$|\delta g_Z|_{\max} = 2\pi G\rho \left(\sqrt{R^2 + H_+^2} - H_+ - \sqrt{R^2 + H_-^2} + H_-\right).$$

Under our assumption that $R \ll H_+ < H_-$, we can write:

$$|\delta g_Z|_{\max} \approx 2\pi G\rho \left(H_+ + \frac{R^2}{2H_+} - H_+ - H_- - \frac{R^2}{2H_-} + H_- \right)$$
$$= \pi G\rho R^2 \frac{H_- - H_+}{H_+H_-} = \frac{GM}{H_+H_-}.$$

Consequently,

$$d = \sqrt{\frac{GM}{|\delta g_Z|_{\max}}} = \sqrt{H_+H_-}.$$

We have to prove that $\sqrt{H_+H_-}$ is closer to H_+ than $\frac{1}{2}(H_+ + H_-)$, which is the depth of the barycenter, i.e., we want to prove that

$$\sqrt{H_+H_-} - H_+ < \frac{1}{2}(H_+ + H_-) - H_+,$$

or $\sqrt{H_+H_-} < \frac{1}{2}(H_+ + H_-)$. In fact, we have

$$H_+H_- < \frac{1}{4}(H_+^2 + H_-^2 + 2H_+H_-),$$

i.e.,

$$H_+^2 + H_-^2 - 2H_+H_- > 0,$$

which is true since $(H_+ - H_-)^2 > 0$.

Exercise 4 By using the approximate gravity anomaly of a semi-infinite vertical homogeneous cylinder with ($R_0 \ll H$) (see (2.36)):

$$\delta g_Z(r) = 2\pi\mu \left(\sqrt{R_0^2 + H^2 + r^2} - \sqrt{H^2 + r^2} \right),$$

prove that the UB (7.44) is worse then the UB (7.45), but (7.53) is even better than (7.45).

Solution
Since the gravitational acceleration of the semi-infinite vertical homogeneous cylinder δg_Z is a function of the planar coordinate r only, we start the exercise by computing the derivative $\frac{d\delta g_Z(r)}{dr} = \delta g_Z'(r)$:

$$\delta g_Z'(r) = 2\pi\mu r \left(\frac{1}{\sqrt{R_0^2 + H^2 + r^2}} - \frac{1}{\sqrt{H^2 + r^2}} \right).$$

In order to find the maximum value of $\delta g'_Z(r)$ we have to look for r for which $\delta g''_Z(r) = 0$:

$$\delta g''_Z(r) = 2\pi\mu \left[\frac{R_0^2 + H^2}{(R_0^2 + H^2 + r^2)^{\frac{3}{2}}} - \frac{H^2}{(H^2 + r^2)^{\frac{3}{2}}} \right] = 0,$$

i.e.,

$$\left(\frac{H^2 + r^2}{R_0^2 + H^2 + r^2} \right)^{\frac{3}{2}} = \frac{H^2}{R_0^2 + H^2}.$$

the above equation can be written as

$$\left(1 - \frac{R_0^3}{R_0^2 + H^2 + r^2} \right)^{\frac{3}{2}} = 1 - \frac{R_0^2}{R_0^2 + H^2}.$$

Now, recalling the binomial approximation $(1+x)^\alpha \approx 1 + \alpha x$, and our assumption that $R_0 \ll H$, we have

$$\frac{3}{2} \frac{R_0^2}{R_0^2 + H^2 + r^2} \approx \frac{R_0^2}{H^2},$$

i.e., $r^2 \approx \frac{1}{2}H^2 - R_0^2 \approx \frac{H^2}{2}$ and therefore $r \approx \frac{H}{\sqrt{2}}$. Substituting this value of r in $\delta g_Z(r)$ and $\delta g'_Z(r)$ we get

$$\delta g_Z \left(\frac{H}{\sqrt{2}} \right) = 2\pi\mu \left(\sqrt{R_0^2 + \frac{3}{2}H^2} - \sqrt{\frac{3}{2}H^2} \right) \approx \frac{2}{3}\pi\mu \sqrt{\frac{3}{2}} \frac{R_0^2}{H} = \frac{2}{\sqrt{6}}\pi\mu \frac{R_0^2}{H}$$

and

$$\delta g'_Z \left(\frac{H}{\sqrt{2}} \right) \approx \frac{2}{3\sqrt{3}}\pi\mu \frac{R_0^2}{H^2},$$

from which we can derive the UB (7.44):

$$\text{UB} = 1.5 \frac{\delta g_Z \left(\frac{H}{\sqrt{2}} \right)}{\delta g'_Z \left(\frac{H}{\sqrt{2}} \right)} \sim \frac{\frac{2}{\sqrt{6}}\pi\mu \frac{R_0^2}{H}}{\frac{2}{3\sqrt{3}}\pi\mu \frac{R_0^2}{H^2}} = 1.5 \frac{3}{\sqrt{2}} H \approx 3.2H$$

If we now consider Eq. (7.45), we have that

$$\mathrm{UB} = 0.86\frac{\delta g_Z(0)}{\delta g'_Z\left(\frac{H}{\sqrt{2}}\right)} \approx \frac{R_0^2}{H}\frac{3\sqrt{3}H^2}{2R_0^2} \approx 2.2H.$$

Finally, we can compute $\Delta\delta g_Z = \delta g''_Z + \frac{1}{r}\delta g'_Z$, so that

$$\Delta\delta g_Z(0) = \delta g''_Z(0) + \frac{1}{r}\delta g'_Z(0) = 4\pi\mu\left[\frac{1}{\sqrt{R_0^2 + H^2}} - \frac{1}{H}\right] \approx \frac{2\pi\mu R_0^2}{H^3}.$$

Therefore, from (7.53) we have:

$$\mathrm{UB} \leq 2.45\sqrt{\frac{|\delta g_Z(0)|}{\Delta\delta g_Z(0)}} = 2.45\sqrt{\frac{[(R_0^2 + H^2) - H]}{\frac{R_0^2}{H^3}}} \approx \frac{2.45}{\sqrt{2}}H = 1.73H.$$

Chapter 8
On the Mathematical Characterization of the Inverse Gravity Problem

8.1 Outline of the Chapter

Before we can enter a more advanced theory of inversion of the gravity data, we need to understand its indeterminacy when the inverse problem is stated in its most general form.

In fact, if we suppose that the anomalous density ρ is square integrable in $\mathbb{R}^3_-$, or in a more restricted domain, it is possible to reconstruct the full set of solutions that produce the given gravity anomaly on S_0.

The theory can go considerably beyond the simple L^2 analysis (see [67, 126]), yet in such a space a much simpler solution can be given, particularly readable if we make use of the Fourier picture, which is permitted by the simple geometric setting in which we work.

This clarifies that, whenever we obtain a "unique" solution to the inverse gravimetric problem, it can happen only on condition that we have applied suitable constraints to restrict the space in which we are looking for such a solution.

We have already met signs of this indeterminacy by examining elementary inverse problems; for instance, in Sect. 7.6 we have seen that, if we restrict so much the space of solutions as to include only homogeneous spheres, of density ρ and radius R, one of the two parameters has to be a priori assigned if we want to find the other.

Even the Bouguer plate displays such a duality between density and width. In fact, a conspicuous part of potential source inverse problems solved in literature concerns the identification of the unknown shape of a homogeneous body with a known density ρ_0. In this case the "unknown parameter" is the boundary of the anomalous body and so we introduce, by this restriction of the model, a nonlinearity between unknowns and data, which somehow complicates the solution. This for instance is the case when we assume that $\delta g_Z(\boldsymbol{x})$ is produced by a two-layer configuration, where the lower boundary is unknown.

F. Sansò, D. Sampietro, *Analysis of the Gravity Field*, Lecture Notes in Geosystems Mathematics and Computing, https://doi.org/10.1007/978-3-030-74353-6_8

For such restricted models it is important to understand whether a solution is unique or not. Suitable theorems will be proved for these cases.

Another obvious characteristic of the inverse gravimetric problem is its ill-posedness, namely a wild behaviour of the solution when we give a small variation to the data.

Since data are always affected by several kinds of errors, from the noise in the discrete observational data to interpolation errors, one cannot hope to derive a sensible inversion without regularizing it, namely imposing some regularity condition. This can be done in a deterministic approach, for instance by using a Tikhonov principle. Other deterministic regularization methods are known in the literature, see [74], yet in our opinion the Tikhonov approach remains the most important.

Therefore in this chapter we include an analysis of nonlinear and linearized inverse gravimetric problems, regularized according to Tikhonov. The results are obtained by applying the theory developed in Appendix C.

Another approach to the regularization problem is based on stochastic methods. This is also presented in the chapter, together with a specific example concerning the estimation of the Moho depth.

The two approaches can come very close to one another, when the discrepancy principle of Morozov [87] is applied. This item is also discussed in Appendix C.

8.2 Non-uniqueness of the Gravity Inversion: The L^2-Decomposition Theorem and the Full Class of Its Solutions

The story starts with a very simple remark, although with far reaching consequences. In general terms, assume to have a certain mass distributed with density ρ in a body B, generating an exterior Newtonian field with potential u, then if we take any other

$$\bar{\rho} = \rho + \Delta\varphi \tag{8.1}$$

with φ a smooth function with support strictly contained in B, we find that $\bar{\rho}$ generates the same exterior potential. In other words, $\Delta\varphi$, with (supp (φ) $\subset B$) generates a zero-potential exterior Newtonian field. This is a simple consequence of Green's identity (1.93): in fact, if S is the boundary of B, P is a point outside B (i.e., $P \in B^c$), and Q a point running on B, one has

$$\begin{aligned} \int_B \frac{1}{\ell_{PQ}} \Delta\varphi \, dB_Q = \int_B \varphi(Q) \Delta_Q \frac{1}{\ell_{PQ}} \, dB_Q + \\ + \int_S \left(\frac{\partial\varphi}{\partial n} \frac{1}{\ell_{PQ}} - \varphi \frac{\partial}{\partial n} \frac{1}{\ell_{PQ}} \right) dS \equiv 0. \end{aligned} \tag{8.2}$$

In fact, if $P \in B^c$, then $\frac{1}{\ell_{PQ}}$ is harmonic in B and

$$\varphi|_s \equiv \left.\frac{\partial\varphi}{\partial n}\right|_S \equiv 0 \tag{8.3}$$

because the support of φ is contained in B, so φ and its normal derivative have to be identically zero close to S.

Based on the above result, a series of functional analysis arguments lead to a nice theorem clarifying completely the situation, when the hypothesis is made that the exterior field is generated by a $\rho \in L^2(B)$ [7, 67, 126].

Theorem 8.1 *Let S, the boundary of the bounded body B, be a sufficiently smooth surface for Green's identities to hold, and let u be the Newtonian potential generated in $\Omega = B^c$ by a mass density $\rho(Q) \in L^2(B)$ Consider the subspace $HL^2(B)$ of $L^2(B)$ defined by*

$$HL^2(B) \equiv \{\rho \in L^2(B);\ \Delta\rho = 0\ in\ B\}, \tag{8.4}$$

also called Bergman Space *with exponent* 2 [6]. *$HL^2(B)$ is closed in $L^2(B)$, i.e., it is a Hilbert subspace of L^2. The projection of ρ on $HL^2(B)$, ρ_h, generates the same exterior potential as ρ, i.e.,*

$$\int_B \frac{1}{l_{PQ}}\rho(Q)dB_Q \equiv \int_B \frac{1}{l_{PQ}}\rho_h(Q)dB_Q, \quad \forall P \in \Omega. \tag{8.5}$$

The orthogonal complement of $HL^2(B)$ in $L^2(B)$

$$HL^2(B)^\perp \equiv \{\rho_0 \in L^2(B);\ \langle\rho_0, \rho_h\rangle_{L^2(B)} = 0 \quad \forall\rho_h \in HL^2(B)\} \tag{8.6}$$

is a closed subspace of $L^2(B)$, too, that can be described by the equivalence

$$\rho_0 \in HL^2(B)^\perp \Longleftrightarrow \rho_0 = \Delta\varphi,\ \varphi|s = \left.\frac{\partial\varphi}{\partial n}\right|_S \equiv 0 \quad (\varphi \in H^{2,2}(B)) \tag{8.7}$$

$HL^2(B)^\perp$ is exactly the subspace of $L^2(B)$ consisting of the mass distributions that generate a null outer potential.

Corollary 8.1 *The solutions of the "inverse problem"*

$$\int_B \frac{1}{\ell_{PQ}}\rho(Q)dB_0 = u(P) \quad \forall P \in S \tag{8.8}$$

have in $L^2(B)$ the form

$$\rho(Q) \equiv \rho_h(Q) + \rho_0(Q); \tag{8.9}$$

ρ_h is the unique harmonic function in $L^2(B)$ satisfying (8.7) *and among all solutions is the one of minimum norm.*

We will not prove here Theorem 8.1 in this general form; rather, we will prove a version of it, adapted to the geometry with which we are working, that allows a systematic use of Fourier transforms, thus simplifying the reasoning.

We start by formulating the problem: find all $\rho(Q) \in L^2(\mathbb{R}^3_-)$, where

$$\mathbb{R}^3_- \equiv \{Q \equiv (\boldsymbol{\xi}, \zeta); \boldsymbol{\xi} \in \mathbb{R}^2, \zeta \leq 0\},$$

such that

$$G \int_{R^3_-} \frac{\zeta}{(r^2_{x\xi} + \zeta^2)^{3/2}} \rho(\boldsymbol{\xi}, \zeta) d_2\xi d\zeta = \delta g_Z(\boldsymbol{\xi}), \quad \forall \boldsymbol{x} \in \mathbb{R}^2. \tag{8.10}$$

Recalling Proposition 3.3, if we introduce the 2D Fourier Transform and put

$$\begin{cases} \delta \hat{g}_Z(\boldsymbol{p}) = \mathscr{F}_{\boldsymbol{x} \to \boldsymbol{p}}\{\delta g_Z(\boldsymbol{x})\}, \\ \hat{\rho}(\boldsymbol{p}, \zeta) = \mathscr{F}_{\boldsymbol{\xi} \to \boldsymbol{p}}\{\rho(\xi, \zeta)\} \end{cases} \tag{8.11}$$

then using the Convolution Theorem 3.1 we obtain

$$- G 2\pi \int_{-\infty}^{0} e^{2\pi p \zeta} \hat{\rho}(\boldsymbol{p}, \zeta) d\zeta = \delta \hat{g}_Z(\boldsymbol{p}), \quad \forall \boldsymbol{p} \in \mathbb{R}^2. \tag{8.12}$$

For the sake of brevity we shall put (8.10) and (8.12) in the operator form

$$\delta g_Z = K(\rho); \text{ and } \delta \hat{g}_Z = \hat{K}(\hat{\rho}). \tag{8.13}$$

Notice that, thanks to Plancherel Theorem (Remark 3.2), the requirement $\rho \in L^2(\mathbb{R}^3_-)$ is translates into

$$\|\rho\|^2_{L^2} = \int_{-\infty}^{0} \int_{\mathbb{R}^2} \rho(\boldsymbol{\xi}, \zeta)^2 d_2\xi d\zeta = \int_{-\infty}^{0} \int_{\mathbb{R}^2} \left|\hat{\rho}(\boldsymbol{p}, \zeta)\right|^2 d_2 p d\zeta + \infty. \tag{8.14}$$

We start our analysis by characterizing in terms of Fourier Transform the Bergman Space $HL^2(\mathbb{R}^3_-)$ and its orthogonal complement.

Proposition 8.1 *Let $\rho_h(\boldsymbol{\xi}, \zeta) \in HL^2(\mathbb{R}^3_-)$. Then the* FT *of ρ_h has the form*

$$\hat{\rho}_h(\boldsymbol{p}, \zeta) = A(\boldsymbol{p}) e^{2\pi p \zeta} \tag{8.15}$$

and the following isometry holds:

$$\|\rho_h\|^2_{L^2(\mathbb{R}^3_-)} = \frac{1}{4\pi}\int_{\mathbb{R}^2} \frac{|A(\boldsymbol{p})|^2}{p} d_2 p. \tag{8.16}$$

Proof If $\rho_h(\boldsymbol{\xi}, \zeta) \in HL^2(\mathbb{R}^3_-)$, then ρ_h is harmonic in $\mathbb{R}^3_-$ and therefore a smooth function in the lower half-space, satisfying the Laplace equation, which we write as

$$\Delta\rho_h(\xi, \zeta) = \Delta_\xi \rho_h(\xi, \zeta) + \rho_h''(\xi, \zeta), \tag{8.17}$$

with

$$\rho_h'' = \frac{\partial^2 \rho_h}{\partial \zeta^2}.$$

If we set

$$\rho_h(\xi, \zeta) = \int e^{-i2\pi \boldsymbol{p}\cdot\boldsymbol{\xi}} \hat{\rho}_h(\boldsymbol{p}, \zeta) d_2 p,$$

then we can easily verify that in the Fourier domain (8.17) becomes

$$-4\pi^2 p^2 \hat{\rho}_h(\boldsymbol{p}, \zeta) + \hat{\rho''}_h(\boldsymbol{p}, \zeta) = 0. \tag{8.18}$$

In Eq. (8.18), $\boldsymbol{p}$ can be considered as a vector parameter, so this is an ordinary differential equation, with the general solution

$$\hat{\rho}_h(\boldsymbol{p}, \zeta) = A(\boldsymbol{p})e^{2\pi p\zeta} + B(\boldsymbol{p})e^{-2\pi p\zeta}. \tag{8.19}$$

Since the function $\int_{\mathbb{R}^2} \left|\hat{\rho}(\boldsymbol{p}, \zeta)\right|^2 d_2 p$ must be integrable in ζ according to (8.14), we see that in (8.19) we have to put $B(\boldsymbol{p}) \equiv 0$ and (8.15) is proved. Moreover, a direct computation gives (8.16), i.e.,

$$\|\rho_h\|^2_{L^2} = \int |A(\boldsymbol{p})|^2 \int_{-\infty}^{0} e^{4\pi p\zeta} d\zeta d_2 p = \frac{1}{4\pi}\int \frac{|A(\boldsymbol{p})^2|}{p} d_2 p. \tag{8.20}$$

□

The meaning of Proposition 8.1 is precisely that $\rho_h \in HL^2(\mathbb{R}^3_-)$ if and only if $\hat{\rho}_h$ has the form (8.15) with a function $A(\boldsymbol{p})$ such that

$$\int \frac{|A(\boldsymbol{p})|^2}{p} d_2 p < +\infty. \tag{8.21}$$

For instance, we shall use the fact that, if we choose $A(\boldsymbol{p}) = \chi_D(\boldsymbol{p})$, the characteristic function of an arbitrary bounded measurable set D, then (8.21) is certainly satisfied.

Moreover, we note that to every $A(\boldsymbol{p})$ satisfying (8.21) there corresponds exactly one function $f(\boldsymbol{p}) \in L^2(\mathbb{R}^2)$ by setting $A(\boldsymbol{p}) = \sqrt{p} f(\boldsymbol{p})$; since $L^2(\mathbb{R}^2)$ is a complete Hilbert space, also the set of functions $\{A(\boldsymbol{p})\}$ satisfying (8.21) and hence $HL^2(\mathbb{R}^3_-)$ (thanks to (8.16)) forms a complete Hilbert space.

We next provide a characterization of the space $HL^2(\mathbb{R}^3_-)^\perp$.

Proposition 8.2 *We have the equivalence*

$$\rho_0 \in HL^2(\mathbb{R}^3_-)^\perp \Longleftrightarrow \int_{-\infty}^{0} \hat{\rho}_0(\boldsymbol{p}, \zeta) e^{2\pi p\zeta} d\zeta \equiv 0 \quad a.e.\ in\ \boldsymbol{p} \in \mathbb{R}^2. \tag{8.22}$$

Furthermore, every function ρ_0 that can be represented as

$$\rho_0 = \Delta\varphi, \ with\ \varphi|_{S_0} = \varphi'|_{S_0} \equiv 0, \tag{8.23}$$

and such that $\varphi, \varphi' \to 0$ when $\zeta \to -\infty$, belongs to $HL^2(\mathbb{R}^3_-)^\perp$.

Proof If $\rho_0 \in HL^2(\mathbb{R}^3_-)^\perp$, then for any $\rho_h \in HL^2(\mathbb{R}^3_-)$

$$0 = \int_{-\infty}^{0} \int_{R^2} \rho_h(\boldsymbol{\xi}, \zeta) \rho_0(\boldsymbol{\xi}, \zeta) d_2\xi d\zeta = \int_{-\infty}^{0} \int_{R^2} \hat{\rho}_h(\boldsymbol{p}, \zeta)^* \hat{\rho}_0(\boldsymbol{p}, \zeta) d_2 p d\zeta, \tag{8.24}$$

so that, using Proposition 8.1, we have too

$$\begin{aligned} 0 &\equiv \int_{-\infty}^{0} \int_{R^2} A^*(\boldsymbol{p}) e^{2\pi p\zeta} \hat{\rho}_0(\boldsymbol{p}, \zeta) d_2 p d\zeta = \\ &= \int A^*(\boldsymbol{p}) \int_{-\infty}^{0} e^{2\pi p\zeta} \hat{\rho}_0(\boldsymbol{p}, \zeta) d\zeta d_2 p \end{aligned} \tag{8.25}$$

for any $A(\boldsymbol{p})$ satisfying (8.21). If we take, as already commented, $A(\boldsymbol{p}) = \chi_D(\boldsymbol{p})$ for an arbitrary bounded set D, we find that

$$0 \equiv \int_D \left[\int_{-\infty}^{0} e^{2\pi p\zeta} \hat{\rho}_0(\boldsymbol{p}, \zeta) d\zeta \right] d_2 p, \tag{8.26}$$

which, by the arbitrariness of D, implies (8.22).

Conversely, let $\hat{\rho}_0$ satisfy (8.22); then (8.25) holds for any $A(\boldsymbol{p})$ satisfying (8.21), i.e., (8.24) holds for any $\rho_h \in HL^2$, i.e., $\hat{\rho}_0 \in HL^2(\mathbb{R}^3_-)^\perp$.

Now, if we go to the Fourier Transform, (8.23) is equivalent to

$$\widehat{\rho_0}(\boldsymbol{p}, \zeta) = \widehat{\varphi}''(\boldsymbol{p}, \zeta) - 4\pi^2 p^2 \widehat{\varphi}(\boldsymbol{p}, \zeta), \quad \widehat{\varphi}(\boldsymbol{p}, 0) \equiv \widehat{\varphi}'(\boldsymbol{p}, 0) \equiv 0. \tag{8.27}$$

Furthermore, we assume that $\widehat{\varphi}(\boldsymbol{p},\zeta)$ and $\widehat{\varphi'}(\boldsymbol{p},\zeta)$ are bounded when $\zeta \to -\infty$, which is implicitly a restriction on the way in which $\varphi(\boldsymbol{\xi},\zeta)$ and $\varphi'(\boldsymbol{\xi},\zeta)$ tend to zero for $\zeta \to -\infty$ and ultimately on the type of ρ_0 generated by (8.23).

In this case we can compute, recalling also (8.15),

$$\forall \rho_h \in HL^2(R^3_-); \quad \int_{-\infty}^0 \int_{\mathbb{R}^2} \rho_0(\boldsymbol{\xi},\zeta)\rho_h(\boldsymbol{\xi},\zeta) d_2\xi d\zeta =$$

$$= \int_{-\infty}^0 \int_{\mathbb{R}^2} \hat{\rho}_0(\boldsymbol{p},\zeta) A^*(\boldsymbol{p}) e^{2\pi p\zeta} d_2p d\zeta = \tag{8.28}$$

$$= \int_{\mathbb{R}^2} A^*(\boldsymbol{p}) \int_{-\infty}^0 [\hat{\varphi}''(\boldsymbol{p},\zeta) - 4\pi^2 p^2 \hat{\varphi}(\boldsymbol{p},\zeta)] e^{2\pi p\zeta} d\zeta d_2p.$$

On the other hand, the above $L^2(\mathbb{R}^3_-)$ product is identically 0 because, due to the behaviour of $\hat{\varphi}(\boldsymbol{p},\zeta)$ at $\zeta = 0$ and for $\zeta \to -\infty$, a double integration by parts gives

$$\int_{\infty-}^0 \hat{\varphi}'' e^{2\pi p\zeta} d\zeta = 4\pi^2 p^2 \int_{-\infty}^0 \hat{\varphi} e^{2\pi p\zeta} d\zeta. \tag{8.29}$$

The proof is complete. □

We are ready now to prove the next fundamental theorem.

Theorem 8.2 *Recalling notation* (8.13)*, one has the following assertions:*

a) Denote by $\mathcal{N}(K)$ the null space of K, i.e.,

$$\mathcal{N}(K) \equiv \{\rho_0; K(\rho_0) = 0\}; \tag{8.30}$$

then

$$\mathcal{N}(K) \equiv HL^2(\mathbb{R}^3_-)^\perp. \tag{8.31}$$

b) Let

$$\delta g_Z = K(\rho), \quad \rho \in L^2(\mathbb{R}^3_-), \tag{8.32}$$

then the class of all solutions of (8.32) *is given by*

$$\mathcal{S}(\delta g_Z) \equiv \{\bar{\rho}; \bar{\rho} = \rho + \rho_0, \rho_0 \in HL^2(\mathbb{R}^3_-)^\perp\}. \tag{8.33}$$

c) The element of minimum norm in $\mathcal{S}(\delta g_Z)$ is given by

$$\rho_h = P_{HL^2}(\rho), \tag{8.34}$$

where P_{HL^2} is the orthogonal projector of L^2 onto HL^2.

d) The class of $\mathscr{S}\{\delta g_Z\}$ generated by some $\rho \in L^2(\mathbb{R}^3_-)$ is a Hilbert space, that we call $\bar{H}^{1/2}(\mathbb{R}^2)$, where the norm is defined by

$$\|\delta g_Z\|^2_{1/2} \equiv \int \left|\delta \hat{g}_Z\right|^2 p d_2 p. \tag{8.35}$$

Proof Let $\rho_0 \in \mathscr{N}(K)$. Then $\hat{\rho}_0$ satisfies (8.22) and $\rho_0 \in HL^2(\mathbb{R}^3_-)^{\perp}$ by Proposition 8.2. Conversely, if $\rho \in HL^2(\mathbb{R}^3_-)^{\perp}$, then $\hat{\rho}$ satisfies the second relation (8.22) in Proposition 8.2, which is equivalent to $\rho_0 \in \mathscr{N}(K)$, because of (8.12).

Point a) is proved by Propositions 8.1 and 8.2.

Point b) is an immediate consequence of the identity

$$K(\bar{\rho}) = K(\rho + \rho_0) = K(\rho), \ \forall \rho_0 \in HL^2(\mathbb{R}^3_-)^{\perp}, \tag{8.36}$$

using point a).

Now let $\rho \in L^2$, then we can establish the orthogonal decomposition

$$\rho = \rho_h + \rho_0, \quad \rho_h \in HL^2(\mathbb{R}^3_-), \quad \rho_0 \in HL^2(\mathbb{R}^3_-)^{\perp}. \tag{8.37}$$

Indeed, $\rho_h \in \mathscr{S}(\delta g_Z)$ by using point b), because $\rho_h = \rho - \rho_0$ and $\rho_0 \in HL^2(\mathbb{R}^3_-)^{\perp}$. Since

$$\|\rho\|^2_{L^2} = \|\rho_h\|^2_{L^2} + \|\rho_0\|^2_{L^2},$$

ρ_h is the element of minimum norm in $\mathscr{S}(\delta g_Z)$. This is indeed a general property of the orthogonal projection. We notice that, since the orthogonal projection is unique, ρ_h is unique too. Point c) is therefore proved. Moreover if $\delta g_Z = K(\rho)$, $\rho \in L^2(\mathbb{R}^3_-)$, by point c) one also has

$$\delta g_Z = K(\rho_h), \rho_h \in HL^2(\mathbb{R}^3_-), \tag{8.38}$$

i.e. (see (8.12)),

$$\delta \hat{g}_Z = \hat{K}(\hat{\rho}_h). \tag{8.39}$$

By Proposition 8.1, there is an $A(\boldsymbol{p})$ satisfying (8.21), such that

$$\delta \hat{g}_Z = -2\pi G \int_{-\infty}^{0} e^{2\pi p \zeta} A(\boldsymbol{p}) e^{2\pi p \zeta} d\zeta, \tag{8.40}$$

namely

$$\delta\hat{g}_Z(\boldsymbol{p}) = -\frac{G}{2}\frac{A(\boldsymbol{p})}{p}. \tag{8.41}$$

Since A has to satisfy (8.21) and viceversa, we note that the space $\bar{H}^{1/2}(\mathbb{R}^2)$ to which δg_Z has to belong, is a Hilbert space because it can be put into a one-to-one, isometric correspondence with $L^2(\mathbb{R}^2)$ by setting

$$\delta\hat{g}_Z(\boldsymbol{p}) = \frac{f(\boldsymbol{p})}{\sqrt{p}}, \quad f \in L^2. \tag{8.42}$$

Point d) is proved. □

Remark 8.1 Let us note that δg_Z can indeed be continued to $\mathbb{R}^3_+$, where it is a harmonic function, regular for $z \to +\infty$. In terms of FT one can write that

$$\delta\hat{g}_Z(\boldsymbol{p}, z) = e^{-2\pi p z}\delta\hat{g}_Z(\boldsymbol{p}, 0), \tag{8.43}$$

so that $\delta\hat{g}_Z(\boldsymbol{p}, 0)$ can be considered as the trace of $\delta\hat{g}_Z$ on $S_0 \equiv \{z = 0\}$.

From potential theory [54, 84], one would expect the trace of the attraction of a body with an L^2 density to be in $H^{1/2}$, corresponding to the requirement of having a finite norm of the form

$$\|\delta g_Z\|^2_{H^{1/2}} = \int_{R^2} |\delta\hat{g}_Z|^2 (1 + p^2)^{1/2} d_2 p < +\infty. \tag{8.44}$$

This is not the case, however, because we are dealing with unbounded domains. In fact, the norm (8.35) is not equivalent to (8.44), as the counterexample

$$\delta\hat{g}_Z = \frac{1}{p}\frac{1}{1 + p^2}$$

demonstrates, because for such a function $\left\|\delta\hat{g}_Z\right\|^2_{H^{1/2}} = \int \frac{1}{p^2}\frac{(1+p^2)^{\frac{1}{2}}}{(1+p^2)^2} d_2 p = +\infty$, while

$$\left\|\delta\hat{g}_Z\right\|^2_{\bar{H}^{1/2}} = \int \frac{1}{p}\frac{1}{(1 + p^2)^2} p d_2 p + \infty$$

due to the different singular behaviour at the origin.

Remark 8.2 Let us recall that harmonic functions obey the so-called *maximum principle* [6], namely, they attain their maximum and minimum values at the boundary.

Since $\rho_h(\boldsymbol{\xi}, \zeta)$ is harmonic in $\mathbb{R}^3_-$ and goes to zero when $\zeta \to -\infty$, we expect that, when it is positive, it attains its maximum on $S_0 = \{z = 0\}$. This obviously does not correspond to any specific geological behaviour of masses; in particular, one should not think that ρ_h inherits any physical characteristic of the density ρ generating δg_Z.

Closing this section we note that formula (8.41) enables us to compute ρ_h from δg_Z. In fact, one has

$$\hat{\rho}_h(\boldsymbol{p}, \zeta) = -\frac{2}{G} p\, \delta \hat{g}_Z(\boldsymbol{p}) e^{2\pi p \zeta}. \tag{8.45}$$

Naturally one would like as well to have the corresponding relation for the original functions $\rho_h(\boldsymbol{\xi}, \zeta)$ and $\delta g_Z(\boldsymbol{x})$.

The Convolution Theorem shows that we can write

$$\rho_h(\boldsymbol{x}, z) = -\frac{2}{G} \int_{\mathbb{R}^2} H(\boldsymbol{x} - \boldsymbol{\xi}, z) \delta g_Z(\boldsymbol{\xi}) d_2\xi \tag{8.46}$$

where, recalling Proposition 3.3,

$$\begin{aligned} H(\boldsymbol{x}, z) &= \int_{R^2} p e^{2\pi p z - i 2\pi \boldsymbol{p}\cdot\boldsymbol{x}} d_2 p = \\ &= \frac{1}{2\pi} \frac{\partial}{\partial z} \int_{R^2} e^{2\pi p z - i 2\pi \boldsymbol{p}\cdot\boldsymbol{x}} d_2 p = \\ &= -\frac{1}{2\pi} \frac{|\boldsymbol{x}|^2 - 2z^2}{(|x|^2 + z^2)^{5/2}} \quad (z \le 0). \end{aligned} \tag{8.47}$$

Another question is whether we could compute ρ_h from ρ. The answer is indeed in the affirmative, but we leave the proof to Exercise 2.

8.3 Formulation and First Functional Properties of Various Restricted Inverse Gravimetric Problems

The behaviour of the solution of the general inverse gravimetric problem, at least when densities are sought in $L^2(\mathbb{R}^3_-)$, was settled in the previous section. In this section we start the analysis of the problem, by restricting the class of possible solutions so much, that we can expect to identify only one of them from data.

We have already seen in the previous section that, if we require the density ρ to be harmonic in $\mathbb{R}^3_-$, we achieve the uniqueness, yet, since we want to work with models that, if not real, are at least plausible from the geological point of view, we will direct our investigation to anomalies created by simpler bodies, e.g., bodies with constant density ρ_0. As we have seen in Chap. 2 and in Sects. 3.5, 3.6, for such simple bodies the form of the gravity signal becomes a function of the boundary shape, in fact a functional mapping, between the surface S, i.e., the boundary of B, and the corresponding anomaly δg_Z.

This mapping is however non linear, making its mathematical analysis more difficult, unless we pass to linearized versions, as we did for instance in Sect. 2.7.

It is for this reason that sometimes in the geophysical literature one reads of a nonlinear gravimetric problem as opposed to a linear one. One should keep in mind though that, rather than an opposition, this is a choice of the analyst to restrict, on the basis of some prior information, the solution space to a nonlinear manifold in a larger linear space.

We will see in the next chapter how to deal with a different kind of prior knowledge in the framework of the linear theory and how to mix the two points of view. So we move here to the formulation, under different restrictive hypotheses, of direct problems, that we have already studied in Part I of the book, and we will specify functional properties of the direct operators, that are essential for understanding how to perform the inversion.

In this section we shall address mainly the problem of identifying the interface of a two-layer configuration, where the upper layer is filled with the constant density ρ_0 (see Fig. 8.1) and the lower with zero density.

As discussed in Sect. 2.7, this is equivalent to considering two layers with density ρ_+ and ρ_-, with contrast $\rho_0 = \rho_+ - \rho_-$.

The interface S is described by the function $z = -H(\boldsymbol{\xi}) = -\tilde{H}(\boldsymbol{\xi}) + \delta H(\boldsymbol{\xi})$, with a reference surface $\tilde{S}$, of equation $z = -\tilde{H}(\boldsymbol{\xi})$ and an anomaly $\delta H(\xi)$, that in some cases we shall assume to be identically zero outside a bounded domain in the plane (x,y), $D \subset S_0$. We will always require that S cannot rise above S, or

$$H(\boldsymbol{\xi}) = \tilde{H}(\boldsymbol{\xi}) - \delta H(\boldsymbol{\xi}) \geq 0; \quad \delta H(\boldsymbol{\xi}) \leq H(\boldsymbol{\xi}). \tag{8.48}$$

Before we enter into the matter, we want to observe that there is a case in which this theory can be also used to determine the shape of a bulky body B. This happens when the boundary S of B is divided in two parts, a known $S_+ = \{z = -H_+(\boldsymbol{\xi}), \boldsymbol{\xi} \in D\}$ (see Fig. 8.2), while the lower surface, $S_- = \{z = -H_-(\boldsymbol{\xi})\}$ is unknown.

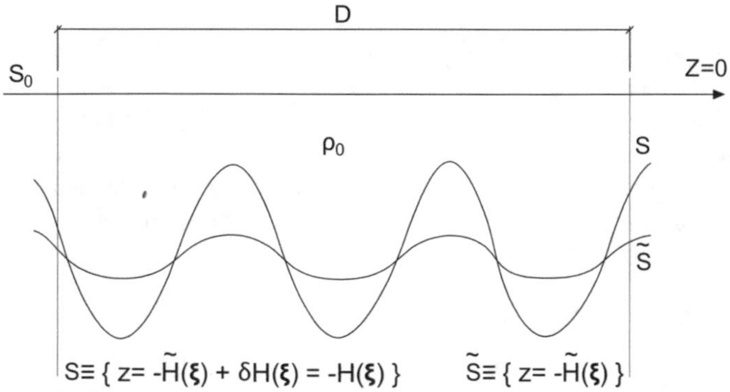

Fig. 8.1 The classical two-layer configuration

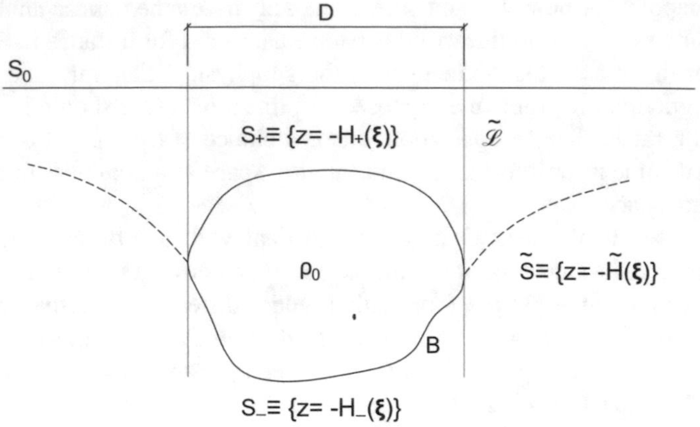

Fig. 8.2 Transforming the attraction of B into the attraction of a layer

In this case we could continue the surface S_+ with a reference surface $\tilde{S}$, arbitrarily defined outside D and in this way form the layer $\mathscr{L}$ with interface

$$\tilde{S} \equiv \{z = -\tilde{H}(\boldsymbol{\xi}), \boldsymbol{\xi} \in D^c; z = -H_+(\boldsymbol{\xi}), \boldsymbol{\xi} \in D\}. \tag{8.49}$$

If we let $\delta\tilde{g}_z$ denote the attraction of $\tilde{\mathscr{L}}$ filled with a density ρ_0, we see that the actually observed δg_Z plus $\delta\tilde{g}_Z$ represents the anomaly generated by the layer $\mathscr{L}$ with interface

$$S \equiv \{z = -\tilde{H}(\boldsymbol{\xi}), \boldsymbol{\xi} \in D^c; z = -H_-(\boldsymbol{\xi}), \boldsymbol{\xi} \in D\}, \tag{8.50}$$

also filled with density ρ_0.

As we see, we are reconducted exactly to the problem defined before, with

$$\delta H(\boldsymbol{\xi}) = H_-(\boldsymbol{\xi}) - H_+(\boldsymbol{\xi}), \boldsymbol{\xi} \in D; \quad \delta H(\boldsymbol{\xi}) = 0, \boldsymbol{\xi} \in D^c. \tag{8.51}$$

Variants of the above formulation, possibly introducing $\tilde{H}(\boldsymbol{\xi})$ also for $\boldsymbol{\xi} \in D$, are easy to obtain.

We proceed by defining three different models, with a decreasing degree of generality. Our purpose is to define function sets $\mathscr{K}$ in the space $L^2(\mathbb{R}^2)$, in such a way that the direct operator

$$\delta g = A(\delta H), \; \delta H \in \mathscr{K}, \tag{8.52}$$

is continuous from $\mathscr{K} \subset L^2(\mathbb{R}^2)$ to $\delta g \in L^2(\mathbb{R}^2)$. In order to do that, we will impose constraints on $\mathscr{K}$ that maybe are not the most general, yet they are acceptable from the geophysical point of view and at the same time they simplify proofs.

8.3.1 Model A (Nonlinear)

The shape of $A(\delta H)$ is given by

$$(\mu = G\rho_0) \quad \delta g_Z = \mu \int_{R^2} \int_{-\tilde{H}+\delta H}^{0} \frac{\zeta}{(r_{x\xi}^2 + \zeta^2)^{3/2}} d\zeta d_2\xi =$$
$$= -\mu \int_{R^2} \left\{ \frac{1}{r_{x\xi}} - \frac{1}{\left(r_{x\xi}^2 + (\tilde{H} - \delta H)^2\right)^{1/2}} \right\} d_2\xi, \tag{8.53}$$

Moreover, we will define $\mathscr{K}$, the class, of possible solutions by the constraints

$$\delta H(\boldsymbol{\xi}) = 0, \quad \boldsymbol{\xi} \in D^c, \ (D \text{ bounded in } S_0), \tag{8.54}$$

$$\delta H(\boldsymbol{\xi}) \leq \tilde{H}(\boldsymbol{\xi}); \quad -\tilde{H}(\boldsymbol{\xi}) + \delta H(\xi) \geq -\bar{H} > -\infty; \tag{8.55}$$

(8.54) implies that δH is identically zero outside D, and (8.55) implies that $\delta H(\xi)$ is bounded above and below.

In fact, if we further assume that $0 \leq \tilde{H}(\boldsymbol{\xi}) < \bar{H}$, which is not a real constraint since $\bar{H}$ is largely arbitrary, we see that

$$-\bar{H} \leq \tilde{H}(\xi) - \bar{H} \leq \delta H(\xi) < \bar{H}. \tag{8.56}$$

This implies that, if it is measurable, $\delta H(\boldsymbol{\xi})$ belongs automatically to $L^2(D)$ and in addition all integrals involving $\delta H(\xi)$ can be extended to D or to the whole $\mathbb{R}^2$ in an equivalent manner. In particular

$$\int_{\mathbb{R}^2} \delta H^2(\boldsymbol{\xi}) d_2\xi = \int_D \delta H^2(\boldsymbol{\xi}) d_2\xi \leq \bar{H}^2 |D| < +\infty. \tag{8.57}$$

Since we will not need to specify the shape and the dimension of D, as long as it is a bounded domain in $\mathbb{R}^2$ (e.g., we could even assume that D is a disk of some finite radius R), nor the value of $\bar{H}$, we think that the constraints (8.56) are not so severe from the geophysical point of view, that they will spoil our conclusions.

After all, one should not forget that even our reference model, with $\delta g_Z(\boldsymbol{x})$ given on the whole S_0 and the Z-axis directed as the normal gravity, is actually an abstraction and an approximation.

We want to prove that:

a) $\mathscr{K}$ is a closed, convex set in $L^2(D)$;
b) $A(\delta H)$ is continuous $A : L^2(D) \to L^2(\mathbb{R}^2)$, i.e.

$$\delta H_n \underset{L^2(D)}{\to} \delta H^* \Longrightarrow ||A(\delta H_n) - A(\delta H^*)||_{L^2(\mathbb{R}^2)} \to 0. \tag{8.58}$$

For a) we need to prove that if $\delta H_n \underset{L^2(D)}{\rightarrow} \delta H^*$ then $\delta H^* \in \mathscr{K}$. But this is elementary, because if $\delta H_n \rightarrow \delta H^*$ in $L^2(D)$, then a subsequence that we denote again as $\{\delta H_n\}$, tends pointwise a.e. in D to δH^* (see [118]), and this implies that the constraints (8.55) hold true for $\delta H^*(\boldsymbol{\xi})$, too.

In particular,

$$|\delta H^*(\xi)| \leq \bar{H} \quad \text{a.e. in } \xi \in D. \tag{8.59}$$

Furthermore, $\mathscr{K}$ is convex, because if $\delta H_1, \delta H_2 \in \mathscr{K}$, then

$$\tilde{H}(\xi) - \bar{H} \leq t\delta H_1 + (1-t)\delta H_2 \leq \tilde{H}(\xi), \quad 0 < t < 1. \tag{8.60}$$

Now, moving to point b), we first have to prove that $A(\delta H) \in L^2(\mathbb{R}^2)$. For this purpose it is convenient to rewrite (8.53) in the form

$$\begin{aligned} \delta g_Z = &- \mu \int_{\mathbb{R}^2} \left\{ \frac{1}{r_{x\xi}} - \frac{1}{[r_{x\xi}^2 + \tilde{H}(\xi)^2]^{1/2}} \right\} d_2\xi + \\ &- \mu \int_{\mathbb{R}^2} \left\{ \frac{1}{[r_{x\xi}^2 + \tilde{H}(\xi)^2]^{1/2}} - \frac{1}{[r_{x\xi}^2 + (\tilde{H} - \delta H)^2]^{1/2}} \right\} d_2\xi. \end{aligned} \tag{8.61}$$

The first integral on the right-hand side of (8.61) is a known function $\delta\tilde{g}_Z(\boldsymbol{x})$, while the second can be restricted to D only, because in D^c, $\delta H = 0$ and the integrand goes to zero, too. So we have

$$\begin{aligned} dg(\boldsymbol{x}) = \delta g(\boldsymbol{x}) - \delta\tilde{g}_Z(\boldsymbol{x}) &= -\mu \int_D \left\{ \frac{1}{[r_{x\xi}^2 + \tilde{H}^2]^{1/2}} - \frac{1}{[r_{x\xi}^2 + (\tilde{H} - \delta H)^2]^{1/2}} \right\} d_2\xi \\ &= -\mu \int_D a(\boldsymbol{x}, \xi | \delta H) d_2\xi = A(\delta H). \end{aligned} \tag{8.62}$$

Since

$$\frac{1}{(r^2 + H^2)^{1/2}} \leq \frac{1}{r}$$

whatever is H, we see that for every $\delta H \in \mathscr{K}$,

$$|a(\boldsymbol{x}, \boldsymbol{\xi} | \delta H)| \leq \frac{1}{r_{x\xi}}. \tag{8.63}$$

Now, assume that D is just a disk of radius R and that $r_x \le 2R$; then D is certainly contained in a disk $D(x, 4R)$ with center in x and radius $4R$, so that

$$\int_D |a(\boldsymbol{x}, \boldsymbol{\xi}|\delta H)| d_2\xi \le \int_{D(x,4R)} \frac{1}{r_{x\xi}} d_2\xi = 8\pi R. \tag{8.64}$$

Furthermore, when $r_x > 2R$, we also have

$$r_{x\xi} \ge r_x - r_\xi \ge r_x - R \quad (r_\xi \le R). \tag{8.65}$$

But then, for $r_x > 2R$, from (8.63), fixing $\boldsymbol{x}$ and $\boldsymbol{\xi}$, for some $0 < \vartheta < 1$, we obtain

$$|a| \le \frac{1}{r_{x\xi}} - \frac{1}{[r_{x\xi}^2 + \bar{H}^2]^{1/2}} \le \frac{1}{2} \frac{\bar{H}^2}{[r_{x\xi}^2 + \vartheta \bar{H}^2]^{3/2}} \le \frac{1}{2} \frac{\bar{H}^2}{r_{x\xi}^3} \le \frac{1}{2} \frac{\bar{H}^2}{(r_x - R)^2}. \tag{8.66}$$

Therefore, when $r_x > 2R$,

$$\int_D |a(\boldsymbol{x}, \boldsymbol{\xi}|\delta H)| d_2\xi \le \frac{1}{2} \frac{\bar{H}^2 |D|}{(r_x - R)^3}. \tag{8.67}$$

Putting together (8.64) and (8.67) we see that there exist a constant C and hence a function

$$F(\boldsymbol{x}) = \begin{cases} \dfrac{C}{R^3}, & \text{if } r_x \le 2R, \\ \dfrac{C}{(r_x - R)^3}, & \text{if } r_x > 2R, \end{cases} \tag{8.68}$$

such that

$$\int_D |a(\boldsymbol{x}, \boldsymbol{\xi}|\delta H)| d_2\xi \le F(\boldsymbol{x}). \tag{8.69}$$

We note that $F(\boldsymbol{x})$ is bounded above and belongs to $L^2(\mathbb{R}^2)$, so that we can assert that

$$|dg(\boldsymbol{x})| \le \frac{C}{R^3} \text{ and } ||dg(\boldsymbol{x})||_{L^2(\mathbb{R}^2)} = ||A(\delta H)||_{L^2(\mathbb{R}^2)} < +\infty. \tag{8.70}$$

Incidentally, with (8.63) and (8.66) we have proved also that there is a constant C such that, for any $\boldsymbol{x}$ and $\boldsymbol{\xi} \in D$

$$|a(\boldsymbol{x}, \boldsymbol{\xi}|\delta H)| \le \frac{C}{r_{x\xi}} \quad \forall \delta H \in \mathscr{K}. \tag{8.71}$$

Since $\frac{1}{r_{x\xi}}$ is integrable over the bounded set D as it is shown by (8.64), we have, considering the sequence $a(\boldsymbol{x}, \boldsymbol{\xi}|\delta H_n)$, that the Lebesgue *dominated convergence theorem* can be applied (see [118]), showing that

$$\{\delta H_n(\boldsymbol{\xi}) \to \delta H^*(\boldsymbol{\xi}) \text{ in } L^2(D) \text{ and a.e.}\} \Rightarrow$$
$$\Rightarrow \lim_{n\to\infty} A(\delta H_n) = \lim_{n\to\infty} \mu \int_D a(\boldsymbol{x}, \boldsymbol{\xi}|\delta H_n) d_2\xi = \tag{8.72}$$
$$= \mu \int_D a(\boldsymbol{x}, \boldsymbol{\xi}|\delta H^*) d_2\xi = A(\delta H^*).$$

The limit (8.72) is pointwise, i.e., for any $x \in \mathbb{R}^2$. So $\delta g_n(\boldsymbol{x}) = A(\delta H_n) \to \delta g^*(\boldsymbol{x}) = A(\delta H^*)$. On the other hand, by (8.69)

$$|dg_n(\boldsymbol{x})|, |dg^*(\boldsymbol{x})| \leq F(\boldsymbol{x}) \in L^2(\mathbb{R}^2),$$

so that

$$|dg_n(\boldsymbol{x}) - dg^*(\boldsymbol{x})|^2 \leq 4F(\boldsymbol{x})^2,$$

which is integrable on $\mathbb{R}^2$. Therefore, invoking again the dominated convergence theorem, we can conclude that

$$\lim_{n\to\infty} \int_{\mathbb{R}^2} |dg_n(\boldsymbol{x}) - dg^*(\boldsymbol{x})|^2 d_2x = \lim_{n\to\infty} ||A(\delta H_n) - A(\delta H^*)||^2_{L^2\mathbb{R}^2} = 0. \tag{8.73}$$

The continuity of $A(\delta H)$ on the set $\mathscr{K}$ is thus proved.

For later reference, we observe that the above reasoning applies to $L^1(\mathbb{R}^2)$, too, because $F(x)$ in (8.68) is also integrable over the whole plane. In other words, $A(\delta H_n)$ is continuous from L^1 (D) to $L^1(\mathbb{R}^2)$ as well, i.e.,

$$\lim_{n\to\infty} ||A(\delta H_n) - A(\delta H^*)||_{L^1} = \lim_{n\to\infty} \int_{\mathbb{R}^2} |dg_n(\boldsymbol{x}) - dg^*(\boldsymbol{x})| |d_2x = 0. \tag{8.74}$$

8.3.2 Model B (Linearized, with Reference $\tilde{H}$)

The linearization of (8.62) with respect to δH is straightforward and gives

$$dg(\boldsymbol{x}) = A\delta H = \mu \int_D \frac{\tilde{H}(\xi)}{[r^2_{x\xi} + \tilde{H}(\xi)^2]^{3/2}} \delta H(\boldsymbol{\xi}) d_2\xi. \tag{8.75}$$

To treat this model, it is convenient to use the Fourier transform of (8.75), namely, always using Proposition 3.3,

$$d\widehat{g}(\boldsymbol{p}) = \widehat{A}\delta H = 2\pi\mu \int_D e^{i2\pi \boldsymbol{p}\cdot\boldsymbol{\xi} - 2\pi p \tilde{H}(\boldsymbol{\xi})} \delta H(\boldsymbol{\xi}) d_2\xi. \tag{8.76}$$

In this case we shall define $\mathscr{K}$ by imposing on $\delta H(\xi)$ the constraints

$$H(\boldsymbol{\xi}) = 0, \qquad \boldsymbol{\xi} \in D^c, \qquad \delta H(\boldsymbol{\xi}) < \tilde{H}(\boldsymbol{\xi}). \tag{8.77}$$

So the lower bound constraint

$$\delta H(\boldsymbol{\xi}) \geq \tilde{H}(\boldsymbol{\xi}) - \bar{H},$$

is relaxed. However, in this case we have to add the following condition on the reference surface $\tilde{S}$:

$$\tilde{H}(\boldsymbol{\xi}) \geq H_0 > 0, \quad \boldsymbol{\xi} \in D. \tag{8.78}$$

Since with these constraints $\tilde{H}$ and H_0 need not have a specific value, we consider that (8.78) is not too restrictive. To verify L^2 continuity we can use (8.76), in two steps: first we use the Schwarz inequality to get

$$\begin{aligned} |d\widehat{g}(\boldsymbol{p})|^2 &\leq \mu^2 4\pi^2 \int_D e^{-4\pi p \tilde{H}(\boldsymbol{\xi})} d_2\xi \cdot \int_D |\delta H(\boldsymbol{\xi})|^2 d_2\xi \leq \\ &\leq 4\pi^2\mu^2 e^{-4\pi p H_0} |D| \, ||\delta H||^2_{L^2(D)}. \end{aligned} \tag{8.79}$$

Then we integrate (8.79) and, recalling Plancherel's Theorem, we get

$$\begin{aligned} ||dg||^2_{L^2(\mathbb{R}^2)} &= ||d\widehat{g}||^2_{L^2(\mathbb{R}^2)} \leq 4\pi^2\mu^2 |D| 2\pi \int_0^{+\infty} e^{-4\pi H_0 p} p dp ||\delta H||^2_{L^2(D)} = \\ &= \mu^2 \frac{\pi}{2} \frac{|D|}{H_0^2} ||\delta H||^2_{L^2(D)}. \end{aligned} \tag{8.80}$$

This proves at once that $\widehat{A}$, and hence A, is a continuous operator from $\mathscr{K} \subset L^2(D)$ to $L^2(\mathbb{R}^2)$.

In fact, we can assert even something more, because (8.78) implies that the kernel of the integral operator (8.76),

$$\widehat{a}(\boldsymbol{p}, \xi) = e^{i2\pi \boldsymbol{p}\cdot\boldsymbol{\xi} - 2\pi p \tilde{H}(\xi)},$$

is such that

$$\int_{R^2} \int_D |\widehat{a}(\boldsymbol{p}, \xi)|^2 d^2\xi d_2 p \leq \frac{1}{8\pi} \frac{|D|}{H_0^2}. \tag{8.81}$$

Therefore, according to a classical theorem of Picard on integral equations of the first kind (see [57]), the operator $\widehat{A}$ is not only continuous but also compact. This means that its inverse, if it exists, cannot be bounded, i.e., there are sequences $\delta H_n \in L^2(D)$ such that $||\delta H_n||_{L^2(D)} = 1$ while $||\delta \widehat{g}_n||^2_{L^2(D)} = ||\widehat{A}\delta H_n||_{L^2(\mathbb{R}^2)} \to 0$. We shall return later to this comment.

8.3.3 Model C (Linearized, with Constant Reference $\bar{H}$)

In this model we assume that the variation $\delta H(\boldsymbol{\xi})$ that we used for the linearization of the gravimetric problem, can be referred to a flat horizon, i.e., we assume that $\tilde{S} = \{z = -\bar{H}\}$, with $\bar{H}$ constant.

In this case the set $\mathscr{K}$ is simply defined, by the constraint

$$\delta H(\boldsymbol{\xi}) \leq \bar{H}; \tag{8.82}$$

apart from that, we must only have $\delta H \in L^2(\mathbb{R}^2)$. What is remarkable in this case is that the Eq. (8.75) becomes a simple convolution equation, namely

$$dg(\boldsymbol{x}) = \mu \int_{R^2} \frac{\bar{H}}{[r^2_{\boldsymbol{x},\boldsymbol{\xi}} + \bar{H}^2]^{3/2}} \delta H(\boldsymbol{\xi}) d_2\xi. \tag{8.83}$$

Accordingly, Eq. (8.76) becomes

$$d\widehat{g}(\boldsymbol{p}) = 2\pi\mu e^{-2\pi p\bar{H}} \delta\widehat{H}(\boldsymbol{p}). \tag{8.84}$$

Since

$$0 < e^{-2\pi p\bar{H}} \leq 1 \quad \forall \boldsymbol{p}, \tag{8.85}$$

(8.84) implies that

$$\begin{aligned} ||dg||^2_{L^2(\mathbb{R}^2)} &= ||d\widehat{g}(\boldsymbol{p})||^2_{L^2(\mathbb{R}^2)} \leq \mu^2 4\pi^2 ||\delta\widehat{H}(\boldsymbol{p})||^2_{L^2(\mathbb{R}^2)} = \\ &= \mu^2 4\pi^2 ||\delta H(\boldsymbol{\xi})||^2_{L^2(\mathbb{R}^2)} \end{aligned} \tag{8.86}$$

so that the continuity of $\widehat{A}$ and then of A in $L^2(\mathbb{R}^2)$ is straightforward.

One has to remark that, on the one hand, we are able with this model to relax all conditions in the definition of $\mathscr{K}$, apart from (8.82); in particular, we don't need to require that $\delta H(\boldsymbol{\xi}) = 0$ ($\boldsymbol{\xi} \in D^C$) to give a meaning to the operator A. On the other hand, we loose one property of A with respect to the model B; in fact, in this case A is clearly selfadjoint in $L^2(\mathbb{R}^2)$ and even positive, because

$$\langle A\delta H, \delta H\rangle = (e\widehat{A}\delta\widehat{H}, \delta\widehat{H}) = 2\pi\mu \int e^{-2\pi p\bar{H}} |\delta\widehat{H}(\boldsymbol{p})|^2 d_2 p \geq 0,$$

but it is no longer compact.

In fact, consider the sequence $\delta\widehat{H}_n(\boldsymbol{p}) = ce^{i2n\alpha - 2\pi p}$, with α the angular anomaly of $\boldsymbol{p}$ in $\mathbb{R}^2$ and c a conveniently small constant; one easily verifies that these are $L^2(\mathbb{R}^2)$ orthogonal functions with constant norm

$$||\delta\widehat{H}_n(\boldsymbol{p})||^2_{L^2(\mathbb{R}^2)} = c^2 \int_0 \int_0^{+\infty} e^{-4\pi p} pd\, pd\alpha = \frac{c^2}{8\pi}.$$

On the other hand, the corresponding functions $d\widehat{g}_n$ are

$$d\widehat{g}_n = c\mu 2\pi e^{i2n\alpha} e^{-2\pi p(1+\bar{H})},$$

which indeed are again orthogonal functions with constant norm given by

$$||d\widehat{g}_n||^2_{L^2(\mathbb{R}^2)} = \frac{c^2\mu^2\pi}{2(1+\bar{H})^2}.$$

So, no subsequence of $\{d\widehat{g}_n\}$ can be strongly convergent and $\widehat{A}$ (and therefore A too) cannot be compact.

Indeed, such a property would be reacquired, if we restricted A to $L^2(D)$, as a subspace of $L^2(\mathbb{R}^2)$. In fact, in this case, repeating the calculations done in (8.81) with $\widehat{a}(\boldsymbol{p}, \boldsymbol{\xi}) = e^{2\pi i \boldsymbol{p}\cdot\boldsymbol{\xi} - 2\pi p\bar{H}}$ one gets

$$\int_{\mathbb{R}^2} \int_D |\widehat{a}(\boldsymbol{p}, \boldsymbol{\xi})|^2 d_2\xi\, d_2 p = \frac{1}{8\pi}\frac{|D|}{H^2} < +\infty,$$

so that now $\widehat{A}$ (and A) is again compact. In spite of the above remark, in general $\widehat{A}$, and therefore A, have in any case an unbounded inverse $\widehat{A}^{-1}$, and A^{-1}. In fact, one deduces from (8.85) that

$$\delta\widehat{H}(\boldsymbol{p}) = \frac{1}{2\pi\mu} e^{2\pi p\bar{H}} \delta\widehat{g}(\boldsymbol{p}), \tag{8.87}$$

and it is clear that $\delta\widehat{H} \in L^2$ only if

$$\int e^{4\pi p\bar{H}} |\delta\widehat{g}(\boldsymbol{p})|^2 d_2 p < +\infty. \tag{8.88}$$

Obviously not all $\delta\widehat{g} \in L^2$ can satisfy the constraint (8.88); try for instance $\delta\widehat{g}(p) = (1 + p^2)^{-1}$ which is in $L^2(\mathbb{R}^2)$, but yields an infinite integral of the type (8.88).

This is basically because the direct operator exponentially smoothes the high frequencies, so that the inverse operator blows up the high frequency spectrum of the data.

So the operator $\widehat{A}^{-1}$ is unbounded and, although in this case it has a simple explicit form (8.87), one cannot guarantee that a small variation in the data $\delta\widehat{g}$ will result in a small variation of the solution $\delta\widehat{H}$.

Concluding this section we can say, for the constant density two-layer models considered, the operator A that links the interface anomaly $\delta H(\xi)$ to the gravity anomaly dg, is always a continuous operator from the suitably restricted set of functions $\mathscr{K} \subset L^2$, to $L^2(\mathbb{R}^2)$; yet the inverse operator A^{-1}, assuming it exists, is not at all bounded from $L^2(\mathbb{R}^2)$ to $\mathscr{K}$, i.e., the inverse problem in all cases is improperly posed, in the given topologies, and, to get approximate solutions we need to find proper regularization tools.

However, the question of the uniqueness of the solution has to be analyzed.

8.4 Some Uniqueness Results for the Two-Layer Model

We address here the uniqueness of the solution for the case of the two layers of constant density, according to models A and B of the previous section. We do not have to deal with model C, since we have an explicit expression, i.e. (8.87), for the inverse of $\widehat{A}$, meaning that the corresponding direct equation has certainly a unique solution, when it exists.

Since uniqueness depends critically on the so-called unique continuation property of harmonic functions, already mentioned in the text, we establish it formally here.

Proposition 8.3 *Let $u(P)$ be a regular harmonic function in an open, connected set Ω and assume that*

$$\begin{aligned} &u(P) = 0, \ \forall P \in B(\bar{P}, \varepsilon) \subset \Omega, \\ &(B(\bar{P}, \varepsilon) \equiv \{P; \mid P\bar{P}| \le \varepsilon\}). \end{aligned} \tag{8.89}$$

Then

$$u(P) = 0 \quad \forall P \in \Omega.$$

We send to the literature for the proof (e.g., [6, 133]), but we complete the reasoning by a Remark.

Remark 8.3 Assume that a certain Newtonian gravitation potential u generates a null attraction in a fixed direction (e.g., in the direction of the Z-axis) in space, in a set Ω which is the complement of the bounded set B, $\Omega = B^c$. This means that the masses generating u are all contained in B and that the total mass is finite, so that

$u(P)$ tends to 0 at infinity; then indeed $u(P)$ is itself zero everywhere in Ω. In fact, u is harmonic in Ω and therefore, in the half-space $\mathbb{R}^3_H = \{z \geq H\}$, with H large enough to leave the body B below the plane $\{z = H\}$; then we have

$$u(\boldsymbol{x}, z) = -\int_z^{+\infty} u_z(\boldsymbol{x}, \zeta) d\zeta \equiv 0. \tag{8.90}$$

But then $u(P)$ is identically zero in $\mathbb{R}^3_H$ as well as in $\Omega \subset \mathbb{R}^3_H$, because of Proposition 8.3.

We can pass now to the proofs of uniqueness.

8.4.1 Model A

The proof is based on Theorem 8.1. In order to use a simple argument, we shall adopt some restrictive hypotheses on the regularity of the interface $S \equiv \{z = -\tilde{H}(\boldsymbol{\xi}) + \delta H(\boldsymbol{\xi})\}$ between the two layers; namely, we shall assume that $H(\boldsymbol{\xi}) = H(\boldsymbol{\xi}) - \delta H(\boldsymbol{\xi})$ is continuous together with its first derivatives. This implies that S is endowed with a continuous normal vector $\boldsymbol{n}$.

Moreover, the vector $\boldsymbol{n}$ is never orthogonal to the z-direction, i.e., S never reaches an inclination of 90°. Let us further recall that, according to Model A, $\delta H(\boldsymbol{\xi})$ is zero outside the bounded set D, and in addition $H(\boldsymbol{\xi}) \leq \bar{H}$, $\forall \boldsymbol{\xi} \in D$, so that essentially the mass (with constant density ρ) generating $\delta g_Z(\boldsymbol{x})$ is all contained into the cylinder with base D and height $\bar{H}$.

Now assume that two layers, with the same density and with interfaces $H_1(\boldsymbol{\xi}) = \tilde{H}(\boldsymbol{\xi}) - \delta H_1(\boldsymbol{\xi})$, $H_2(\boldsymbol{\xi}) = \tilde{H}(\boldsymbol{\xi}) - \delta H_2(\boldsymbol{\xi})$, generate the same $\delta g_Z(\boldsymbol{x})$ on S_0, and therefore on the whole upper halfspace $\mathbb{R}^3_+$. We note that indeed both $H_1(\boldsymbol{\xi})$ and $H_2(\boldsymbol{\xi})$ coincide with $\tilde{H}(\boldsymbol{\xi})$ when $\boldsymbol{\xi} \in D^c$, so if we take the difference between the two mass distributions we are left with a body constituted of a collection of "meatballs" $\{B_k\}$, as in Fig. 8.3, with density $\pm\rho$ in each of the B_k, depending whether S_1 or S_2 stays above the other. So we see that the body $\delta B = \bigcup_k B_k$ with density

$$\boldsymbol{\xi} \in B_k \quad \bar{\rho}(\boldsymbol{\xi}) = \begin{cases} +\rho, \text{if} - H_1(\boldsymbol{\xi}) > -H_2(\boldsymbol{\xi}) \text{ in } B_k, \\ -\rho, \text{if} - H_2(\boldsymbol{\xi}) > -H_1(\boldsymbol{\xi}) \text{ in } B_k, \end{cases} \tag{8.91}$$

generates a null attraction on S_0 and above. Since δB is bounded, thanks to Remark 8.3, $\bar{\rho}(\boldsymbol{\xi})$ generates a null attraction everywhere in δB^c, and so also a null potential in the same set. But then, according to Theorem 8.1, $\bar{\rho}(\boldsymbol{\xi})$ must be orthogonal in $L^2(\delta B)$ to all distributions $\rho_h(\boldsymbol{\xi})$ harmonic in δB and square integrable there, namely

$$\int_{\delta B} \bar{\rho}(\boldsymbol{\xi}) \rho_h(\boldsymbol{\xi}, \zeta) dB = 0, \quad \forall \rho_h \in HL^2(\delta B). \tag{8.92}$$

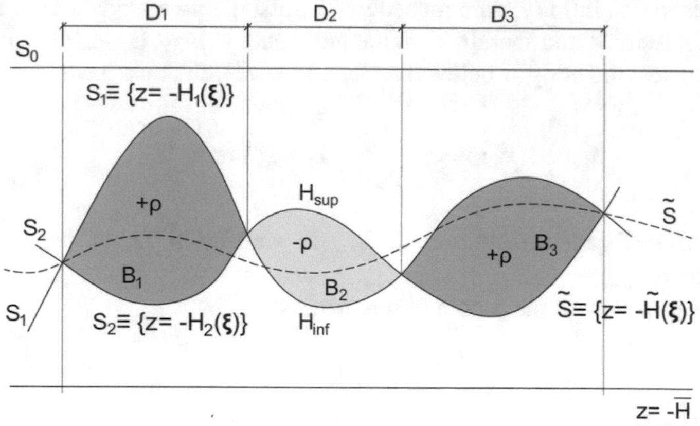

Fig. 8.3 An example of the difference between two bodies of constant density ρ with the same base D, lower, $z = -\bar{H}$ and reference $z = -\bar{H}(\xi)$

Since δB is composed of sets disjoint in terms of volume, i.e., $m(\delta B_k \cap \delta B_j) = 0$, $k \neq j$, a function ρ_h harmonic in δB has to be separately harmonic in each of the components B_k. In this sense $\bar{\rho}(\boldsymbol{\xi})$ itself is harmonic in δB, so that we can take $\rho_h = \bar{\rho}$ in (8.92) showing that we must have

$$\bar{\rho}^2 m(\delta B) = 0 \quad \Longrightarrow \quad m(\delta B) = 0;$$

since, by using the notation of Fig. 8.3

$$m(\delta B) = \int_D [H_{\text{sup}}(\boldsymbol{\xi}) - H_{\text{inf}}(\boldsymbol{\xi})] d_2\xi = 0$$

we must have

$$H_{\text{sup}}(\boldsymbol{\xi}) = H_{\text{inf}}(\boldsymbol{\xi}). \tag{8.93}$$

Since $H_{\text{Sup}} = (H_1 \text{ or } H_2)$ and $H_{\text{inf}} = (H_2 \text{ or } H_1)$, (8.93) implies

$$H_1(\boldsymbol{\xi}) = H_2(\boldsymbol{\xi}),$$

i.e., uniqueness of the solution is proved. We want to emphasize that, in order to prove the uniqueness of the solution, we had to put restrictive conditions on the function $H(\boldsymbol{\xi})$ and its gradient. Such conditions could be relaxed, for instance by allowing $H(\boldsymbol{\xi})$ to have a finite number of jumps, in which case an adaptation of the above reasoning following an old idea by Novikov and others [95, 126, 162] can still prove uniqueness.

Yet, to our knowledge, the reasoning cannot be pushed so far as to cover the case $H(\boldsymbol{\xi}) \in L^2(D)$. This is a warning we will take up again when discussing the Tikhonov regularization in the next section.

8.4.2 Model B

In this case we have to prove the uniqueness of the solution of Eq. (8.75), where the operator A is now linear, continuous and even compact, $L^2(D) \to L^2(\mathbb{R}^2)$, as shown by (8.81).

Here we assume that the reference surface, $\tilde{S} = \{z = -\tilde{H}(\boldsymbol{\xi})\}$, is continuous with a continuous and bounded gradient, namely with a continuous normal field $\tilde{\boldsymbol{n}}$ which is never orthogonal to the z-axis, at least for $\boldsymbol{\xi} \in D$. Note that this time the restriction is not so tough, as it concerns only the reference surface $\tilde{S}$, for the choice of which we have a certain amount of freedom.

To proceed, following [127], we first consider the function

$$dg(\boldsymbol{x}, z) = \mu \int_D \frac{\tilde{H}(\boldsymbol{\xi}) + z}{[r_{\boldsymbol{x}\boldsymbol{\xi}}^2 + (\tilde{H}(\boldsymbol{\xi}) + z)^2]^{3/2}} \delta H(\boldsymbol{\xi}) d_2\xi, \tag{8.94}$$

which is obviously harmonic as a function of $P \equiv (\boldsymbol{x}, z)$, everywhere in $\mathbb{R}^3$, outside the bounded piece of surface $\tilde{S}$ that projects on D (see Fig. 8.4), let us call it $\tilde{S}_D$.

We notice that the operator A can be written as

$$dg(\boldsymbol{x}) = A(\delta H) \equiv dg(\boldsymbol{x}, z)|_{z=0}, \tag{8.95}$$

i.e., the known term of (8.75) is just $dg(\boldsymbol{x}, z)$ restricted to $S_0 = \{z = 0\}$. Therefore, if we have a $\delta H(\xi)$ that gives $dg(\boldsymbol{x}) = 0$, the same $\delta H(\xi)$ gives a harmonic $dg(\boldsymbol{x}, z)$ which is zero on S and therefore on $\mathbb{R}^3_+$ as well.

Notice that, as we shall see below, the integral over D in (8.94) can as well be viewed as an integral over $\tilde{S}_D$.

So, as a consequence of Proposition (8.3) and Remark 8.3, $dg(\boldsymbol{x}, z)$ must vanish everywhere in $\mathbb{R}^3$ outside $\tilde{S}_D$, and even the potential $u(\boldsymbol{x}, z)$ such that

$$\frac{\partial}{\partial z} u(\boldsymbol{x}, z) = -dg(\boldsymbol{x}, z),$$

namely

$$u(\boldsymbol{x}, z) = \mu \int_D \frac{\delta H(\boldsymbol{\xi})}{[r_{x\xi}^2 + (\tilde{H}(\xi) + z)^2]^{3/2}} d_2\xi, \tag{8.96}$$

must vanish in $\mathbb{R}^3 \backslash \tilde{S}_D$.

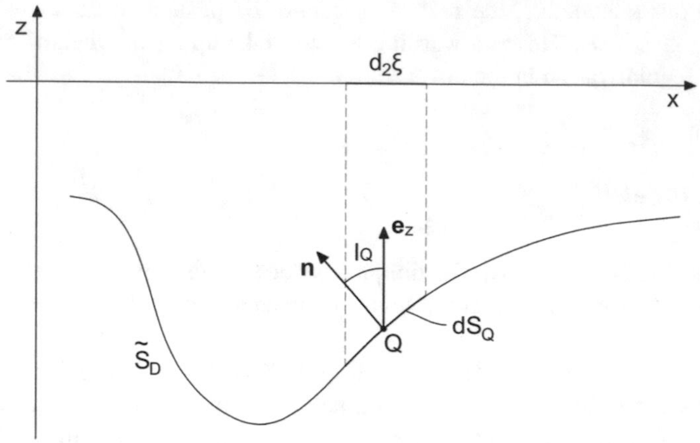

Fig. 8.4 The geometry of the single layer on $\tilde{S}_D$

Now recall that (see Fig. 8.4), letting I_ξ denote the inclination of dS on the (x, y)-plane,

$$d_2\boldsymbol{\xi} = dS_\xi \cos I_\xi, \tag{8.97}$$

and notice that to any $\boldsymbol{\xi}$ in D corresponds one and only one $Q \in \tilde{S}_D$, for which it holds $\zeta_Q = -\bar{H}(\boldsymbol{\xi})$ too. Therefore, (8.96) can be re-written as

$$\begin{gathered} u(P) = \mu \int_{\tilde{S}_D} \frac{\alpha(Q)}{\ell_{PQ}} dS_Q \\ (\alpha(Q) = \delta H(Q) \cos I_Q). \end{gathered} \tag{8.98}$$

Since

$$|\alpha(Q)| \leq |\delta H(Q)|, \tag{8.99}$$

as $\delta H \in L^2(D)$ and therefore also $\delta H \in L^1(D)$, because D is bounded, we see that $\alpha(Q) \in L^1(\tilde{S}_D)$, too, so that (8.98) represents in a legitimate form a single-layer potential (see Remark 1.5).

This statement comes from

$$\begin{aligned} &\int_{\tilde{S}_D} |\alpha(Q)| dS_Q \leq \int_{\tilde{S}_D} |\delta H(\boldsymbol{\xi})| \cos I_Q dS_Q = \\ &= \int_D |\delta H(\boldsymbol{\xi})| d_2\xi \leq \left(\int_D \delta H^2(\boldsymbol{\xi}) d_2\xi \right)^{1/2} |D|^{1/2}. \end{aligned}$$

At this point, again referring to Remark 1.5, we find that, almost everywhere on $\tilde{S}_D$,

$$-4\pi\alpha(Q) = \frac{\partial u}{\partial n}(Q_+) - \frac{\partial u}{\partial n}(Q_-). \tag{8.100}$$

But, since u is zero in $\mathbb{R}^3 \backslash \tilde{S}_D$, the right-hand side of (8.100) is zero too, i.e.,

$$\alpha(Q) = \delta H(Q) \cos I_Q = 0 \quad \text{a.e. } Q \in \tilde{S}_D;$$

on the other hand, since, according to our hypotheses,

$$\cos I_Q \geq c > 0$$

we find that

$$\delta H(Q) \equiv \delta H(\boldsymbol{\xi}) = 0 \quad \text{a.e. on } D. \tag{8.101}$$

The uniqueness of the solution of (8.75) is thus proved, under the above hypotheses. This concludes our discussion about the uniqueness of solutions in Models A, B and C.

8.5 Tikhonov Regularized Solutions for the Inversion of a Two-Layer model

Since the inverse gravimetric problem is always improperly posed, particularly for the models A, B and C discussed in Sect. 8.2, the only hope to derive sensible information on the solution from an imperfect knowledge of the data is to accept to find an approximate solution to the direct problem by applying some regularization technique, especially the Tikhonov regularization discussed in Appendix C.

We will do that separately for each of the three models.

8.5.1 *Model A*

This is the nonlinear model (8.62) and we want to apply to it the Theorem (C.1) on the existence of the minimum of the Tikhonov principle for the functional,

$$J_\lambda(dg, \delta H) = \|dg - A(\delta H)\|_Y^2 + \lambda \|\delta H\|_X^2 . \tag{8.102}$$

The set $\mathscr{K}$ and the Hilbert space Y for the continuity of $A(\delta H)$ have already been established in (8.54), (8.55) and (8.58). In particular, in (8.58) we proved that the nonlinear operator $A(\delta H)$ is continuous from $L^2(D)$ to $L^2(\mathbb{R}^2)$, so that it is only natural to choose as "large" space $Y \equiv L^2(D)$.

We have therefore only to define what is the space X such that (see (C.5))

$$\mathscr{K} \subset X \subset L^2(D) \tag{8.103}$$

and the embedding of X in $L^2(D)$ is compact.

A simple choice in this case is to look at one of the Sobolev spaces $H^{n,2}(D)$ consisting of functions square integrable on D together with their derivatives up to the order n. Indeed, the higher n is, the more regular the functions belonging to $H^{n,2}$ are. For instance, in [115] one can find a nice numerical example with the choice $\delta H \in H^{2,2}(D)$; here we prefer the choice $\delta H \in H^{1,2}(D)$, namely

$$\|\delta H\|_X^2 = \int_D \left(\delta H^2(\boldsymbol{\xi}) + |\nabla \delta H(\boldsymbol{\xi})|^2\right) d_2\xi < \infty, \tag{8.104}$$

because such a choice guarantees on the one hand the compact embedding of $X = H^{1,2}(D)$ into $L^2(D)$ (Rellich's theorem) and on the other hand it allows δH to be a little more irregular, namely closer to a geophysically acceptable behaviour.

So in this case the Tikhonov functional looks as follows:

$$J_\lambda(dg, \delta H) = \int_{\mathbb{R}^2} \left| dg(\boldsymbol{x}) - \mu \int_D \left[\frac{1}{\left[r_{x\xi}^2 + \tilde{H}\right]^{1/2}} - \frac{1}{\left[r_{x\xi}^2 + \left(\tilde{H} - \delta H\right)^2\right]^{1/2}} \right] \right|^2 d_2\xi\, d_2x$$
$$+ \lambda \int_D \left(\delta H^2 + |\nabla \delta H|^2\right) d_2\xi; \tag{8.105}$$

the choice of λ according to Morozov's discrepancy principle (see Appendix C) requires some knowledge of the possible error present in the data $dg(\boldsymbol{x})$. The numerical solution of the minimum problem (8.105) could be obtained by discretization and a further application of the classical Newton–Raphson method, or, perhaps more easily, by linearization and simple iteration, although in this case there is no guarantee of convergence.

Remark 8.4 We make this remark of negative nature because it highlights a global characteristic of the inverse gravimetric problem that we need not misinterpret.

Sometimes, in the theory of solutions of nonlinear equations, a role is played by operators enjoying a monotonicity property. This, however, is not our case, because while the direct nonlinear operator

$$dg = A(\delta H) \tag{8.106}$$

is indeed monotone in δH, namely $\delta H_1 > \delta H_2 \Rightarrow dg_1 < dg_2$, if $\rho_0 > 0$ (and viceversa if $\rho_0 < 0$) as a simple drawing can show. The converse statement, i.e., $dg_1 < dg_2 \Rightarrow \delta H_1 > \delta H_2$, is false.

It is enough to build a counterexample, based on Fig. 8.5. The first interface S_1 is just a plane $z = -\bar{H}_1 = -\left(\bar{H}_2 - \delta H_1\right)$ with a constant $\delta H_1 = \bar{H}_2 - \bar{H}_1$; the second interface S_2 is a composition of a flat Bouguer plate, $z = -\bar{H}_2$, with a perturbation δH_2 that is just a column of density 0 with radius ε, jumping up to zero. The attraction $d\bar{g}_1$ is just a constant; the attraction dg_2 is the difference between a constant $d\bar{g}_2$ and the attraction of the matter missing in the column. Since this tends to 0 when $\varepsilon \to 0$, we see that for a sufficiently small ε we will have $dg_2 > dg_1$ without having $\delta H_1 > \delta H_2$, i.e., monotonicity is not verified. This remark, though elementary, is important to avoid misinterpretations when inverting gravity data.

8.5.2 *Model B*

As we have already stated, this model is important because this is what one finds when trying to solve iteratively the nonlinear problem. In fact, even if we start

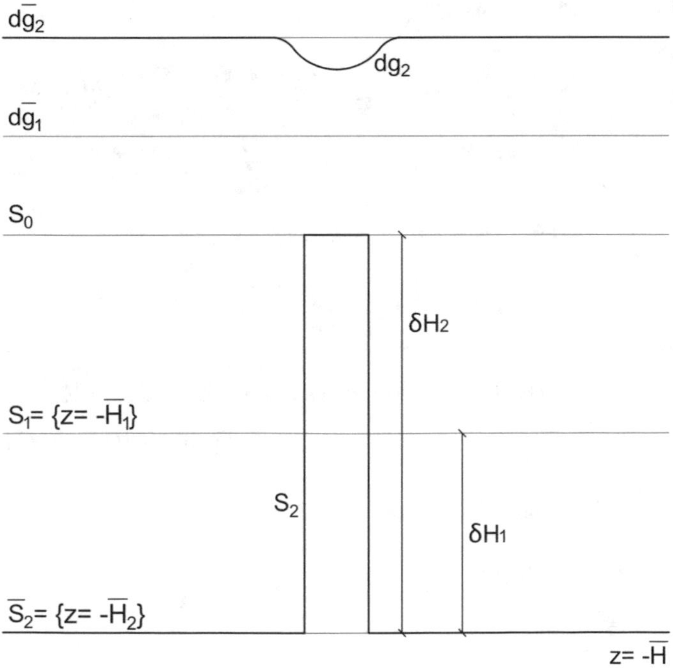

Fig. 8.5 A two-layer configuration where $\Delta g_2 > dg_1 = d\bar{g}$: 1, but $\delta H_2 \equiv \left\{0,\ \xi \in D^c_\varepsilon; = \bar{H}_2 \xi \in D^c_\varepsilon\right\}$ is not smaller than $\delta H_1 = \bar{H}_2 - \bar{H}_1$

linearizing with a flat horizon (Model C), soon after the first solution the new approximate reference $\tilde{S}$ can no longer be taken flat.

In order to write the Tikhonov normal equation for this model, we prove first a proposition that is a combination of Propositions 3.3 and 3.2.

Proposition 8.4 *One has*

$$\int \frac{H_1}{\left[r_{x\xi}^2 + H_1^2\right]^{3/2}} \frac{H_2}{\left[r_{x\eta}^2 + H_2^2\right]^{3/2}} d_2x = 2\pi \frac{H_1 + H_2}{\left[r_{\xi\eta}^2 + (H_1 + H_2)^2\right]^{3/2}} \tag{8.107}$$

Proof Let us first recall that

$$\int_{R^2} e^{-i2\pi \boldsymbol{p}\cdot\boldsymbol{x}} d_2x = \delta(\boldsymbol{x}).$$

Therefore, using Proposition 3.3 twice i.e.,

$$\frac{H}{\left[r_{x\xi}^2 + H^2\right]^{3/2}} = 2\pi \int e^{-i\pi \boldsymbol{p}\cdot(\boldsymbol{x}-\boldsymbol{\xi}) - 2\pi p H} d_2p$$

in the first term in (8.107), we get

$$4\pi^2 \int_{\mathbb{R}^2} \int_{\mathbb{R}^2} e^{-i2\pi \boldsymbol{p}\cdot(\boldsymbol{x}-\boldsymbol{\xi}) - 2\pi p H_1} \int_{\mathbb{R}^2} e^{-i2\pi \boldsymbol{q}\cdot(\boldsymbol{x}-\boldsymbol{\eta}) - 2\pi q H_2} d_2p d_2q d_2x =$$

$$= 4\pi^2 \int_{\mathbb{R}^2} \int_{\mathbb{R}^2} e^{i2\pi \boldsymbol{p}\cdot\boldsymbol{\xi} + i2\pi \boldsymbol{q}\cdot\boldsymbol{\eta} - 2\pi p H_1 - 2\pi q H_2} \int_{\mathbb{R}^2} e^{-i2\pi(\boldsymbol{p}+\boldsymbol{q})\cdot\boldsymbol{x}} d_2x d_2q d_2p =$$

$$4\pi^2 \int_{\mathbb{R}^2} \int_{\mathbb{R}^2} e^{i2\pi \boldsymbol{p}\cdot\boldsymbol{\xi} + i2\pi \boldsymbol{q}\cdot\boldsymbol{\eta} - 2\pi p H_1 - 2\pi q H_2} \delta(\boldsymbol{p} + \boldsymbol{q}) d_2q d_2p =$$

$$4\pi^2 \int_{\mathbb{R}^2} e^{i2\pi \boldsymbol{p}\cdot\boldsymbol{\xi} - i2\pi \boldsymbol{p}\cdot\boldsymbol{\eta} - 2\pi p H_1 - 2\pi p H_2} d_2p =$$

$$= 4\pi^2 \int_{\mathbb{R}^2} e^{-i2\pi \boldsymbol{p}\cdot(\boldsymbol{\eta}-\boldsymbol{\xi})} e^{-2\pi p(H_1+H_2)} d_2p = 2\pi \frac{H_1 + H_2}{\left[r_{\xi\eta}^2 + (H_1 + H_2)^2\right]^{3/2}}.$$

We observe that if we put $H_1 = \tilde{H}(\boldsymbol{\xi})$ and $H_2 = \tilde{H}(\boldsymbol{\eta})$, then (8.107) reads

$$\int_{\mathbb{R}^2} \frac{\tilde{H}(\boldsymbol{\xi})}{\left[r_{x\xi^2 + \tilde{H}(\xi)^2}\right]^{3/2}}, \frac{\tilde{H}(\boldsymbol{\eta})}{\left[r_{x\eta}^2 + \tilde{H}(\boldsymbol{\eta})^2\right]^{3/2}} d_2x = \tag{8.108}$$

$$= 2\pi \frac{\tilde{H}(\boldsymbol{\xi}) + \tilde{H}(\boldsymbol{\eta})}{\left[r_{\xi\eta}^2 + \left(\tilde{H}(\boldsymbol{\xi}) + \tilde{H}(\boldsymbol{\eta}) \right)^2 \right]^{3/2}}. \tag{8.109}$$

□

As we have seen in Sect. 8.2, the operator (8.73) is continuous from $L^2(D)$ to $L^2(D^2)$, when $\tilde{H}(\xi)$ satisfies the further constraint (8.78).

This allows us to apply Theorem C.2 (see Appendix C), choosing $X \equiv L^2(D)$ and $Y \equiv L^2(\mathbb{R}^2)$.

So in this case our Tikhonov functional writes

$$J_\lambda(dg, \delta H) = \|dg - A\delta H\|_{L^2(R^2)} + \lambda \|\delta H\|_{L^2(D)}^2 =$$

$$= \int_{R^2} \left[dg(x) - \int_D \mu \frac{\tilde{H}(\xi)}{\left[r_{x\xi}^2 + \tilde{H}(\xi)^2 \right]^{3/2}} \delta H(\xi)\, d_2\xi \right]^2 d_2x +$$

$$+ \lambda \int_D \delta H(\xi)^2 d_2\xi. \tag{8.110}$$

We note that with the above definition of X and Y, the transpose $A^\top$ of the operator A is also an integral operator from $L^2(\mathbb{R}^2)$ to $L^2(D)$, given by

$$A^\top dg = \mu \int_{R^2} \frac{\tilde{H}(\xi)}{\left[r_{x\xi}^2 + \tilde{H}(\xi)^2 \right]^{3/2}} dg(x)\, d_2x, \tag{8.111}$$

namely we exchange the roles of $\boldsymbol{x}$ and $\boldsymbol{\xi}$ in the kernel of the operator A.

After this remark, we find that the normal equation of Tikhonov's principle becomes

$$A^\top A(\delta H) + \lambda\, \delta H = A^\top dg \tag{8.112}$$

with $A^\top \boldsymbol{dg}$ given by (8.111) and $A^\top A(\delta H)$, thanks to (8.109), given by

$$\boldsymbol{\xi} \in D, \quad A^\top A(\delta H) = \mu^2 \int_D \frac{\tilde{H}(\boldsymbol{\xi}) + \tilde{H}(\boldsymbol{\eta})}{\left[r_{\xi\eta}^2 + \left(\tilde{H}(\boldsymbol{\xi}) + \tilde{H}(\boldsymbol{\eta}) \right)^2 \right]^{3/2}} \delta H(\boldsymbol{\eta})\, d_2\eta. \tag{8.113}$$

Equation (8.112) is therefore an integral equation that can be solved algebraically by discretization.

We have to stress here that no attention has been paid to the constraints (8.77). Concerning for the first, however, we do not need to worry any further, because it is automatically accounted for by defining $X = L^2(D)$, namely a space of functions different from 0 only in D. As a matter of fact, (8.112) is an equation from $L^2(D)$ to $L^2(D)$, so its solution is by definition a function δH defined in D only.

The second constraint in (8.77), though, is more difficult to treat. However, if we consider the discussion in Remark 8.2, particularly observing Fig. 8.5, we understand that there are two possibilities: either the solution δH of (8.112) satisfies, so to say spontaneously, the inequality (8.77) and then we have reached the minimum, or we are on the boundary of $\mathcal{K}$ and the obstacle $\delta H < \tilde{H}$ has to modify our solution.

We are satisfied here with the statement that the constrained solution does in fact exist, because its numerical treatment, once the problem is discretized, belongs to the matter of linear programming, which is beyond the scope of this text.

We only mention here that in case the free solution of (8.112) presents too large areas where $\delta H > \tilde{H}$, namely an interface "emerging" from $S_0 \equiv \{z = 0\}$, this means that we have been too optimistic in fixing a "low" value of the error in the data and consequently a too low value of λ. Therefore, it is suggested to repeat the inversion increasing the value of λ namely the degree of regularization. When however the area where $\delta H > \tilde{H}$ is "small", this can be ignored and the solution can be truncated at $-\tilde{H} + \delta H = 0$.

In this case in fact, though we do not reach the optimum, we should be close to it. Since the mass density generating an anomalous effect, i.e. ρ_0 in the volume where the constraint is violated, should be small and hence its gravitational effect should be small too.

Remark 8.5 We have to recall that sometimes a solution of Model B is just a step of an iterative solution of Model A. In this case, although Model B requires only an $L^2(D)$ regularization of the Tikhonov functional, when Model B is derived from Model A we expect it to minimize a functional of the form

$$\bar{J}_\lambda\,(dg,\,\delta H) = \|dg - A\delta H\|^2_{L^2(\mathbb{R}^2)} + \lambda \int_D \left(\delta H^2 + |\nabla \delta H|^2\right) d_2\xi, \tag{8.114}$$

with $\delta H \in H^{1,2}\,(D)$.

Since δH appears together with its gradient in (8.114), when we perform the variation of $\bar{J}_\lambda$, namely when we study the equation

$$\frac{d}{dt}\bar{J}_\lambda\,(dg,\,\delta H + th)\bigg|_{t=0} = 0, \quad \forall h \in X, \tag{8.115}$$

we need to put some constraints on δH. In fact, (8.115) becomes

$$2\int_D h\,(\xi)\left[A^\top dg - A^\top A\delta H\right] d_2\xi + 2\lambda \int_D [h\,(\xi)\,\delta H\,(\xi) + \nabla h\,(\xi)\cdot \nabla\delta H\,(\xi)]\,d_2\xi = 0. \tag{8.116}$$

To conclude our reasoning we need to reverse the gradient operator from ∇h onto $\nabla\delta H$, namely we have to use the first Green identity (see (1.91))

$$\int_D \nabla h\,(\xi)\cdot\nabla\delta H\,(\boldsymbol{\xi})\,d_2\xi = \int_C h\,(\boldsymbol{\xi})\,\frac{\partial\delta H\,(\boldsymbol{\xi})}{\partial \boldsymbol{n}}d\xi - \int_D h\,(\boldsymbol{\xi})\,\Delta\delta H\,(\boldsymbol{\xi})\,d_2\xi, \tag{8.117}$$

where C is boundary of D, $C = \partial D$.

Since we want to assume that $\delta H\,(\boldsymbol{\xi}) = 0$ in D^c, it is only natural that, since $\nabla\delta H$ has to be a measurable function, we assume also that

$$\delta H\,(\boldsymbol{\xi})|_C = 0 \tag{8.118}$$

In this case also the variation of δH in X, $h\,(\boldsymbol{\xi})$, has to satisfy a similar relation (i.e., has to be zero on C), so that (8.117) reduces to

$$\int_D \nabla h\,(\boldsymbol{\xi})\cdot\nabla\delta H\,(\boldsymbol{\xi})\,d_2\xi = -\int_D h\,(\boldsymbol{\xi})\,\Delta\delta H\,(\boldsymbol{\xi})\,d_2\xi \tag{8.119}$$

Combining (8.116) and (8.119), we arrive at Tikhonov's normal equation

$$\begin{cases} A^\top A\delta H + \lambda\delta H + \lambda\Delta\delta H = A^\top dg, \\ \delta H|_C = 0 \end{cases} \tag{8.120}$$

Indeed, the solution of (8.120) is somehow more complicated than that of (8.112), yet (8.120) too lends itself to a discretization if the $\Delta\delta H$ is replaced by the usual 5-point scheme $\delta H_{i+1,k} + \delta H_{i-1,k} + \delta H_{i,k+1} + \delta H_{i,k-1} - 4\delta H_{i,k}$, see Fig. 8.6) and the domain D is chosen conveniently simple, e.g., a rectangle in $\mathbb{R}^2$, in order to easily include the conditions $\delta H = 0$ on the boundary.

Remark 8.6 The problem of the best choice of the space X to be used in the Tikhonov regularization (8.105) gained recently some interest [110]. For instance, it is easy to prove, by numerical examples, that the choice of $X = L^2(D)$ is somehow too weak and tends to produce improper oscillation in the solution δH. On the other hand, the choice $X = H^{1,2}(\sigma)$ has the disadvantage that cannot allow any sharp discontinuity in the solution δH, while such features are quite common in real world. It is then suggested that an intermediate space, namely that of functions of bounded variations [3, 55] has the advantage of retaining the compact embedding in $L^1(D)$ and, at the same time, to include discontinuous functions, although not too wildly discontinuos. We report here just the definition of the bounded variation norm, i.e.,

$$\|f\,(x)\|_{BV(D)} = \|f\|_{L^1(D)} + \sup_{\boldsymbol{\varphi}\,\in\, C^1(D)} \int_D \nabla\cdot\boldsymbol{\varphi}\,(x)\,f\,(x)\,d_2x, \ |\boldsymbol{\varphi}| \le 1, \tag{8.121}$$

leaving the proof of the existence of a Tikhonov minimum to Exercise 4. The disadvantage in using a regularizing functional like $\|\delta H\|_{BV(D)}$ is in that the corresponding variational equation is no longer linear, even in the case of Model

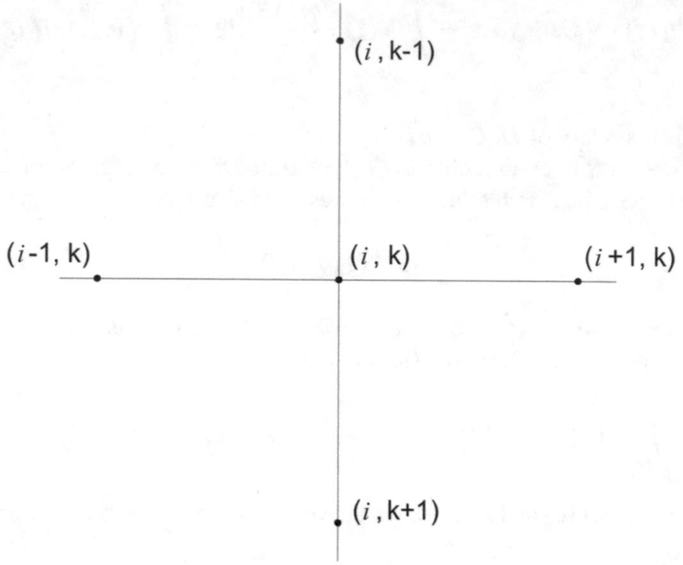

Fig. 8.6 The 5-point discretization of the Laplacian

B where $A\delta H$ is linear. So the numerics involved in the minimization of (8.121) is always more complicated than that implied by a quadratic functional $J(dg, \delta H)$.

Still another weaker functional has been proposed in [110], namely

$$\Phi(f) = \|f\|_{L^2} + \int_D \frac{|\nabla f|^2}{|\nabla f|^2 + \varepsilon^2} d_2x. \tag{8.122}$$

Numerical examples seem to be promising. One can prove that the topology induced by (8.122) is that of almost everywhere pointwise convergence of f and of its gradient. However, to our knowledge, the corresponding mathematical analysis, has never been explored.

8.6 Exercises

Exercise 1 The normal Tikhonov equation of the type (8.120), for a square domain $D \equiv ([0, 1] \times [0, 1])$, reads:

$$\mu^2 \int \frac{\tilde{H}(\xi) + \tilde{H}\left(\underline{\eta}\right)}{\left[r_{\xi\eta}^2 + \left(\tilde{H}(\xi) + \tilde{H}\left(\underline{\eta}\right)^2\right)\right]^{\frac{3}{2}}} \delta H\left(\underline{\eta}\right) d_2\eta + \tag{8.123}$$
$$+\lambda \delta H(\xi) + \lambda \Delta \delta H(\xi) = f(\xi),$$

with $\boldsymbol{\xi}, \underline{\eta} \in D$, and f given by (8.111). We want to transform it into a Single integral equation for δH.

Solution To this aim we need the Green function of Δ on the square $[01] \times [01]$. This is a function $G(\boldsymbol{x}, \boldsymbol{\xi}) = G(x_1, x_2, \xi_1, \xi_2)$ satisfying the following relations:

$$\Delta G(\boldsymbol{x}, \boldsymbol{\xi}) = \delta(\boldsymbol{x} - \boldsymbol{\xi}),$$
$$G(\boldsymbol{x}, \boldsymbol{\xi}) = G(\boldsymbol{\xi}, \boldsymbol{x}),$$
$$G(\boldsymbol{x}, \boldsymbol{\xi}) = 0 \quad \boldsymbol{x} \in C \text{ (boundary of } D).$$

Assume $\varphi_k(t)$ is an orthonormal, L^2 complete sequence on $[01]$, i.e.,

$$f(t) - \sum_{k=1}^{+\infty} \varphi_k(t) \langle f, \varphi_k \rangle_{L^2(0,1)}. \tag{8.124}$$

Then we can write symbolically

$$\delta\left(t - t'\right) = \sum_{k=1}^{+\infty} \varphi_k(t) \varphi_k\left(t'\right). \tag{8.125}$$

Assume further that $\{\varphi_k\}$ are smooth and such that

$$\varphi''_k(t) = -a_k^2 \varphi_k(t),$$
$$\varphi_k(0) = \varphi_k(1) = 0.$$

Such relations are clearly satisfied by $\varphi_k(t) = \sqrt{2} \sin k\pi t$ with $a_k = k\pi$, $k = 1, 2, \ldots$ That the system $\{\varphi_k\}$ is orthonormal in $L^2(0, 1)$, i.e.,

$$\int_0^1 \varphi_k(t) \varphi_j(t)\, dt = \delta_{kj} = \begin{cases} 1, & \text{if } k = j, \\ 0, & \text{if } k \neq j, \end{cases}$$

is easily verified by direct calculation of the integrals. That it is complete in $L^2(0, 1)$ is established as follows. Let $f \in L^2(0, 1)$ be such that

$$\langle f, \varphi_k \rangle_{L^2(0,1)} = 0 \quad \forall k > 0;$$

if we can prove that $f = 0$ a.e. in [0, 1], then $\{\varphi_k\}$ is complete. Extend $f(t)$ to the interval $[-1, 1]$ as an odd function:

$$\bar{f}(t) = \begin{cases} f(t), & \text{if } t \geq 0, \\ -f(-t), & \text{if } t < 0. \end{cases}$$

It is clear that $\bar{f} \in L^2(-1, 1)$, because

$$\int_{-1}^{1} \bar{f}^2(t)\, dt = 2\int_0^1 f^2(t)\, dt < +\infty.$$

Therefore $\bar{f}(t)$ admits a Fourier series expansion

$$\bar{f}(t) = a_0 + \sum_{n=1}^{+\infty} (a_n \cos \pi n t + b_n \sin \pi n t);$$

since $\bar{f}(t)$ is odd, one has $a_0 = a_n = 0$ for all $n \geq 1$. So

$$\bar{f}(t) = \sum_{n=1}^{+\infty} b_n \sin \pi n t,$$

with

$$b_n = \langle \bar{f}(t), \sin \pi n t \rangle_{L^2(-1,\, 1)}.$$

But

$$\int_{-1}^{1} \bar{f}(t) \sin \pi n t\, dt = 2\int_0^1 f(t) \sin \pi n t\, dt = 0 \quad \forall n > 0,$$

so

$$\bar{f}(t) = 0 \Longrightarrow f(t) = 0.$$

Now put

$$G(\boldsymbol{x}, \boldsymbol{\xi}) = \sum_{k,j=1}^{+\infty} \frac{\varphi_k(x_1)\,\varphi_j(x_2)\,\varphi_k(\xi_1)\,\varphi_j(\xi_2)}{\pi^2\left(k^2 + j^2\right)}.$$

Clearly,

$$G(\boldsymbol{x}, \boldsymbol{\xi}) = G(\boldsymbol{\xi}, \boldsymbol{x}) \text{ and } G(\boldsymbol{x}, \boldsymbol{\xi})|_{x \in C} = 0.$$

Moreover,

$$
\begin{aligned}
\Delta_x G(\boldsymbol{x}, \boldsymbol{\xi}) &= \sum_{k,j} \frac{\pi^2 k^2 + \pi^2 j^2}{\pi^2 \left(k^2 + j^2\right)} \varphi_k(x_1)\, \varphi_j(x_2)\, \varphi_k(\xi_1)\, \varphi_j(\xi_2) = \\
&= \sum_k \varphi_k(x_1)\, \varphi_k(\xi_1) \sum_j \varphi_j(x_2)\, \varphi_j(\xi_2) = \\
&= \delta(x_1 - \xi_1)\, \delta(x_2 - \xi_2) = \delta(\boldsymbol{x} - \boldsymbol{\xi}).
\end{aligned}
$$

Thus, $G(\boldsymbol{x}, \boldsymbol{\xi})$ is the sought-for Green function. Now observe that, using the second Green identity 1.95 and the fact that

$$
\begin{aligned}
\Delta_x G(\boldsymbol{x}, \boldsymbol{\xi}) = \Delta_\xi G(\boldsymbol{x}, \boldsymbol{\xi}) &= \delta(\boldsymbol{x} - \boldsymbol{\xi}), \\
G(\boldsymbol{x}, \boldsymbol{\xi}) &= 0, \ \boldsymbol{\xi} \in C, \\
\delta H(\boldsymbol{\xi}) &= 0, \boldsymbol{\xi} \in C,
\end{aligned}
$$

one gets

$$
\begin{aligned}
\int_D G(\boldsymbol{x}, \boldsymbol{\xi})\, \Delta \delta H(\xi)\, d_2\xi &= \int_D \Delta_\xi G(\boldsymbol{x}, \boldsymbol{\xi})\, \delta H(\boldsymbol{\xi})\, d_2\xi = \\
&= \int_D \delta(\boldsymbol{x} - \boldsymbol{\xi})\, \delta H(\xi)\, d_2\xi = \delta H(\boldsymbol{x}).
\end{aligned}
$$

Therefore, multiplying the normal equation by $G(\boldsymbol{x}, \boldsymbol{\xi})$ and integrating, we obtain

$$
\begin{aligned}
&\mu^2 \int_D \delta H(\boldsymbol{\eta}) \cdot \left\{ \int_D d_2\boldsymbol{\xi} G(\boldsymbol{x}, \boldsymbol{\xi}) \frac{\tilde{H}(\boldsymbol{\xi}) + \tilde{H}(\boldsymbol{\eta})}{\left[r_{\xi\eta}^2 + \left(\tilde{H}(\xi) + \tilde{H}(\eta) \right)^2 \right]^{\frac{3}{2}}} d_2\xi \right\} d_2\eta + \\
&+ \lambda \int_D G(\boldsymbol{x}, \boldsymbol{\xi})\, \delta H(\boldsymbol{\eta})\, d_2\eta + \lambda \delta H(\boldsymbol{x}) = \int_D G(\boldsymbol{x}, \boldsymbol{\xi})\, f(\boldsymbol{\xi})\, d_2\boldsymbol{\xi},
\end{aligned}
$$

which is the sough-for integral equation for δH.

Exercise 2 Let $\rho(\boldsymbol{x}, z) \in L^2\left(\mathbb{R}^3_-\right)$ be a given mass density; find the equivalent harmonic $\rho_{\text{h}}(\boldsymbol{x}, z)$, generating the same attraction on $S_0 = \{z = 0\}$.

Solution Recall from Corollary 8.1 that

$$
\delta \hat{g}_z(\boldsymbol{p}) = -2\pi G \int_{-\infty}^{0} e^{2\pi p \zeta}\, \hat{\rho}(\boldsymbol{p}, \zeta)\, d\zeta,
$$

with

$$\hat{\rho}(\boldsymbol{p}, \zeta) = -\frac{2}{G} p e^{2\pi p z} \delta \hat{g}_z(\boldsymbol{p}),$$

so that

$$\hat{\rho}(\boldsymbol{p}, z) = 4\pi \int_{-\infty}^{0} p e^{2\pi p (z+\zeta)} \delta \hat{g}_z(\boldsymbol{p})\, d\zeta.$$

Recalling that (see 8.47)

$$H(\boldsymbol{x}, z) = \mathscr{F}^* \left(p e^{2\pi p z} \right) = -\frac{1}{2\pi} \frac{|\boldsymbol{x}|^2 - 2z^2}{\left(|\boldsymbol{x}|^2 + z^2 \right)^{\frac{5}{2}}},$$

we find

$$\rho_h(\boldsymbol{x}, z) = -2 \int_{-\infty}^{0} \int_{e^2} \frac{|\boldsymbol{x} - \boldsymbol{\xi}|^2 - 2(z+\zeta)^2}{\left[|\boldsymbol{x} - \boldsymbol{\xi}|^2 + (z+\zeta)^2 \right]^{\frac{5}{2}}} \rho(\boldsymbol{\xi}, \zeta)\, d_2\xi\, d\zeta.$$

Exercise 3 Let B be a ball of radius R and assume we are given on its boundary surface the anomaly $\delta g(R, \sigma) = -\left.\frac{\partial u(r,\sigma)}{\partial r}\right|_{r=R}$ with $\sigma = \sigma(\lambda, \varphi)$ the solid angle and λ, φ the spherical longitude and latitude respectively. Find the unique (minimum L^2 norm) internal density ρ, that generates the given attraction.

Solution Note that if ρ has to be harmonic in B, it must have the form

$$\rho_h = \sum_{n=0}^{+\infty} \sum_{m=-n}^{n} \rho_{nm} \left(\frac{r}{R} \right)^n Y_{nm}(\sigma).$$

Moreover, for $P \in B^{\mathrm{c}}$ $(r_P > R)$ and $Q \in B$ $(r_Q < R)$,

$$\ell_{PQ}^{-1} = \sum_{n=0}^{+\infty} \sum_{m=-n}^{n} \frac{r_Q^n}{r_P^{n+1}} \frac{1}{2n+1} Y_{nm}(\sigma_P) Y_{nm}(\sigma_Q),$$

where $\{Y_{nm}(\sigma)\}$ are orthonormalized spherical harmonics. Therefore

$$u\left(r_P,\ \sigma_P\right)=G\int_0^R r_Q^2\int_\sigma \rho_h\left(Q\right)\frac{1}{\ell_{PQ}}d\sigma dr_Q=$$

$$=G\sum_{n=0}^{+\infty}\frac{1}{2n+1}\sum_{m=-n}^{n}\frac{Y_{nm}\left(\sigma_Q\right)}{r_P^{n+1}}\rho_{nm}\int_0^R r_Q^{n+2}dr_Q=$$

$$=G\sum_{n=0}^{+\infty}\frac{1}{(2n+1)\ (n+3)}\sum_{m=-n}^{n}\rho_{nm}Y_{nm}\left(\sigma_Q\right)\frac{R^{n+3}}{r_P^{n+1}}.$$

Accordingly

$$-\left.\frac{\partial u\left(r,\sigma\right)}{\partial r}\right|_{r_P=R}=GR\sum_{n=0}^{+\infty}\sum_{m=-n}^{n}\frac{n+1}{(2n+1)\ (n+3)}\rho_{nm}Y_{nm}\left(\sigma\right).$$

Therefore, if

$$\delta g\left(R,\sigma\right)=\sum_{n=0}^{+\infty}\sum_{m=-n}^{n}\delta g_{nm}Y_{nm}\left(\sigma\right)$$

the solution is given by

$$GR\frac{n+1}{(2n+1)\ (n+3)}\rho_{nm}=\delta g_{nm},$$

i.e.,

$$\rho_{nm}=\frac{1}{GR}\frac{(2n+1)\ (n+3)}{n+1}\delta g_{nm},$$

to be used in the series defining ρ_h.

Exercise 4 Prove that the Tikhonov functional (see (8.121) for the definition of $|\cdot|_{BV(D)}$)

$$J_\lambda\left(dg,\ \delta H\right)=|dg-A\delta H|_{L^1(R^2)}+\lambda\left|\delta H\right|_{BV(D)}$$

has a minimum in the closed convex set $\mathscr{K}$

$$\mathscr{K}=\left\{\delta H\in BV\left(D\right);\quad\left|\delta H\left(x\right)\right|\leq\bar{H},\quad x\in D\quad\text{a.e.}\right\}.$$

Solution That $\mathscr{K}$ is closed and convex in L^1(D) can be shown by the same reasoning leading to (8.59), (8.60). In fact, if $\{\delta H_n\}$ is a Cauchy sequence in L^1 and $|\delta H_n\left(\xi\right)|\leq\bar{H}$, then $\{\delta H_n\}$ is Cauchy also in $L^2(D)$ and the reasoning of Sect. 8.2 holds.

Referring to the Theorem B.1 of Appendix B, we can put

$$Y = L^1\left(\mathbb{R}^2\right), \quad X = BV(D), \quad H = L^1(D);$$

then $\mathscr{K} \subset X$ and $X \subset H$, with compact embedding, as it is proved in [55]. So we need only the continuity of

$$A(\delta H) : L^1(D) \to L^1\left(\mathbb{R}^2\right),$$

in order to be able to apply the conclusions of Theorem B.1. But this, already claimed in (8.74), comes from the following remarks: setting $dg(x) = \int_D a(\boldsymbol{x}, \boldsymbol{\xi} | \delta H)\, d_2x = A(\delta H), dg_n(x) = A(\delta H_n)$ one has, recalling (8.63), that $a(\boldsymbol{x}, \boldsymbol{\xi} | \delta H)$ is dominated by a function integrable on D. So pointwise

$$\forall x \quad \lim dg_n(x) = \int_D \lim a(\boldsymbol{x}, \boldsymbol{\xi} | \delta H_n)\, d_2\xi = A(\delta H) = dg(x).$$

Furthermore, recalling (8.68), we have

$$|dg_n(x)|, |dg(x)| \leq F(x) \in L^1\left(\mathbb{R}^2\right);$$

so we can pass to the limit under the integral and claim that

$$\lim |A(\delta H_n) - A(\delta H)|_{L'(\mathbb{R}^2)} = \lim \int_{\mathbb{R}^2} |dg_n(x) - dg(x)|\, d_2x = 0,$$

as we needed to prove.

Chapter 9
General Inversion Approaches

9.1 Outline of the Chapter

In the previous chapter, also with the help of Appendices B and C, we learned how to regularize by a Tikhonov approach some inverse gravimetric problems. In particular, we have seen applications to different simplified models, specifically those that can be described by two homogeneous layers with an unknown interface.

In this chapter we return to the more general inverse problem where an unknown density $\delta\rho\,(\boldsymbol{\xi}, z)$ is sought in the layer $\{-H \leq z \leq 0\}$, when the gravity anomaly $\delta g(x)$ is given on the upper plane S_0. In this case we already know that a large intrinsic non-uniqueness plagues the solution [46]; nevertheless, one could expect that, e.g., the use of a regularizing functional, according to the Tikhonov approach, would remedy such a drawback. Nevertheless, as we shall see, this leads invariantly to some harmonic mass distribution that is known to place most of masses close to the upper plane S_0: certainly a non-realistic solution. It is for this reason that several authors (e.g., [77]) have proposed to introduce weights into the regularizing function with the purpose of "pushing down" the $\delta\rho$ solution.

There is clearly a lot of arbitrariness in such an approach, yet, after deriving the analytic solution of the Tikhonov normal equations with depth-dependent weights, we give an argument in Sect. 9.3 that justifies the choice of a weight going like z^{-2} with depth. An analogous solution is then derived for a discretized model in Sect. 9.4. This is not a simple discretization of the solution formula previously found; rather, it is the solution for a model discretized from the beginning by imposing that $\delta\rho$ is constant in prisms of a regular raster.

The nice feature here is that the elementary formula for the gravity generated by a prism spontaneously introduces the DFT (Discrete Fourier Transform) of the unknown field and the data, with a relation that strongly resembles the one of the continuous case.

Another factor that helps in providing plausible solutions is to introduce a prior distribution $\delta\rho_0\,(\boldsymbol{\xi}, z)$, and force $\delta\rho$ to be close to $\delta\rho_0(\boldsymbol{\xi}, z)$.

F. Sansò, D. Sampietro, *Analysis of the Gravity Field*, Lecture Notes in Geosystems Mathematics and Computing, https://doi.org/10.1007/978-3-030-74353-6_9

A stochastic treatment of this scenario once more leads to a solution equivalent to Tikhonov's, if the regularizing parameter is properly chosen in relation to the variances of the various signals, yet the provision of a sensible prior information is typically more easily done in terms of the geological materials that produce the gravimetric signal. This compels the introduction of a field of labels describing such materials as a further unknown of the problem.

The necessary additional information supplied is a prior distribution of labels (see [126]). The model that emerges from the above considerations is similar to the one that is used in image analysis. It is therefore only natural to resort to the use of the so-called Monte Carlo–Markov Chain (MCMC) methods, that have been successfully applied in such an area.

So, after recalling the Bayesian paradigm in Sect. 9.5, we describe briefly in Sect. 9.6 the MCMC methods which, combined with the Simulated Annealing, provide an approximate solution to a specific kind of optimization problems. This general approach is adapted to the inverse gravimetric problem in Sect. 9.7.

Concerning the inverse gravimetric problem and more generally the analysis and the numerics aspects of inverse problems one can consult [13, 44–46, 60, 75, 81, 82, 90–92].

Finally, in Sect. 9.8 some realistic examples are treated by the above approach to provide a feeling for its effectiveness.

9.2 The Tikhonov Principle with Depth Weighted Regularizers

Let us first define precisely the model we want to invert.

We assume that data are the anomalies $\delta g_Z(\boldsymbol{x})$, $\boldsymbol{x} \in S_0 = \mathbb{R}^2$, and that, apart from errors, δg_Z (namely the errorless anomalies) are generated by an anomalous mass distribution with density $\delta\rho(\boldsymbol{\xi}, z)$, in the layer $\mathscr{L} \equiv \left\{\boldsymbol{\xi} \in \mathbb{R}^2; -H \le z \le 0\right\}$, such that the usual, see (2.3), Newton formula holds:

$$\delta g_Z(\boldsymbol{x}) = G \int_{-H}^{0} \int_{R^2} \frac{z}{\left[|\boldsymbol{x} - \boldsymbol{\xi}|^2 + z^2\right]^{3/2}} \delta\rho(\boldsymbol{\xi}, z)\, d_2\xi dz. \tag{9.1}$$

Let us recall that (see (3.104) and (3.105)) such observation equation can be conveniently written in terms of the FT $\delta\hat{g}_Z$ of δg_Z and $\delta\hat{\rho} \equiv \delta\hat{\rho}(\boldsymbol{p}, z)$ which is the FT of $\delta\rho(\boldsymbol{\xi}, z)$, with respect to $\boldsymbol{\xi}$, yielding

$$\delta\hat{g}_Z(\boldsymbol{p}) = -2\pi G \int_{-H}^{0} e^{2\pi p z} \delta\hat{\rho}(\boldsymbol{p}, z)\, dz. \tag{9.2}$$

For the sake of brevity, this equation is written in operator form as

$$\delta \hat{g}_Z = \hat{A} \delta \hat{\rho}. \tag{9.3}$$

If we assume $\delta \hat{\rho}$ to be in L^2 in the layer $\mathscr{L}$, then we can think of setting up a suitable Tikhonov principle to estimate $\delta \hat{\rho}$, namely, to consider the functional

$$J\left(\delta \hat{\rho}\right) = \left\| \delta \hat{g}_Z - \hat{A} \delta \hat{\rho} \right\|^2_{L^2(\mathbb{R}^2)} + \lambda \left\| \delta \hat{\rho} \right\|^2_{L^2(\mathscr{L})} \tag{9.4}$$

and look for its minimum. This is the simplest form of Tikhonov's principle.

Before we continue, it is worth stressing that the regularizer $\left\| \delta \hat{\rho} \right\|^2_{L^2(\mathscr{L})}$ does in fact correspond perfectly to $\|\delta \rho\|^2_{L^2(\mathscr{L})}$, since

$$\int_{-H}^{0} \int_{\mathbb{R}^2} \left| \delta \hat{\rho}\left(\boldsymbol{p}, z\right) \right|^2 d_2 p dz = \int_{-H}^{0} \int_{\mathbb{R}^2} \delta \rho (\xi, z)^2 d_2 \xi dz, \tag{9.5}$$

thanks to Parseval's identity (3.45).

Let us first solve explicitly (9.4) to explain why we propose to modify it, following the work of some authors (e.g., [77]).

The Euler equation for the principle $\min \Phi\left(\delta \hat{\rho}\right)$ is

$$\left(\hat{A}^* \hat{A} + \lambda I\right) \delta \hat{\rho} = \hat{A} * \delta \hat{g}_Z \tag{9.6}$$

or, in explicit form,

$$4\pi^2 G^2 e^{2\pi p z} \int_{-H}^{0} e^{2\pi p z'} \delta \hat{\rho}\left(\boldsymbol{p}, z'\right) dz' + \lambda \delta \hat{\rho}\left(\boldsymbol{p}, z\right) = 2\pi G e^{2\pi p z} \delta \hat{g}_Z\left(\boldsymbol{p}\right) \tag{9.7}$$

Denote

$$\varphi\left(\boldsymbol{p}\right) = \int_{-H}^{0} e^{2\pi p z'} \delta \hat{\rho}\left(\boldsymbol{p}, z'\right) dz'; \tag{9.8}$$

then (9.7) yields

$$\delta \hat{\rho} = -\frac{2\pi G}{\lambda} \left[\delta \hat{g}_Z\left(\boldsymbol{p}\right) + 2\pi G \varphi\left(\boldsymbol{p}\right)\right] e^{2\pi p z}. \tag{9.9}$$

Substituting back (9.9) into (9.8), one readily finds $\varphi\left(\boldsymbol{p}\right)$ and, ultimately, as shown in Exercise 1, one gets the explicit solution

$$\delta \hat{\rho} = -\frac{\frac{2\pi G}{\lambda} \delta \hat{g}_Z\left(\boldsymbol{p}\right)}{1 + \frac{4\pi^2 G^2}{\lambda} \psi\left(p\right)} e^{2\pi p z}, \tag{9.10}$$

where

$$\psi(p) = \int_{-H}^{0} e^{4\pi p z'} dz' = \frac{1 - e^{-4\pi p H}}{4\pi p}. \tag{9.11}$$

Examining (9.10) we realize that $\delta\hat{\rho}$ has the form

$$\delta\hat{\rho}(\boldsymbol{p}, z) = C(\boldsymbol{p})\, e^{2\pi p z}, \tag{9.12}$$

i.e., it is once more harmonic in $\mathscr{L}$ (recall Remark 8.2).

Apart from functional considerations in Remarks 8.5, and 8.6, we have already commented in §8 that such a solution is not realistic for geological reasons because most of the mass with density $\delta\rho$ is pushed close to S_0. This is particularly evident in the form (9.2) as

$$\int_{\mathbb{R}^2} |\delta\rho(\xi, z)|^2 d_2\xi \equiv \int_{\mathbb{R}^2} |C(\boldsymbol{p})|^2 e^{4\pi p z} d_2 p, \tag{9.13}$$

so that, for small $z(< 0)$, $e^{4\pi p z}$ has little effect, at least at low frequencies, while for large negative z the integral of $\delta\rho$ squared becomes small. To counteract such a drawback we can make two choices that we will combine.

First of all, we can assume that we have some prior information on $\delta\rho$, namely a $\delta\rho_0$ that can serve as a reference distribution. This already tends to waive the non-uniqueness of the inverse gravimetric problem, which acts through the minimization of the L^2 norm implicit in (9.4), giving rise to a harmonic solution. This is obtained by regularizing $\delta\hat{\rho} - \delta\hat{\rho}_0$, rather than $\delta\hat{\rho}$. Furthermore, we can introduce a weight in the regularization, $w(z)$, modifying $J(\delta\hat{\rho})$ to

$$\begin{aligned} J(\delta\hat{\rho}) = & \int_{\mathbb{R}^2} \left|\delta\hat{g}_Z(\boldsymbol{p}) - \hat{A}\delta\hat{\rho}\right|^2 d_2 p + \\ & + \lambda \int_{-H}^{0} w(z) \int_{\mathbb{R}^2} \left|\delta\hat{\rho}(\boldsymbol{p}, z) - \delta\hat{\rho}_0(\boldsymbol{p}, z)\right|^2 d_2 p\, dz. \end{aligned} \tag{9.14}$$

In this case it is clear that, if $w(z)$ is decreasing with depth, more weight is put on deviations of $\delta\hat{\rho}$ from $\delta\hat{\rho}_0$ close to $z = 0$ than at lower level, so that the minimizer will more easily accept larger differences $\delta\hat{\rho} - \delta\hat{\rho}_0$ at depth rather than close to S_0.

Put in this way, it is clear that there is a large indeterminacy in the choice of $w(z)$ and then in the corresponding minimizer $\delta\hat{\rho}$; we will discuss this issue in the next section, arriving at some justification of the choice that is already present in literature.

Remark 9.1 Sometimes one is willing to use a Tikhonov regularizer that tends to smooth estimators. For instance, in [148] the regularizing functional takes on the form

$$\bar{J}\left(\delta\hat{\rho}\right) = \int_{\mathbb{R}^2} \left|\delta\hat{g}_Z\left(\boldsymbol{p}\right) - \hat{A}\delta\hat{\rho}\right|^2 d_2p +$$
$$+\lambda \int_{-H}^{0} w\left(z\right)\left\{\int_{\mathbb{R}^2} \left|\delta\hat{\rho} - \delta\hat{\rho}_0\right|^2 d_2p + \alpha \int_{\mathbb{R}^2} p^2 \left|\delta\hat{\rho}\left(\boldsymbol{p}, z\right)\right|^2 d_2p\right\} dz, \tag{9.15}$$

implementing a simultaneous minimization of $\left|\nabla_\xi \delta\rho\left(\boldsymbol{\xi}, z\right)\right|^2$. Sometimes also the minimization of $\left|\frac{\partial}{\partial z}\delta\rho\left(\boldsymbol{\xi}, z\right)\right|$ is sought, or even of higher-order derivatives.

We close this section by solving the minimum principle $\min_{\delta\hat{\rho}} J\left(\delta\hat{\rho}\right)$ when $J\left(\delta\hat{\rho}\right)$ is as in (9.14).

The Euler equation for the functional (9.14) is

$$4\pi^2 G^2 e^{2\pi pz} \int_{-H}^{0} e^{2\pi pz'} \delta\hat{\rho}\left(\boldsymbol{p}, z\right) dz' + \lambda w\left(z\right)\left[\delta\hat{\rho}\left(\boldsymbol{p}, z\right) - \delta\hat{\rho}_0\left(\boldsymbol{p}, z\right)\right] =$$
$$= -2\pi G e^{2\pi pz} \delta\hat{g}_Z\left(\boldsymbol{p}\right). \tag{9.16}$$

Reasoning exactly as in (9.7), one gets the solution

$$\delta\hat{\rho}\left(\boldsymbol{p}, z\right) = \delta\hat{\rho}_0\left(\boldsymbol{p}, z\right) - \frac{\frac{2\pi G}{\lambda}\left[2\pi G\varphi_0\left(\boldsymbol{p}\right) + \delta\hat{g}_Z\left(\boldsymbol{p}\right)\right]}{1 + \frac{4\pi^2 G^2}{\lambda}\psi_w\left(p\right)} \frac{e^{2\pi pz}}{w\left(z\right)}, \tag{9.17}$$

where

$$\varphi_0\left(\boldsymbol{p}\right) = \int_{-H}^{0} e^{2\pi pz} \delta\hat{\rho}_0\left(\boldsymbol{p}, z\right) dz \tag{9.18}$$

and

$$\psi_w\left(p\right) = \int_{-H}^{0} \frac{e^{4\pi pz'}}{w\left(z'\right)} dz'. \tag{9.19}$$

The calculation is explained in more detail in Exercise 1 where also the variational principle for the functional $\bar{J}\left(\delta\hat{\rho}\right)$ (see (9.15)) is solved.

9.3 Discussion of the Weight $w(z)$

We are seeking arguments that can suggest a "good" shape for $w(z)$. To do that, we first of all re-write the observation equations, modelling their discrepancies as Generalized Random Fields, rather than as deterministic functions, as customary in Tikhonov's regularization theory.

In fact, on the basis of a remark in Appendix B, we know that stochastic models provide solutions equivalent to Tikhonov's, but without undetermined parameters, like λ.

So we write

$$\begin{cases} \delta g_Z = A\delta\rho + \nu, \\ \delta\rho_0 = \delta\rho + \eta, \end{cases} \tag{9.20}$$

or, in Fourier transformed form,

$$\begin{cases} \delta\hat{g}_Z = \hat{A}\delta\hat{\rho} + \hat{\nu}, \\ \delta\hat{\rho}_0 = \delta\hat{\rho} + \hat{\eta}. \end{cases} \tag{9.21}$$

Here we assume $\boldsymbol{\nu} = \boldsymbol{\nu}(\boldsymbol{x})$ to be a white noise on $\mathbb{R}^2$, with zero mean and covariance operator $C_\upsilon \equiv \sigma_\nu^2 I$. According to Appendix B, the GRF $\hat{\boldsymbol{\nu}} = \hat{\boldsymbol{\nu}}(\boldsymbol{p})$ turns out to be also a white noise on $\mathbb{R}^2$, with covariance operator $C_{\hat{\upsilon}} = \sigma_\nu^2 I$, or

$$E\left\{\hat{\boldsymbol{\nu}}(\boldsymbol{p})\,\hat{\nu}^*\left(\boldsymbol{p}'\right)\right\} = \sigma_\nu^2\delta\left(\boldsymbol{p}-\boldsymbol{p}'\right). \tag{9.22}$$

As for the error $\eta = \eta(\boldsymbol{\xi}, z)$ of the prior reference model $\delta\rho_0(\boldsymbol{\xi}, z)$, we assume again to be uncorrelated between two different points $(\boldsymbol{\xi}, z) \neq \left(\boldsymbol{\xi}', z'\right)$, but we want it to have a variance density depending on z. Namely, assuming that η has zero mean, we state

$$E\left\{\eta(\boldsymbol{\xi}, x)\,\eta\left(\boldsymbol{\xi}', z'\right)\right\} = \sigma_\eta^2(z)\,\delta\left(z-z'\right)\delta\left(\boldsymbol{\xi}-\boldsymbol{\xi}'\right). \tag{9.23}$$

Notice that this is just the kernel of the operator C_η that multiplies each function $f \in L^2(\mathscr{L})$ by $\sigma_\eta^2(z)$.

Let $g = C_\eta f$, i.e.,

$$g(\boldsymbol{\xi}, z) = \sigma_\eta^2(z)\,f(\boldsymbol{\xi}, z).$$

Then by taking the Fourier Transform on the variable $\boldsymbol{\xi}$ only, we get

$$\hat{g} = \sigma_\eta^2(z)\,\hat{f}.$$

We conclude that $C_{\hat{\eta}}$ is again the same operator as C_η, but acting on functions of $\boldsymbol{p}$, namely

$$C_{\hat{\eta}}\left(\boldsymbol{p}, z; \boldsymbol{p}', z'\right) = E\left\{\hat{\eta}(\boldsymbol{p}, z)\,\hat{\eta}^*\left(\boldsymbol{p}', z'\right)\right\} = \sigma_\eta^2(z)\,\delta\left(z-z'\right)\delta\left(\boldsymbol{p}-\boldsymbol{p}'\right). \tag{9.24}$$

Moreover, we shall assume that $\hat{\upsilon}$ and $\hat{\eta}$ are uncorrelated to one another.

Let us underline that here, contrary to the more general case treated in Appendix B, $\delta\hat{\rho}$ is considered as a deterministic parameter (function) that we want to estimate, and only the error terms $\hat{\nu}$, $\hat{\eta}$ are GRF's. As for the stochastic models (9.22), (9.24), in reality they are dictated by the need of deriving an estimation principle equivalent to (9.13). Note that the shape of (9.24) allows to represent a situation where the prior information $\delta\hat{\rho}_0$ has a different variance at various depths; in particular it seems reasonable to expect that $\delta\hat{\rho}_0$ might be more accurate at the boundary S_0 and that its variance be increasing with depth.

At this point, to estimate $\delta\hat{\rho}$ we invoke the Least Squares principle in the classical Markov form, namely we seek to minimize

$$\Phi\left(\delta\hat{\rho}\right) = \left\langle \delta\hat{g}_Z - \hat{A}\delta\hat{\rho},\ C_\nu^{-1}\left(\delta\hat{g}_Z - \hat{A}\delta\hat{\rho}\right)\right\rangle_{L^2(\mathbb{R}^2)} +$$

$$+\left\langle \delta\hat{\rho} - \delta\hat{\rho}_0,\ C_\eta^{-1}\left(\delta\hat{\rho} - \delta\hat{\rho}_0\right)\right\rangle_{L^2(\mathscr{L})} = \tag{9.25}$$

$$= \frac{1}{\sigma_\nu^2}\int_{\mathbb{R}^2}\left|\delta\hat{g}_Z - \hat{A}\delta\hat{\rho}\right|^2 d_2p+$$

$$+\int_{-H}^{0}\frac{1}{\sigma_\eta^2(z)}\int_{\mathbb{R}^2}\left|\delta\hat{\rho} - \delta\hat{\rho}_0\right|^2 d_2p\,dz. \tag{9.26}$$

All this is somewhat heuristic, however it could be fully justified by applying the GRF theory presented in Appendix B. Nonetheless, it is sufficient for the purpose of this section, which is only to establish a parallel between Tikhonov's principle, $\min J\left(\delta\hat{\rho}\right)$, with the stochastic least squares principle, $\min \Phi\left(\delta\hat{\rho}\right)$.

In fact, it is enough to glance at formulas (9.14) and (9.25) to realize that the two principles are equivalent, and therefore produce the same estimators, if

$$\lambda w(z) = \frac{\sigma_\nu^2}{\sigma_\eta^2(z)}. \tag{9.27}$$

In other words, (9.27) together with the models (9.22) and (9.24) provides a stochastic interpretation of Tikhonov's formulas.

Given such an interpretation, we can formulate a requirement on the prior distribution error $\eta(\boldsymbol{\xi}, z)$ imposing that, although η can be very rough (noise like) specially in depths, it should be such as to generate a gravimetric signal, we call it χ,

$$\chi(\boldsymbol{x}) = G\int_{-H}^{0}\int_{\mathbb{R}^2}\frac{z}{\left[|\boldsymbol{x}-\boldsymbol{\xi}|^2 + z^2\right]^{3/2}}\eta(\boldsymbol{\xi}, z)\,d_2\xi\,dz, \tag{9.28}$$

that has at least finite real values. Since $\chi(\boldsymbol{x})$ is obviously a homogeneous, zero mean RF, this is usually translated into the sufficient condition that

$$\sigma^2(\chi(\boldsymbol{x})) = C_\chi(0) < +\infty. \tag{9.29}$$

The question is studied in depth in Exercise 2, where it is proved that (9.29) is equivalent to the condition

$$\int_{-H}^{0} \frac{\sigma_\eta^2(z)}{z^2} dz < +\infty. \tag{9.30}$$

So we have now a characterization of $\sigma_\eta^2(z)$ (and then of $w(z)$) that requires this function to be, on the one hand, increasing with negative z, and on the other hand going to 0 at S_0 faster than $|z|$, to allow the integral (9.30) to converge. If we hypothesize a reasonably simple law like

$$\sigma_\eta^2(z) = C\,|z|^\alpha, \tag{9.31}$$

we see that the first integer α complying with the above conditions is

$$\alpha = 2 \Rightarrow \sigma_\eta^2(z) = Cz^2, \quad w(z) = \frac{k}{z^2}. \tag{9.32}$$

This is what is found in literature (see [77, 148]) and we can quote another good reason for choosing $\alpha = 2$ and not higher exponents. In fact, $\frac{1}{z^2}$ is roughly the weight of the influence of masses at depth z on the overlaying gravimetric signal on S_0; so, quoting Li and Oldenburg [77], such a choice "essentially allows equal chance for cells at different depths to be non-zero".

All in all, we consider the choice (9.32) as a physically reasonable compromise between simplicity and the need to "push down" the estimated mass distribution in the inversion process.

Remark 9.2 In the quoted paper Li and Oldenburg suggest a weight $w(z)$ of the form

$$w(z) = \frac{C}{(z + z_0)^2}, \tag{9.33}$$

with z_0 a parameter to be empirically adjusted. In their numerical experiments though, they use a discretized version of the Tikhonov principle and the value of z_0 they find to provide suitable results if $z_0 \sim 20$ m, which is very close to one half of the side of the cell (voxel) they use, that is 50 m.

Since we will discuss Tikhonov's discretization in the next section, with a proposal that automatically accounts for half the width of the voxel, we do not think there is contradiction between the two.

9.4 The Discrete Tikhonov Solution

Recall that in a realistic setting our input data are the values of $\delta g_Z(\boldsymbol{x}_{nm})$, where $\{\boldsymbol{x}_{nm}\} \equiv \{x_n, y_m\}$ are the centers of squares that compose a regular grid on the plane S_0; namely (see Fig. 9.1),

$$\begin{cases} \{\boldsymbol{x}_{nm}\} = \{(x_n, y_m)\} = \{(n - 1/2)\Delta, (m - 1/2)\Delta\}, \\ n = 1, 2, \ldots, N; \ m = 1, 2, \ldots, M \end{cases} \tag{9.34}$$

Now to process such a data set one could think to apply the continuous Tikhonov formulas (9.17), (9.18), (9.19) by discretizing them, as discussed in Sect. 3.7, namely, by constraining the wavelength vector $\boldsymbol{p}$ to a grid similar to that of Fig. 9.1, transforming all integrals into sums and applying to the resulting formulas the algorithms of the DFT. Here, however, we prefer to follow another way, namely to discretize the problem from scratch, reducing the space of solution $\rho(\boldsymbol{x}, z)$ to a finite-dimensional space where ρ is constant in each volume element $V_{n,m,\ell}$ (voxel) of a regular raster composed by L layers of prisms, having a square base (Δ, Δ) and a height D as shown in Fig. 9.2.

To simplify the notation, we shall often collect the two indices referring to the x-and y-directions, into a unique integer vector index, for instance, putting

$$\boldsymbol{n} = (n, m), \quad \boldsymbol{j} = (j, k), \tag{9.35}$$

and so on. We will also call $\mathscr{G}$ the grid of integers $(n = 1, \ldots, N, m = 1, \ldots, M)$ that $\boldsymbol{n}$ is supposed to sweep.

So every voxel $V_{\boldsymbol{n},\ell}$ will have a barycenter

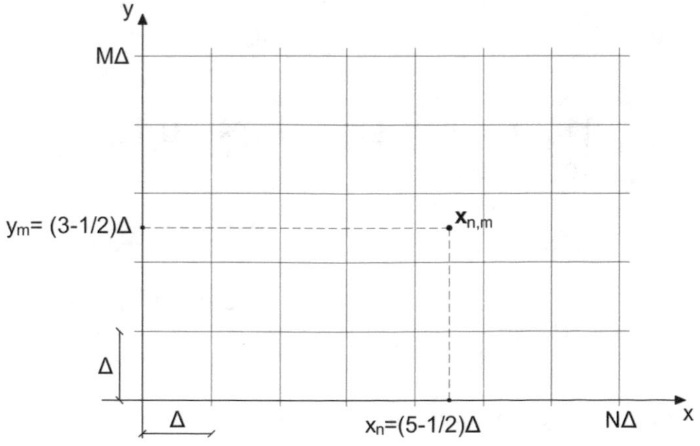

Fig. 9.1 The reference data grid on S_0; note that data are referred to the centers of the pixels $\boldsymbol{x}_{n,m}$

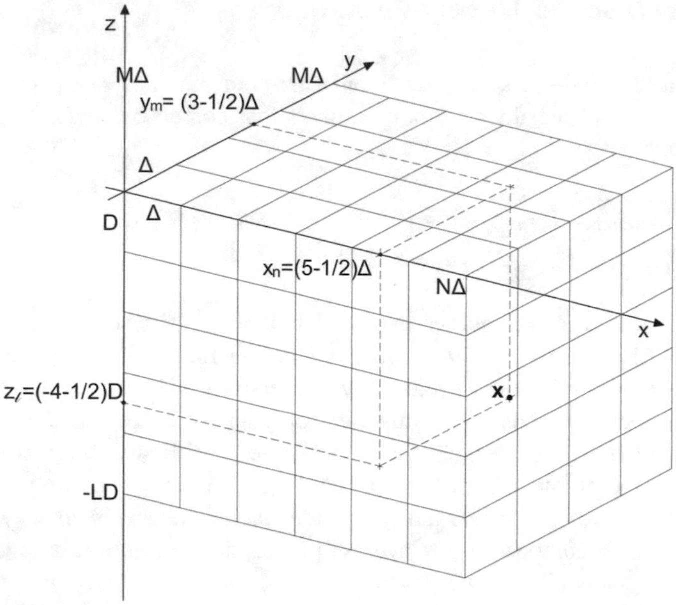

Fig. 9.2 The regular raster of voxels for a piecewise constant density; the centered of the voxel V_{534} is $(\boldsymbol{x}_{53}, z_4) = (x_5, y_3, z_4)$

$$\boldsymbol{C}_{\boldsymbol{n},\ell} = (\boldsymbol{x}_{\boldsymbol{n}}, z_\ell) \equiv (x_n, y_m, z_\ell) \equiv \left(\left(n - \frac{1}{2}\right)\Delta, \left(m - \frac{1}{2}\right)\Delta, \left(\ell - \frac{1}{2}\right)D\right) \tag{9.36}$$

and a corresponding indicator function

$$\chi_{\boldsymbol{n},\ell}(\boldsymbol{x}, z) = \chi_{\boldsymbol{n}}(\boldsymbol{x})\chi_\ell(z), \tag{9.37}$$

where

$$\chi_{\boldsymbol{n}}(x) = \begin{cases} 1, & \text{if } (n-1)\Delta \le x < n\Delta\ ;\ (m-1)\Delta \le y < m\Delta \\ 0, & \text{otherwise} \end{cases}$$

and

$$\chi_\ell(z) = \begin{cases} 1, & \text{if } -\ell D \le z < -(\ell - 1)D \\ 0, & \text{otherwise} \end{cases}$$

On $V_{\boldsymbol{n},\ell}$ the density $\delta\rho(\boldsymbol{x}, z)$ is constant by assumption:

$$\delta\rho(V_{\boldsymbol{n},\ell}) = \delta\rho_{\boldsymbol{n},\ell}. \tag{9.38}$$

Outside the large prism

$$B_0 = \{0 \le x \le N\Delta; 0 \le y \le M\Delta; -LD \le z \le 0\} \tag{9.39}$$

we suppose that

$$\delta\rho(\boldsymbol{x}, z) \equiv 0, \ \ (\boldsymbol{x}, z) \in B_0^c. \tag{9.40}$$

As we see, the solution space is now reduced to a presumably large, but finite number of dimensions, $N \times M \times L$.

Before coming to the observation equations and the subsequent application of Tikhonov's principle, we warn the reader that here, as suggested in Appendix C, if we have a prior mass distribution $\delta\rho_0$ (naturally this is supposed to be discretized too) we subtract it from $\delta\rho$, thus defining a new unknown

$$d\rho = \delta\rho - \delta\rho_0; \tag{9.41}$$

likewise we define a new observation function

$$dg = \delta g_Z - \delta g_{Z0} = \delta g_Z - A\delta\rho_0, \tag{9.42}$$

so that the observation equation of gravity "anomaly" on S_0 is now written as

$$dg = Ad\rho. \tag{9.43}$$

Notice that we have not used this trick in Sect. 9.3, because there we were focusing on the modelling of the error $\eta = \delta\rho - \delta\rho_0$ and so it was useful to keep explicit the dependence of the solution on the prior information.

Then, our observation equation is here only (9.43), or better its FT namely

$$d\widehat{g}(\boldsymbol{p}) = \widehat{A}d\widehat{\rho} = -2\pi G \int_{-L}^{0} e^{2\pi p z} d\widehat{\rho}(\boldsymbol{p}, z) dz. \tag{9.44}$$

If we start from the discretized unknown

$$d\rho = \sum_{\ell=1}^{L} \sum_{\boldsymbol{n}\in\mathcal{G}} d\rho_{\boldsymbol{n},\ell}, \ \chi_{\boldsymbol{n},\ell}(\boldsymbol{x}, z) \tag{9.45}$$

and take its FT with respect to $\boldsymbol{x}$, we get (see Example 3.5)

$$d\widehat{\rho}(\boldsymbol{p}) = \sum_{\ell=1}^{L} \sum_{\boldsymbol{n}\in\mathcal{G}} d\rho_{\boldsymbol{n},\ell}, \ e^{i2\pi \boldsymbol{p}\cdot\boldsymbol{x}_{\boldsymbol{n}}} F(\boldsymbol{p})\chi_\ell(z), \tag{9.46}$$

where

$$\boldsymbol{p} = \left(p_x, p_y\right), \; F(\boldsymbol{p}) = \text{sinc}\,(\pi p_x \Delta)\,\text{sinc}\,(\pi p_y \Delta) \cdot \Delta^2, \tag{9.47}$$

so that the corresponding gravimetric signal, found by computing $\widehat{A}$, $d\widehat{\rho}$ from (9.46), is

$$d\widehat{g}(\boldsymbol{p}) = -2\pi G \sum_{\ell=1}^{L} \sum_{\boldsymbol{n}\in\mathscr{G}} e^{i2\pi \boldsymbol{p}\cdot\boldsymbol{x_n}+2\pi p z_\ell} \bar{F}(p) d\rho_{\boldsymbol{n},\ell}, \tag{9.48}$$

where

$$\bar{F}(p) = F(p)\frac{\sinh(\pi p D)\cdot D}{\pi p D}. \tag{9.49}$$

Now, driven by the theory of DFT, we reckon $\widehat{g}(\boldsymbol{p})$ for $\boldsymbol{p}$ belonging to a conjugate grid on the Fourier plane, namely we put

$$\begin{cases} \boldsymbol{p} = \boldsymbol{p_j}, & \boldsymbol{j} = (j,k) \in \mathscr{G}, \\ \boldsymbol{p}_j = (j f_X,\, k f_Y), & f_X = \frac{1}{N\Delta},\; f_Y = \frac{1}{M\Delta}. \end{cases} \tag{9.50}$$

With (9.50) one can compute

$$\begin{cases} i\boldsymbol{p_j}\cdot\boldsymbol{x_n} + p_j z_\ell = i\left(\frac{jh}{N} + \frac{km}{M}\right) - \varphi_{j,\ell}, \\ \varphi_{j,\ell} = \sqrt{\frac{j^2}{N^2} + \frac{k^2}{M^2}}\left(\ell - \frac{1}{2}\right)\frac{D}{\Delta}, \end{cases} \tag{9.51}$$

so that (9.48) can be written as

$$d\widehat{g}(\boldsymbol{p_j}) = -2G\pi \sum_{\ell=1}^{L} H_{j\ell}\delta\widehat{\rho}_{\boldsymbol{j},\ell}, \tag{9.52}$$

where

$$\begin{aligned} H_{j,\ell} &= e^{-2\pi\varphi_{j,\ell}}\bar{F}(p_j), \\ \delta\widehat{\rho}_{j,\ell} &= \sum_{\boldsymbol{n}\in\mathscr{G}} e^{2\pi i\left(\frac{jn}{N}+\frac{km}{M}\right)}\delta\rho_{\boldsymbol{n},\ell}. \end{aligned} \tag{9.53}$$

Let us observe that, (9.51), (9.53), show that $H_{\boldsymbol{j},\ell}$ is a matrix of complex numbers and that $\delta\widehat{\rho}_{\boldsymbol{j},\ell}$ turns out to be just the DFT of $\delta\rho_{\boldsymbol{n},\ell}$. So we know that from d $\delta\widehat{\rho}_{\boldsymbol{j},\ell}$ we can retrieve $\delta\rho_{\boldsymbol{n},\ell}$ by the inverse DFT formula, namely

$$\delta\rho_{n,\ell} = \sum_{j\in\mathscr{G}} e^{-2\pi i\left(\frac{jn}{N}+\frac{km}{M}\right)}\delta\widehat{\rho}_{j,\ell}. \tag{9.54}$$

The Tikhonov principle now writes

$$\min \sum_{j\in\mathscr{G}} \left| d\widehat{g}_j + 2\pi G \sum_{\ell=1}^{L} H_{j,\ell}\delta\widehat{\rho}_{j,\ell} \right|^2 + \lambda \sum_{\ell=1}^{L} w_\ell \sum_{j\in\mathscr{G}} |\delta\widehat{\rho}_j, \ell|^2, \tag{9.55}$$

where the weights w_ℓ are chosen according to the rule discussed in Sect. 9.3, namely

$$w_\ell = \frac{1}{z_\ell^2} = \frac{1}{\left(\ell - \frac{1}{2}\right)^2 D^2}. \tag{9.56}$$

Notice that, with the choice (9.56) a possible proportionality constant of the weights is absorbed by the other constant λ. Furthermore, observe that the regularizing term in (9.55) has this form because we know that the following identity holds:

$$\sum_{n\in\mathscr{G}} \delta\rho_{n,\ell}^2 = \sum_{j\in\mathscr{G}} |\delta\, \rho_{j,\ell}|^2. \tag{9.57}$$

The variational equation of the minimum principle (9.55) is

$$2\pi G H_{j,\ell}^* \left(d\widehat{g}_j + 2\pi G \sum_{\ell'=1}^{L} H_{j,\ell'}\delta\widehat{\rho}_{j,\ell'} \right) + \lambda w_\ell \delta\widehat{\rho}_{j,\ell} = 0. \tag{9.58}$$

The solution of (9.58) can be obtained by an argument already applied in Sect. 9.2 and explicitly developed in Exercise 4; the result is

$$\begin{cases} \delta\widehat{\rho}_{j\ell} = -\frac{2\pi G}{\lambda w_\ell} \frac{H_{j\ell}^* \delta\widehat{g}(\boldsymbol{p}_j)}{1+\frac{2\pi G}{\lambda}\psi(\boldsymbol{p}_j)}, \\ \psi(\boldsymbol{p}_j) = \sum_{\ell'=1}^{L} \frac{2\pi G}{w_{\ell'}} H_{j\ell'}^* H_{j\ell}. \end{cases} \tag{9.59}$$

Though expected, the similarity of (9.59) to (9.10) is remarkable.

Remark 9.3 Let us note that the solution (9.59) can be completely computed only after a value has been given to λ. For this purpose we can use the Morozov principle (see Appendix C). In this case, once a mean quadratic level δ^2 of the admissible error of the observation equations has been fixed, λ is the solution of the equation

$$\frac{1}{NM}\sum_{j\in\mathscr{G}}\left|1-\frac{4\pi^2G^2}{w_\ell}\frac{H_{j\ell}^*}{\lambda+2\pi G\psi(p_j)}\right|^2\left|d\widehat{g}(p_j)^2\right|=\delta^2. \tag{9.60}$$

Indeed (9.60) has to be solved by some standard numerical method, e.g., Newton's method.

Remark 9.4 (Multiresolution) Before closing this section we would like to get a look at a different regularization approach that for lack of space we will not be able to develop in the book. Yet, since a significant literature is growing in this direction [12, 29], namely that of the so-called *multiresolution approach*, we deem it useful to introduce at least the general concept with the purpose of explaining the logical connection with the line of thought pursued in the book.

Let us start by observing that whatever approach we take, when we come to computing a solution of a general improperly posed linear problem

$$y = Ax \tag{9.61}$$

we need to transform such an equation into a discrete one, where y, x are vectors and A is just a matrix.

The same is true if we consider the normal equation

$$A^T Ax = A^T y, \tag{9.62}$$

which can be derived from (9.61) if one applies a simple least squares method, with the only difference that now the operator to be inverted, namely the normal operator $A^T A$, is positive definite. The discretization of (9.61) or (9.62) is essentially an approximation process by projection of both the observation field and the unknown onto finite-dimensional subspaces.

Namely, if we put

$$x \in X;\, X_N = \text{span}\{\xi_1, \xi_2, \dots, \xi_N\},$$

$$y \in Y;\, Y_M = \text{span}\{\eta_1, \eta_2, \dots, \eta_M\}$$

we replace, for instance, (9.61) with

$$X \cong \sum_{n=1}^{N} x_n\xi_n,$$

$$Y \cong \sum_{m=1}^{M} y_m\eta_m,$$

and then we write

$$y_m = \sum_{n=1}^{N} x_n \left\langle \eta_m^*, A\xi_n \right\rangle_Y, \tag{9.63}$$

where $\{\eta_m^*\}$, the singular elements associated to A, are characterized by

$$\left\langle \eta_m^*, \eta_k \right\rangle_Y = \delta_{mk}. \tag{9.64}$$

If the sequence $\{\eta_m\}$ is orthonormal, one has $\eta_m^* = \eta_m$.

In an analogous way (9.62) becomes

$$\sum_{n=1}^{N} \left\langle \xi_m^*, A^T A\xi_n \right\rangle_X x_n = \sum_{k=1}^{M} \left\langle \xi_m^*, A^T \eta_k \right\rangle y_k = y_m, \tag{9.65}$$

with ξ_m^* characterized by

$$\left\langle \xi_m^*, A^T \eta_k \right\rangle_X = \left\langle A\xi_m^*, \eta_k \right\rangle_Y = \delta_{mk}. \tag{9.66}$$

A scheme like this, possibly with some variants, is known as the *Riesz–Galerkin method* (see [74]). The method is known to converge to the exact solution when $N, M \to \infty$, if the operator $A^T A$ is strictly positive, so that $\left(A^T A\right)^{-1}$ exists and it is bounded. But this is not the case when the original problem (9.61) is improperly posed, because $\left(A^T A\right)^{-1}$ is unbounded and so the solvability of the discretized normal system (9.65) is guaranteed only for fixed finite N, M, although the system becomes more and more unstable by increasing N, M because some of the eigenvalues of the normal matrix $A^T A$ tend to zero.

To remedy this drawback, the Tikhonov approach modifies the normal system by adding λI to the normal matrix, so that the equation to be solved becomes

$$\left(A^T A + \lambda I\right) X = A^T Y.$$

This shifts up the spectrum of $A^T A$ by λ, so that the numerical instability is avoided. On the contrary, a multiresolution scheme would adopt layers of basis functions $\{\xi_n\}_\ell$ such that from a layer to the next the "resolution" of the basis is increased. For each layer the number of unknowns, $\{x_n\}_\ell$, is increased and the corresponding least squares problem is solved. The process is stopped when numerical instability shows up (see [29]). A number of clever things are done in multiresolution analysis to orthogonalize the different layers, to produce a faster synthesis of the solution with a lower numerical burden. In this case the base functions are called *wavelets*. Just to give a flavour of the idea we provide in Fig. 9.3 a graphic example in one dimension for the so-called *Haar basis/Haar wavelets*.

Notice that the Haar functions have been used in this section to discretize $\delta\rho\,(\boldsymbol{\xi}, z)$, although with a fixed level of resolution only.

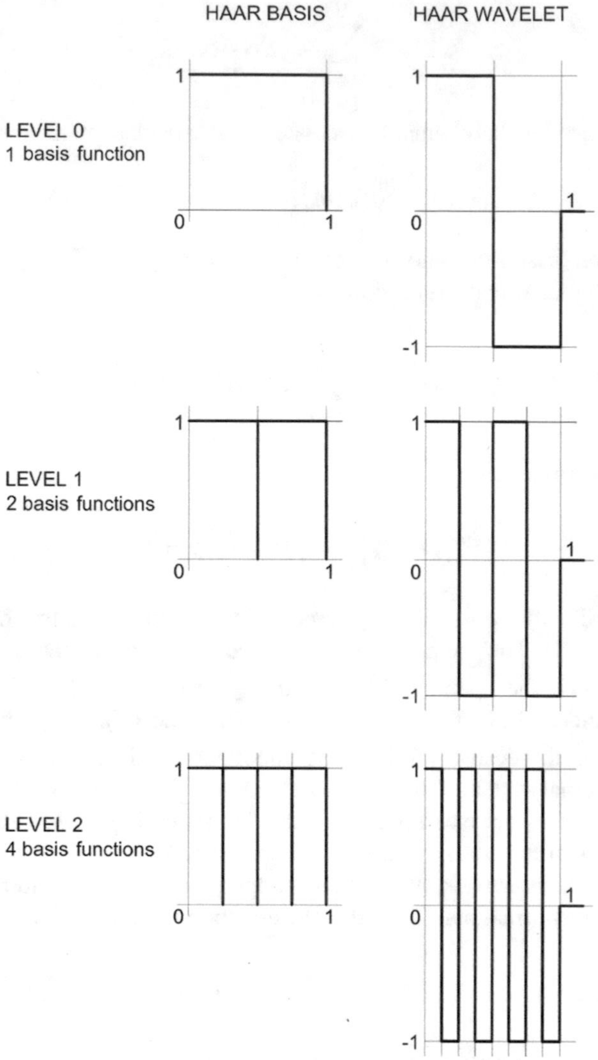

Fig. 9.3 Haar basis and Haar wavelets in 1D; notice that difference levels of Haar functions are not orthogonal, while Haar wavelets are

This shows that resorting to Haar wavelets for the unknown density in an inverse gravimetric problem could be beneficial and in fact has been done in numerical examples (see [47]).

The question of the choice of a convenient sequence for $\{\eta_k^*\}$ or $\{\xi_k^*\}$ requires many more mathematical tools and we send the interested reader to literature (see for instance [29]).

9.5 The Bayesian Paradigm

One unpleasant feature of the deterministic approach for inverting a gravity anomaly field is that, due to the large indeterminacy of its solution, either one has to strongly restrict the space of solutions (e.g., by assuming that the anomaly is generated only by one specific body of constant density) or one has to resort to some mathematical regularization which has a large degree of arbitrariness, leaving us with the feeling that the elaborated solution is an artefact. If, on the contrary, one examines the routine approach of a company operator working to interpret a gravity map, one realizes that the starting point is, so to say, a geological scenario, derived from some prior knowledge of the area or just guessed on the basis of geological experience. Then by successive small changes of densities in the various parts of the volume under analysis, an attempt is made to go closer and closer to the available gravity data.

This practical approach, despite the coarseness of the geological information, seems to work in most cases and therefore there is a need to find a method to include this prior data set into our solution.

We will build the proper model for the gravity context in one of the next sections, yet we anticipate here that the crucial step is to introduce a parallel field $L(\boldsymbol{\xi}, z)$ of geological labels qualifying the material at a point $(\boldsymbol{\xi}, z)$ as belonging to a specifical class of rocks. Among other things, this allows the density function $\rho(\boldsymbol{\xi}, z)$ to undergo sudden jumps passing from one material to another, without the artificial smoothing typically present in regularized solutions.

There is another scientific branch in which a similar problem has been thoroughly studied, namely image analysis, where a solution was provided to the need of breaking an image into patches, smooth inside each patch, but presenting jumps between one patch and another. There the introduction of labels (a method inferred from physical models of ferromagnetism [50]) has been crucial. The methodology applied to get estimates from data belong to the large field of Bayesian methods, which have the conceptual advantage of mimicking the human stepwise learning in the presence of an increasing information. Of course such methods have already been introduced into geophysics (classical is the book of Jeffreys [70]), sometimes with a quite original elaboration like the one by Tarantola [147]; however, the specific use of labels has been mostly developed in image analysis and it is that literature that inspired our presentation.

In this section we start summarizing the general Bayesian approach to the analysis of a system described in terms of a vector of parameters $\boldsymbol{x}$ and some prior knowledge on it and by a vector of observable quantities $\boldsymbol{y}$ about which some new information is introduced, in terms of a numerical vector of observations $\boldsymbol{y}_0$. The purpose of the method is to show how the innovative information on $\mathbf{y}$ modifies our knowledge of $\boldsymbol{x}$; this is based mathematically just on the concept of conditional probability, but accompanied by a thorough interpretation of the formulas.

In the Bayesian approach everything is uncertain and therefore everything is stochastic. Even what we could consider as constant c is in fact modelled as a RV

C, with a singular distribution, $P(C = c) = 1$. So the physical world where x and y live is described by a general distribution, $f_{XY}(x, y)$, which represents our state of knowledge of (x, y), before new information is introduced.

Notice that (x, y) can be vector variates, that can assume continuous or discrete values, or they can be even infinite-dimensional (random fields), although in this case considerable caution should be applied before we could speak about a probability density, as the symbol $f(\,)$ suggests.

As a warning, let us agree that here and in the next section, often we will use, for the sake of brevity, the term distribution of a RV X for the probability density function, $f_X(x)$; this should not be confused with the probability distribution function that we will denote as $F_X(x)$.

In any event, although we consider only the finite-dimensional case, we will use $f_{XY}(x, y)$ even when x and/or y are partly continuous and partly discrete.

Of course, depending on the case, f will have the meaning of a probability density or of a proper probability. Now we start with a formula that could be taken as definition of conditional probability, namely

$$f_{XY}(x, y) = f_{Y|X}(y|x) f_X(x). \tag{9.67}$$

We interpret (9.67) in the following way: $f_X(x)$ summarizes what we know a priori about x, while $f_{Y|X}$ describes the residual variability in Y, when we assume that X, i.e., the state of the system, is perfectly known. Usually in the Bayesian literature the notation

$$f_X(x) = p(x) \tag{9.68}$$

is used, to recall that this is the prior distribution of X, while $f_{Y|X}$, describing the observation process, is denoted as the likelihood of the data

$$f_{Y|X}(x, y) = \mathscr{L}(x, y). \tag{9.69}$$

Often $\mathscr{L}(x, y)$ is derived from observation equations

$$E\{Y|X\} = E\{Y|X = x\} + \nu = g(x) + \nu \tag{9.70}$$

by assigning some distribution to the "noise" ν, $f_\nu(\nu)$, so that in this case

$$\mathscr{L}(x, y) = f_\nu[y - g(x)]. \tag{9.71}$$

Notice that, since we have not specified a dependence of ν on x in $f_\nu(\nu)$, this means that we are implicitly assuming that ν and X are stochastically independent. This is the most commonly adopted model, though not necessarily the only one.

On the other hand, our focus is on finding the distribution of X when we get a sample y_0 from the RV Y. In other words, we are interested in $f_{X|Y}(x|Y = y_0)$, which we will call the *posterior distribution of* X. Notice that it is typical of the Bayesian approach to consider y_0 as a constant, contrary to the classical (or frequentist) concept that views it as a member of the collection of possible values of the random variable Y.

Indeed, we have the formula

$$f_{XY}(x, y) = f_{Y|X}(X|Y) f_Y(y) \tag{9.72}$$

which, compared with (9.67), provides the result

$$f_{X|Y}(x|y = y_0) = \frac{f_{Y|X}(y_0, x) f_X(x)}{f_Y(y_0)}. \tag{9.73}$$

Since we have

$$f_Y(y_0) = \int f_{Y|X}(y_0, \xi) f_X(\xi) d\xi, \tag{9.74}$$

also recalling the notation (9.68), (9.69), we can cast (9.73) in the form

$$f_{X|Y}(x|y_0) = \frac{\mathscr{L}(x, y_0) p(x)}{\int \mathscr{L}(\xi, y_0) p(\xi) d\xi} \tag{9.75}$$

which is also known as the *Bayes formula*.

Formula (9.75) is the solution of Bayes problem, which is constituted by the full knowledge of the posterior probability density.

Let us underline that from the Bayesian viewpoint

$$C(y_0) = \frac{1}{\int \mathscr{L}(\xi, y_0) p(\xi) d\xi} \tag{9.76}$$

is just a normalization constant, although one has to acknowledge that when we come to numbers the real computation of $C(y_0)$ is often a very hard task. Since knowing a probability distribution means knowing how to sample it, as we will see in the next section, many methods have been investigated on how to sample a distribution, given a part from a normalization constant.

Remark 9.5 (The MAP) Although the spirit of the Bayesian theory is to provide the posterior distribution of X, namely (9.75), often we need to summarize this information by indices that characterize such a distribution. Classical indices would be the posterior mean μ_X and covariance C_X as localization and dispersion of the distribution. Nevertheless μ_X is particularly meaningful if the posterior distribution is unimodal, with a bulk of probability and monotonous tails, like for instance a Gaussian density.

This however is not the case when the posterior distribution is multimodal, so that μ_X could even fall in a region of low probability density. In such cases it is more meaningful to appeal to the maximum likelihood principle, substituting μ_X

with the mode of the distribution, $\widehat{x}$. In Bayesian literature $\widehat{x}$ is called the *maximum a posteriori* (MAP) of $f_{XY}(x, y)$.

Analytically, the necessary condition $\widehat{x}$ has to satisfy is

$$\nabla_X f_{X|Y}(x|y_0) = C(y_0) \cdot \nabla_X[\mathscr{L}(x, y_0)p(x)] = 0. \tag{9.77}$$

Traditionally, in statistics (also due to the importance of the distributions of exponential families), instead of maximizing $f_{X|Y}$ one prefers to use the log-likelihood function

$$\begin{aligned} \ell(x, y_0) &= -\log f_{X|Y}(x|y_0) = \\ &= -\log C(y_0) - \log \mathscr{L}(x, y_0) - \log p(x). \end{aligned} \tag{9.78}$$

This time $\widehat{x}$ has to be such that

$$\widehat{x} = \arg\min \ \ell(x, y_0) \tag{9.79}$$

and (9.77) is substituted by the score equation

$$\nabla_X \ell(\widehat{x}, y_0) = 0. \tag{9.80}$$

The name is related to the fact that in statistics the random variable

$$U(X) = \nabla_X \ell(X) \tag{9.81}$$

is usually called the *score variable.*

We note that the analytic solution of equation (9.80) does not depend on $C(y_0)$; yet, as we shall see in the next section, $\widehat{x}$ is not really derived from the minimum condition, but rather as the stochastic limit of a suitable sequence $\{x_n\}$. Nevertheless here we close just noticing that an "approximate" index of dispersion of X around $\widehat{x}$ is provided by the classical likelihood theory as

$$\begin{aligned} &C_X \cong [D^2\ell(\widehat{x}, y_0)]^{-1} \\ &\left(D^2\ell(\widehat{x}, y_0) \equiv \left[\frac{\partial}{\partial x_i}\frac{\partial}{\partial x_k}\ell(\widehat{x}, y_0)\right]\right). \end{aligned} \tag{9.82}$$

This matrix $D^2\ell(\widehat{x}, y_0)$ is known as the (observed) *information matrix* (see [145], §2.4).

The result (9.82) is obtained by searching for a normal approximation of $f_{X|Y}$ in the high density region containing $\widehat{x}$. This in turn derives from a second-order expansion of $\ell\,(X, Y_0)$ around $\widehat{x}$, namely

$$\ell(X, Y_0) \cong \ell(\widehat{x}, y_0) + (X - \widehat{x})^T \nabla \ell(\widehat{x}, y_0) + \\ + \frac{1}{2}(X - \widehat{x})^T D^2 \ell(\widehat{x}, y_0)(X - \widehat{x}), \tag{9.83}$$

so that observing that $\ell(\widehat{x}, y_0)$ is constant, since $\widehat{x}$ is function of y_0 only, $\nabla \ell(\widehat{x}, y_0) = 0$ by definition of $\widehat{x}$, it turns out that $\ell = - \log f_{X|Y}$ has approximately a normal shape with $\ell(\widehat{x}, y_0)$ as average and $[D^2 \ell(\widehat{x}, y_0)]^{-1}$ as covariance function.

Also in this case some information on C_X, whenever such matrix retains a statistical meaning, will be in practice derived not from an analytic form, but as a stochastic limit.

9.6 Stochastic Optimization: The Simulated Annealing

To keep things easier, let us agree that from now on, the elements $\boldsymbol{x}$ and $\boldsymbol{y}$ are finite-dimensional vectors. In the last section we have introduced the Bayesian architecture leading to the search of the MAP, namely to

$$\widehat{\boldsymbol{x}} = \arg\max f_{X|Y}(\boldsymbol{x}, \boldsymbol{y}_0) \tag{9.84}$$

or equivalently to the minimization problem

$$\begin{cases} \widehat{\boldsymbol{x}} = \arg\min \ell(\boldsymbol{x}, \boldsymbol{y}_0), \\ \ell(\boldsymbol{x}, \boldsymbol{y}_0) = - \log f_{X|Y}(\boldsymbol{x}, \boldsymbol{y}_0). \end{cases} \tag{9.85}$$

We shall assume that we are able to identify a bounded region Q of the space where $\boldsymbol{x}$ lies, such that we know that $\widehat{\boldsymbol{x}}$ belongs to the interior of Q. We also assume that $f_{X|Y}(\boldsymbol{x}, \boldsymbol{y}_0)$ is continuous and strictly positive on Q, so that $\ell(\boldsymbol{x}, \boldsymbol{y}_0)$ is also regular. Therefore, our Bayesian estimation approach is reduced to the optimization problem (9.85) with $\widehat{\boldsymbol{x}} \in Q$. Readers interested in optimization problems can consult [136, 139] and the references therein. Conversely, assume that a continuous, positive and bounded function $\mathscr{E}(\boldsymbol{x})$ ($m \le \mathscr{E}(\boldsymbol{x}) \le M, \boldsymbol{x} \in Q$) is given on Q and we wish to solve

$$\hat{\boldsymbol{x}} = \arg\min_{x \in Q} \mathscr{E}(\boldsymbol{x}). \tag{9.86}$$

Then we can as well consider this as a MAP problem if we assume that for any positive real T,

$$x \in Q, \quad f_T(x) = \frac{e^{-\frac{\mathscr{E}(x)}{T}}}{\int_Q e^{-\frac{\mathscr{E}(\xi)}{T}} d\xi} = C_T e^{-\frac{\mathscr{E}(x)}{T}} \tag{9.87}$$

is just the posterior distribution of some random variable X, namely we assume that $f_{X|Y}(x, y_0) = f_T(x)$ (or $\ell(x, y_0) = -\frac{\mathscr{E}(x)}{T} + \log C_T$), which is strictly positive on Q. In fact, on Q

$$C_T \geq \frac{1}{|Q|} e^{\frac{m}{T}} > 0$$

and

$$f_T(x) \geq C_T e^{-\frac{M}{T}} > 0. \tag{9.88}$$

Therefore every optimization problem can be as well interpreted as the search of the MAP of a Bayesian system. Functions with the shape of $f_T(x)$, given in (9.87), are well-known in thermodynamics where $\mathscr{E}(x)$ has the meaning of an energy and T (apart from a constant) that of a temperature; this explains the symbols used. In the thermodynamics literature C_T^{-1} is also known as the *energy partition function*; its knowledge implies an impossible numerical task when the number of variables increases, as it happens in the application we have in mind.

In any event, let us assume the function $\mathscr{E}(x)$ is given and, without loss of generality, let us define the coordinates in the parameter space so that the unique minimum of $\mathscr{E}$ in Q is attained at $\widehat{x} = 0$. Furthermore, assume that Q is just a prism centered at $\widehat{x} = 0$, namely

$$Q = \{-a_1 \leq x_1 \leq a_1, \ldots, -a_n \leq x_n \leq a_n\}. \tag{9.89}$$

The first result will be obtained by considering the family of distributions (9.87) and assuming that one is able to draw independent samples from any of them. Then, consider any sequence $\{T_k\}$ tending monotonically to 0 and let X_k be a RV with probability density $f_{T_k}(x)$. We assume that each X_k is stochastically independent of X_j when $j \neq k$. Then the following theorem holds.

Theorem 9.1 *The sequence $\{X_k\}$ is stochastically convergent to $\widehat{x} = 0$.*
Moreover, if we consider the space

$$Q^{\infty} = Q \otimes Q \otimes \ldots$$

of independent sample sequences drawn from $\{X_k\}$, then we have that

$$P\{\lim_{k \to \infty} x_k = 0\} = 1 \tag{9.90}$$

$$(x_k \sim X_k; \ \{x_k\} \in Q^{\infty}).$$

Proof Since the second statement implies the first, we need only to prove (9.90). We note that the complement of the set $\{\{\boldsymbol{x}_k\} \in Q^\infty | \lim_{k\to\infty} \boldsymbol{x}_k = 0\}$ is the set NZL (no zero limit) defined as follows

$$\text{NZL} \equiv \{\{\boldsymbol{x}_k\} \in Q^\infty; \exists \varepsilon > 0, \exists \{j\} \subset \{k\}, |\boldsymbol{x}_j| > \varepsilon \,\forall j\}. \tag{9.91}$$

In (9.91) the subsequence $\{j\}$ of $\{k\}$ is understood to be infinite, so that if we take the subsequence $\{T_j\}$ of $\{T_k\}$, we have $T_j \to 0$, is $j \to \infty$.

We claim that

$$P\{NZL\} = 0 \tag{9.92}$$

Indeed, take any integer J in the subsequence $\{j\}$:

$$P\{NZL\} = P\{|\boldsymbol{X}_j| > \varepsilon, \forall j\} \leq P\{|\boldsymbol{X}_J| > \varepsilon\}. \tag{9.93}$$

But, for any constant $\bar{\mathscr{E}}$,

$$\begin{aligned} P\{|\boldsymbol{X}_J| > \varepsilon\} &= \frac{\int_{|\boldsymbol{x}|>\varepsilon} e^{-\frac{\mathscr{E}(x)}{T_J}} dx}{\int_Q e^{-\frac{\mathscr{E}(x)}{T_I}} dx} \equiv \\ &\equiv \frac{\int_{|\boldsymbol{x}|>\varepsilon} e^{-\frac{\mathscr{E}(x)-\bar{E}}{T_J}} dx}{\int_Q e^{-\frac{\mathscr{E}(x)-\bar{\mathscr{E}}}{T_J}} dx} \end{aligned} \tag{9.94}$$

Now, choose $\bar{\mathscr{E}}$ so close to $\mathscr{E}(0)$ that (see Fig. 9.4)

$$Q_{\bar{\mathscr{E}}} \equiv \{\mathscr{E}(0) \leq \mathscr{E}(\boldsymbol{x}) \leq \bar{\mathscr{E}}\} \subset \{|X| \leq \varepsilon\}; \tag{9.95}$$

then

$$Q^{\mathrm{c}}_{\bar{\mathscr{E}}} \equiv \{\mathscr{E}(\boldsymbol{x}) > \bar{\mathscr{E}}\} \supset \{|X| > \varepsilon\}. \tag{9.96}$$

Therefore, since the integrand is everywhere positive,

$$\int_{|\boldsymbol{x}|>\varepsilon} e^{-\frac{\mathscr{E}(x)-\bar{\mathscr{E}}}{T_J}} dx \leq \int_{Q^{c}_{\bar{\mathscr{E}}}} e^{-\frac{\mathscr{E}(x)-\bar{\mathscr{E}}}{T_J}} dx. \tag{9.97}$$

Moreover

$$\int_Q e^{-\frac{\mathscr{E}(x)-\bar{\mathscr{E}}}{T_J}} dx = \int_{Q_{\bar{\mathscr{E}}}} e^{-\frac{\mathscr{E}(x)-\bar{\mathscr{E}}}{T_J}} dx + \int_{Q^{c}_{\bar{\mathscr{E}}}} e^{-\frac{\mathscr{E}(x)-\bar{\mathscr{E}}}{T_J}} dx. \tag{9.98}$$

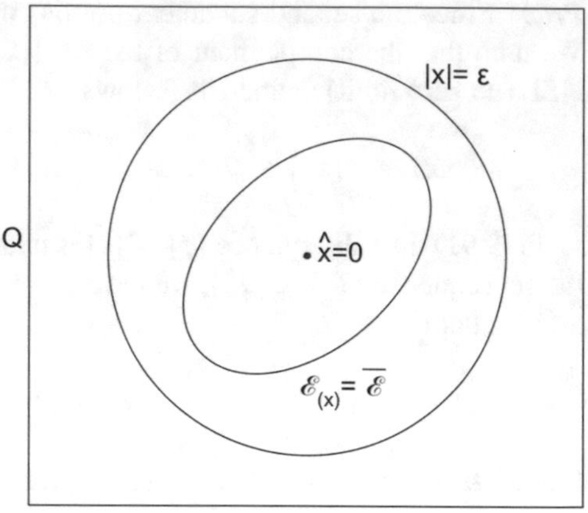

Fig. 9.4 The set Q, the ball $\{|X| < \varepsilon\}$ and the boundary of the sets $Q_{\bar{\mathscr{E}}}$, $Q^c_{\bar{\mathscr{E}}}$

Since

$$e^{-\frac{\mathscr{E}(x)-\bar{\mathscr{E}}}{T_J}} > 1, \quad x \in Q_{\bar{\mathscr{E}}},$$

we have

$$\int_Q e^{-\frac{\mathscr{E}(x)-\bar{\mathscr{E}}}{T_j}} dx > |Q_{\bar{\mathscr{E}}}|, \tag{9.99}$$

with $|Q_{\bar{\mathscr{E}}}|$ denoting the volume of $Q_{\bar{\mathscr{E}}}$. By using (9.97), (9.99) in (9.94) and (9.93) we obtain

$$P\{\text{NZL}\} \leq \frac{1}{|Q_{\bar{\mathscr{E}}}|} \int_{Q^c_{\bar{\mathscr{E}}}} e^{-\frac{\mathscr{E}(x)-\bar{\mathscr{E}}}{T_J}} dx. \tag{9.100}$$

Now, since $\mathscr{E}(x) - \bar{\mathscr{E}} > 0$ in $Q^c_{\bar{\mathscr{E}}}$, we have $e^{-\frac{\mathscr{E}(x)-\bar{\mathscr{E}}}{T_J}} < 1$ in this region, and then $\lim_{J \to \infty} e^{-\frac{\mathscr{E}(x)-\bar{\mathscr{E}}}{T_J}} = 0 \; \forall x \in Q^c_{\bar{\mathscr{E}}}$.

Therefore, considering that (9.100) holds $\forall J$ and taking the limit for $J \to \infty$, by the dominated convergence theorem (9.100) implies

$$P(\text{NZL}) = 0,$$

as it was claimed. □

What Theorem 9.1 says is that, since when $T_k \to 0$ the distributions $f_{T_k}(x)$ become more and more peaked around $\widehat{x}$ (see Fig. 9.5), it is enough to sample (independently) each $X_k \sim f_{T_k}$ to obtain a sequence $\{x_k\}$ that tends to $\widehat{x}$ with probability 1.

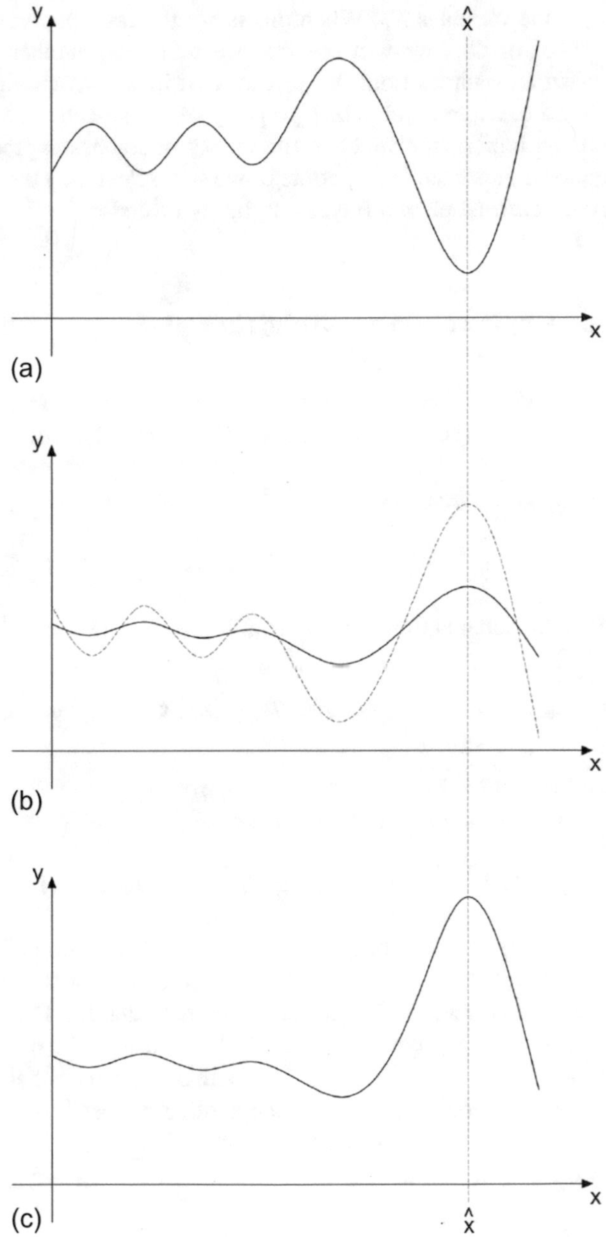

Fig. 9.5 (**a**) a function $\mathscr{E}(x)$ with an absolute maximum at $\widehat{\boldsymbol{x}}$, (**b**) $f_{T=10}(x)$, (**c**) $f_{T=0.1}(x)$; $\widehat{x}$ is a maximum of both, but $f_{10}(x)$ is considerably flatter than $f_{0.1}(x)$

According to the theorem, then, one is inclined to take a T_k tending very fast to zero in order to avoid drawing too many samples. However, this is true only if we can really sample the variables $\boldsymbol{X}_k$. When this is not the case, in particular because we cannot get hold of C_{T_k}, we will see that we will need another machine that allows us to construct samples from the f_{T_k}, at least in an asymptotic way. In any event we will need to exploit a classical property of the so-called *Markov chains*, which constitute an important chapter of the theory of stochastic processes. Since this is too large to be presented in this context, we will select only those results that are useful for our construction of a Bayesian gravity inversion.

9.7 Markov Chains and the Gibbs Sampler

We consider a process $\{\boldsymbol{X}_k\}$ with discrete times, $k = 1, 2, \ldots$ and state space a (bounded) prism Q in $\mathbb{R}^n$ (i.e. $\boldsymbol{x}_k \sim \boldsymbol{X}_k$; $\boldsymbol{x}_k \in Q$). Assume the process is described by a family of *probability density functions* (pdf). Namely, for any N put $\boldsymbol{\xi}_N^T = (\boldsymbol{X}_1^T, \boldsymbol{X}_2^T, \ldots, \boldsymbol{X}_N^T)$, then we know

$$f_{\xi_N}(\boldsymbol{x}_1, \boldsymbol{x}_2, \ldots, \boldsymbol{x}_N), \tag{9.101}$$

where f_{ξ_N} satisfy the famous Kolmogorov conditions, i.e., they are pdf's and

$$\int_Q f_{\xi_N}(\boldsymbol{x}_1, \boldsymbol{x}_2, \ldots, \boldsymbol{x}_{N-1}, \boldsymbol{x}_N) d\boldsymbol{x}_N \equiv f_{\xi_{N-1}}(\boldsymbol{x}_1, \boldsymbol{x}_2, \ldots, \boldsymbol{x}_{N-1}); \tag{9.102}$$

see for instance [52], Ch. 1, §1.

The process $\{\boldsymbol{X}_k\}$ is said to be a *Markov process if*

$$f_{\boldsymbol{X}_N|\boldsymbol{X}_{N-1}\ldots\boldsymbol{X}-1}(\boldsymbol{x}_N, \boldsymbol{x}_{N-1}, \ldots, \boldsymbol{x}_1) \equiv f_{\boldsymbol{X}_N|\boldsymbol{X}_{N-1}}(\boldsymbol{x}_N|\boldsymbol{x}_{N-1}), \tag{9.103}$$

i.e., by fixing the value of $\boldsymbol{X}_{N-1}$ only, the conditional distribution of $\boldsymbol{X}_N$, relative to the past history $(\boldsymbol{x}_1, \boldsymbol{x}_2, \ldots, \boldsymbol{x}_{N-1})$, is fixed, or, in other words, $\boldsymbol{X}_N$ depends on the process at previous time $N - 1$, but not on the path that has led the process to $\boldsymbol{x}_{N-1}$. Strictly speaking, a Markov chain is a Markov process with discrete time and discrete state space, however, also in agreement with our approach that "continuous" and "discrete" theories should be close to one another, we shall use irrespectively the term Markov chain (MC) in our context.

Due to the well-known identity

$$f_{\boldsymbol{X}_N,\boldsymbol{X}_{N-1}}(\boldsymbol{x}_n, \boldsymbol{x}_{N-1}) = f_{\boldsymbol{X}_N|\boldsymbol{X}_{N-1}}(\boldsymbol{x}_N|\boldsymbol{x}_{N-1}) f_{\boldsymbol{x}_{N-1}}(\boldsymbol{x}_{N-1}), \tag{9.104}$$

we see, by integrating with respect to $\boldsymbol{x}_{N-1}$, that the evolution of the (marginal) distribution of the process from time $N - 1$ to time N follows the law

$$
\begin{aligned}
f_{X_N}(\boldsymbol{x}_N) &= \int_Q f_{X_N, X_{N-1}}(\boldsymbol{x}_N, \boldsymbol{x}_{N-1}) d\boldsymbol{x}_{N-1} = \\
&= \int_Q f_{X_N|X_{N-1}}(\boldsymbol{x}_N|\boldsymbol{x}_{N-1}) f_{X_{N-1}}(\boldsymbol{x}_{N-1}) d\boldsymbol{x}_{N-1}.
\end{aligned} \tag{9.105}
$$

To simplify the notation we will write

$$f_N(\boldsymbol{x}_N) = f_{X_N}(\boldsymbol{x}_N), \tag{9.106}$$

$$K_N(\boldsymbol{x}_{N-1}, \boldsymbol{x}_N) = f_{X_N|X_{N-1}}(\boldsymbol{x}_N|\boldsymbol{x}_{N-1}). \tag{9.107}$$

In particular, $K_N(\boldsymbol{x}, \boldsymbol{y})$, which represents the conditional probability density that the process goes from x at time $N - 1$ to any y at time N, is called the *transition kernel*.

The following properties of a transition kernel are self-evident:

$$K_N(\boldsymbol{x}, \boldsymbol{y}) \geq 0, \tag{9.108}$$

$$\int_Q K_N(\boldsymbol{x}, \boldsymbol{y}) dy = 1. \tag{9.109}$$

We will say that K_N is strictly positive if

$$K_N(\boldsymbol{x}, \boldsymbol{y}) \geq k_0 > 0, \quad \forall \boldsymbol{x}, \boldsymbol{y} \in Q_k. \tag{9.110}$$

The MC $\{\boldsymbol{X}_N\}$, is said to be homogeneous in time if its transition kernel $K(\boldsymbol{x}, \boldsymbol{y})$ is the same for all N.

For the so-called irreducible and aperiodic, homogeneous MC's a fundamental ergodic theorem claims that the sequence of pdf's $f_N(\boldsymbol{x})$ tends to a pdf $f(\boldsymbol{x})$ for $N \to \infty$; $f(\boldsymbol{x})$ is also a stable equilibrium function of the chain.

Rather than discussing the general properties mentioned above, we will adapt the ergodic theorem to our present context, by the following enunciation.

Theorem 9.2 (Ergodic) *Let $\{\boldsymbol{X}_N\}$ be a homogenous* MC *with strictly positive transition kernel* (*see* (9.110)). *Then, for every initial pdf $f_1(x)$, there is a pdf $f(x)$ such that*

$$\lim_{N\to\infty} \| f_N(\boldsymbol{x}) - f(\boldsymbol{x})\|_{L^1} = 0; \tag{9.111}$$

furthermore, $f(\boldsymbol{x})$ is the unique equilibrium distribution of the chain.

Proof Call L_0^1 the subspace of L^1 consisting of the functions g such that

$$\int_Q g(\boldsymbol{x}) dx = 0; \tag{9.112}$$

it is obvious that L_0^1 is a closed subspace of L^1.

Notice that for any two pdf's $f(\boldsymbol{x})$, $f'(\boldsymbol{x})$ the function

$$g(\boldsymbol{x}) = f(\boldsymbol{x}) - f'(\boldsymbol{x})$$

belongs to L_0^1.

Define the integral operator K with kernel $K(\boldsymbol{x}, \boldsymbol{y})$, the transition kernel of $\{\boldsymbol{X}_N\}$, by

$$h(\boldsymbol{y}) = \int_Q K(\boldsymbol{x}, \boldsymbol{y}) g(\boldsymbol{x}) d\boldsymbol{x} = Kg; \tag{9.113}$$

then K acts from L_0^1 to L_0^1, i.e., $\int_Q h(\boldsymbol{y}) d\boldsymbol{y} = 0$ as it is obvious from property (9.109). Furthermore, K is continuous and contracting in L_0^1, namely, there is a number $c_0 < 1$ such that

$$\|h\|_{L^1} = \|Kg\|_{L^1} \leq c_0 \|g\|_{L^1} \qquad \forall g \in L_0^1. \tag{9.114}$$

The proof of (9.114) is to be found in Exercise 5.

Then we have

$$\begin{aligned} \left\| f_{N+p} - f_N \right\|_{L^1} &= \left\| K(f_{N+p-1} - f_{N-1}) \right\|_{L^1} = \\ &= \cdots = \left\| K^{N-1}(f_{p+1} - f_1) \right\|_{L^1} \leq \\ &\leq c_0^{N-1} \left\| f_{p+1} - f_1 \right\|_{L^1} \end{aligned} \tag{9.115}$$

because $f_{p+1} - f_1 \in L_0^1$. Furthermore,

$$\left\| f_{p+1} - f_1 \right\|_{L^1} = \int_Q |f_{p+1} - f_1| dx \leq \int_Q |f_{p+1}| dx + \int_Q |f_1| dx = 2, \tag{9.116}$$

because f_{p+1} and f_1 are pdf's. By (9.115), for $N \to \infty$ and uniformly in p,

$$\left\| f_{n+p} - f_N \right\|_{L^1} < 2c_0^{N-1} \to 0, \tag{9.117}$$

so $\{f_N(\boldsymbol{x})\}$ is a Cauchy sequence in L^1. Therefore, there is $f(\boldsymbol{x}) \in L^1$ such that $f_N \to f$ in L^1. Moreover, since $1 = \|f_N\|_{L^1} \to \|f\|_{L^1}$ we have that $\int_Q f(\boldsymbol{x}) dx = 1$ and since a subsequence of $f_N(\boldsymbol{x})$ tends to $f(\boldsymbol{x})$ almost everywhere (a.e.), we have that $f(\boldsymbol{x}) \geq 0$, a.e. i.e., f is a pdf.

Since K is a continuous operator in L_0^1, if $f_N - f \to 0$ ($f_N - f \in L_0^1$) then $K(f_N - f) \to 0$, i.e., $K f_N \to K f$.

Hence, letting $N \to \infty$ in the identity

$$f_N(\boldsymbol{y}) = \int_Q K(\boldsymbol{x}, \boldsymbol{y}) f_{N-1}(\boldsymbol{x}) d\boldsymbol{x},$$

we get

$$f(\boldsymbol{y}) = \int_Q K(\boldsymbol{x}, \boldsymbol{y}) f(\boldsymbol{x}) d\boldsymbol{x}, \tag{9.118}$$

i.e., f is an equilibrium function of the chain; in fact, if we start the chain with $X_1 \sim f(\boldsymbol{x})$, then we obtain $f(\boldsymbol{x})$ as the distribution of $\boldsymbol{X}_N$ at any time.

Since K is a contraction in L_0^1, it is clear that f is unique, because

$$f = Kf, \; \bar{f} = K\bar{f} \Rightarrow (f - \bar{f}) = K(f - \bar{f})$$

and so

$$\|f - \bar{f}\|_{L^1} \leq c_0 \|f - \bar{f}\|_{L^1} \quad (c_0 < 1),$$

which is possible only if $f = \bar{f}$.

Therefore, $\forall f_1(\boldsymbol{x})$, the sequence f_N always tends to $f(\boldsymbol{x})$, i.e. the equilibrium distribution is stable □

Remark 9.6 (sampling a MC) Let us note that the repeated application of (9.103) yields the telescoping formula

$$\begin{aligned} f_{\xi_N}(\boldsymbol{x}_N, \boldsymbol{x}_{N-1}, \ldots, \boldsymbol{x}_1) &= f_{\boldsymbol{X}_N|\boldsymbol{X}_{N-1}\ldots\boldsymbol{X}_1}(\boldsymbol{x}_N|\boldsymbol{x}_{N-1}, \ldots, \boldsymbol{x}_1)\, f_{\xi_{N-1}}(\boldsymbol{x}_{N-1}, \ldots, \boldsymbol{x}_1) = \\ &= f_{\boldsymbol{X}_N|\boldsymbol{X}_{N-1}}(\boldsymbol{x}_N|\boldsymbol{x}_{N-1})\, f_{\boldsymbol{X}_{N-1}|\boldsymbol{X}_{N-2}}(\boldsymbol{x}_{N-1}|\boldsymbol{x}_{N-2}) \\ &\cdots f_{\boldsymbol{X}_2|\boldsymbol{X}_1}(\boldsymbol{x}_2|\boldsymbol{x}_1)\, f_{\boldsymbol{X}_1}(\boldsymbol{x}_1) \end{aligned} \tag{9.119}$$

It follows that to obtain a sample for $\boldsymbol{\xi}_N$ we can apply the following elementary mechanism:

$$\boldsymbol{x}_1 \sim \boldsymbol{X}_1 \sim f_{\boldsymbol{X}_1}(\boldsymbol{x}_1)$$

$$\boldsymbol{x}_2 \sim \boldsymbol{X}_2\big|_{\boldsymbol{X}_1=\boldsymbol{x}_1} \sim f_{\boldsymbol{X}_2|\boldsymbol{X}_1}(\boldsymbol{x}_2|\boldsymbol{x}_1), \boldsymbol{x}_1 \text{ from above}$$

$$\ldots$$

$$\boldsymbol{x}_N \sim \boldsymbol{X}_N\big|_{\boldsymbol{X}_{N-1}=\boldsymbol{x}_{N-1}} \sim f_{\boldsymbol{X}_N|\boldsymbol{X}_{N-1}}(\boldsymbol{x}_N|\boldsymbol{x}_{N-1}) \;,\; \boldsymbol{x}_{N-1} \text{ from above}$$

The sequence $(\boldsymbol{x}_1, \boldsymbol{x}_2, \ldots, \boldsymbol{x}_N)$ is a sample from $\boldsymbol{\xi}_N$. The construction of the sequence, according to the above scheme, is said to run the MC starting from x_1.

Now assume we are given a pdf $f_X(\boldsymbol{x})$ on Q that is strictly positive and bounded above:

$$A_0 \geq f_X(\boldsymbol{x}) \geq a_0 \,; \tag{9.120}$$

note that this is exactly the case for all pdf's belonging to the family $\{f_T(\boldsymbol{x})\}$ given in (9.87).

It is our purpose to build from $f_X(\boldsymbol{x})$ a MC that admits such a pdf as limit distribution, so that running the MC, after a transient time we derive samples from it. Not only this but we want to do that in a way that is independent of any unpleasant normalization constant contained in $f_X(\boldsymbol{x})$, so that in the case of $f_T(\boldsymbol{x})$ we will avoid the computation of the constant C_T.

Our solution to the above problem will come from a combination of the next two lemmas, which together give rise to the so-called *Gibbs Algorithm* [51].

Since to build a MC, which in our case will be a homogeneous MC, we need a transition kernel $K(\boldsymbol{x}, \boldsymbol{y})$, we start from this. To simplify formulas we adopt a particular notation, namely let

$$\boldsymbol{x}^T = (x_1, x_2, \ldots, x_n)$$

we put, for

$$m > \ell, \quad (\boldsymbol{x}_\ell^m)^T = (x_\ell, x_{\ell+1}, \ldots, x_m),$$

so that, for instance, we can write

$$\boldsymbol{x}^T = ((\boldsymbol{x}_1^{i-1})^T, x_i, (\boldsymbol{x}_{i+1}^n)^T) = ((\boldsymbol{x}_1^{i-1})^T, (\boldsymbol{x}_i^n)^T)\,; \tag{9.121}$$

moreover, we will put

$$\boldsymbol{x}_{-i}^T = ((\boldsymbol{x}_1^{i-1})^T, (\boldsymbol{x}_{i+1}^n)^T),$$

namely $\boldsymbol{x}_{-i}$ is the vector $\boldsymbol{x}$ with the component x_i erased.

With this notation we have for instance the relation between marginal distributions

$$f_{X_{-i}}(\boldsymbol{x}_{-i}) = f_{X_{-i}}(\boldsymbol{x}_1^{i-1}, \boldsymbol{x}_{i+1}^n) = \int_{-a_i}^{a_i} f_X(\boldsymbol{x}_1^{i-1}, x_i, \boldsymbol{x}_{i+1}^n) dx_i, \tag{9.122}$$

assuming that Q is just the product of intervals as in (9.89). Notice that, due to (9.120), the marginals (9.122) satisfy the inequalities

$$A_0(2a_i) \geq f_{X_{-i}}(\boldsymbol{x}_{-i}) \geq a_0(2a_i) > 0. \tag{9.123}$$

Moreover, the conditional distributions of $\boldsymbol{X}_i$ conditional to $\boldsymbol{X}_{-i}$ are given by

$$K_i(\boldsymbol{x}_1^{i-1}, x_i, \boldsymbol{x}_{i+1}^n) = f_{X_i|X_{-i}}(x_i|\boldsymbol{x}_1^{i-1}, \boldsymbol{x}_{i+1}^n) = \\ = \frac{f_X(\boldsymbol{x}_1^{i-1}, x_i, \boldsymbol{x}_{i+1}^n)}{f_{X_{-i}}(\boldsymbol{x}_1^{i-1}, \boldsymbol{x}_{i+1}^n)}. \tag{9.124}$$

From (9.123) we see that such conditional distribution are bounded too, below and above

$$\frac{A_0}{a_0(2a_i)} \geq K_i(\boldsymbol{x}_1^{i-1}, x_i, \boldsymbol{x}_{i+1}^n) \geq \frac{a_0}{A_0(2a_i)} > 0. \tag{9.125}$$

We also note that, if f_X contains some constant factor, then all X_{-i} will contain the same constant too, so that K_i given by (9.124) will be independent of it. We are ready now to formulate the first lemma.

Lemma 9.1 *Define*

$$K(\boldsymbol{x}, \boldsymbol{y}) = \prod_{i=1}^n K_i(\boldsymbol{y}_1^{i-1}, y_i, \boldsymbol{x}_{i+1}^n); \tag{9.126}$$

then $K(\boldsymbol{x}, \boldsymbol{y})$ is a strictly positive, bounded, transition kernel.

Proof That $K(\boldsymbol{x}, \boldsymbol{y})$ is strictly positive comes from the same property of each K_i, see (9.125). Then one needs to prove that

$$\int_Q K(\boldsymbol{x}, \boldsymbol{y})d\boldsymbol{y} = 1; \tag{9.127}$$

the proof is given in Exercise 6 for $n = 2$; the generalization to any n is straightforward. □

With particular applications in mind, we note that $K_i(\boldsymbol{y}_1^{i-1}, y_i, \boldsymbol{x}_{i+1}^n)$ is nothing but the conditional distribution $K_i(\boldsymbol{x})$ given in (9.124), where the first i components are taken from the vector $\boldsymbol{y}$, while the remaining components are taken from the vector $\boldsymbol{x}$.

The next lemma explains why we have introduced a $K(\boldsymbol{x}, \boldsymbol{y})$ as in (9.126).

Lemma 9.2 *Let $\{X_N\}$ be the* MC *generated by the transition kernel $K(\boldsymbol{x}, \boldsymbol{y})$, given by formula* (9.126)*; then $\{X_N\}$ admits $f_X(\boldsymbol{x})$ as limit distribution.*

Proof That $\{X_N\}$ admits a unique stable limit distribution comes from Theorem 9.2 and the remark that $K(\boldsymbol{x}, \boldsymbol{y})$ is strictly positive. That $f_X(\boldsymbol{x})$ is the limit distribution is proved in Exercise 7, for $n = 2$. The generalization to any n is straightforward. □

The combination of Lemmas 9.1 and 9.2 suggests a procedure to estimate the MAP $\hat{\boldsymbol{x}}$.

In fact, let us first fix a value of the "temperature" T, namely the corresponding pdf $f_T(\boldsymbol{x})$; then, by using (9.126) we can build the transition kernel $K_T(\boldsymbol{x}, \boldsymbol{y})$ relative to $f_T(\boldsymbol{x})$ and, according to Remark 9.6, starting from an arbitrary $\boldsymbol{x}_1$, we can run a MC $\{\boldsymbol{x}_{kT}\}$ that has $f_T(\boldsymbol{x})$ as limit distribution. Therefore we arrive at a time N along the MC when we can consider $\boldsymbol{x}_{NT}$ as a sample from $f_T(\boldsymbol{x})$. Now let us define a sequence $T_k \to 0$ and then for each k let us run the above algorithm, so that for each newly generated MC we can extract a sample $\boldsymbol{x}_{N_k T_k}$.

Each MC is generated independently (in fact all the sampling procedures and the initial state are generated anew) so that $\{\boldsymbol{X}_{N_k T_k}\}$ is a stochastically independent sequence, and each $\boldsymbol{x}_{N_k T_k}$ is sampled from $f_{T_k}(\boldsymbol{x})$; therefore Theorem 9.1 applies and we can claim that

$$\lim_{k\to\infty} \boldsymbol{X}_{N_k T_k} = \hat{\boldsymbol{x}}, \tag{9.128}$$

with probability 1.

On a pure rational ground we seem to have hit the target, but it is clear that the procedure outlined above is numerically extremely demanding, because for each k we have to generate an MC and wait until it has reached a stationary distribution. To fasten the construction of the sequence, one might be tempted to decide to change the temperature T_k after each sweep of the components of x by the Gibbs algorithm, so once $\boldsymbol{x}_{T_k}$ has been sampled, one new sample x_{T_k+1} is sampled from the changed conditional distribution $K_{T_{k+1}}(\boldsymbol{x}, \boldsymbol{y})$, starting form the previous $\boldsymbol{x}_{T_k}$, namely

$$\boldsymbol{x}_{T_{k+1}} \sim K_{T_{k+1}}(\boldsymbol{x}_{T_k}, \boldsymbol{y})\,. \tag{9.129}$$

This approach can in fact produce the wanted result and is known in statistical literature as *Gibbs Sampler* (see for instance [49]).

As a matter of fact, it is not difficult to see that the sequence $\{\boldsymbol{X}_k\}$ generated by the Gibbs Sampler is an MC, though not anymore homogenous because the kernels K_{T_k} change from one step to the next.

It is clear though that critical to the practical application of the Gibbs Sampler is the choice of the cooling schedule $\{T_k\}$. In fact, one might be willing to have a fast cooling to decrease the numerical burden, yet if the distribution f_T is flat at high temperature we can expect to be able to draw samples from every zone of the probability space, but if we go to zero too fast the MC cannot have the time to approach the area where the absolute maximum is, and it remains trapped in one of the lobes of f_T around one of the local maxima, rather then jumping to the subset of highest probability density.

The question was settled by a theorem of Geman and Geman in their seminal paper on the Gibbs Sampler application to image restoration. We report here the formulation of the theorem, sending the interested reader to the literature [50].

Theorem 9.3 *The sequence of variables* $\{\boldsymbol{X}_{T_k}\}$ *constructed by the algorithm* (9.129) *constitutes a non-homogeneous* MC; *if the cooling schedule* $\{T_k\}$ *is such that*

$$T_k \geq \frac{C}{\log(1+k)}, \tag{9.130}$$

then the limit distribution to which $f_{T_k}(\boldsymbol{x})$ *tend is*

$$\delta(\boldsymbol{x} - \widehat{\boldsymbol{x}}), \tag{9.131}$$

i.e., the sequence $\{\widehat{\boldsymbol{X}}_k\}$ *is stochastically convergent to the constant* $\widehat{\boldsymbol{x}}$*.*

Let us observe that the "safe" value for C found in [50], namely $C = n \cdot (\max \mathscr{E}(\boldsymbol{x}) - \min \mathscr{E}(\boldsymbol{x}))$, is by far too large to be practical.

In the literature one can find many examples where more favourable constants and even a faster schedule, e.g., $T_k = \frac{C}{k+1}$, have been used successfully.

Remark 9.7 One may think the task of constructing the Gibbs Sampler based on conditional distributions as in (9.124) might be as complicated as it was to sample from the original distribution. This is not the case because, as we see in (9.124), the function K_i of x_i, which is the distribution to be sampled, is in fact proportional to $f_X(\boldsymbol{x}_i^{i-1}, x_i, \boldsymbol{x}_{i+1}^n)$ where $\boldsymbol{x}_1^{i-1}, \boldsymbol{x}_{i+1}^n$ are fixed. So, to transform this into a real probability distribution we only need to normalize our function with respect to the individual variable x_i, which is, as we shall see in our application, more easily feasible, specially when the part of f_X that contains x_i is factored from the rest of the function, so that the rest is cancelled during normalization.

9.8 Bayesian Gravity Inversion: Construction of the Model

In this section we will apply the concepts presented in the last three sections to the problem of inverting a set of gravity anomalies, given on the $S_0 \equiv \{z = 0\}$ plane, retrieving the densities $\rho(\boldsymbol{\xi}, z)$ contained in the layer $\{H \leq z \leq 0\}$, which can reproduce (or closely reproduce) the data.

What we present here is one possible Bayesian model that the authors have exploited in the last years; other choices however are possible, always in a Bayesian framework.

The context is the same as the one we illustrated in Sect. 9.4, namely the gravity data dg_0 are collected on a grid on S_0; only, here we shall reorder them into a unique vector $d\boldsymbol{g}_0$, namely we create some correspondence between the index i of dg_{0i} and the double index $\boldsymbol{n}$ of $d\boldsymbol{g}_{0\boldsymbol{n}}$.

The dimension of the vector $\boldsymbol{y}$ will therefore be $N_y = N \cdot M$ (see Fig. 9.1).

Moreover, the variables ρ are discretized as constant densities inside the voxels $V_{n,\ell}$ (cf. Fig. 9.2) that decompose the set B where $\rho \neq 0$ into

$$N_V = N \cdot M \cdot L, \tag{9.132}$$

elements. The values of ρ are organized into a unique vector $\boldsymbol{\rho}$, associating the index i of ρ_i to the position of the corresponding voxel V_i.

The vector $\boldsymbol{\rho}$ will be part of the vector of unknowns $\boldsymbol{x}$ that, beyond $\boldsymbol{\rho}$, will contain also another vector $\boldsymbol{L}$ (labels) such that the component L_i, referring to the voxel V_i with density ρ_i, can assume an integer value in the set

$$L_i = \{1, 2, \dots, c\}. \tag{9.133}$$

To each integer k in (9.133) there corresponds a certain geological material, taken from a list compiled by an expert geologist that has an a priori knowledge of the area where the inversion takes place. Also, $\boldsymbol{L}$ is considered as a vector RV as $\boldsymbol{\rho}$ is, so that $\boldsymbol{x}$ is just the argument of the RV

$$\boldsymbol{X} = \begin{bmatrix} \boldsymbol{\rho} \\ \boldsymbol{L} \end{bmatrix}, \tag{9.134}$$

as typical of Bayesian models.

To proceed with the theory of Sect. 9.5, we need to define the joint distribution of observables and parameters (9.67), namely

$$\begin{aligned} f_{X,Y}(\boldsymbol{x}, \boldsymbol{y}) &= f(\boldsymbol{dg}, \boldsymbol{\rho}, \boldsymbol{L}) \\ &= \mathscr{L}(\boldsymbol{dg}|\boldsymbol{\rho}, \boldsymbol{L}) p(\boldsymbol{\rho}, \boldsymbol{L}) = \\ &= \mathscr{L}(\boldsymbol{dg}|\boldsymbol{\rho}, \boldsymbol{L}) p(\boldsymbol{\rho}|\boldsymbol{L}) p(\boldsymbol{L}), \end{aligned} \tag{9.135}$$

or, alternatively, the "energy" function

$$\begin{aligned} \mathscr{E}(\boldsymbol{dg}, \boldsymbol{\rho}, \boldsymbol{L}) &= -\log f(\boldsymbol{dg}, \boldsymbol{\rho}, \boldsymbol{L}) = \\ &= -\log \mathscr{L}(\boldsymbol{dg}, |\boldsymbol{\rho}, \boldsymbol{L}) - \log p(\boldsymbol{\rho}|\boldsymbol{L}) - \log p(\boldsymbol{L}) \\ &= \mathscr{E}_{dg}(\boldsymbol{dg}|\boldsymbol{\rho}, \boldsymbol{L}) + \mathscr{E}_{0\rho}(\boldsymbol{\rho}|\boldsymbol{L}) + \mathscr{E}_{0L}(\boldsymbol{L}), \end{aligned} \tag{9.136}$$

which is the target function to be minimized according to the methods explained in Sects. 9.6, 9.7.

The easiest part is that of the likelihood, which, given the observation model

$$\boldsymbol{dg} = A\boldsymbol{\rho} + \boldsymbol{\nu}, \tag{9.137}$$

when we assume the noise vector to be normally distributed, is given by

$$\begin{aligned} \mathscr{L}(\boldsymbol{dg}|\boldsymbol{\rho}, \boldsymbol{L}) &= \mathscr{L}(\boldsymbol{dg}|\boldsymbol{\rho}) = \\ &= \frac{1}{(2\pi)^{\frac{NM}{2}} (\det C_\nu)^{1/2}} e^{-\frac{1}{2}(\boldsymbol{dg}-A\boldsymbol{\rho})^T C_\nu^{-1}(\boldsymbol{dg}-A\boldsymbol{\rho})}. \end{aligned} \tag{9.138}$$

Therefore, the corresponding energy is, apart from an irrelevant constant,

$$\mathcal{E}_{dg}(d\boldsymbol{g}|\boldsymbol{\rho}) = \frac{1}{2}(d\boldsymbol{g} - A\boldsymbol{\rho})^T C_\nu^{-1}(d\boldsymbol{g} - A\boldsymbol{\rho}) . \tag{9.139}$$

To proceed with the prior distribution energy, it is convenient first to introduce a system of neighborhoods, $\{\Delta_i\}$, of the voxels $\{V_i\}$. Each neighborhood is geometrically defined for one of the internal voxels and then replicated for all the others.

The simplest neighborhood of V_i is that of the so-called nearest neighbours, constituted by the 6 voxels having a face in common with V_i (see Fig. 9.6).

Other neighborhood systems could be used, like the neighborhood consisting of all the 26 voxels V_i that have at least one point in common with V_i.

The chosen dimension of the neighborhood is an index representing the extent of the spatial interaction among the variables as well as among the variables $\{L_i\}$, as we shall see.

The neighbourhood of V_i has to be cut in an obvious way when V_i is at the boundary of the body B_0 (see Fig. 9.2).

Now we can address the construction of $\mathcal{E}_{0\rho}(\boldsymbol{\rho}|\boldsymbol{L})$. Through this energy function we want to express two prior conditions: one is that each ρ_i, when the label of V_i is $L_i = k$, will have to respect the geological prior knowledge of the material that corresponds to the k label.

The second condition is that all the ρ_j, when $j \in \Delta_i$, and $L_j = k$ too, have to be close to ρ_i. So we will split $\mathcal{E}_{0\rho}$ into the sum of two terms, one of geological nature, the other related to smoothness, i.e.,

$$\mathcal{E}_{0\rho} = \mathcal{E}_{0\rho G}(\boldsymbol{\rho}|\boldsymbol{L}) + \mathcal{E}_{0\rho s}(\boldsymbol{\rho}|\boldsymbol{L}). \tag{9.140}$$

As for the first one, we put

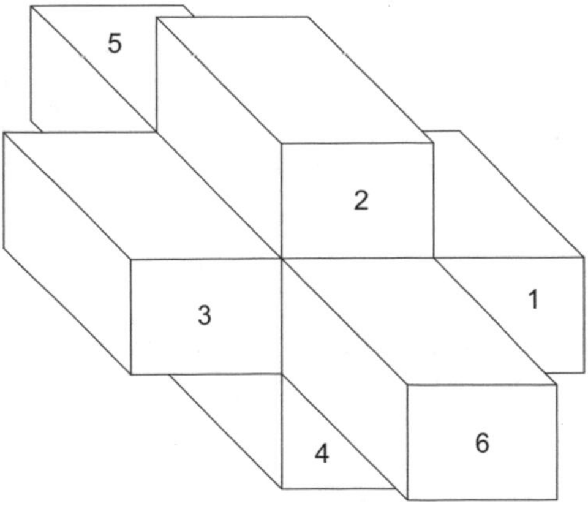

Fig. 9.6 The 6-voxel neighborhood of the (hidden) central voxel

$$\mathscr{E}_{0\rho G}(\boldsymbol{\rho}|\boldsymbol{L}) = \sum_{i=1}^{N_V} \mathscr{E}_{0\rho_i}(\rho_i|L_i) \tag{9.141}$$

and

$$\mathscr{E}_{0\rho_i}(\rho_i|L_i = k) = \frac{1}{2}\frac{(\rho_i - \mu_k)^2}{\sigma_k^2} - \log \chi_3\left(\frac{\rho_i - \mu_k}{\sigma_k}\right), \tag{9.142}$$

with

$$\chi_a(t) = \begin{cases} 1, & \text{if } |t| < a, \\ 0, & \text{if } |t| > a. \end{cases} \tag{9.143}$$

Note that here we consider that $\log 0 = -\infty$, so that, when $\frac{\rho_i - \mu_i}{\sigma_i} > 3$, we have $e^{-\mathscr{E}_{0\rho_i}(\rho_i|L_i=k)} = 0$.

To explain (9.142), we first consider Table 9.1, borrowed from [149], but other similar tables could be used. Now imagine that in the area where we want to perform the inversion, the geological information is that the materials identified by some labels, corresponding to a subset of the lines in Table 9.1, are present, perhaps adding one more label for water if we are in a marine area.

We note first of all that the resulting table defines the sample space of $\{(\rho_i, L_i),\ i = 1, \ldots, N\}$, in the sense that for $L_i = k$ the corresponding ρ_i can assume only values in an interval with upper bound $\rho_{\mathrm{U}}(k)$ and lower bound $\rho_{\mathrm{L}}(k)$, which can be read out of Table 9.1,

$$L_i = k \Rightarrow \rho_{\mathrm{L}}(k) \le \rho_i \le \rho_{\mathrm{U}}(k). \tag{9.144}$$

Moreover, we can define the average on the above interval as

$$\mu(L_i) = \mu_k = \frac{1}{2}(\rho_{\mathrm{L}}(k) + \rho_{\mathrm{U}}(k)) \tag{9.145}$$

and the standard deviation

$$\sigma(L_i) = \sigma_k = \frac{1}{6}(\rho_{\mathrm{U}}(k) - \rho_{\mathrm{L}}(k)), \tag{9.146}$$

so that the interval (9.144) can also be written as $[\mu_k - 3\sigma_k, \mu_k + 3\sigma_k]$. We will assume to have for every ρ_i in the distribution $f(\boldsymbol{\rho}|\boldsymbol{L})$ a factor of the form

$$f(\rho_i|L_i = k) = \frac{1.0027}{\sqrt{2\pi}\sigma_k} e^{-\frac{1}{2}\left(\frac{\rho_i - \mu_k}{\sigma_k}\right)^2} \chi_3\left(\frac{\rho_i - \mu_k}{\sigma_k}\right). \tag{9.147}$$

Table 9.1 Variation of rock density for different rocks. [149]

Rock type	Density range [g/cm^3]	Density average [g/cm^3]
Sediments (wet)		
Soil	1.2–2.4	1.92
Clay	1.6–2.6	2.21
Gravel	1.7–2.4	2.00
Sand	1.7–2.3	2.00
Sandstone	1.6–2.8	2.35
Shale	1.7–3.2	2.40
Limestone	1.9–2.9	2.55
Dolomite	2.3–2.9	2.70
Igneous rocks		
Rhyolite	2.3–2.7	2.52
Andesite	2.4–2.8	2.61
Granite	2.5–2.8	2.64
Granodiorite	2.7–2.8	2.73
Porphyry	2.6–2.9	2.74
Quartz diorite	2.6–3.0	2.79
Diorite	2.7–3.0	2.85
Lavas	2.8–3.0	2.90
Diabase	2.5–3.2	2.91
Basalt	2.7–3.3	2.99
Gabbro	2.7–3.5	3.03
Peridotite	2.8–3.4	3.15
Acid igneous	2.3–3.1	2.61
Basic igneous	2.1–3.2	2.79
Metamorphic rocks		
Quartzite	2.5–2.7	2.60
Schists	2.4–2.9	2.64
Graywacke	2.6–2.7	2.65
Marble	2.6–2.9	2.75
Serpentine	2.4–3.1	2.78
Slate	2.7–2.9	2.79
Gneiss	2.6–3.0	2.80
Amphibolite	2.9–3.0	2.96
Eclogite	3.2–3.5	3.4
Metamorphic	2.4–3.1	2.74

Notice that (9.147) is a distribution exactly on the interval $[\mu_k - 3\sigma_k, \mu_k + 3\sigma_k]$. which explains the normalising factor 1.0027.

When the above distributions are multiplied for all values of i, we arrive to the form

$$f_{\rho G}(\boldsymbol{\rho}|\boldsymbol{L}) = \frac{1.0027^{N_V}}{(2\pi)^{N_V/2}\Pi_{i=1}^{N_V}\sigma(L_i)}\times$$
$$\times \exp\{-\frac{1}{2}(\boldsymbol{\rho} - \boldsymbol{\mu}(\boldsymbol{L}))^T \Sigma(\boldsymbol{L})^{-1}(\boldsymbol{\rho} - \boldsymbol{\mu}(\boldsymbol{L})\}\times \tag{9.148}$$
$$\times \prod_{i=1}^{N_V} \chi_3\left(\frac{\rho_i - \mu(L_i)}{\sigma(L_i)}\right),$$

where indeed

$$\mu(L_i = k) = \mu_k, \ \sigma(L_i = k) = \sigma_k, \tag{9.149}$$

and

$$\Sigma_{i\ell}(\boldsymbol{L}) = \sigma(L_i = k)\delta_{i\ell} = \sigma_k\delta_{i\ell}. \tag{9.150}$$

We can observe that if we model the prior of $\boldsymbol{\rho}|\boldsymbol{L}$ putting $\mathscr{E}_{0\rho s} \equiv 0$ in (9.140), then $f_{\rho G}$ given by (9.148) is exactly this prior. In this case, suppose we want to sample $\boldsymbol{\rho}|\boldsymbol{L}$ from (9.148). Since the components of $\boldsymbol{\rho}|\boldsymbol{L}$ are independent, they can be extracted separately and, if z_k is a sample from a standardized normal, we can just put

$$\rho_k|_{L_i=k} = \mu_k + \sigma_k z_k \text{ if } |z_k| < 3\,; \tag{9.151}$$

all samples of z_k with $|z_k| > 3$ are simply rejected. Numerical experience, however, shows that with such a simple mechanisms, exactly due to the independence of ρ_i from $\rho_j (j \neq i)$, even two neighbouring voxels with the same label could have quite different sampled densities.

For instance, looking at Table 9.1, two close voxels of limestone, could have one a density of 2 g/cm^3, the other of 2.7 g/cm^3; a contrast hardly justifiable from a geological standpoint. It is true that all ρ_i are related one to the other also through the likelihood (9.139), when we come to the posterior of ρ given the observations; but this connection is rather weak, specially for deep voxels, so the above objection is justified. Then, it is important to introduce in $\mathscr{E}$ a term that helps smoothing the density variations between neighbouring voxels with the same labels. In this way the field ρ will be smooth in volumes with the same label and discontinuous across surfaces that are boundaries from one label to a different one. Such an effect could be achieved in two ways: either by introducing a correlation between densities with the same label, namely by making the matrix $\sum(\boldsymbol{L})$ block diagonal with each block, internally non-diagonal, depending on the value of L, or by expressing directly that the differences of ρ between close voxels are small. Here we pursue this second approach by stating that

$$\begin{cases} f_{\rho s}(\boldsymbol{\rho}|\boldsymbol{L}) = e^{-\mathscr{E}_{0\rho s}(\boldsymbol{\rho},\boldsymbol{L})}, \\ \mathscr{E}_{0\rho s}(\boldsymbol{\rho},\boldsymbol{L}) = \frac{1}{2}\sum_{k=1}^{N_V}\sum_{j\in\Delta_k}\left(\frac{\rho_k-\mu_k}{\sigma_k} - \frac{\rho_j-\mu_j}{\sigma_j}\right)^2 \delta_{L_k L_j}. \end{cases} \tag{9.152}$$

As we see, the value of $\mathscr{E}_{0\rho s}$ is small if neighbouring voxels V_k, V_j, having the same label so that $\mu_k = \mu_j$ and $\sigma_k = \sigma_j$, also have densities ρ_k, ρ_j close to one another.

The decision to use the normalized densities in (9.152), namely

$$\bar{\rho}_k = \frac{\rho_k - \mu_k}{\sigma_k}, \tag{9.153}$$

is based on experience with the method and on the remark that, in the end if we assume the field $\bar{\boldsymbol{\rho}} = \Sigma^{-1}(\boldsymbol{\rho} - \boldsymbol{\mu})$, with the appropriate $\Sigma(\boldsymbol{L})$, $\boldsymbol{\mu}(\boldsymbol{L})$, to be smooth even across the boundaries of regions with different labels, this could be less unrealistic than the same hypothesis applied directly to $\boldsymbol{\rho}$, as it is implicit for instance in the Tikhonov principle, when a term with the modulus squared of the gradient of ρ is introduced.

Moreover, it is intuitive that a material with larger σ_k can in principle exhibit larger density variations than one with a smaller σ_k, so the use of the σ's in (9.152) provides a natural weighting of the smoothness condition among the various materials.

We also remark here that since in the end $\mathscr{E}_{0\rho s}$ given by (9.152) is a homogeneous, symmetric, positive quadratic form in the components of $\bar{\boldsymbol{\rho}}$, there must be a symmetric positive matrix $\mathscr{S}(\boldsymbol{L})$ (we can call it a selection matrix) such that

$$\mathscr{E}_{0\rho s} = \frac{1}{2}\bar{\boldsymbol{\rho}}^T \mathscr{S}(\boldsymbol{L})\bar{\boldsymbol{\rho}} = \frac{1}{2}(\boldsymbol{\rho} - \boldsymbol{\mu})^T \Sigma^{-1}\mathscr{S}(\boldsymbol{L})\Sigma^{-1}(\boldsymbol{\rho} - \boldsymbol{\mu}). \tag{9.154}$$

Summarizing (9.136), (9.148), (9.154) we find the shape of the prior of $\boldsymbol{\rho}$ given $\boldsymbol{L}$, apart from a "constant" C, which generally depends on $\boldsymbol{L}$, but which we do not make explicit here,

$$f_{0\rho}(\boldsymbol{\rho},\boldsymbol{L}) = Ce^{-E_{0\rho}(\boldsymbol{\rho}|\boldsymbol{L})}, \tag{9.155}$$

$$\mathscr{E}_{0\rho}(\boldsymbol{\rho},\boldsymbol{L}) = \frac{1}{2}\bar{\boldsymbol{\rho}}^T\bar{\boldsymbol{\rho}} + \frac{1}{2}\bar{\boldsymbol{\rho}}^T\mathscr{S}(\boldsymbol{L})\bar{\boldsymbol{\rho}} - r(\bar{\boldsymbol{\rho}}), \tag{9.156}$$

$$r(\bar{\boldsymbol{\rho}}) = \sum_{i=1}^{N_V} r(\bar{\rho}_i) = -\sum_{i=1}^{N_V} \log \chi_3(\bar{\rho}_i). \tag{9.157}$$

Now we can pass to specifying the prior $p(\boldsymbol{L})$ or (see (9.136)) the corresponding energy $\mathscr{E}_{0\boldsymbol{L}}(\boldsymbol{L})$. Since this will include the information on the prior field of labels, $\boldsymbol{L}_0$, it is important to specify how this can be supplied by the geologist that provides the input to the inversion. Essentially this is a 3D geological map specifying the volumes occupied by the different labels. An example is shown in Fig. 9.7,

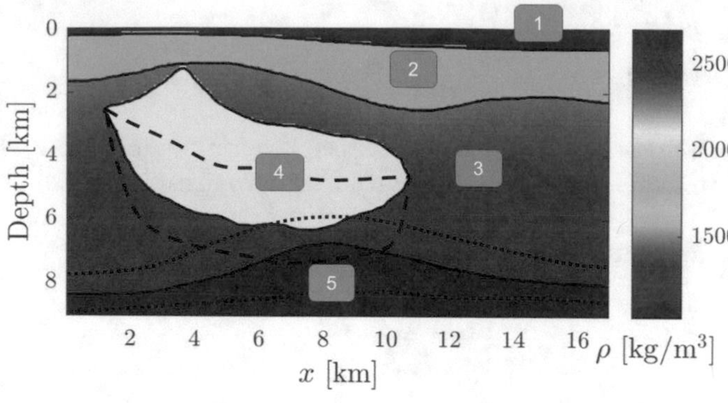

Fig. 9.7 A hypothetical simplified geological section with the presence of (1) water, (2) sediments of first type, (3) sediments of second type, (4) salt dome, (5) basement, with densities as in Table 9.2.—delimiting the area of uncertainty of label 4, …delimiting the area of uncertainty of label 5

Table 9.2 Densities of the model in Fig. 9.7

Material	Label	Density [kg/m^3]
Water	1	1030
Sediments 1	2	2200
Sediments 2	3	2300 (top)–2600 (bottom)
Salt dome	4	2100
Basement	5	2700

representing a salt dome embedded in a sedimentary background. Very important, too, is that the map should show the zones where the labels are practically certain and the zones (buffers) where the labels could correspond to one or another material.

Another important aspect that we learn form Fig. 9.7 is that, given a certain label $L_{0i} = k$ of the voxel V_i in one of the buffers, the label L_i can perhaps have another value, but not any value corresponding to all the materials. For instance, in the buffer of label 4 in Fig. 9.7, we could possibly find labels of the sediment 3 or for a few voxels, of the basement 5, but not labels 1 and 2. This will have to be reflected in the prior probability of L_i. Moreover, for those voxels that the geologist claims to have a sure label (i.e., no uncertainty), we will have to put $P(L_i = L_{0i}) = 1$, and then L_i is, so to say, no longer a variable, but just a constant, given number.

Considering that in a realistic case one can have up to, say, $0.5 \cdot 10^6$ voxels, i,.e., up to 10^6 unknowns, in terms of densities and labels, it is not negligible to cancel some 10^5 of those, that will remain fixed in the estimation procedure.

In any event, we propose a prior $p(\boldsymbol{L})$ of the form

$$\begin{cases} p(\boldsymbol{L}) = ce^{-\mathscr{E}_{0L}(L)}, \\ \mathscr{E}_{0L}(\boldsymbol{L}) = \Sigma_{i=1}^{N_V} d_i(L_i, L_{0i}) + \Sigma_{i=1}^{N_V} \Sigma_{j \in \Delta_i} D_i(L_i, L_j). \end{cases} \tag{9.158}$$

The "distance" functions $d_i(\cdot, \cdot)$ and $D_i(\cdot, \cdot)$ represent the penalty we have to pay, in the total value of $\mathscr{E}(\boldsymbol{\rho}, \mathbf{L})$, in the first case if we choose an $L_i \neq L_{0i}$, in the second case if we have a label L_i different from the surrounding labels.

So, let us define

$$d_i(L_i, L_{0i}) = \begin{cases} 0, & \text{if } L_i = L_{0i} = k, \\ q_{ij}, & \text{if } L_i = j \neq k; \end{cases} \tag{9.159}$$

the values q_{ij} can be tuned in such a way that they are larger if the probability that $L_i = j$ is smaller and they can also take into account the position of V_i in the buffer, assigning a larger probability (a smaller q_{ij}) to the event that L_i changes from $L_{0i} = k$ to $L_i = j$.

Note that we call the quantities $e^{-d_i(L_i, L_{0i})}$ probabilities, which strictly speaking is not correct because they do not add to 1; however, with a suitable choice of the normalization constant C, they do become genuine probabilities.

Let us also notice that if for a certain voxel V_i the label j is forbidden, then we have just to put $q_{ij} = +\infty$, so that

$$P(L_i = j) = Ce^{-q_{ij}} = 0. \tag{9.160}$$

So, if for instance the voxel V_i is in a position where the geologist has decided that $L_{0i} = k$ surely, then we will simply have

$$d_i(L_i, L_{0i}) = \begin{cases} 0, & \text{if } L_i = k, \\ +\infty, & \text{if } L_i \neq k. \end{cases} \tag{9.161}$$

Similarly, we have for the function $D_i(L_i, L_j)$

$$D_i(L_i, L_j) = \begin{cases} 0, & \text{if } L_i = L_j (= k) \quad (j \in \Delta_i), \\ Q_{im}, & \text{if } L_i = m \neq k, \end{cases} \tag{9.162}$$

where the Q_{im} could be in general different from the q_{im} defined in (9.159). Nevertheless, Q_{im}, similarly to q_{im}, will be large if the probability that $L_i = m \neq L_j$ is small, and viceversa. Therefore, the meaning of the second term in $\mathscr{E}_{0L}$ given by (9.158) is that the labels of neighbouring voxels are forced to homogenize without too many oscillations.

So we are finally able to collect all our results to find the joint distribution of observables (i.e., $d\boldsymbol{g}$) and parameters (e.g., $(\boldsymbol{\rho}, \mathbf{L})$), namely

$$f(d\boldsymbol{g}, \boldsymbol{\rho}, \mathbf{L}) = Ce^{-\mathscr{E}(d\boldsymbol{g}, \boldsymbol{\rho}, \mathbf{L})}, \tag{9.163}$$

$$\mathscr{E}(d\boldsymbol{g}, \boldsymbol{\rho}, \mathbf{L}) = \frac{1}{2}(d\boldsymbol{g} - A\boldsymbol{\rho})^T C_\nu^{-1}(d\boldsymbol{g} - A\boldsymbol{\rho}) +$$

$$+\frac{1}{2}[\bar{\boldsymbol{\rho}}^T\bar{\boldsymbol{\rho}}+\bar{\boldsymbol{\rho}}^T\mathscr{S}(\mathbf{L})\bar{\boldsymbol{\rho}}]-r(\bar{\boldsymbol{\rho}})+ \tag{9.164}$$
$$+\sum_i d_i(L_i, L_{0i})+\sum_i\sum_{j\in\Delta_i} D_i(L_i, L_j).$$

At this point we could consider the construction of the model as accomplished and we could pass to the next argument, namely the application of the Gibbs Sampler for the minimization of $\mathscr{E}$, which will be summarily presented in the next section.

Nevertheless, we warn the reader that still another change in (9.164) is desirable. In fact, one can remark that, if we increase the number of unknowns without increasing the number of observation equations, for instance by taking layers of voxels with a smaller height, the terms relative to the prior distribution, for instance $\bar{\boldsymbol{\rho}}^T\bar{\boldsymbol{\rho}}$, will prevail on the term containing the observed gravity anomalies. This is certainly an unwanted effect, tending to underweight the contribution of the observations to the estimation process. So we found it useful to introduce some empirical constants that at least weight the different terms as function of the number of equations/components that they contain. Therefore, we will proceed with an energy function of the form

$$\begin{aligned}\mathscr{E}(d\boldsymbol{g},\boldsymbol{r},\mathbf{L}) &= \frac{1}{2}(d\boldsymbol{g}-\boldsymbol{\rho})^T C_v^{-1}(d\boldsymbol{g}-\boldsymbol{\rho})+\\ &+\frac{\eta}{2}[\bar{\boldsymbol{\rho}}^T\bar{\boldsymbol{\rho}}+\mathscr{S}(\mathbf{L})\bar{\boldsymbol{\rho}}]-r(\bar{\boldsymbol{\rho}})\\ &+\frac{\gamma}{2}\sum_i d_i(L_i, L_{0i})+\frac{\lambda}{2}\sum_i\sum_{j\in\Delta_i} D_i(L_i, L_j).\end{aligned} \tag{9.165}$$

The constants η, γ, λ have to be calibrated by repeated trials, adapting to the given observation vector $d\boldsymbol{g}_0$ and a priori labels vector $\boldsymbol{L}_0$, also taking into account (cf. [119]) the remarks made above. Indeed, a change of the expression of $\mathscr{E}$ implies at the same time a change of the normalizing constant C in (9.163); nevertheless this is not worrisome for us because a lot of effort is made precisely in order to obtain our estimate without knowing C.

A final remark is that the term in the square brackets in (9.165) has a meaning similar to the term $\lambda\boldsymbol{x}^T\boldsymbol{x}$ in the Tikhonov method and in particular it regularizes, given the labels $\boldsymbol{L}$, the estimation of $\boldsymbol{\rho}$; therefore, a first rough guess for the value of η could be derived by applying the same concept that we have seen in the first part of the chapter, for a deterministic approach.

9.9 Bayesian Gravity Inversion: the Gibbs Sampler

In this section we want to study how to apply the Gibbs Sampler concept to the minimization of the energy $\mathscr{E}$ given by expression (9.165).

We start from the pdf of the posterior

$$f_T(\boldsymbol{\rho}, \boldsymbol{L}|d\boldsymbol{g}_0) = \frac{C_T e^{-\frac{\mathscr{E}(\boldsymbol{\rho}, \boldsymbol{L}, d\boldsymbol{g}_0)}{T}}}{C_T \sum_{(L)} \int e^{-\frac{\mathscr{E}(\boldsymbol{\rho}, \boldsymbol{L}, d\boldsymbol{g}_0)}{T}} d\boldsymbol{\rho}}; \tag{9.166}$$

here $\Sigma_{(L)}$ means that the sum is taken over all possible values of all the components of $\boldsymbol{L}$. We first of all eliminate C_T and then observe that the denominator will be function of T and $d\boldsymbol{g}_0$, which is a constant vector in the Bayesian concept.

So we can rewrite (9.166) as

$$f_T(\boldsymbol{\rho}, \boldsymbol{L}|d\boldsymbol{g}_0) = C_T(d\boldsymbol{g}_0) e^{-\frac{\mathscr{E}(\boldsymbol{\rho}, \boldsymbol{L}, d\boldsymbol{g}_0)}{T}}, \tag{9.167}$$

considering that the important here is to know that $C_T(d\boldsymbol{g}_0)$ (this will be different from C_T in (9.166)) does not depend on $\boldsymbol{\rho}$ and $\boldsymbol{L}$, because when we apply the Gibbs algorithm we can sample from our distribution, simplifying the constant $C_T(d\boldsymbol{g}_0)$ too.

Since from here on all distributions and energies considered will be conditional to $d\boldsymbol{g}_0$, we will no longer write this variable in the conditioning list.

The Gibbs algorithm is here defined sweeping the grid of voxels from $i = 1$, to $i = N_V$. For each i we want to draw a sample from

$$\begin{aligned} & f_T(\rho_i, L_i = k|\boldsymbol{\rho}_{-i} \boldsymbol{L}_{-i}) = \\ & = \frac{f_T(\rho_i, L_i = k, \boldsymbol{\rho}_{-i}, \boldsymbol{L}_{-i})}{\sum_m \int f_T(\rho_i = t, L_i = m, \boldsymbol{\rho}_{-i}, \boldsymbol{L}_{-i}) dt} = \\ & = \frac{e^{-\frac{1}{T}\mathscr{E}(\boldsymbol{\rho}, \boldsymbol{L})}}{\sum_m \int e^{-\frac{1}{T}\mathscr{E}(\rho_i = t, L_i = m, \boldsymbol{\rho}_{-i}, \boldsymbol{L}_{-i})} dt}. \end{aligned} \tag{9.168}$$

The first step to be performed here is to simplify (9.168) considering that, in such a distribution, $\boldsymbol{\rho}_{-i}$, $\boldsymbol{L}_{-i}$ act like constants, so if we can split $\mathscr{E}$ into the sum of two pieces, one of which depends on $\boldsymbol{\rho}_{-i}$, $\boldsymbol{L}_{-i}$ only, namely

$$\mathscr{E} = H(\rho_i, L_i|\boldsymbol{\rho}_{-i}, \boldsymbol{L}_{-i}) + F(\boldsymbol{\rho}_{-i}, \boldsymbol{L}_{-i}), \tag{9.169}$$

then $e^{-\frac{1}{T}F}$ becomes a factor that can be eliminated from the numerator and denominator.

The analysis of the various pieces of $\mathscr{E}$ (see (9.165)) is performed in Exercise 8, where it is proved that

Table 9.3 Notation used in the Gibbs Sampler section

$A =$ forward matrix $= [\boldsymbol{a}_1, \boldsymbol{a}_2 \dots \boldsymbol{a}_i \dots \boldsymbol{a}_{N_V}]$	$(\boldsymbol{a}_k =$ columns f)
$A_i = \lvert\boldsymbol{a}_i\rvert^2$	
$d\boldsymbol{g}_{0i} = d\boldsymbol{g}_0 - \Sigma_{k \neq i} \boldsymbol{a}_k \rho_k$	
$B_i = \boldsymbol{a}_i^T d\boldsymbol{g}_{0i}$	
$\Gamma_i = \{j \in \Delta_i; L_j = L_i\}$	
$N_i =$ number of elements of Γ_i	
$M_i = \frac{1}{N_i} \sum_{j \in \Gamma_i} \frac{\rho_j - \mu_j}{\sigma_j}$	

$$H(\rho_i, L_i | \boldsymbol{\rho}_{-i}, \boldsymbol{L}_{-i}) = \frac{1}{2\sigma_v^2}(A_i \rho_i^2 - 2B_i \rho_i) + \frac{\eta}{2}\bar{\rho}_i^2 +$$

$$+ \eta N_i(\bar{\rho}_i^2 + 2M_i \bar{\rho}_i) - r(\bar{\rho}_i) + \frac{\gamma}{2} d_i(L_i, L_{0i}) + \frac{\lambda}{2} \sum_{j \in \Gamma_i} [D_i(L_i, L_j) + D_j(L_j, L_i)]; \tag{9.170}$$

in (9.170) and in the subsequent reasoning we adopt the notation specified in Table 9.3.

Let us notice that A is a constant, $d\boldsymbol{g}_{oi}$ and B_i are functions of $\boldsymbol{\rho}_{-i}$; Γ_i, N_i are functions of L_i, $\boldsymbol{L}_{-i}$; M_i is a function of L_i, $\boldsymbol{L}_{-i}$, ρ_{-i}.

At this point we can claim that

$$f_T(\rho_i, L_i | \rho_{-i}, \boldsymbol{L}_{-i}) = C(\boldsymbol{\rho}_{-i}, \boldsymbol{L}_{-i}) e^{-\frac{H(\rho_i, L_i)}{T}} \tag{9.171}$$

here in H we have neglected the dependence on $\boldsymbol{\rho}_{-i}$, $\boldsymbol{L}_{-i}$, that are fixed in sampling (ρ_i, L_i), while the dependence of C on such variables is just recalled. Obviously, we do not know a priori the form of $C(\boldsymbol{\rho}_{-i}, \boldsymbol{L}_{-i})$; on the other hand, we will be able to get it a posteriori by an easily computable normalization formula.

Before we accomplish this, however, we want to pass from the couple (ρ_i, L_i) to $(\bar{\rho}_i, L_i)$ in view of a simplification in the writing of H. Recalling the definition (9.153), of $\bar{\rho}_i$, we have

$$d\rho_i = \sigma_i d\bar{\rho}_i;$$

accordingly, the pdf for the new variables becomes

$$\bar{f}_T(\bar{\rho}_i, L_i | \boldsymbol{\rho}_{-i}, \boldsymbol{L}_{-i}) = f_T(\mu_i + \sigma_i \bar{\rho}_i, L_i | \boldsymbol{\rho}_{-i}, \boldsymbol{L}_{-i}) \sigma_i, \tag{9.172}$$

with f_T given by (9.171).

Therefore, lifting $\sigma_i = \sigma(L_i)$ to the exponent, and substituting $\rho_i = \mu_i + \sigma_i \bar{\rho}_i$ in (9.170) and reordering, we arrive at the new definition

$$\begin{aligned}\bar{H}(\bar{\rho}_i, L_i|\bar{\boldsymbol{\rho}}_{-i}\bar{\boldsymbol{L}}_{-i}) &= \alpha_i\bar{\rho}_i^2 - 2\beta_i\bar{\rho}_i - r(\bar{\rho}_i) + \\ &+ \frac{A_i\mu_i^2}{2\sigma_\nu^2} - \frac{B_iM_i}{\sigma_\nu^2} - \log\sigma_i + \frac{\gamma}{2}d_i(L_i, L_{-i}) + \\ &+ \frac{\lambda}{2}\sum_{j\in\Gamma_i}[D_i(L_i, L_j) + D_j(L_j, L_i)],\end{aligned} \tag{9.173}$$

where

$$\alpha_i = \frac{A_i\sigma_i^2}{\sigma_\nu^2} + \eta\left(N_i + \frac{1}{2}\right)\ ;\ \beta_i = -\frac{A_i\sigma_i\mu_i}{2\sigma_\nu^2} + \frac{B_i\sigma_i}{\sigma_\nu^2} + \eta N_i\mu_i. \tag{9.174}$$

Notice that in the conditioning variables we have switched from $(\boldsymbol{\rho}_{-i}, \boldsymbol{L}_{-i})$ to $(\bar{\boldsymbol{\rho}}_{-i}, \boldsymbol{L}_{-i})$; this is indeed allowed, because fixing one or the other is the same thing.

As we see, the first three terms in (9.173) are the only ones to include $\bar{\rho}_i$, while all the others depend on L_i only, obviously except for $\boldsymbol{\rho}_{-i}, \boldsymbol{L}_{-i}$.

We want to modify these first terms by completing the square of a binomial, so we use the identity

$$\alpha_i\bar{\rho}_i^2 - 2\beta_i\bar{\rho}_i = \alpha_i\left(\bar{\rho}_i - \frac{\beta_i}{\alpha_i}\right)^2 - \frac{\beta_i^2}{\alpha_i}, \tag{9.175}$$

and we denote

$$\begin{cases}\alpha_i = \frac{1}{2\bar{\sigma}_i^2}, \\ \frac{\beta_i}{\alpha_i} = \bar{\mu}_i,\end{cases} \tag{9.176}$$

which define the new variables $\bar{\sigma}_i^2$, $\bar{\mu}_i$. Finally, we can write (9.173) as

$$\begin{aligned}\bar{H}(\bar{\rho}_i, L_i|\bar{\boldsymbol{\rho}}_{-i}, \boldsymbol{L}_{-i}) &= \frac{1}{2}\left(\frac{\bar{\rho}_i - \bar{\mu}_i}{\bar{\sigma}_i}\right) - r(\bar{\rho}_i) + \\ &- \frac{\beta_i^2}{\alpha_i} + \frac{A_i\mu_i^2}{2\sigma_\nu^2} - \frac{\beta_i\mu_i}{\sigma_\nu^2} - \log\sigma_i + \\ &+ \frac{\gamma}{2}d_i(L_i, L_{-i}) + \frac{\lambda}{2}\sum_{j\in\Gamma_i}[D_i(L_i, L_j) + D_j(L_j, L_i)].\end{aligned} \tag{9.177}$$

Let us put

$$J(\bar{\rho}_i, L_i) = \frac{1}{2}\left(\frac{\bar{\rho}_i - \bar{\mu}_i}{\bar{\sigma}_i}\right)^2 - r(\bar{\rho}_i)$$

and

$$G(L_i) = -\frac{\beta_i^2}{\alpha_i} + \frac{A_i \mu_i^2}{2\sigma_v^2} - \frac{\beta_i \mu_i}{\sigma_v^2} - \log \sigma_i + \\ + \frac{\gamma}{2} d_i(L_i, L_{-i}) + \frac{\lambda}{2} \sum_{j \in \Gamma_i} [D_i(L_i, L_j) + D_j(L_j, L_i)]. \tag{9.178}$$

Then we obtain

$$f_T(\bar{\rho}_i, L_i | \bar{\boldsymbol{\rho}}_{-i}, \boldsymbol{L}_{-i}) = C e^{-\frac{\bar{H}(\bar{\rho}_i, L_i)}{T}} = \\ = C e^{-\frac{J(\bar{\rho}_i, L_i)}{T}} \cdot e^{-\frac{G(L_i)}{T}}. \tag{9.179}$$

The point is that now we are able to compute explicitly the integral

$$\int e^{-\frac{J(\bar{\rho}_i, L_i)}{T}} d\bar{\rho}_i = g_T(L_i), \tag{9.180}$$

where

$$g_T(L_i) = \sqrt{2\pi T} \bar{\sigma}_i \left[\operatorname{erf}\left(\frac{3 - \bar{\mu}_i}{\bar{\sigma}_i \sqrt{T}} \right) - \operatorname{erf}\left(-\frac{3 + \bar{\mu}_i}{\bar{\sigma}_i \sqrt{T}} \right) \right], \tag{9.181}$$

as proved in Exercise 9.

Let us remind the reader that the error function is defined as

$$\operatorname{erf}(\bar{z}) = \int_{-\infty}^{\bar{z}} \frac{1}{\sqrt{2\pi}} e^{-\frac{z^2}{2}} dz \tag{9.182}$$

and is well known and numerically computable (see [111]). Therefore, integrating in $d\bar{\rho}_i$ (9.179) we obtain

$$p(L_i) = \int f_T(\bar{\rho}_i, L_i | \bar{\boldsymbol{\rho}}_{-i}, \boldsymbol{L}_{-i}) d\bar{\rho}_i = \\ = C g_T(L_i) e^{-\frac{G(L_i)}{T}} \tag{9.183}$$

where $g_T(L_i)$ and $G(L_i)$ are given functions. Since the prior $p(L_i)$ has to be normalized, (9.183) yields

$$1 = \sum_k p(L_i) = C \sum_k g_T(L_i = k) e^{-\frac{G(L_i = k)}{T}}; \tag{9.184}$$

the sum in (9.184) is easily computable because the number of labels is small (usually less than 10), therefore (9.184) determines C numerically.

Once C is known, so is the discrete distribution $p(L_i)$ from (9.183), and it is easy to make a random sampling from it.

Finally, rewriting (9.179) in the form

$$\begin{aligned} g_T(\bar{\rho}_i, L_i|\bar{\boldsymbol{\rho}}_{-i}\boldsymbol{L}_{-i}) &= \frac{1}{g_T(L_i)} e^{-\frac{J(\bar{\rho}_i, L_i)}{T}} \cdot C g(L_i) e^{-\frac{G(L_i)}{T}} \\ &= \frac{1}{g_T(L_i)} e^{-\frac{J(\bar{\rho}_i, L_i)}{T}} p(L_i), \end{aligned} \tag{9.185}$$

we readily see that

$$g_T(\bar{\rho}_i|L_i, \bar{\boldsymbol{\rho}}_{-i}, \boldsymbol{L}_{-i}) = \frac{1}{g_T(L_i)} e^{-\frac{J(\bar{\rho}_i, L_i)}{T}}, \tag{9.186}$$

so that if we are able to sample from this pdf we have achieved the capability of sampling $(\bar{\rho}_i, L_i)$ (or (ρ_i, L_i)) by sampling L_i first and then $\bar{\rho}_i$ from (9.186).

This is exactly what is required to run the Gibbs algorithm and therefore to obtain, approximately, the minimization of $\mathscr{E}$ as required.

So, returning to (9.186) and observing that

$$e^{-\frac{r(\bar{\rho}_i)}{T}} = [\chi_3(\bar{\rho}_i)]^{1/T} \equiv \chi_3(\bar{\rho}_i), \tag{9.187}$$

because $\chi_3(\bar{\rho}_i)$ can assume only the values 0 or 1, we have

$$f_T(\bar{\rho}_i|L_i, \bar{\boldsymbol{\rho}}_{-i}, \boldsymbol{L}_{-i}) = \frac{e^{-\frac{1}{2}\left(\frac{\bar{\rho}_i - \mu_i}{\bar{\sigma}_i\sqrt{T}}\right)}}{\sqrt{2\pi T}\bar{\sigma}_i k_{Ti}} \chi_3(\bar{\rho}_i), \tag{9.188}$$

with

$$k_{Ti} = \left[erf\left(\frac{-3 - \bar{\mu}_i}{\bar{\sigma}_i}\right) - erf\left(-\frac{2 + \bar{\mu}_i}{\bar{\sigma}_i}\right)\right] < 1. \tag{9.189}$$

If we let $u(\bar{\rho}_i; \bar{\mu}_i, \sqrt{T}\bar{\sigma}_i)$ denote the normal pdf with mean $\bar{\mu}_i$ and standard deviation $\sqrt{T}\bar{\sigma}_i$, we see that

$$f_T(\bar{\rho}_i|L_i, \bar{\boldsymbol{\rho}}_{-i}, \boldsymbol{L}_{-i}) = u(\bar{\rho}_i; \bar{\mu}_i, \sqrt{T}\bar{\sigma}_i) \frac{\chi_3(\bar{\rho}_i)}{k_{Ti}} \tag{9.190}$$

so that

$$\forall \bar{\rho}_i, \qquad \frac{f_T(\bar{\rho}_i|L_i, \bar{\boldsymbol{\rho}}_{-i}, \boldsymbol{L}_{-i})}{u(\bar{\rho}_i; \bar{\mu}_i, \sqrt{T}\bar{\sigma}_i)} \equiv \frac{\chi_3(\bar{\rho}_i)}{k_{Ti}} \leq \frac{1}{k_{Ti}}. \tag{9.191}$$

In this situation there is a statistical tool to draw samples from f_T, namely the so-called *acceptance/rejection method* (see [145], §3.3.3.).

This works as follows: let us draw independently two samples,

$$\bar{\rho}_i^* \sim N(\bar{\rho}_i\,;\,\bar{\mu}_i, \sqrt{T}\bar{\sigma}) \tag{9.192}$$

and u^* from a uniform distribution on [0,1],

$$u^* \sim U(0,1)\,; \tag{9.193}$$

then

$$\begin{aligned}
&\text{if } \chi_3(\bar{\rho}_i^*) = 1,\ \ u^* \le k_{Ti} \Rightarrow \text{accept } \bar{\rho}_i^* \\
&\text{if } \chi_3(\bar{\rho}_i^*) = 0,\ \ u^* > k_{Ti} \Rightarrow \text{repeat the draw.}
\end{aligned}$$

The explanation of the algorithm is in the given reference.

Here we just comment that, since k_{Ti} is smaller, but typically not much smaller than 1, and anyway $k_{Ti} \to 1$ in the annealing phase, because $T_i \to 0$, the rate of rejection of the draws from the normal (9.192) is usually also small and the algorithm becomes swiftly efficient.

Once we know how to sample from (9.185), the Gibbs Sampler rules give an MC sequence $(\boldsymbol{\rho}, \boldsymbol{L})_k$ tending to the solution.

Before passing to a final section, where numerical examples will be shown, we warn the reader that, although theoretically clean, the recipe of this section requires a lot of experiments, simplifications and numerical refinements, before it can be successfully applied. Yet this can be done, with satisfactory results, as we shall see in the next section.

9.10 Bayesian Gravity Inversion: Numerical Examples

In this section we will present a set of numerical examples, based on synthetic as well as on real data, to show the potentiality of the Bayesian Inversion Method introduced in Sects. 9.5–9.9.

Example 9.1 (Two-Layer Model) In this first example we will use the Bayesian algorithm to invert for a two-layer model such as the one presented in Sect. 8.4. We will invert a grid of gravity anomalies at ground level (constant altitude equal to 0 m above sea level) to recover the depth of the interface between two layers, namely the discontinuity between the lower crust and the upper mantle (Moho). In order to have a realistic dataset we select a Moho depth model from literature [58] in an area of 10° × 10°, in the easternmost part of the Mediterranean Sea. Then, assuming a constant density of 2670 kg m^{-3} for the crust and of 3300 kg m^{-3} for the mantle, we compute, by means of the algorithm presented in Sec. 6.1, the first

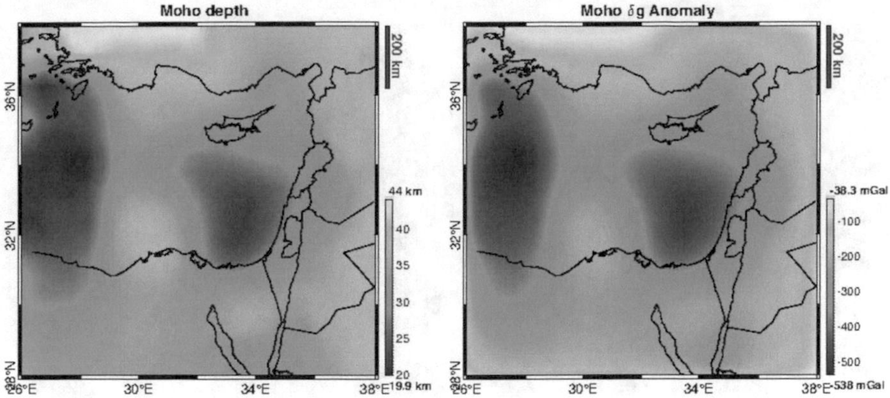

Fig. 9.8 The reference Moho model used in the two-layer example and the corresponding δg anomaly

vertical derivative of the gravitational potential due to the Moho discontinuity. A remark is necessary at this point: even if very simple, the above model can represent a real-life inversion problem for the determination of the shape of the Moho once all the density variations within the crust and the upper mantle have been removed (e.g., by filling the oceans or the sedimentary layers to 2670 $\mathrm{kg\,m^{-3}}$ by means of terrain correction procedure). In any case, this model, with just two layers of known densities and unknown geometry, presents the important advantage of guaranteeing the uniqueness of the solution of the inverse gravimetric problem, i.e., there is just one surface that will generate the given gravity anomaly.

The reference Moho model used to generate the synthetic observations, and its gravitational anomaly are shown in Fig. 9.8. A 1 mGal white noise has been added to the gridded gravity signal.

In order to perform the inversion we modelled the area down to a depth of 46 km, using $120 \times 140 \times 70$ voxels in the x, y and z direction respectively, for a total of about $1.2 \cdot 10^6$ prisms. Before applying the Bayesian inversion we need to introduce the prior information on the labels and on the densities. The former is defined, here, by supposing that all the voxels above a depth of 30 km are crust, while the others are mantle. The latter, i.e., the density prior probability, is introduced supposing to know the exact density of the two layers. Note that this simple test, where densities are fixed and the uniqueness of the solution is guaranteed, is reported here to understand the capability of the proposed Gibbs-Sampler and Simulated Annealing to find the minimum of the objective function. It should be also observed that the above problem can be solved, once it has been suitably regularized, by means of (8.84) exploiting FFT algorithms (for instance, the well known Parker–Oldenburg inversion [96]).

Since we are working with a realistic, yet synthetic model, we are able to compute the difference with respect to the true reference model and the corresponding mean square error.

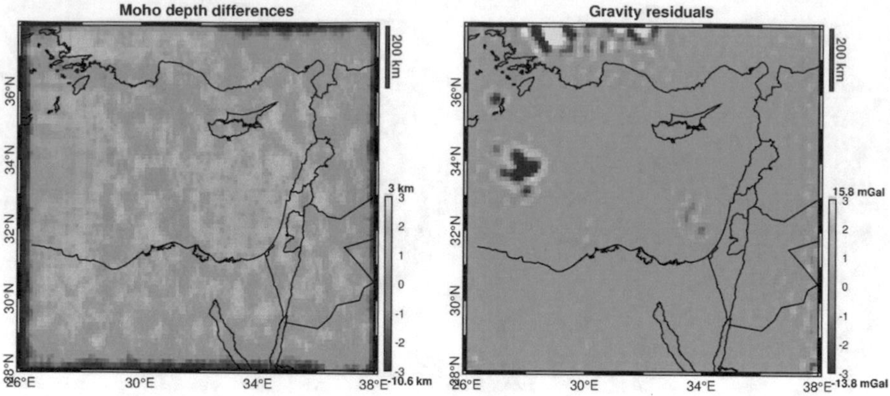

Fig. 9.9 Result of the Bayesian inversion in terms of difference between the reference Moho and the estimated one (left), and in terms of gravity residuals (right)

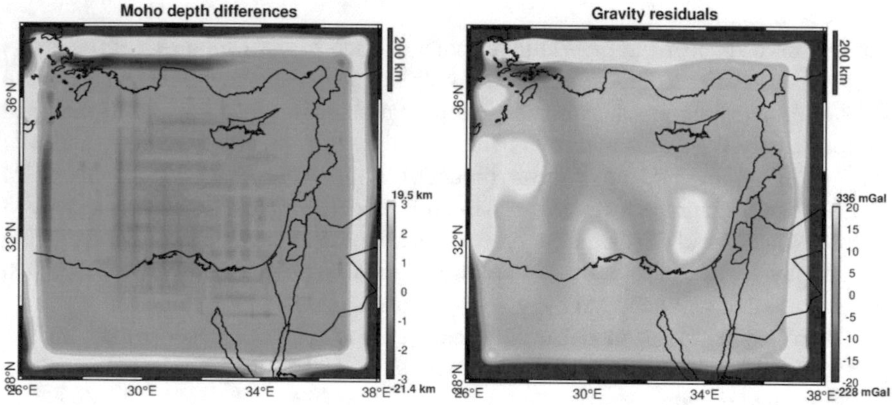

Fig. 9.10 Result of the Parker–Oldenburg inversion in terms of difference between the reference Moho and the estimated one (left), and in terms of gravity residuals (right)

In Fig. 9.9 the results of the inversion are shown both in terms of difference between the inverted Moho depth and the reference one (used to generate the gravitational signal), and in terms of gravity residuals.

The RMSE of the recovered Moho for this closed-loop test is of about 1.6 km, with an average error of 0.8 km. The largest error (of the order of 10 km) is obtained in correspondence of the border where the inversion is less constrained by the observations (the average error without considering the borders decreases to 0:4 km). As for the residual gravity anomaly, it can be seen that the algorithm has been able to fit the signal, at least at the level of the observation accuracy (RMS of the residual is 0:9 mGal). For the sake of completeness we compare the Bayesian inversion results with those obtained by a classical Parker–Oldenburg inversion (see Fig. 9.10).

It can be seen that the Parker–Oldenburg algorithm, which makes use of the FFT, has a much more evident border effect both in terms of retrieved Moho depth and gravity anomaly. The RMSE of the solution in terms of Moho depth is of 3 km, which drops to 0.5 if the borders are not considered. Apart from the borders, the error, which is reflected also in the residual gravity anomaly, is basically due to the regularization introduced by the inversion algorithm (and required in order to obtain the convergence of the solution), which has the effect of removing the high frequencies (wavelength shorter than about 60 km in this example) of the signal.

Coming to the computational burden, for this experiment the Bayesian inversion takes about 10 hours, while the FFT inversion can be performed in almost real time. As said before, this example is reported here only to show the capability of the Bayesian inversion to perform at least at the same level (in terms of results) of a more classical method. In the next few examples we will see how the flexibility of the Bayesian inversion allows to deal also with much more complex (and realistic) scenarios, which cannot be solved with the Parker–Oldenburg inversion. Concluding this first numerical example, we can say that the Bayesian algorithm, even if requiring much more computation time, performs similarly to a classical Fourier-based inversion when dealing with the standard two-layer problem, at least when no high frequencies are present in the discontinuity surface. If high frequency details are present in the discontinuity surface, the regularization of the Fourier inversion tends to over-smooth the result, while the Bayesian inversion is able to recover also these details.

Example 9.2 (3D Mass Density Estimation) We consider again the synthetic model introduced in Example 9.1. However, in contrast with what has been done in the previous test, we will now consider known and fixed, the depth of the discontinuity surface between the two layers and invert for the mass density distributions inside the crust and the upper mantle. We recall here that, when we built the synthetic model we considered a constant density for the crust of 2670 $\mathrm{kg\,m^{-3}}$ and of 3300 $\mathrm{kg\,m^{-3}}$ for the mantle. Note that, even if in principle the above problem again has a unique solution (at least when searching for a constant density contrast between two layers of known geometry), in practice we will invert for a large set of densities (one for each voxel in which the model has been discretized), only imposing a certain degree of regularity on the density distribution, thus loosing in fact the uniqueness of the solution.

Similarly to the previous example, this inversion, too, can be performed by means of a classical inversion method, namely the algorithm described by Li and Oldenburg [77]. As already stated, in this test, the prior on the labels is given by the exact position of the Moho. As for the prior in terms of density distribution we will assume that the density of the crust is $\bar{\rho}_{\mathrm{C}} = 2750\ \mathrm{kg\,m^{-3}}$ with a standard deviation $\bar{\sigma}_{\mathrm{C}} = 122\ \mathrm{kg\,m^{-3}}$, while for the mantle we have $\bar{\rho}_{\mathrm{M}} = 3000\ \mathrm{kg\,m^{-3}}$ and $\bar{\sigma}_{\mathrm{M}} = 122\ \mathrm{kg\,m^{-3}}$. Note that here intentionally a prior information quite different from the synthetic model has been considered.

The inversion results, in terms of densities, are shown along an East-West profile in central part of the model in Fig. 9.11, which is quite representative of the general

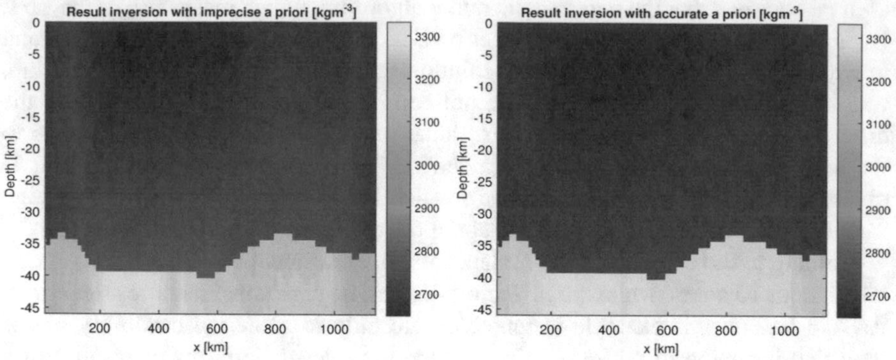

Fig. 9.11 Result of the Bayesian inversion for different values of the a-priori mass density. $\bar{\rho}_C = 2750\ \text{kg m}^{-3}$, $\bar{\rho}_M = 3000\ \text{kg m}^{-3}$ (left); $\bar{\rho}_C = 2713\ \text{kg m}^{-3}$, $\bar{\rho}_M = 3240\ \text{kg m}^{-3}$ (right). True values are $\rho_C = 2670\ \text{kg m}^{-3}$ and $\rho_M = 3300\ \text{kg m}^{-3}$

situation. It can be seen that the algorithm tends to correctly decrease the density in the upper part of the model, and to increase the one in the lower part. The resulting RMSE decreases from 80 kg m^{-3} in the a-priori model to 44 kg m^{-3} for the crust and from 300 kg m^{-3} to 59 kg m^{-3} for the mantle. Basically this RMSE is due to the fact that the proposed algorithm, and in general the inversion of the gravitational field, hardly changes the average value of the densities due to the removal of the normal field from the observations, which essentially has the effect of removing the average value of the gravitational accelerations. In order to alleviate this problem, a simple Least Squares adjustment can be performed to estimate an average density value for each layer. Once estimated, these values can be used as prior knowledge in the subsequent inversion. In the current example, the Least Squares adjustment yields to an average a-priori density of $\bar{\rho}_C = 2713\ \text{kg m}^{-3}$ for the crust and $\bar{\rho}_M = 3240\ \text{kg m}^{-3}$ for the mantle. With this new, more correct, a-priori knowledge, the RMSE of the inversion result is reduced to 22 kg m^{-3} for the crust and to 29 kg m^{-3} for the mantle density. As for the fitting of the gravitational field, it does not depend, as expected, on the accuracy of the a-priori knowledge being of the order of 0.1 mGal for both the performed inversions.

Example 9.3 (Inversion of Complex Bodies) In the first two examples we have analyzed simple cases which can be solved also by exploiting classical techniques. The aim of this example is to investigate a scenario typical of oil exploration, namely, the presence of a complex salt dome in a sedimentary background. In this specific situation seismic prospecting hardly obtains a reliable solution, especially for the lower surface of the body. Note that here both the density distribution and the geometries are unknown and therefore classical methods such as the Parker–Oldenburg or the Li–Oldenburg methods cannot be used. The extension of the model is of about 45 km in the horizontal plane (both in the x and y directions) and about 13 km in the vertical one. The number of voxels used is 53×49 (horizontal) and 64 (vertical), assuming a voxel resolution of 900 m in the horizontal

Table 9.4 A-priori density information for the Example 9.3

Material [km]	Mean density [kg m^{-3}]	Density std [kg m^{-3}]	Density gradient [kg m^{-4}]
Filled water	2000	0	0
Sediments	2210	22	0.0096
Salt	2100	7	0

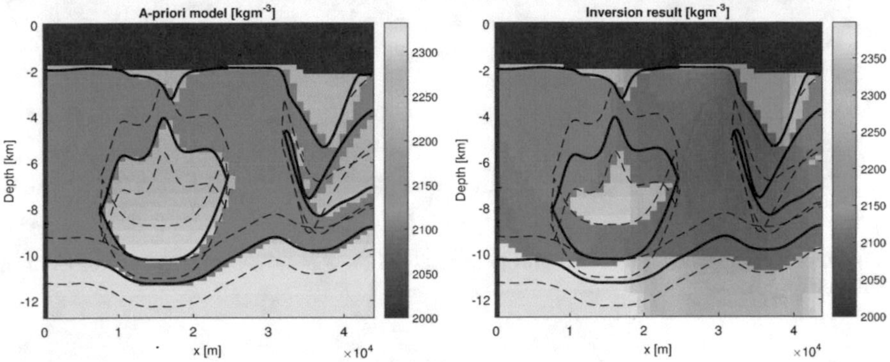

Fig. 9.12 A priori model (left), and results of the inversion (right). Black solid line represents the boundaries between geological units from the a-priori model. Black dashed lines represents the uncertainty regions for the boundaries between geological units

plane, and varying between 100 and 300 m in the vertical direction. The geological formations present in the model are water, sediments and the salt dome. The salt density is assumed to be constant in space, while for the sedimentary layer a vertical gradient has been assumed. An overview of the a-priori knowledge on the density distribution is summarized in Table 9.4, while the labels geometry, together with their accuracies, as given in the a-priori model, are shown in Fig. 9.12(left). Looking at this profile, one can clearly see the complexity of the salt dome formation, which presents holes and arcs that in general cannot be recovered with classical inversion algorithms. Considering the inversion results, Fig. 9.12(right), it can be observed that the algorithm correctly modifies the geometry by enlarging the salt dome when a mass surplus is found in the a-priori model (thus effectively lightening the model) or contracting the salt when dealing with a lack of mass. Moreover, the density distribution in the salt has been kept almost constant according to the a-priori information, while lateral density variations have been added within the sedimentary layer. The improvements in fitting the gravitational field can be seen in Fig. 9.13, where it can be observed how the residuals decrease (in terms of standard deviation) from 6.4 mGal for the a-priori model to 0.7 mGal for the final solution.

Example 9.4 (A Real-Case Study) After a set of synthetic tests, we report now a real-case study with the aim of illustrating the capability of the Bayesian inversion to deal with complex scenarios. The investigated region covers an area of about 150 km $\times$ 80 km, and it is modelled by a set of 10 labels representing 10 different geological materials. The volume is in this case discretized into 152 $\times$ 79 $\times$ 74

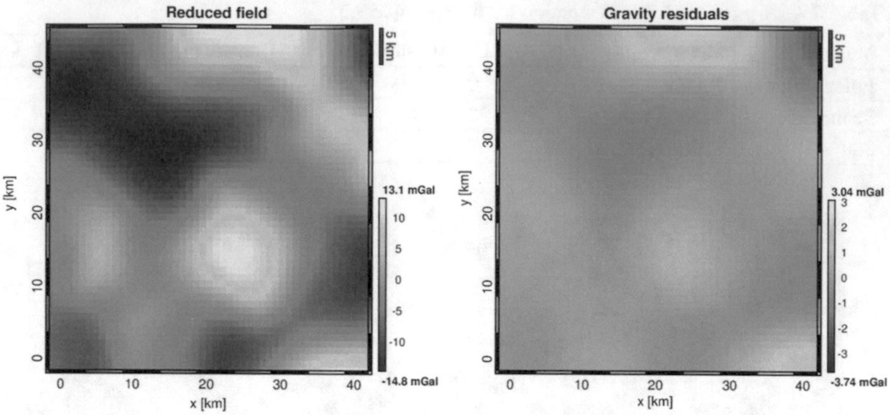

Fig. 9.13 Residuals in terms of the gravity anomaly of the a-priori model (left), and of the final solution (right)

voxels, each with a size of about 1 km in the horizontal directions and thickness increasing with depth varying from 100 m to 700 m in the vertical direction. The region is modelled up to a depth of 25 km. As it can be seen from Fig. 9.14, within the considered volume, apart from a set of 9 layered geological formations, also a body is present.

The prior knowledge on the model consists of the depth of each mass discontinuity (solid line in Fig. 9.14) and of the corresponding accuracy (dash lines in Fig. 9.14), defining the region in which the discontinuity between two geological units can move. For each geological formation the mean density and its accuracy in terms of standard deviation are also defined. Note that, already in the a-priori model a linear gradient in the density distribution of some specific geological units has been included.

The inversion result is shown in Fig. 9.14b where it can be seen that the algorithm modified the shape of each geological formation according to the a-priori accuracy. For instance the bottom of the body has moved upward maintaining its shape, while the top of the body has modified and again moved upward. The density of the body was slightly increased from 2375 kg m^{-3} to 2394 kg m^{-3} in an almost uniform way. On the contrary high frequency density variations were added in the upper layers. Considering the Orange layer, it can be observed that its thickness was notably reduced and that a smooth density variation was also added. Figure 9.15 shows the residuals, with respect to the observed gravity (here in terms of Bouguer anomaly at zero level), of the a-priori and inverted models. It can be seen that the algorithm has been able to improve the fitting of the gravity field from 18.2 mGal to 0.7 mGal in terms of standard deviations. A major misfit, of −4 mGal, can be found in the South-East corner. This residual is probably due to an overestimation of the accuracy of the a-priori model of the body which, together with the smoothness of the density variation inside the model itself, constitute a too strong constraint to allow for a

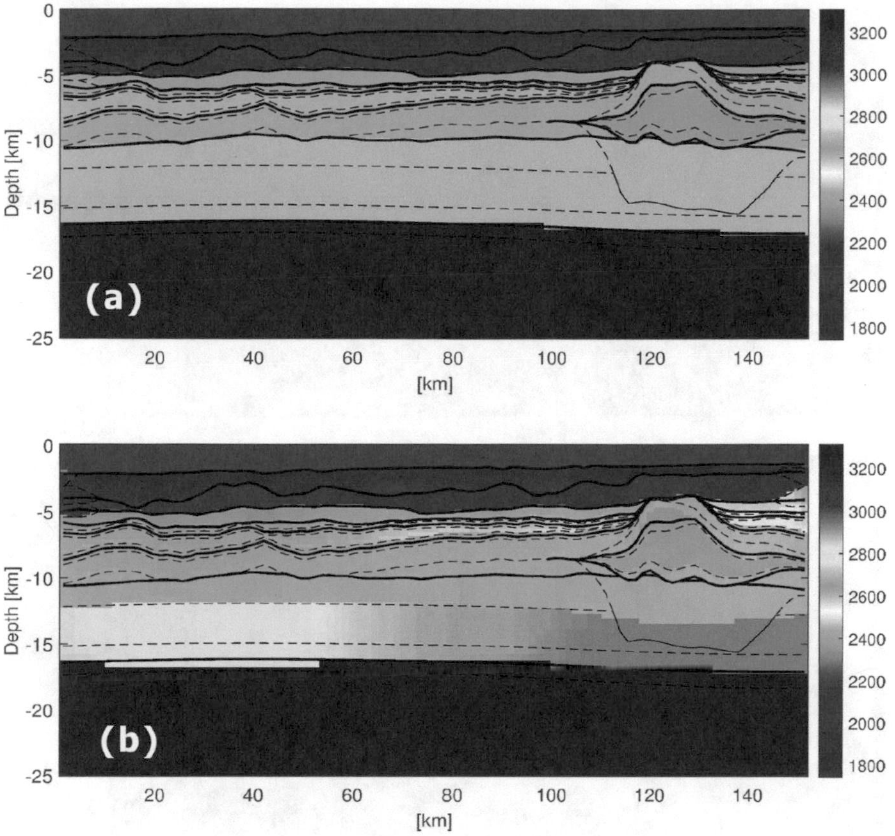

Fig. 9.14 A-priori model (**a**), results from the Bayesian inversion (**b**), in terms of mass density (units $\mathrm{kg\,m^{-3}}$). Solid lines represent the a priori discontinuities between geological formations, and dash lines represent the uncertainty related to each discontinuity

perfect gravity fitting. Other important discrepancies (smaller than 4 mGal) are due to border effects which are inevitable when dealing with gravity.

Example 9.5 (A Real Case Study at Regional Scale) Here we demonstrate the use of the Bayesian inversion to study the crustal structure at regional scale. The aim of the study, which is taken from [112], was to increase the knowledge of the crustal structure of Guangdong Province (south China), to investigate the geoneutrino flux from the inner Earth layers at the Jiangmen Underground Neutrino Observatory (more details can be found in [112]). The extent of the considered region is $6^\circ \times 4^\circ$ in the East and North directions, respectively, to which a border of 3° has been added in order to decrease border effects during the inversion. The spatial resolution of the model, which is dictated by the specific application of geoneutrino studies, is of 0.5° in the (x, y)-plane, with a 100 m resolution in the vertical direction. The volume has been modelled down to a depth of 50 km. The observations to be

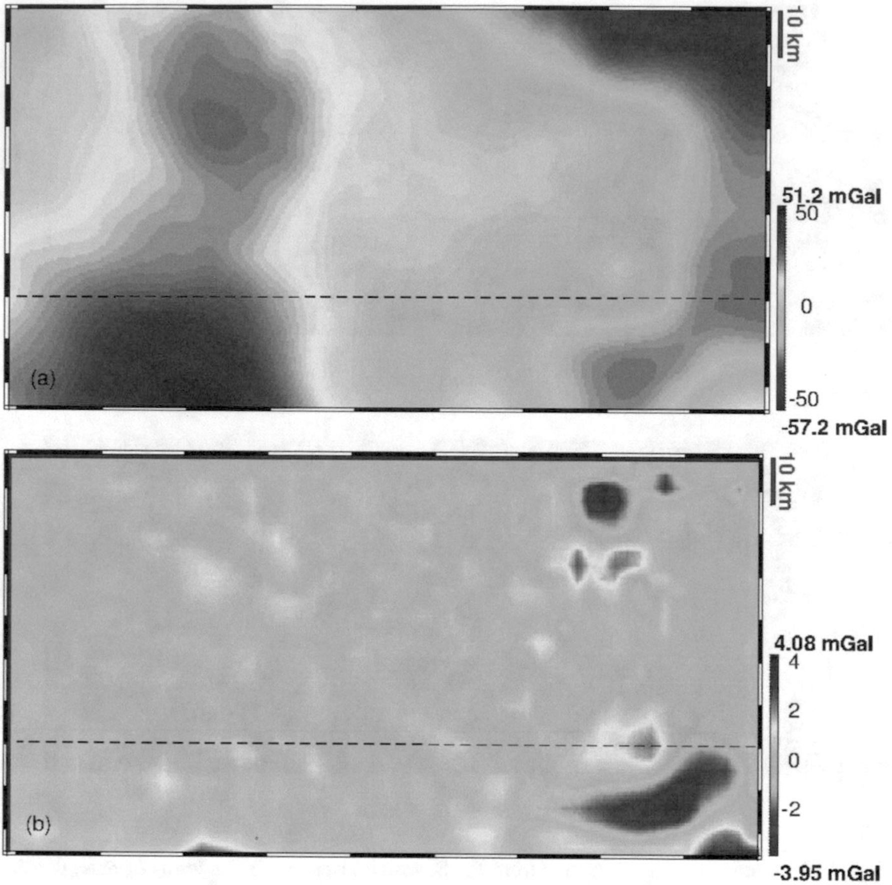

Fig. 9.15 Difference between the observed gravity field and the a-priori model (**a**) or the inversion solution (**b**). Black dash lines represents the profile shown in Fig. 9.14

inverted are the gravity anomalies synthesized up to spherical harmonic degree and order 200 from the GOCE only GO_CONS_GCF_2_SPW_R5 model [113]. They have been computed with a horizontal resolution of 50 km and at an ellipsoidal height of 600 m, namely, the altitude closest to the terrain, but outside masses. The accuracy of the chosen observed signal, predicted from the model itself, is of 1 mGal.

The a-priori model, and the corresponding constraints, are generated from published studies, including deep seismic sounding profiles, receiver functions, teleseismic P-wave velocity models and Moho depth maps. The final model is described by three layers overlying the mantle, namely lower, middle and upper crusts. Again, this simplified modelling is related to the final aim of the study, since in order to predict the geoneutrino flux at a given point, only the volume of these three layers is required. Information derived from the CRUST1.0 [76] model is used to have at least as a raw information where no local measurements are present. In details the basement and the discontinuity between the upper, middle and the lower crust are obtained by means of a regularized least squares adjustment, starting from the available data, such that the obtained surface is smooth and lies inside its admissibility boundary. The accuracy of the above datasets and the admissibility boundary (i.e., the regions on uncertainty for each discontinuity surface between two layers) have been obtained by analyzing the intersection between different available profiles. Again we refer to [112] for details. An example of the discontinuity and corresponding admissibility regions for the profile AA′ of Fig. 9.16 are shown in Fig. 9.17.

The value of the mean density is taken directly from the densities given in [66], interpolated over the horizontal grid by means of a nearest neighbour interpolation. Its variability is assumed to be constant for all the layers and fixed to a value to be tuned, choosing among four values in the range $10 \div 40\ \mathrm{kg\,m^{-3}}$. The final result of this step is, similarly to the previous example, a set of a priori geometries and densities together with their accuracies.

Moreover, some deterministic constraints are imposed to the final solution:

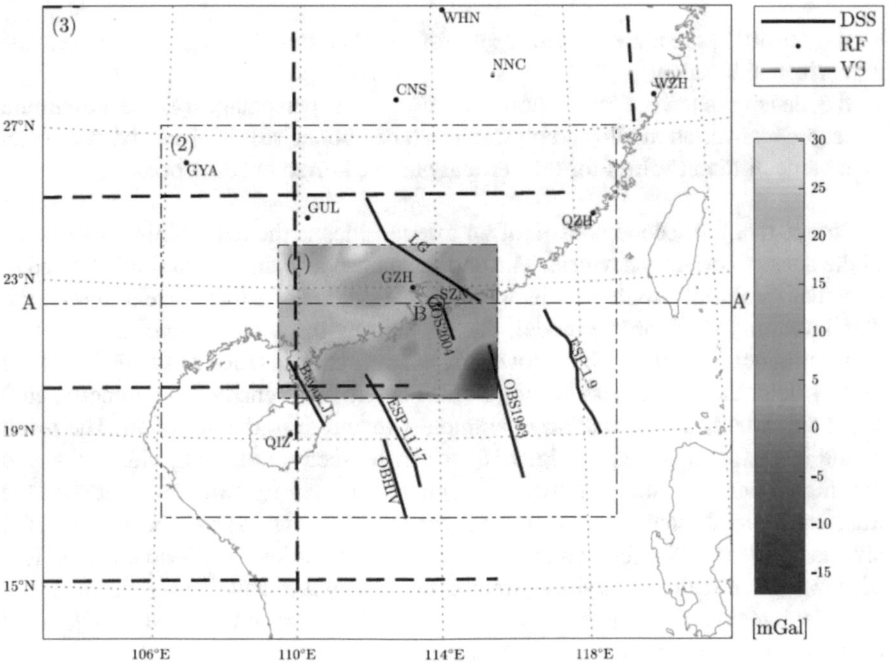

Fig. 9.16 Input geophysical data used for the construction of the 3D model. Deep seismic sounding profiles (DSS), P-wave velocity models profiles (VS) and locations of seismograph stations (RF) reported in the figure correspond to the input seismic data used to build the a-priori model. The observed gravity anomalies (area 1) and the border regions (areas 2, 3)

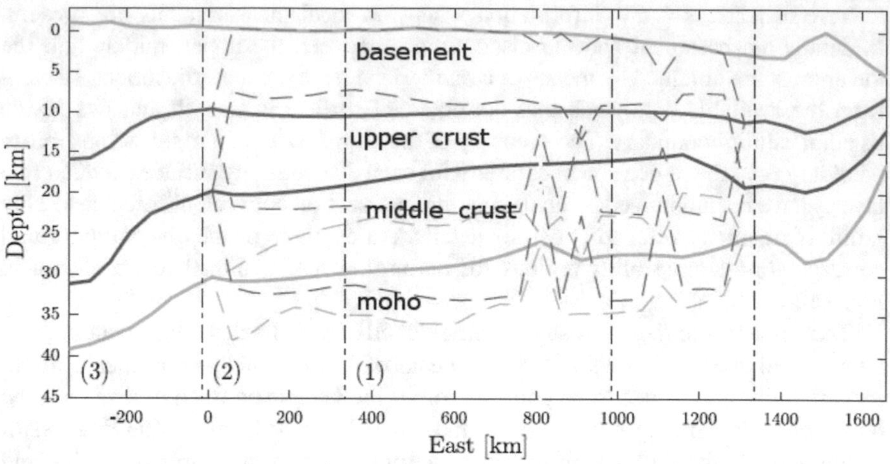

Fig. 9.17 Section AA′ from the given geophysical input (depth magnification 20×). Dashed lines are the admissibility ranges while continuous lines are the prior reference surfaces. Upper and middle crust refers to the depth of the end of the upper and middle crust, respectively

- the density has to increase with depth for each "column" of voxels with the same horizontal location;
- the density allows a maximum variation as a percentage of the maximum theoretical variation allowed between two neighbour voxels. This constraint can assume different values for the vertical and the horizontal directions.

Notice that the second constraint on the densities is the deterministic "version" of the density correlation matrix. The choice to introduce this constraint is related to computational time problems arising when use is done of a true correlation matrix. The inversion of the above model, due to the low resolution of the model itself, requires about 5 minutes. This allows, given some criteria (such as the fitting of the gravity field, the smoothness of the solution in terms of density and geometry, etc.) to set empirically all the ancillary parameters required for the inversion. The result of the inversion is shown in Fig. 9.18. It can be seen that the algorithm changed the initial density model, but always keeping sharp density variations between the different layers. Moreover, inside each layer density variations have been added, but always consistent with the defined constraints. Geometries have been changed too, but always within the admissibility region defined by the available geophysical data and keeping smooth surfaces. The obtained solution fits the observed gravitational field with a standard deviation of 1.1 mGal.

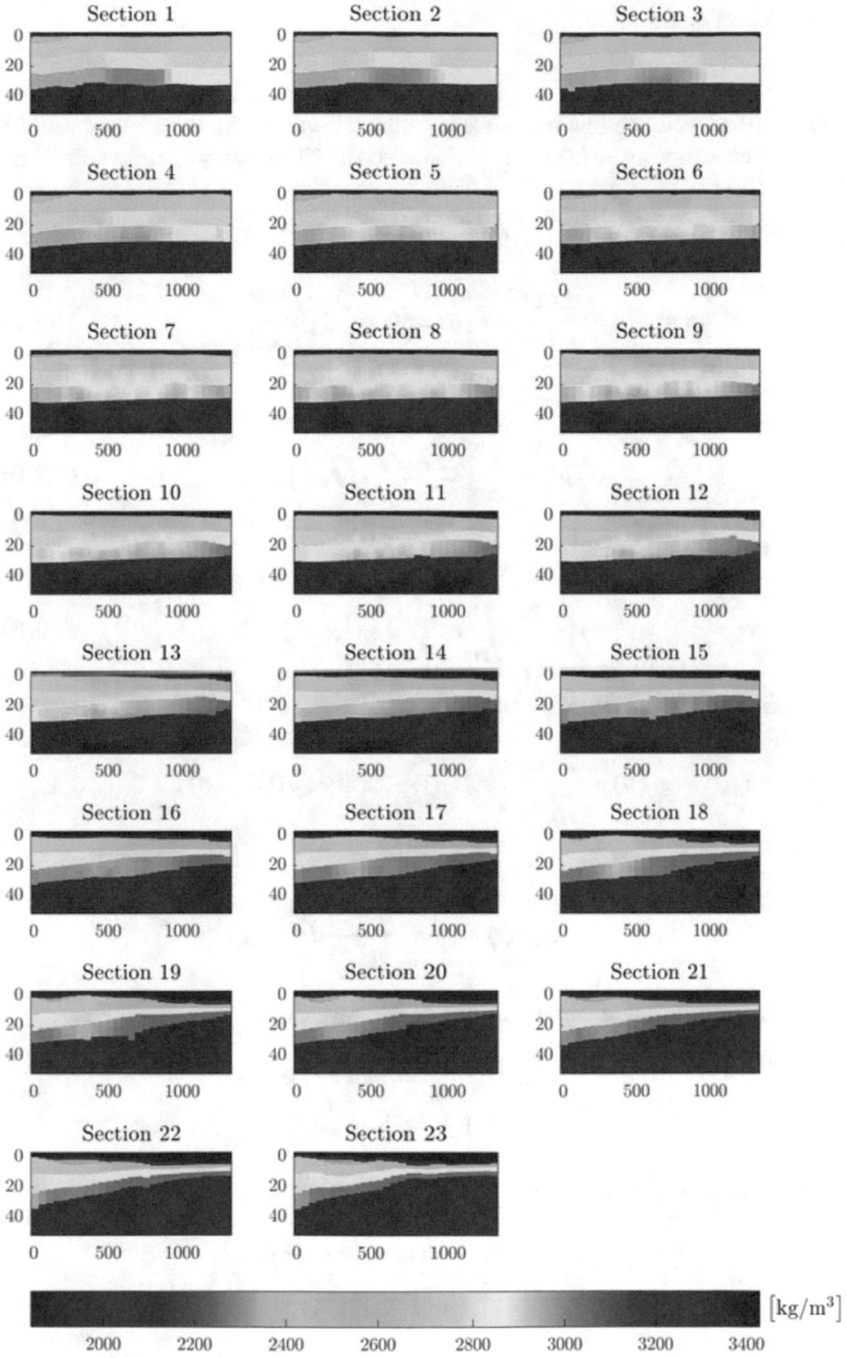

Fig. 9.18 Vertical cross-sections estimated density of the best solution. Horizontal axis represents the East coordinate, while the vertical one the depth with a magnification of 20×, both expressed in [km]. The sections are at each step of the horizontal grid, starting from north (1) to south (23)

9.11 Exercises

Exercise 1 Solve the Euler equation of Tikhonov principle min $J\left(\delta\hat{\rho}\right)$, with J given by (9.14). Notice that the unweighted solution is just a particular case of the above, when choosing $w(z) = 1$. Solve also the Tikhonov principle with $\bar{J}\left(\delta\hat{\rho}\right)$ given by (9.15).

Solution The Euler equation (9.16) can be written in the form

$$\delta\hat{\rho} = \delta\hat{\rho}_0 - \frac{2\pi G}{\lambda}\left[\delta\hat{g}_z + 2\pi G\varphi\left(p\right)\right]\frac{e^{2\pi pz}}{w\left(z\right)}, \tag{9.194}$$

where

$$\varphi\left(\boldsymbol{p}\right) = \int_{-H}^{0} e^{2\pi pz'}\delta\hat{\rho}\left(\boldsymbol{p}, z'\right)dz'; \tag{9.195}$$

we also denote

$$\varphi_0\left(\boldsymbol{p}\right) = \int_{H}^{0} e^{2\pi pz'}\delta\hat{\rho}_0\left(\boldsymbol{p}, z'\right)dz'. \tag{9.196}$$

Substituting (9.194) in (9.195) one gets

$$\varphi\left(\boldsymbol{p}\right) = \varphi_0\left(\boldsymbol{p}\right) - \frac{2\pi G}{\lambda}\left[\delta g_Z\left(\boldsymbol{p}\right) + 2\pi G\varphi\left(\boldsymbol{p}\right)\right]\psi_w\left(p\right), \tag{9.197}$$

where

$$\psi_w\left(p\right) = \int_{-H}^{0}\frac{e^{4\pi pz'}}{w\left(z'\right)}dz'.$$

The algebraic solution of (9.197) for $\varphi\left(\boldsymbol{p}\right)$ is

$$\varphi\left(\boldsymbol{p}\right) = \frac{\varphi_0\left(\boldsymbol{p}\right) - \frac{2\pi G}{\lambda}\delta\hat{g}_Z\left(\boldsymbol{p}\right)\psi_w\left(p\right)}{1 + \frac{4\pi^2G^2}{\lambda}\psi_w\left(p\right)}. \tag{9.198}$$

Substituting this expression in (9.194) we finally obtain

$$\delta\hat{\rho}\left(\boldsymbol{p}, z\right) = \delta\hat{\rho}_0 - \frac{2\pi G}{\lambda}\frac{2\pi G\varphi_0\left(\boldsymbol{p}\right) + \delta\hat{g}_Z\left(\boldsymbol{p}\right)}{1 + \frac{4\pi^2G^2}{\lambda}\psi_w\left(p\right)}\frac{e^{2\pi pz}}{w\left(z\right)}.$$

We move now to solve the principle min $\bar{J}\left(\delta\hat{\rho}\right)$ with $\bar{J}$ given by (9.15). The Euler equation in this case is

$$-\hat{A}^*\left(\delta\hat{g}_Z - \hat{A}\delta\hat{\rho}\right) + \lambda w(z)\left[\delta\hat{\rho} - \delta\hat{\rho}_0 + \alpha p^2\delta\hat{\rho}\right] = 0,$$

that is

$$4\pi^2 G^2 e^{2\pi pz}\int_{-H}^{0} e^{2\pi pz'}\delta\hat{\rho}\left(\boldsymbol{p}, z'\right)dz' + \lambda w(z)\left[\left(1+\alpha p^2\right)\delta\hat{\rho} - \delta\hat{\rho}_0\right] =$$
$$= -2\pi G e^{2\pi pz}\delta\hat{g}_Z. \tag{9.199}$$

Putting, as before,

$$\varphi(\boldsymbol{p}) = \int_{-H}^{0} e^{2\pi pz'}\delta\hat{\rho}(\boldsymbol{p})\,dz', \tag{9.200}$$

$$\varphi_0(\boldsymbol{p}) = \int_{-H}^{0} e^{2\pi pz'}\delta\hat{\rho}_0(\boldsymbol{p})\,dz', \tag{9.201}$$

and

$$\psi_w(p) = \int_{-H}^{0}\frac{e^{4\pi pz'}}{w(z')}dz', \tag{9.202}$$

we obtain at first from (9.199)

$$\delta\hat{\rho} = \frac{\delta\hat{\rho}_0}{1+\alpha p^2} - \frac{2\pi G e^{2\pi pz}}{\lambda w\left(1+\alpha p^2\right)}\delta\hat{g}_Z - \frac{4\pi^2 G^2}{\lambda w}e^{2\pi pz}\varphi(\boldsymbol{p}), \tag{9.203}$$

which substituted in (9.200) gives

$$\varphi(\boldsymbol{p}) = \frac{\varphi_0(\boldsymbol{p})}{1+\alpha p^2} - \frac{2\pi G}{\lambda\left(1+\alpha p^2\right)}\psi_\omega(p)\,\delta\hat{g}_Z(\boldsymbol{p}) - \frac{4\pi^2 G^2}{\lambda}\psi_\omega(p)\,\varphi(\boldsymbol{p}). \tag{9.204}$$

Finally, substituting back in (9.203) and simplifying we find

$$\delta\hat{\rho}(\boldsymbol{p}, z) = \frac{\delta\hat{\rho}_0(\boldsymbol{p}, z)}{1+\alpha p^2} - \frac{4\pi^2 G^2 e^{2\pi pz}}{\lambda w(z)\left(1+\alpha p^2\right)\left(1+\frac{4\pi^2 G^2}{\lambda}\psi_w(p)\right)}\varphi_0(\boldsymbol{p}) +$$
$$- \frac{2\pi G e^{2\pi pz}}{\lambda w(z)\left(1+\alpha p^2\right)\left(1+\frac{4\pi^2 G^2}{\lambda}\psi_w(p)\right)}\delta g_Z(\boldsymbol{p}). \tag{9.205}$$

Exercise 2 Consider a noise-like mass density distribution $\eta(\xi, z)$ in the layer $\mathscr{L} = \left\{\mathbb{R}^2 \times [-H, 0]\right\}$ and the corresponding gravimetric signal, which in general form reads

$$\chi(\boldsymbol{x}) = A\eta = \int_{-H}^{0} \int_{R^2} a(\boldsymbol{x}, \boldsymbol{\xi}, z)\, \eta(\boldsymbol{\xi}, z)\, d_2\xi dz, \tag{9.206}$$

or, in the Fourier domain,

$$\chi(\boldsymbol{p}) = \hat{A}\hat{\eta} = \int_{-H}^{0} \int_{\mathbb{R}^2} \hat{\alpha}(\boldsymbol{p}, \boldsymbol{q}, z)\, \hat{\eta}(\boldsymbol{q}, z)\, d_2 p dz. \tag{9.207}$$

If $E\{\boldsymbol{\eta}(\boldsymbol{\xi}, z)\} = 0$, and $E\{\eta(\boldsymbol{\xi}, z)\, \eta(\boldsymbol{\xi}', z')\} = \sigma_\eta^2(z)\, \delta(z - z')\, \delta(\boldsymbol{\xi} - \boldsymbol{\xi}')$, compute the covariances $C_\chi(\boldsymbol{x} - \boldsymbol{x}')$ and $C_{\hat{\chi}}(\boldsymbol{p} - \boldsymbol{p}')$ and show that χ has a finite variance, namely $C_\chi(0) < +\infty$, if

$$\int_{-H}^{0} \frac{\sigma_\eta^2(z)}{z^2} dz < +\infty. \tag{9.208}$$

Observe that, in particular, if η is a pure white noise, i.e., $\sigma_\eta^2(z) = \text{const}$, then χ does not have a finite variance, but still has a finite covariance for $\boldsymbol{x} \neq \boldsymbol{x}'$.

Solution The formula for $C_{\hat{\chi}}(\boldsymbol{p} - \boldsymbol{p}')$ could be obtained directly by recalling the rules of covariance propagation for a modulated white noise, but we re-derive it for didactic reasons. Let us first of all observe that the relation between the Fourier transform of a gravimetric signal $\delta\hat{g}_z$ and that of the density distribution that generates it, $\delta\hat{\rho}$, is

$$\delta\hat{g}_Z(\boldsymbol{p}) = -2\pi G \int_{-H}^{0} e^{2\pi p z} \delta\hat{\rho}(\boldsymbol{p}, z)\, dz. \tag{9.209}$$

A comparison with (9.207) shows that

$$\hat{a}(\boldsymbol{p}, \boldsymbol{q}, z) = -2\pi G e^{2\pi p z} \delta(\boldsymbol{p} - \boldsymbol{q}). \tag{9.210}$$

Furthermore, since

$$\hat{\eta}(\boldsymbol{p}, z) = \int_{\mathbb{R}^2} e^{i2\pi \boldsymbol{p}\cdot\boldsymbol{\xi}} \boldsymbol{\eta}(\boldsymbol{\xi}, z)\, d_2\xi, \tag{9.211}$$

we have

$$
\begin{aligned}
C_{\hat{\eta}}\left(\boldsymbol{p}, \boldsymbol{p}', z, z'\right) &= E\left\{\hat{\eta}\left(\boldsymbol{p}, z\right) \hat{\eta}^*\left(\boldsymbol{p}', z'\right)\right\} = \\
&= \int_{R^2} e^{i2\pi\left(\boldsymbol{p}\cdot\boldsymbol{\xi} - \boldsymbol{p}'\cdot\boldsymbol{\xi}'\right)} C_\eta\left(\boldsymbol{\xi}, \boldsymbol{\xi}', z, z'\right) d_2\xi d_2\xi' = \\
&= \sigma_\eta^2(z)\,\delta\left(z - z'\right) \iint_{\mathbb{R}^2} e^{i2\pi\left(\boldsymbol{p}\cdot\boldsymbol{\xi} - \boldsymbol{p}'\cdot\boldsymbol{\xi}'\right)} \delta\left(\boldsymbol{\xi} - \boldsymbol{\xi}'\right) d_2\xi d_2\xi' = \\
&= \sigma_\eta^2(z)\,\delta\left(z - z'\right)\delta\left(\boldsymbol{p} - \boldsymbol{p}'\right).
\end{aligned}
\tag{9.212}
$$

Now, propagating the covariance of (9.207) and taking (9.213) into account, we find

$$
\begin{aligned}
C_{\hat{\chi}}\left(\boldsymbol{p}, \boldsymbol{p}'\right) &= \int_{-H}^{0} dz \int_{-H}^{0} dz' \iint_{\mathbb{R}^2} \hat{a}\left(\boldsymbol{p}, \boldsymbol{q}, z\right) \hat{a}^*\left(\boldsymbol{p}', \boldsymbol{q}', z'\right) \cdot \\
&\cdot \sigma_\eta^2(z)\,\delta\left(z - z'\right)\delta\left(\boldsymbol{p} - \boldsymbol{p}'\right) dq dq' = \\
&= 4\pi^2 G^2 \int_{-H}^{0}\int_{-H}^{0} \sigma_\eta^2(z)\, e^{2\pi p z} e^{2\pi p' z} \delta\left(\boldsymbol{p} - \boldsymbol{p}'\right)\delta\left(z - z'\right) dz dz' = \\
&= 4\pi^2 G^2 \left(\int_{-H}^{0} \sigma_\eta^2(z)\, e^{4\pi p z} dz\right) \delta\left(\boldsymbol{p} - \boldsymbol{p}'\right),
\end{aligned}
\tag{9.213}
$$

showing that in any event $\hat{\chi}$ is a modulated white noise, as it has always to be for the Fourier Transform of a homogeneous signal (see Appendix B). Therefore, $\hat{\chi}$ never has a finite variance, which however is not true for χ. In fact,

$$
\begin{aligned}
C_\chi\left(\boldsymbol{x}, \boldsymbol{x}'\right) &= C_\chi\left(\boldsymbol{x} - \boldsymbol{x}'\right) = \iint_{\mathbb{R}^2} e^{i2\pi\left(\boldsymbol{p}\cdot\boldsymbol{x} - \boldsymbol{p}'\cdot\boldsymbol{x}'\right)} \cdot \\
&\cdot\; 4\pi^2 G^2 \delta\left(\boldsymbol{p} - \boldsymbol{p}'\right)\left(\int_{-H}^{0} \sigma_\eta^2(z)\, e^{4\pi p z} dz\right) d_2p\, d_2p' = \\
&= 4\pi^2 G^2 \int_{\mathbb{R}^2} e^{i\pi \boldsymbol{p}\cdot\left(\boldsymbol{x} - \boldsymbol{x}'\right)} \left(\int_{H}^{0} \sigma_\eta^2(z)\, e^{4\pi p z} dz\right) d_2p.
\end{aligned}
\tag{9.214}
$$

Already at this point we can draw an important conclusion: if we require χ to have a finite variance, we must have

$$
\sigma^2(\chi) = C_\chi(0) = 4\pi^2 G^2 \int_{\mathbb{R}^2}\int_{-H}^{0} \sigma_\eta(z)\, e^{4\pi p z} dz\, d_2p < +\infty. \tag{9.215}
$$

Considering that, with the substitution $t = -4\pi p z$, we can write

$$
4\pi^2 \int_{\mathbb{R}^2} e^{4\pi p z} d_2p = \frac{\pi}{2z^2} \int_0^{+\infty} t e^{-t} dt = \frac{\pi}{2z^2}, \tag{9.216}
$$

we see that (9.125) is verified if

$$\frac{\pi G^2}{2}\int_{-H}^{0}\frac{\sigma_\eta^2(z)}{z^2}dz < +\infty, \tag{9.217}$$

and (9.208) is proved. We note that (9.208) is an interesting condition, showing that if we want $\chi = A\eta$ to have a finite variance, we are forced to require that $\sigma_\eta^2(z)$ tends to zero for $z \to 0$ faster than z. If, for instance, we assume that

$$\sigma_\eta^2(z) = cz^\alpha \tag{9.218}$$

(remember that we expect $\sigma_\eta^2(z)$ to increase with z), we have to require that $\alpha > 1$, and if we further assume that α is an integer, we are forced to take $\alpha \geq 2$. In particular, if on the contrary we take $\alpha = 0$, namely we assume η to be a pure white noise, then necessarily $\sigma^2(\chi) = +\infty$; this is true even if η is a white noise in a thin layer close to S_0. To complete the exercise let us write (9.125) in the form

$$C_\chi(\boldsymbol{x}-\boldsymbol{x}') = 4\pi^2G^2\int_{-H}^{0}\sigma_\eta^2(z)\int_{R^2}e^{i2\pi\boldsymbol{p}\cdot(\boldsymbol{x}-\boldsymbol{x}')}e^{4\pi pz}d_2p\,dz. \tag{9.219}$$

Recalling that

$$2\pi\int_{R^2}e^{i2\pi\boldsymbol{p}\cdot\boldsymbol{\xi}}e^{4\pi pz}d_2p = \frac{-2z}{\left[|\boldsymbol{\xi}|^2+4z^2\right]^{\frac{3}{2}}}, \tag{9.220}$$

from formula (9.219) we find

$$C_\chi(\boldsymbol{x}-\boldsymbol{x}') = 2\pi G^2\int_{-H}^{0}\frac{-2z}{\left[|\boldsymbol{\xi}|^2+4z^2\right]^{\frac{3}{2}}}\sigma_\eta^2(z)\,dz. \tag{9.221}$$

The form (9.221) cannot be further developed unless we specify $\sigma_\eta^2(z)$. For instance, if we put $\sigma_\eta^2(z) = \sigma_0^2$, namely, we consider the pure white noise case, we obtain

$$\begin{aligned} C_\chi(\boldsymbol{x}-\boldsymbol{x}') &= -4\pi G^2\sigma_0^2\int_{-H}^{0}\frac{z}{\left[|\boldsymbol{\xi}|^2+4z^2\right]^{\frac{3}{2}}}dz = \\ &= -\pi G^2\sigma_0^2\left\{\frac{1}{|\boldsymbol{x}-\boldsymbol{x}'|}-\frac{1}{\left[|\boldsymbol{x}-\boldsymbol{x}'|^2+4H^2\right]^{\frac{1}{2}}}.\right\} \end{aligned} \tag{9.222}$$

It is interesting to observe that, despite the fact that $C_\chi(0) = +\infty$, formula (9.222) provides a legitimate covariance function, namely a symmetric positive-definite function of $\boldsymbol{x} - \boldsymbol{x}'$.

Exercise 3 Referring to the notation introduced in the previous exercise, compute the covariance functions $C_{\hat{\chi}}(\boldsymbol{p} - \boldsymbol{p}')$ and $C_\chi(\boldsymbol{x} - \boldsymbol{x}')$ of the gravimetric signal generated by a density distribution $\eta(\boldsymbol{\xi}, z)$ in $\mathscr{L}$ with covariance

$$C_\eta(\boldsymbol{\xi}, z, \boldsymbol{\xi}', z') = \sigma_\eta^2(z)\,\delta(z - z')\,\delta(\boldsymbol{\xi} - \boldsymbol{\xi}'), \tag{9.223}$$

$$\sigma_\eta^2(z) = cz^2. \tag{9.224}$$

Solution In practice we have to apply (9.213) and (9.219) with $\sigma_\eta^2(z)$ given by (9.224). Since

$$f(p) = \int_{-H}^{0} z^2 e^{4\pi p z} dz = \frac{2}{(4\pi p)^3} - \frac{e^{-4\pi p H}}{(4\pi p)^3}\left[(4\pi p)^2 H^2 + 8\pi p H + 2\right], \tag{9.225}$$

as one can easily verify by taking the derivative with respect to H, we obtain

$$C_{\hat{\chi}}(\boldsymbol{p} - \boldsymbol{p}') = 4\pi^2 G f(p)\,\delta(\boldsymbol{p} - \boldsymbol{p}'). \tag{9.226}$$

Notice that $C_{\hat{\chi}}$, which is also the spectral density of χ, $S_{\chi(\boldsymbol{p})}$, is always positive as it has to be, i.e., $f(p)$ is always positive, so that

$$\iint_{\mathbb{R}^2} \varphi(\boldsymbol{p})\varphi^*(\boldsymbol{p}')\,C_{\hat{\chi}}(\boldsymbol{p} - \boldsymbol{p}')\,d_2p\,d_2p' =$$

$$= 4\pi^2 G^2 \int_{\mathbb{R}^2} f(p)\,|\varphi(\boldsymbol{p})|^2 d_2p \geq 0 \qquad \forall \varphi \in \mathscr{D}\left(\mathbb{R}^2\right).$$

As for $C_\chi(\boldsymbol{x} - \boldsymbol{x}')$, one has:

$$C_\chi(\boldsymbol{x} - \boldsymbol{x}') = -4\pi G^2 c \int_{-H}^{0} \frac{z^3}{[(\boldsymbol{x} - \boldsymbol{x}')^2 + 4z^2]^{\frac{3}{2}}} dz =$$

$$= \frac{\pi G^2 c}{4}\left[\left(|\boldsymbol{x} - \boldsymbol{x}'|^2 + 4H^2\right)^{\frac{1}{2}} + \frac{|\boldsymbol{x} - \boldsymbol{x}'|^2}{(|\boldsymbol{x} - \boldsymbol{x}'|^2 + 4H^2)^{\frac{1}{2}}} - 2\,|\boldsymbol{x} - \boldsymbol{x}'|\right] \tag{9.227}$$

The integral in (9.227) is computed via the substitution $\zeta = -z$ and $t = \zeta^2$. From (9.227) we see that

$$C_\chi(0) = \frac{\pi G^2 c H}{2},$$

i.e., χ has finite variance. When $r = |z - x'| \to +\infty$ one has

$$C_\chi(r) = O\left(\frac{1}{r^3}\right),$$

so that the function is integrable on $\mathbb{R}^2$.

Exercise 4 Find the solution of the normal Tikhonov equation (9.58), namely

$$4\pi^2 G^2 H^*_{j,\ell} \sum_{\ell'=1}^{L} H_{j,\ell'}\, \delta\hat{\rho}_{j,\ell'} + \lambda w_\ell \delta\hat{\rho}_{j,\ell} = -2\pi G H^*_{j,\ell} d\hat{g}\left(\boldsymbol{p}_j\right). \tag{9.228}$$

Solution Put $d\hat{g}_j = d\hat{g}\left(\boldsymbol{p}_j\right)$ and observe that (9.228) can be written in the form

$$\delta\hat{\rho}_{j,\ell} = -\frac{2\pi G}{\lambda w_\ell} H^*_{j,\ell} \bar{H}_j, \tag{9.229}$$

with

$$\bar{H}_j = d\hat{g}_j + 2\pi G \sum_{\ell'=1}^{L} H_{j,\ell'}\, \delta\hat{\rho}_{j,\ell'}. \tag{9.230}$$

Inserting (9.229) into (9.230), one finds that

$$\bar{H}_j = d\hat{g}_j - \frac{2\pi G}{\lambda} \psi_j \bar{H}_j, \tag{9.231}$$

where

$$\psi_j = \sum_{\ell'=1}^{L} \frac{2\pi G}{w'_\ell} H^*_{j,\ell'} H_{j,\ell'}. \tag{9.232}$$

To be more explicit we observe that in reality ψ_j is only a function of p_j, since

$$H^*_{j,\ell'} H_{j,\ell'} = \left|H_{j\ell'}\right|^2 = \bar{F}\left(p_j\right) e^{-4\pi\sqrt{\frac{j^2}{N^2}+\frac{k^2}{H^2}}\left(\ell' - \frac{1}{2}\right)\frac{D}{\Delta}} \tag{9.233}$$

where the factor $\bar{F}\left(p_j\right)$, purely depending on the shape of the voxels, is

$$\bar{F}\left(p_j\right) = \operatorname{sinc}\left(\pi p_x \Delta\right) \operatorname{sinc}\left(\pi p_y \Delta\right) \frac{\sinh\left(\pi p D\right)}{\pi p D} \Delta^2 D. \tag{9.234}$$

Therefore, (9.231) takes on the form

$$\bar{H}_j = \frac{d\hat{g}_j}{1 + \frac{2\pi G}{\lambda}\psi_j} \tag{9.235}$$

and finally (9.229) yields

$$\delta\hat{\rho}_{j,\ell} = -\frac{2\pi G}{\lambda w_\ell}\frac{H^*_{j,\ell} d\hat{g}_j}{1 + \frac{2\pi G}{\lambda}\psi_j}. \tag{9.236}$$

Exercise 5 Prove that if $K(\boldsymbol{x}, \boldsymbol{y})$ is a strictly positive transition kernel, $K(\boldsymbol{x}, \boldsymbol{y}) \geq K_0 > 0$, continuous in $\boldsymbol{x}$ and $\boldsymbol{y}$ on Q, then putting

$$h(\boldsymbol{y}) = \int_Q K(\boldsymbol{x}, \boldsymbol{y})\, g(\boldsymbol{y})\, d\boldsymbol{y} \qquad \left(h, g \in L^1_0\right) \tag{9.237}$$

one has

$$\| h \|_{L^1} \leq c_0 \| g \|_{L^1} \tag{9.238}$$

for some $c_0 < 1$.

Solution Since $h, g \in L^1_0$, i.e., $\int_Q g(\boldsymbol{x})\, d\boldsymbol{x} = \int_Q h(\boldsymbol{x})\, d\boldsymbol{x} = 0$, one can write

$$h(\boldsymbol{x}) = h_+(\boldsymbol{x}) - h_-(\boldsymbol{x}), \tag{9.239}$$

with

$$h_+(\boldsymbol{x}) = \begin{cases} > 0 & \text{on } I_+, \\ = 0 & \text{on } I_-, \end{cases} \qquad h_-(\boldsymbol{x}) = \begin{cases} = 0 & \text{on } I_+, \\ > 0 & \text{on } I_-, \end{cases} \qquad (I_+ \cup I_- = Q); \tag{9.240}$$

similarly, $g(\boldsymbol{x}) = g_+(\boldsymbol{x}) - g_-(\boldsymbol{x})$,

$$g_+(\boldsymbol{x}) = \begin{cases} > 0 & \text{on } J_+, \\ = 0 & \text{on } J_-, \end{cases} \qquad g_-(\boldsymbol{x}) = \begin{cases} = 0 & \text{on } J_+, \\ > 0 & \text{on } J_-, \end{cases} \qquad (J_+ \cup J_- = Q). \tag{9.241}$$

Therefore,

$$\int_Q h_+(\boldsymbol{x})\,dx = \int_{I_+} h_+(\boldsymbol{x})\,dx = \int_{I_-} h_-(\boldsymbol{x})\,dx = \int_Q h_-(\boldsymbol{x})\,dx, \tag{9.242}$$

$$\int_Q g_+(\boldsymbol{x})\,dx = \int_{J_+} g_+(\boldsymbol{x})\,dx = \int_{J_-} g_-(\boldsymbol{x})\,dx = \int_Q g_-(\boldsymbol{x})\,dx, \tag{9.243}$$

Moreover

$$\|h\|_{L^1} = \int_Q |h(\boldsymbol{x})|\,dx = \int_{I_+} h_+(\boldsymbol{x})\,dx + \int_{I_-} h_-(\boldsymbol{x})\,dx, \tag{9.244}$$

$$\|g\|_{L^1} = \int_Q |g(\boldsymbol{x})|\,dx = \int_{I_+} g_+(\boldsymbol{x})\,dx + \int_{J_-} g_-(\boldsymbol{x})\,dx. \tag{9.245}$$

From (9.242) and (9.243) we have

$$\begin{aligned} h_+(\boldsymbol{y}) = h(\boldsymbol{y}) &= \int_Q K(\boldsymbol{x},\boldsymbol{t})\,[g_+(\boldsymbol{x}) - g_-(\boldsymbol{x})]\,dx- \\ &= \int_{J_+} K(\boldsymbol{x},\boldsymbol{y})\,g_+(\boldsymbol{x})\,dx - \int_{J_-} K(\boldsymbol{x},\boldsymbol{y})\,g_-(\boldsymbol{x})\,dx, \end{aligned} \tag{9.246}$$

for $y \in I_+$, and

$$h_-(\boldsymbol{y}) = -h(\boldsymbol{y}) = -\int_{J_+} K(\boldsymbol{x},\boldsymbol{y})\,g_+(\boldsymbol{x})\,dx + \int_{J_-} K(\boldsymbol{x},\boldsymbol{y})\,g_-(\boldsymbol{x})\,dx, \tag{9.247}$$

for $y \in I_-$.

Now, since $|h(\boldsymbol{y})| = h_+(\boldsymbol{x}) + h_-(\boldsymbol{x})$, from (9.244) and (9.246), (9.247), we find

$$\begin{aligned} \|h\|_{L^1} = &\int_{I_+}\int_{J_+} K(\boldsymbol{x},\boldsymbol{y})\,g_+(\boldsymbol{x})\,dxdy - \int_{I_+}\int_{J_-} K(\boldsymbol{x},\boldsymbol{y})\,g_-(\boldsymbol{x})\,dxdy- \\ &- \int_{I_-}\int_{J_+} K(\boldsymbol{x},\boldsymbol{y})\,g_+(\boldsymbol{x})\,dxdy + \int_{I_-}\int_{J_-} K(\boldsymbol{x},\boldsymbol{y})\,g_-(\boldsymbol{x})\,dxdy. \end{aligned} \tag{9.248}$$

Exchanging the integrals and rearranging we get

$$\begin{aligned}\|h\|_{L^1} &= \int_{J_+} g_+(\boldsymbol{x})\left[\int_{I_+} K(\boldsymbol{x},\boldsymbol{y})\,dy - \int_{I_-} K(\boldsymbol{x},\boldsymbol{y})\,dy\right]dx- \\ &-\int_{J_-} g_-(\boldsymbol{x})\left[\int_{I_+} K(\boldsymbol{x},\boldsymbol{y})\,dy - \int_{I_-} K(\boldsymbol{x},\boldsymbol{y})\,dy\right]dx = \\ &= \int_{Q} g(\boldsymbol{x})\left[\int_{I_+} K(\boldsymbol{x},\boldsymbol{y})\,dy - \int_{I_-} K(\boldsymbol{x},\boldsymbol{y})\,dy\right]dx \le \\ &\le \int_{Q} |g(\boldsymbol{x})|\left|\int_{I_+} K(\boldsymbol{x},\boldsymbol{y})\,dy - \int_{I_-} K(\boldsymbol{x},\boldsymbol{y})\,dy\right|dx.\end{aligned} \tag{9.249}$$

Notice that, since $K(\boldsymbol{x},\boldsymbol{y})$ is a transition kernel, letting $a(\boldsymbol{x}) = \int_{I_+} K(\boldsymbol{x},\boldsymbol{y})\,dy$, $b(\boldsymbol{x}) = \int_{I_-} K(\boldsymbol{x},\boldsymbol{y})\,dy$, the functions $a(\boldsymbol{x})$, $b(\boldsymbol{x})$ are continuous on Q, and

$$a(\boldsymbol{x}) + b(\boldsymbol{x}) = 1. \tag{9.250}$$

Moreover, from the condition of strict positivity of $K(\boldsymbol{x},\boldsymbol{y})$

$$\begin{cases} a(\boldsymbol{x}) > K_0\,|I_+| > 0 \\ b(\boldsymbol{x}) > K_0\,|I_-| > 0 \end{cases} \quad \forall \boldsymbol{x} \in Q. \tag{9.251}$$

Therefore, for some $c_0 < 1$, $|a(\boldsymbol{x}) - b(\boldsymbol{x})| \le c_0$, $\forall \boldsymbol{x} \in Q$. Returning to (9.250), we conclude $\|h\|_{L^1} \le c_0 \int_Q g(\boldsymbol{x})\,dx = c_0\,\|g\|_{L^1}$ as it was to be proven.

Exercise 6 Prove that in the case dim $\boldsymbol{x}$ = dim $\boldsymbol{y}$ = 2 the kernel $K(\boldsymbol{x},\boldsymbol{y})$ described in (9.124) satisfies the normalization condition

$$\int_Q K(\boldsymbol{x},\boldsymbol{y})\,dy = 1. \tag{9.252}$$

Solution With $n = 2$ the formula (9.126) gives:

$$\begin{aligned}K(\boldsymbol{x},\boldsymbol{y}) &= K(x_1, x_2, y_1, y_2) = \\ &= K_1(y_1|\,x_2)\cdot K_2(y_2|\,y_1) = \\ &= f_{x_1|x_2}(y_1|\,x_2)\,f_{x_2|x_1}(y_2|\,y_1) = \\ &= \frac{f_{\mathbf{x}}(y_1, x_2)}{\int_{-a_1}^{a_1} f_{\mathbf{x}}(\xi, x_2)\,d\xi}\cdot\frac{f_{\mathbf{x}}(y_1, y_2)}{\int_{-a_2}^{a_2} f_{\mathbf{x}}(y_1, \eta)\,d\eta}.\end{aligned} \tag{9.253}$$

Therefore

$$\int_Q K(\boldsymbol{x}, \boldsymbol{y})\, d\boldsymbol{y} = \int_{-a_2}^{a_2} dy_2 \int_{-a_1}^{a_1} K(x_1, x_2, y_1, y_2)\, dy_1 =$$
$$= \int_{-a_1}^{a_1} \frac{f_x(y_1, x_2)}{\int_{-a_1}^{a_1} f_x(\xi, x_2)\, d\xi} \frac{\int_{-a_2}^{a_2} f_x(y_1, y_2)\, dy_2}{\int_{-a_2}^{a_2} f_x(y_1, \eta)\, d\eta} dy_1 = \tag{9.254}$$
$$= \frac{\int_{-a_1}^{a_1} f_x(y_1, x_2)\, dy_1}{\int_{-a_1}^{a_1} f_x(\xi, x_2)\, d\xi} = 1,$$

as we needed to prove.

Exercise 7 Prove, for the case $n = 2$, that the equilibrium distribution of $\{\boldsymbol{x}_k\}$ with transition kernel given by (9.124) is $f_x(\boldsymbol{x})$; namely, that,

$$\int_Q K(\boldsymbol{x}, \boldsymbol{y})\, f_{\boldsymbol{x}}(\boldsymbol{x})\, d_2x = f_{\boldsymbol{x}}(\boldsymbol{y}). \tag{9.255}$$

Solution Recalling (9.253) one has

$$\int_Q K(\boldsymbol{x}, \boldsymbol{y})\, f_{\boldsymbol{x}}(\boldsymbol{x})\, d_2x = \int_{-a_2}^{a_2} \int_{-a_1}^{a_1} \frac{f_{\boldsymbol{x}}(y_1, x_2)}{\int_{-a_1}^{a_1} f_{\boldsymbol{x}}(\xi, x_2)\, d\xi} \frac{f_{\boldsymbol{x}}(y_1, y_2)}{\int_{-a_2}^{a_2} f_{\boldsymbol{x}}(y_1, \eta)\, d\eta} f_{\boldsymbol{x}}(x_1, x_2)\, dx_2 dx_1 =$$

$$= \int_{-a_2}^{a_2} \frac{f_{\boldsymbol{x}}(y_1, x_2)}{\int_{-a_1}^{a_1} f_{\boldsymbol{x}}(\xi, x_2)\, d\xi} \frac{f_{\boldsymbol{x}}(y_1, y_2)}{\int_{-a_2}^{a_2} f_{\boldsymbol{x}}(y_1, \eta)\, d\eta} \int_{-a_1}^{a_1} f_{\boldsymbol{x}}(x_1, x_2)\, dx_1 dx_2 =$$

$$= f_{\boldsymbol{x}}(y_1, y_2) \int_{-a_2}^{a_2} \frac{f_{\boldsymbol{x}}(y_1, x_2)}{\int_{-a_2}^{a_2} f_{\boldsymbol{x}}(y_1, \eta)\, d\eta} dx_2 \equiv f_{\boldsymbol{x}}(y_1, y_2),$$

i.e., (9.255) holds.

Exercise 8 Prove that the decomposition (9.169) of the energy function $\mathscr{E}$ and the expression (9.170), are correct.

Solution Remember that the energy function $\mathscr{E}$ can be written (consider also the notation in Table 9.3)

$$\mathscr{E} = \frac{1}{2\sigma_v^2} |\mathbf{dg}_0 - A\boldsymbol{\rho}|^2 + \frac{\eta}{2} \sum_k \bar{\rho}_k^2 + \frac{\eta}{2} \sum_k \sum_{j \in \Gamma_k} \left(\bar{\rho}_k - \bar{\rho}_j\right)^2 -$$
$$- \sum_k r(\bar{\rho}_k) + \frac{\gamma}{2} \sum_k d_k(L_k, L_{0k}) + \frac{\lambda}{2} \sum_k \sum_{j \in \Gamma k} D_k\left(L_k, L_j\right) \tag{9.256}$$

and let us analyze the decomposition for each element. Consider that any piece of function of $(\boldsymbol{\rho}_{-i}, \boldsymbol{L}_{-i})$ only can be included into the definition of a front constant $C_T(\boldsymbol{\rho}_{-i}, L_{-i})$ that here is of no interest. Therefore, any two quantities differing by a constant of this type will be regarded as equivalent. Using the decomposition of A by columns (see Table 9.3), we get

$$|\mathbf{dg}_0 - A\boldsymbol{\rho}|^2 = \left|\mathbf{dg}_0 - \sum_{k\neq i}\mathbf{a}_k\rho_k - \mathbf{a}_i\rho_i\right|^2 = |\mathbf{dg}_{0,i} - \mathbf{a}_i\rho_i|^2 =$$
$$= |\mathbf{a}_i|^2\rho_i^2 - 2dg_{0,i}^T\mathbf{a}_i\rho_i + |dg_{0,i}|^2 = A_i\rho_i^2 - 2B_i\rho_i + |\mathbf{dg}_{0,i}|^2; \tag{9.257}$$

since $|\mathbf{dg}_{0,i}|^2$ is a function of $\boldsymbol{\rho}_{-i}$ only (see Table 9.3), it can be incorporated into the C_T constant. Moreover,

$$\sum_k \bar{\rho}_k^2 = \bar{\rho}_i^2 + \sum_{k\neq i}\bar{\rho}_k^2 = \bar{\rho}_i^2 + F(\rho_{-i}, L_{-i}); \tag{9.258}$$

in this formula and in the following F is used as symbol of a general function (recall that $\bar{\rho}_i$ is a function of L_i, too). Since F is a function of (ρ_{-i}, L_{-i}), it can be included in C_T. Furthermore

$$\sum_k\sum_{j\in\Gamma_k}(\bar{\rho}_k - \bar{\rho}_j) = \sum_k \bar{\rho}_k^2 N_k - 2\sum_k\bar{\rho}_k\sum_{j\in\Gamma_k}\bar{\rho}_j + \sum_k\sum_{j\in\Gamma_k}\bar{\rho}_j^2. \tag{9.259}$$

Then we can write

$$\sum_k \bar{\rho}_k^2 N_k = \bar{\rho}_i^2 N_i + F(\rho_{-i}, L_{-i}). \tag{9.260}$$

Notice that when k runs over all the voxel indexes, in $\Sigma_k\bar{\rho}_k\Sigma_{j\in\Gamma_k}\bar{\rho}_j$, the index i will appear when $k = i$ and when $j = i$ and $k \in \Gamma_i$ (because in this case $j \in \Gamma_k$); this second case happens for all $k \in \Gamma_i$, so that we can write

$$2\sum_k\bar{\rho}_k\sum_{j\in\Gamma_k}\bar{\rho}_j = 4\bar{\rho}_i\sum_{j\in\Gamma_i}\bar{\rho}_j + F(\rho_{-i}, L_{-i}) = 4\bar{\rho}_i N_i M_i + F(\rho_{-i}, L_{-i}). \tag{9.261}$$

Similarly, we have

$$\sum_k\sum_{j\in\Gamma_k}\bar{\rho}_j^2 = N_i\bar{\rho}_i^2 + F(\rho_{-i}, L_{-i}). \tag{9.262}$$

Therefore, combining (9.259), (9.260), (9.261), (9.262) together we obtain

$$\sum_k\sum_{j\in\Gamma_k}(\bar{\rho}_j - \bar{\rho}_j)^2 = 2N_i\bar{\rho}_i^2 - 4\bar{\rho}_i N_i M_i + F(\rho_{-i}, \mathbf{L}_{-i}). \tag{9.263}$$

Moreover,

$$\sum_k r(\bar{\rho}_k) = r(\bar{\rho}_i) + F(\rho_{-i}) \tag{9.264}$$

and

$$\sum_k d_k(L_k, L_{0k}) = d(L_i, L_{0i}) + F(L_{-i}). \tag{9.265}$$

Finally, always considering the mechanism of formula (9.260), (9.261), we obtain

$$\sum_k \sum_{j\in\Gamma_k} D_k(L_k, L_j) = \sum_{j\in\Gamma_i} D_i(L_i, L_j) + \sum_{k\in\Gamma_i} D_k(L_k, L_i) + F(\mathbf{L}_{-i}). \tag{9.266}$$

Changing the name of the summation index from k to j in (9.266), and using all relations from (9.258) to (9.266), we finally get

$$\begin{aligned}\mathscr{E} &= \frac{1}{2\sigma_v^2}(A_i\rho_i^2 - 2B_i\rho_i) + \frac{\eta}{2}\bar{\rho}_i^2 + \eta N_i(\bar{\rho}_i^2 - 2M_i\bar{\rho}_i) - \\ &- r(\bar{\rho}_i) + \frac{\gamma}{2}d_i(L_i, L_{0i}) + \frac{\lambda}{2}\sum_{j\in\Gamma_i}|D_i(L_i, L_j) + D_j(L_j, L_i)| + F(\rho_{-i}, L_{-i});\end{aligned} \tag{9.267}$$

$F(\rho_{-i}, \mathbf{L}_{-i})$ is as always irrelevant as it can be incorporated into C_T, the rest of the formula is essentially (9.170) we needed to prove.

Exercise 9 Prove that

$$\begin{aligned}I &= \int e^{-\frac{J(\bar{\rho}_i, L_i)}{T}} d\bar{\rho}_i = \int e^{-\frac{1}{2T}\left(\frac{\bar{\rho}_i - \bar{\mu}_i}{\bar{\sigma}_i}\right)^2 - \frac{r(\bar{\rho}_i)}{T}} d\bar{\rho}_i \\ &= \sqrt{2\pi T}\bar{\sigma}_i\left[\operatorname{erf}\left(\frac{3-\bar{\mu}_i}{\bar{\sigma}_i}\right) - \operatorname{erf}\left(-\frac{3+\bar{\mu}_i}{\bar{\sigma}_i}\right)\right].\end{aligned} \tag{9.268}$$

Solution First of all notice that:

$$e^{-\frac{1}{T}r(\bar{\rho}_i)} = e^{-\frac{1}{T}\log\chi_3(\bar{\rho}_i)} = [\chi_3(\bar{\rho}_i)]^{\frac{1}{T}} \equiv \chi_3(\bar{\rho}_i), \tag{9.269}$$

because $\chi_3(\bar{\rho}_i)$ can assume only the values 0 and 1. Hence, (9.268) writes

$$I = \int_{-3}^{3} e^{\frac{1}{2}\left(\frac{\bar{\rho}_i - \bar{\mu}_i}{\sqrt{T}\bar{\sigma}_i}\right)^2} d\bar{\rho}_i. \tag{9.270}$$

Now put $\bar{\rho}_i = \bar{\mu}_i + \sqrt{T}\bar{\sigma}_i z$, $d\bar{\rho}_i = \sqrt{T}\bar{\sigma}_i dz$ and consider that

$$\begin{aligned}I &= \sqrt{2\pi T}\bar{\sigma}_i \int_{-\frac{3+\bar{\mu}_i}{\sqrt{T}\bar{\sigma}_i}}^{\frac{3-\bar{\mu}_i}{\sqrt{T}\bar{\sigma}_i}} \frac{e^{-\frac{1}{2}z^2}}{\sqrt{2\pi}} dz = \\ &= \sqrt{2\pi T}\bar{\sigma}_i\left[\operatorname{erf}\left(\frac{3-\bar{\mu}_i}{\bar{\sigma}_i}\right) - \operatorname{erf}\left(-\frac{3+\bar{\mu}_i}{\bar{\sigma}_i}\right)\right],\end{aligned} \tag{9.271}$$

as we needed to prove.

Exercise 10 Write a Matlab function to perform gravity inversion by the Parker–Oldenburg algorithm.

Solution

```
function [Moho] = Parker_inv(dg,deltax,deltay,h,z0,lambdaMin, lamb-
daMax)
% Compute depth of discontinuity between two layers by means of
% Parker-Oldenburg method Oldenburg (1974).
% Input:
% dg [mGal], observed gravity
% deltax [m]
% deltay [m]
% h [m], height of gravity observations
% z0 [m], average depth of discontinuity
% lambdaMin, lambdaMax [m] minimum and maximum wavelengths
% Output: % Moho [m], depth of the discontinuity surface

G = 6.67408e-11; dg = dg.* 1e-5; % density contrast in kg/m.^3
deltaRho = 670;
z0 = z0 + h;
z0 = z0 ./ 1000;
G = G .* 1e3;
[r c] = size(dg);
if (mod (r, 2) ~ = 0)
    r = r - 1;
    dg (end, :) = [];
end
if (mod (c, 2) ~= 0)
    c = c - 1;
    dg (:, end) = [];
end

Delta_x = deltax * c;
Delta_y = deltay * r;
% set convergence criterion in km
threshold = 0.02;
WH = 1 / lambdaMax;
SH = 1 / lambdaMin;
truncation = 0.2;
dg = dg - mean( dg(:) );
%cosine Tukey window with a truncation of 20% is applied
```

```
wr = tukeywin (r, truncation);
wc = tukeywin (c, truncation);
w2 = wr(:) * wc(:)';
dg = dg .* w2;
dg_hat = fft2 (dg);

%compute frequencies matrix
for f = 1: abs( (r / 2) + 1);
    for g = 1: abs( (c / 2) + 1);
        frequency(f, g) = sqrt( ( (f - 1) / Delta_y)..^2 + ( (g - 1) / Delta_x).^2);
    end;
end;

if (mod (r, 2) == 0)
    frequency(1, :) = [];
end;
if (mod (c, 2) == 0)
    frequency(:, 1) = [];
end;
frequencytotal = [frequency fliplr(frequency); ...
flipud([frequency fliplr(frequency)])];
filter = frequencytotal .*0;
filter(frequencytotal < WH) = 1;
filter(frequencytotal < SH & frequencytotal > WH) = 0.5 .* (1 + cos((((2 * pi)
* frequencytotal(frequencytotal < SH & frequencytotal > WH)) - (2 * pi *
WH)) / (2 * (SH - WH))));
filter(frequencytotal > SH) = 0;
frequencytotal = frequencytotal .* (2 * pi);

% Compute first term of the series
up = -(dg_hat) .* exp (z0 .* (frequencytotal) );
down = (2 * pi * G * deltaRho);
DeltaD_hat(:,:,1) = up ./ down .* filter;
DeltaD = real (ifft2 (DeltaD_hat(:,:,1)));

% Compute second term of the series
DeltaD_hat(:,:,2) = ((frequencytotal ./ factorial(2)) .* (fft2(DeltaD.^2)));
DeltaD_hat(:,:,2) = DeltaD_hat(:,:,2) .* filter;
DeltaD_hat(:,:,2) = DeltaD_hat(:,:,1) - DeltaD_hat(:,:,2);
```

```
DeltaD(:,:,2) = real (ifft2 (DeltaD_hat(:,:,2)));
diff2 = (DeltaD(:,:,2) - DeltaD(:,:,1)).^2;
rms = sqrt (sum (diff2(:)) / (2 * (c * r)));
i=2;
rms =100;

while rms > = threshold & i < 10 %start iterations
    temp = zeros (size (frequencytotal));
for j = 2 : i+1
        temp = temp + (frequencytotal).^(j) ./ factorial(j + 1) .* fft2
(DeltaD(:,:,i).^(j + 1));
    end;
    DeltaD_hat(:,:,i + 1) = DeltaD_hat(:,:,1) - temp .* filter;
    DeltaD(:,:,i+1) = real (ifft2 (DeltaD_hat(:,:,i + 1)));
    diff2 = (DeltaD(:,:,i + 1) - DeltaD(:,:,i)).^2;
    rms = sqrt (sum (diff2(:)) / (2 * (c * r)));
    i = i+1
end;
Moho = (DeltaD(:,:,i) + z0) .* 1000; end;
```

Chapter 10
Some Conclusions

The line of research of gravity inversion and interpretation is not at a dead end: on the contrary, the matter is still hot and attracting new efforts in deepening our understanding of the problem and exploring new numerical algorithms. This is because the information on crustal processes contained in the knowledge of the gravity field outside the masses, although far from giving a univocal picture of the internal distribution, is undeniably the most abundant at all scales (from 10^5 to 1 km) depending on the areas, and so precise that no geophysical-geological model can be acceptable when discordant with available gravity data. Further on the acquisition of new data, specially by spaceborne, airborne and shipborne gravimetry and gradiometry, and the development of new powerful computing machines have stirred the activity on the subject. The matter of how to homogenize and integrate the information on the gravity field coming from different sources (and techniques) at different scales, without resorting to the knowledge of the distribution of internal masses, received in the last decades major attention in Geodesy, leading to the construction of global gravity models with a maximum spatial resolution of about 10 km, and local gravity models with a spatial resolution of the order of 1 km and an accuracy of 1 mGal. From the above results we have learnt that the power of the gravitational signal distributes in a monotonically decreasing manner among the different scales, from planetary to very local. The plot of the so called harmonic degree variances of the potential (see Fig. 4.4) is very convincing and it is clear that this reflects a physical property of the mass distribution inside the Earth, saying the power at larger scales dominates the power at lower scales. Now if one wants to build a consistent theory of the inversion of the gravity field, one is unavoidably faced with a first choice: go global or local. This is because, although the gravity field is one, and conceptually its inversion has the same mathematical characteristics at all scales, the main tools for the interpretation and analysis of the field are the spherical harmonics expansion at a global level and the Fourier Transform at a local level; each of the two has its own theory, theorems and properties. In this book we chose to focus on the local level, which has obvious implications in applied

F. Sansò, D. Sampietro, *Analysis of the Gravity Field*, Lecture Notes in Geosystems Mathematics and Computing, https://doi.org/10.1007/978-3-030-74353-6_10

geophysics, corresponding also to the authors' experience of the last 10–15 years in both teaching and research. But, as claimed above, no local analysis can be conducted on the gravity field without a good understanding of global models. This explains why we introduce those in Chap. 4 and then discuss in detail the issue of Terrain and Residual Terrain Correction, accounting for the external information that can reduce the gravity field to a really "local level". All that in agreement with the idea that we should write a book "useful" to the reader in the sense of providing a clear justification of each step leading from data (observations) to results (mass anomalies). So, apart from the first part, which is essentially a handbook of integration exercises applied to the context of gravity calculations, and the second part just recalled above, the core of the matters is in the third part, where we first illustrate the mathematical setup of the inverse gravimetric problem and then present the most important approaches (in our opinion) followed to reduce the inherent non-uniqueness of the solution. Of course, the literature on the matter is extremely large, so we had to make choices. Thus, along the line of a deterministic approach we follow that of the Tikhonov theory [150], leading to the solution of a classical variational principle, while along the line of a stochastic approach we adopt Bayesian concepts and the Wiener-Kolmogorov principle of minimizing the mean square estimation error [126, 128]. As for the analysis of the abstract problem, this accounts primarily to the identification of the null space of Newton's operator in various functional spaces and eventually the proof of the existence of a minimum-norm solution of the inversion problem. This leads directly to the Tikhonov approach. The analysis of such a problem is well known already from many years in an L^2 context, for the mass density ρ, and it has been recently generalized to more general spaces, like L^P or Sobolev spaces (see [7, 22, 130]). Let us mention though that a mathematical theory for what is perhaps the most immediate choice, namely the L^∞ space, including natural constraints on two mass density (it has to be non-negative and it has a natural upper bound) is still missing. In the book the L^2 theory is presented, although translated from the usual "spherical picture" into a "Cartesian setup" typical of the applied geophysical point of view. The subsequent Tikhonov approach is then discussed with the $H^{1,2}$ norm as regularizing functional, which seems to the authors the most reasonable compromise between the physical reality of a certainly discontinuous mass density and the requirements of a mathematical approximation problem. The corresponding mathematical background is presented in Appendix C, where certainly not the most general point of view is taken, but rather the theory is adapted to the applied geophysical context. Naturally the Tikhonov line of thought is not the only one used to solve improperly posed problems. For instance, the Landweber approach, just to mention another important one [74], provides an alternative strategy. It has to be mentioned that the performance of the approximation, does indeed depend on the choice of the functional basis used in the application. In this respect the use of wavelets derived from multiresolution analysis and in particular of harmonic wavelets [43] is conceptually very appealing and it is probably one of the most important lines not explored in this work, that in our opinion also deserves more numerical experimentation in future. Our choice to go with Tikhonov's approach, is justified, beyond its wide diffusion, by its strong compatibility with our stochastic approach. This is very much agreeing with our belief that the methods that provide reasonable results for the same problem, must be mapped, to a certain extent, with one another. As for the stochastic line a first choice had to be made, whether to go through the relevant infinite-dimensional prior and likelihood distributions, and then look for the distribution of the gravity potential conditional to the observation field, according to the pure Bayesian rules, or to content ourselves with the search of linear predictors (in the observations) based on a second-order analysis of the stochastic fields and on the principle of minimizing the mean square prediction error. All that is similar to the choice between the use of the Bayes formula or that of least squares as ordinary parameter estimation method. The pros and cons are also similar even for the infinite-dimensional case; the authentic Bayesian approach is more consistent and can account for nonlinear prediction problems, but it requires the knowledge of the probabilistic distributions that are by far not available to the analyst. On the contrary, the least squares, as in its "best linear unbiased estimation" version, is less general, but it is based only on the knowledge of the first two moments of the random variables involved; these are usually estimable from the data. The only problem in transferring such concepts to the infinite-dimensional context is that the simple theory of stochastic processes is not sufficient to give a sound mathematical basis to the prediction process, especially when random fields that do not have point values, like a white noise, enter the picture. What is needed to systematize such a matter is the theory of Generalized Random Fields [120], the basis of which are provided in the book in Appendix B. Once rigorously formulated, the method can then easily run by exploiting suitable numerical methods, well known in geophysics, the so-called Monte Carlo Markov Chain approach. But still another feature makes the Bayesian approach so attractive, namely the possibility of introducing into the prediction process a field of "labels", i.e., a discrete qualitative variable describing the geological nature of mass anomalies, together with their density. The field of labels, with a prior geological distribution as input, is then predicted in the output providing a geological scenery of the anomalies, updated by the inclusion of gravity observations. This establishes a strong systematic interaction between the pure inverse gravimetric problem and its geological context. The theory has been validated by realistic and real cases, some of which reported in the text.

In practice the book reports a successful merging of geological and gravimetric information, based on a sound mathematical theory. This is not the end of the story, but rather a fortunate starting point that can potentially provide a method to integrate into a consistent picture the geological setting of specific cases, with all available geophysical information including gravimetry, magnetometry and seismics. This is in our opinion an open way for future research.

Appendix A
Mathematical Auxiliary Material

In this Appendix we briefly recall a number of items concepts and facts from mathematical analysis that are needed in the book. In many cases we will only mention items and definitions, dwelling a little more on critical concepts/theorems that might be less commonly known.

The reminder of linear algebra hereafter is presented in a form that is particularly useful for generalization to infinite-dimensional linear spaces.

A.1 Vector—Matrix Algebra

We assume the reader is acquainted with the concept of linear space. Let us agree that, unless specifically stated, the scalars used in linear operations are just real numbers.

Definition A.1 Let $\boldsymbol{a}_1, \dots, \boldsymbol{a}_k$ be k *nonzero* vectors belonging to the linear space L; they are said to be linearly independent if

$$\sum_{j=1}^{k} \lambda_j \boldsymbol{a}_j = 0 \Longleftrightarrow \lambda_j = 0, \quad j = 1 \dots k; \tag{A.1}$$

the set of elements

$$L_k = \{\sum_{j=1}^{k} \lambda_j \boldsymbol{a}_j ; \lambda_j \in \mathbb{R}, \quad j = 1, \dots, k\} \tag{A.2}$$

is called the span of $(\boldsymbol{a}_j, j = 1, \dots, k)$ and is denoted

$$\text{span}\,\{\boldsymbol{a}_1, \dots, \boldsymbol{a}_k\} = L_k; \tag{A.3}$$

F. Sansò, D. Sampietro, *Analysis of the Gravity Field*, Lecture Notes in Geosystems Mathematics and Computing, https://doi.org/10.1007/978-3-030-74353-6

L_k is a linear subspace of L.

Definition A.2 We say that L has dimension N if there are N linearly independent vectors $(\boldsymbol{a}_i, i = 1, \ldots, N)$, such that

$$L = \text{span}\,\{\boldsymbol{a}_1, \ldots, \boldsymbol{a}_N\}; \tag{A.4}$$

and any $N + 1$ vectors in L cannot be linearly independent, i.e. N is the maximal number of linearly independent vectors in L. We say that $(\boldsymbol{a}_1, \ldots, \boldsymbol{a}_N)$, is a basis of L.

Example A.1 Among linear spaces we can consider the (real) Euclidean spaces $\mathbb{R}^N$, whose elements are vectors in algebraic sense, namely

$$\boldsymbol{\lambda} \in \mathbb{R}^N \Longleftrightarrow \boldsymbol{\lambda} \begin{bmatrix} \lambda_1 \\ \vdots \\ \lambda_N \end{bmatrix}, \quad \lambda_k \in \mathbb{R}_I.$$

We note that a basis in L is a tool that creates a one-to-one correspondence between vectors $\boldsymbol{a} \in L$ and vectors $\boldsymbol{\lambda} \in \mathbb{R}^N$, by

$$\boldsymbol{a} = \sum_{k-1}^{N} \lambda_k \boldsymbol{a}_k; \tag{A.5}$$

$\boldsymbol{\lambda}$ is called the vector of components of $\boldsymbol{a}$ with respect to (or in) the basis $\{\boldsymbol{a}_k\}$.

Definition A.3 A scalar product in L is a map of $L \times L \to \mathbb{R}$ with the following properties:

a) $\boldsymbol{a} \cdot \boldsymbol{b}$ is linear in $\boldsymbol{a}$ and $\boldsymbol{b}$ (i.e., it is bilinear);
b) $\boldsymbol{a} \cdot \boldsymbol{b} = \boldsymbol{b} \cdot \boldsymbol{a}$;
c) $\boldsymbol{a} \cdot \boldsymbol{a} = |\boldsymbol{a}|^2 \geq 0 (|\boldsymbol{a}| = 0 \Longleftrightarrow \boldsymbol{a} = 0)$;

Two vectors $\boldsymbol{a}$ and $\boldsymbol{b}$ are said to be orthogonal if $\boldsymbol{a} \cdot \boldsymbol{b} = 0$.

Observe that, according to Definition A.3, we have $|\boldsymbol{a}+\boldsymbol{b}|^2 = (\boldsymbol{a}+\boldsymbol{b}) \cdot (\boldsymbol{a}+\boldsymbol{b}) = |\boldsymbol{a}|^2 + |\boldsymbol{b}|^2 + 2\boldsymbol{a} \cdot \boldsymbol{b}$. So, when $\boldsymbol{a}$ and $\boldsymbol{b}$ are orthogonal, the Pythagorean rule holds: $|\boldsymbol{a} + \boldsymbol{b}|^2 = |\boldsymbol{a}|^2 + |\boldsymbol{b}|^2$.

Proposition A.1 *The following Cauchy–Schwarz inequality holds:*

$$|\boldsymbol{a} \cdot \boldsymbol{b}| \leq |\boldsymbol{a}| \cdot |\boldsymbol{b}|. \tag{A.6}$$

In fact, if $\boldsymbol{a} = 0$ *or* $\boldsymbol{b} = 0$, (A.6) *is trivially true; otherwise*

$$\left| \frac{\boldsymbol{a}}{|\boldsymbol{a}|} \pm \frac{\boldsymbol{b}}{|\boldsymbol{b}|} \right|^2 = 2 \pm 2 \frac{\boldsymbol{a} \cdot \boldsymbol{b}}{|\boldsymbol{a}|\ |\boldsymbol{b}|} \geq 0,$$

which readily yields (A.6). *Notice that the above inequality can be an equality if only if*

$$\frac{|\boldsymbol{a}|}{|\boldsymbol{a}|} = \pm\frac{\boldsymbol{b}}{|\boldsymbol{b}|},$$

i.e., if there is a real λ *such that*

$$\boldsymbol{b} = \lambda\boldsymbol{a}. \tag{A.7}$$

Definition A.4 A basis $\{\boldsymbol{e}_1, \ldots, \boldsymbol{e}_N\}$ of L is called orthonormal if

$$\boldsymbol{e}_k \cdot \boldsymbol{e}_j = \delta_{kj} \tag{A.8}$$

i.e., any two distinct basis vectors are orthogonal and each basis vector has modulus 1.

Definition A.5 The orthogonal projection of a vector $\boldsymbol{a}$ on a unit vector $\boldsymbol{e}$ is defined as

$$\boldsymbol{P}_e(\boldsymbol{a}) = (\boldsymbol{a} \cdot \boldsymbol{e})\boldsymbol{e}; \tag{A.9}$$

given a subspace S spanned by the orthonormal basis $(\boldsymbol{e}_k, k = 1, \ldots, M)$, the orthogonal projection of $\boldsymbol{a}$ on S is defined as

$$\boldsymbol{P}_s(\boldsymbol{a}) = \sum_{k=1}^{M} (\boldsymbol{a} \cdot \boldsymbol{e}_k)\boldsymbol{e}_k. \tag{A.10}$$

Proposition A.2 *Every base of a subspace* S $(\boldsymbol{a}_k, k = 1, \ldots, N)$ *can be transformed into an orthonormal basis by the so-called Gram-Schmidt procedure; the same holds for the whole space.*

Namely, we put

$$\begin{aligned}
\boldsymbol{a}_1 &= \lambda_{11}\boldsymbol{e}_1,\\
\boldsymbol{a}_2 &= \lambda_{21}\boldsymbol{e}_1 + \lambda_{22}\boldsymbol{e}_2,\\
\boldsymbol{a}_3 &= \lambda_{31}\boldsymbol{e}_1 + \lambda_{32}\boldsymbol{e}_2 + \lambda_{33}\boldsymbol{e}_3,\\
&\ldots\ldots\ldots\ldots\ldots\ldots\ldots\ldots
\end{aligned}$$

and we determine the $\boldsymbol{e}_k$ *and* $\lambda_{nn}(n \geq k)$ *as follows:*

- *from the requirement* $|\boldsymbol{e}_1| = 1$ *we get*

$$\lambda_{11} = |\boldsymbol{a}_1|, \ \boldsymbol{e}_1 = \frac{\boldsymbol{a}_1}{\lambda_{11}};$$

- *from the requirements* $\boldsymbol{e}_2 \cdot \boldsymbol{e}_1 = 0$, $|\boldsymbol{e}_2| = 1$ *we get*

$$\boldsymbol{a}_2 \cdot \boldsymbol{e}_1 = \lambda_{21}, \quad |\boldsymbol{a}_2 - \lambda_{21}\boldsymbol{e}_1| = \lambda_{22},$$

$$\boldsymbol{e}_2 = \frac{1}{\lambda_{12}}(\boldsymbol{a}_2 - \lambda_{21}\boldsymbol{e}_1),$$

and so on.

When $M = N$, *i.e.,* $(\boldsymbol{e}_k;\ k = 1, \ldots, N)$ *is an orthonormal basis of the whole space* L, (A.10) *becomes*

$$\boldsymbol{a} = \sum_{k=1}^{N} (\boldsymbol{a} \cdot \boldsymbol{e}_k)\boldsymbol{e}_k. \tag{A.11}$$

Remark A.1 We notice that (A.11) induces an isometric isomorphism between L and $\mathbb{R}^n$, when the Euclidean space $\mathbb{R}^n$ is endowed with the scalar product and modulus given by

$$\boldsymbol{\lambda} \cdot \boldsymbol{\gamma} = \boldsymbol{\lambda}^T \boldsymbol{\gamma} = \sum_{k=1}^{N} \lambda_k \lambda_k, \quad |\boldsymbol{\lambda}|^2 = \boldsymbol{\lambda} \cdot \boldsymbol{\lambda} = \sum_{k=1}^{N} \lambda_k^2, \quad \boldsymbol{\lambda}, \boldsymbol{\gamma} \in \mathbb{R}^n.$$

In fact, given $\boldsymbol{\lambda}$ we have $\boldsymbol{a} = \Sigma_{k=1}^{N} \lambda_k \boldsymbol{e}_k$, and given $\boldsymbol{a}$ we have $\lambda_k = \boldsymbol{a} \cdot \boldsymbol{e}_k$. Moreover, from (A.8)

$$\begin{aligned} |\boldsymbol{a}|^2 &= \left(\sum_{k=1}^{N} \lambda_k \boldsymbol{e}_k\right) \cdot \left(\sum_{j=1}^{N} \lambda_j \boldsymbol{e}_j\right) = \sum_{k,j=1}^{N} \lambda_k \lambda_j \boldsymbol{e}_k \cdot \boldsymbol{e}_j = \\ &= \sum_{k}^{N} \lambda_k^2 = |\boldsymbol{\lambda}|^2, \end{aligned} \tag{A.12}$$

whence the *isometry*.

Proposition A.3 *Given any subspace* S *of* L *and a vector* $\boldsymbol{a} \notin S$, *there is in* S *exactly one element,* $\hat{\boldsymbol{a}}$, *that lies at minimum distance from* $\boldsymbol{a}$, *namely*

$$\hat{\boldsymbol{a}} = P_s(\boldsymbol{a}). \tag{A.13}$$

Indeed, from (A.10) *and* (A.8), *it easily follows that*

$$P^2(\boldsymbol{a}) = P[P(\boldsymbol{a})] = P(\boldsymbol{a});$$

but then one has $\boldsymbol{a} - P(\boldsymbol{a}) = (I - P)(\boldsymbol{a}) \perp S$*; in fact,*

$$P(\boldsymbol{a} - P(\boldsymbol{a})) = P(\boldsymbol{a}) - P(\boldsymbol{a}) = 0.$$

Therefore, by the Pythagorean rule, for any $a \in S$,

$$\begin{aligned} |\boldsymbol{a} - \hat{\boldsymbol{a}}|^2 &= |(I - P)(\boldsymbol{a}) + P(\boldsymbol{a}) - \hat{a}|^2 = \\ &= |(I - P)\boldsymbol{a}|^2 + |P(\boldsymbol{a}) - \hat{a}|^2. \end{aligned} \tag{A.14}$$

Since $P(\boldsymbol{a}) \in S$*, we can choose* $\hat{a} = P(\boldsymbol{a})$ *in* (A.14) *and thus attain the minimum of* $|\boldsymbol{a} - \hat{a}|^2$.

Definition A.6 A homogeneous linear mapping of $\mathbb{R}^n$ in $\mathbb{R}^m$ is represented by a matrix A, i.e., an $m \times n$ table

$$A = \begin{bmatrix} a_{11} \dots a_{1n} \\ \dots\dots\dots \\ a_{m1} \dots a_m \end{bmatrix}; \tag{A.15}$$

then we define the product $A\boldsymbol{x}$ ($\boldsymbol{x} \in \mathbb{R}^n$) by the usual rows by columns product. $\mathbb{R}^n$ is called the *domain* of A, the set

$$\mathscr{R}(A) = \{\boldsymbol{y};\ \boldsymbol{y} = A\boldsymbol{x}, \forall \boldsymbol{x} \in \mathbb{R}^n\} \subset \mathbb{R}^m. \tag{A.16}$$

is called the *range* of A.

Definition A.7 The null space $\mathscr{N}$ of a A is the linear subspace

$$\mathscr{N} = \{\boldsymbol{x};\ A\boldsymbol{x} = 0\} \subset \mathbb{R}^n \tag{A.17}$$

If $\mathscr{N} = \{0\}$ i.e., $A\boldsymbol{x} = 0 \Longrightarrow \boldsymbol{x} = 0$, we say that A is of full rank or that it induces a transformation "into" $\mathbb{R}^m$. Since, using the notation $\boldsymbol{a}_k$ for the k-th column of A, we can write

$$\boldsymbol{y} = A\boldsymbol{x} = \sum_{k=1}^{n} x_k \boldsymbol{a}_k, \tag{A.18}$$

A is into if $\{\boldsymbol{a}_k,\ k = 1, \dots, n\}$ are linearly independent (see (A.1)). We observe that indeed if $n > m$ this cannot happen.

Let us note that (A.16) implies

$$\mathscr{R}(A) = \text{span}\ \{\boldsymbol{a}_k, k = 1, \dots, n\}. \tag{A.19}$$

Definition A.8 The transformation induced by A is said to be "onto" $\mathbb{R}^m$ if

$$\mathscr{R}(A) = \mathbb{R}^m; \tag{A.20}$$

we observe that (A.20) may hold only if $n \geq m$.

Definition A.9 The transpose of A, denoted A^T, is defined as

$$B = A^T, b_{ki} = a_{ik}; \tag{A.21}$$

so the transpose has domain $\mathbb{R}^m$ and range in $\mathbb{R}^n$. It is easy to verify that

$$\forall \boldsymbol{\lambda} \in \mathbb{R}^n, \forall \boldsymbol{\gamma} \in \mathbb{R}^m, \quad \boldsymbol{\gamma}^T(A\boldsymbol{\lambda}) = (A^T\boldsymbol{\gamma})^T\boldsymbol{\lambda}, \tag{A.22}$$

$$\text{or} \quad \boldsymbol{\gamma} \cdot (A\boldsymbol{\lambda}) = (A^T\boldsymbol{\gamma}) \cdot \boldsymbol{\lambda}. \tag{A.23}$$

Proposition A.4 *By using* (A.22) *it is easy to see that*

$$\mathscr{R}(A^T) = \mathscr{N}(A)^\perp \text{ and } \mathscr{R}(A) \equiv \mathscr{N}(A^T)^\perp. \tag{A.24}$$

Definition A.10 If $m = n$ we say that A is square;
if $A^T = A$ we say that A is symmetric;
if $a_{ik} = \delta_{ik}$, A is called the identity and denoted I;
if $n = m$, and $\mathscr{N}(A) = \{0\}$ we say that A is regular (otherwise A is singular).

If A is regular, then there exists a matrix A^{-1} such that

$$\boldsymbol{y} = A\boldsymbol{x} \Longleftrightarrow \boldsymbol{x} = A^{-1}\boldsymbol{y}, \tag{A.25}$$

$$A^{-1}A = AA^{-1} = I. \tag{A.26}$$

As we know, using the concept of determinant of A as an indicator, we can say that A is regular if and only if

$$\exists A^{-1} \Longleftrightarrow \det A \neq 0. \tag{A.27}$$

Definition A.11 If A is symmetric, we associate to it the quadratic form $\boldsymbol{x}^T(A\boldsymbol{x})$; A is said to be positive definite if

$$\boldsymbol{x}^T(A\boldsymbol{x}) \geq 0, \ \forall \boldsymbol{x}; \tag{A.28}$$

if

$$\boldsymbol{x}^T A\boldsymbol{x} > 0 \qquad \forall \boldsymbol{x} \neq 0, \tag{A.29}$$

we say that A is strictly positive definite.

It is obvious that a strictly positive matrix is regular, i.e. A^{-1} exists.

Definition A.12 We say that the pair $(\lambda, \ \boldsymbol{v})$ consists of an eigenvalue and an eigenvector of the square matrix A if

$$\boldsymbol{v} \neq 0 \text{ and } A\boldsymbol{v} = \lambda \boldsymbol{v}. \tag{A.30}$$

In this case it is clear that λ is a root of the secular equation

$$\det(A - \lambda I) = 0 \tag{A.31}$$

It is also obvious that if $\boldsymbol{v}$ is an eigenvector, then the same is for $\alpha \boldsymbol{v}$, $\forall \alpha \neq 0$. So we can agree conventionally that eigenvectors are always chosen to have modulus 1.

Proposition A.5 *Assume A is square and symmetric, $A^T = A$; then there are n real roots of* (A.31) *i.e., n eigenvalues $\lambda_k, k = 1, \ldots, n$ of A. some of the λ_k can coincide.*

To each λ_k corresponds an eigenvector v_k i.e.,

$$A\boldsymbol{v}_k = \lambda_k \boldsymbol{v}_k. \tag{A.32}$$

In case of repeated eigenvalues the corresponding eigenvectors can be chosen to be orthogonal to one another; eigenvectors corresponding to different eigenvalues are automatically orthogonal to one another. Since there are n eigenvectors, they constitute an orthonormal basis of $\mathbb{R}^n$. If $\lambda = 0$ is not an eigenvalue, then $\mathcal{N}(A) = \{0\}$ and A is regular.

Proposition A.6 *If A is square and symmetric, with the k eigenvalue–eigenvector pair $(\lambda_k, \ \boldsymbol{v}_k)$, then the identity holds*

$$A = \sum_{k=1}^{n} \lambda_k \boldsymbol{v}_k \boldsymbol{v}_k^T; \tag{A.33}$$

if $\lambda_k \neq 0$ for all k, then A is regular and

$$A^{-1} = \sum_{k=1}^{n} \lambda_k^{-1} \boldsymbol{v}_k \boldsymbol{v}_k^T. \tag{A.34}$$

In fact, note that if $\{\boldsymbol{v}_k\}$ is an orthonormal basis of $\mathbb{R}^n$, then the identity

$$\boldsymbol{a} \equiv \sum_{k=1}^{n} \boldsymbol{v}_k (\boldsymbol{v}_k^T \boldsymbol{a}). \tag{A.35}$$

implies that

$$\sum_{k=1}^{n} \boldsymbol{v}_k \boldsymbol{v}_k^T = I. \tag{A.36}$$

Hence, if we multiply (A.35) on the right by A and use (A.32), we get (A.33).

As for (A.34), we have, by the orthogonality of $\{\boldsymbol{v}_k\}$,

$$A^{-1}A = \sum_{k=1}^{n} \lambda_k^{-1} \boldsymbol{v}_k \boldsymbol{v}_k^T \sum_{j=1}^{n} \lambda_j \boldsymbol{v}_k \boldsymbol{v}_k^T =$$

$$= \sum_{k,j=1}^{n} \lambda_k^{-1} \lambda_j \delta_{kj} \boldsymbol{v}_k \boldsymbol{v}_k^T = \sum_k \boldsymbol{v}_k \boldsymbol{v}_k^T = I.$$

Remark A.2 If A is square, symmetric and positive definite, then

$$\lambda_k \geq 0. \tag{A.37}$$

Conversely, if (A.37) holds, then A is positive definite.
If A is strictly positive (A.37) holds without equality.
The relation (A.37) is just an application of Definition A.11 to the identity

$$\lambda_k = \boldsymbol{v}_k^T A \boldsymbol{v}_k. \tag{A.38}$$

Remark A.3 (A.34) shows that the inverse of a symmetric matrix is symmetric and the inverse of a symmetric positive matrix is symmetric and positive.

Definition A.13 The triples $(\lambda_k, \boldsymbol{v}_k, \boldsymbol{u}_k)$, with $\lambda_k \in k$, $\boldsymbol{v}_k \in \mathbb{R}^m$, $\boldsymbol{u}_k \in \mathbb{R}^n$constitute the singular elements of a matrix $A(\mathbb{R}^n \to \mathbb{R}^m)$, if $(\lambda_k, \begin{bmatrix} \boldsymbol{v}_k \\ \boldsymbol{u}_k \end{bmatrix})$ are eigenvalues and eigenvectors of the extended symmetric matrix

$$A = \begin{bmatrix} 0 & A \\ A^T & 0 \end{bmatrix}; \tag{A.39}$$

this implies

$$\begin{cases} A\boldsymbol{u}_k = \lambda_k \boldsymbol{v}_k, \\ A^T \boldsymbol{v}_k = \lambda_k \boldsymbol{u}_k. \end{cases} \tag{A.40}$$

Remark A.4 Notice that (A.40) immediately entails

$$\begin{cases} A^T A\boldsymbol{u}_k = \lambda_k^2 \boldsymbol{v}_k, \\ AA^T \boldsymbol{v}_k = \lambda_k^2 \boldsymbol{v}_k; \end{cases} \tag{A.41}$$

this shows that the $\{\boldsymbol{u}_k\}$ can be taken as orthonormal in $\mathbb{R}^n$ and $\{v_k\}$ as orthonormal in $\mathbb{R}^m$.

Since the above relations hold for any m, n, let us assume for the sake of definiteness that $m \geq n$.

Then, if we further assume that A is into, the first relation in (A.41) shows that $A^T A$ is a square symmetric strictly positive matrix in $\mathbb{R}^n$and $\lambda_k^2 \neq 0$, $k = 1 \ldots n$.

The second relation in (A.41) says that $\{v_k\}$ are orthonormal too but, since AA^T is a matrix in $\mathbb{R}^m$, when $m > n$, there must be other $m - n$ eigenvalues of AA^T equal to zero, corresponding to $\mathcal{N}(A^T)$.

Moreover, since (A.40) can be also written in the form

$$\begin{cases} A(-\boldsymbol{u}_k) = (-\lambda_k)\boldsymbol{v}_k, \\ A^T \boldsymbol{v}_k = (-\lambda_k)(-\boldsymbol{u}_k), \end{cases} \tag{A.42}$$

we see that the eigenvalues come in couples, and if $(\lambda_k, \boldsymbol{v}_k, \boldsymbol{u}_k)$ are singular elements with $\lambda_k > 0$, then $(-\lambda_k, \boldsymbol{v}_k, -\boldsymbol{u}_k)$ are singular elements, too.

So, if we organize the indices k so that for $k = 1, \ldots, n$ we have $\lambda_k > 0$, usually taken in decreasing order, we find from the first relation in (A.40) that

$$A \sum_{k=1}^{n} \boldsymbol{u}_k \boldsymbol{u}_k^T = \sum_{k=1}^{n} \lambda_k \boldsymbol{v}_k \boldsymbol{u}_k^T. \tag{A.43}$$

Since $\{\boldsymbol{u}_k\}$ are n orthonormal vectors,

$$\sum_{k=1}^{n} \boldsymbol{u}_k \boldsymbol{u}_k^T = I$$

and (A.43) becomes

$$A = \sum_{k=1}^{n} \lambda_k \boldsymbol{v}_k \boldsymbol{u}_k^T. \tag{A.44}$$

which is the spectral representation of a general matrix $A; \mathbb{R}^n \to \mathbb{R}^m$.

A.2 Calculus

We assume the reader to be acquainted with the basic notions of calculus: limits, functions of one or more variables, differential calculus with its rules, classical Riemann integral calculus on bounded or unbounded sets.

The same operations on vector fields are also assumed to be known, together with multiple integrals, surface and line integrals. However the king theorem of vector calculus, i.e., the Gauss Theorem, is so important to our subject, that an independent proof will be provided in the text. We only remind here that the following definition will be useful in the sequel.

Definition A.14 Let $\{a_n\}$ be a bounded sequence of numbers; then the two classes of upper and lower bounds, $\{M\}$ and $\{m\}$, of $\{a_n\}$ cannot be empty. We define the upper limit of a_n as

$$\bar{a} = \overline{\lim}\, a_n = \min\{M; M \geq a_n, \forall n\}, \tag{A.45}$$

and the lower limit of $\{a_n\}$ as

$$\underline{a} = \underline{\lim}\, a_n = \max\{m; m \leq a_n, \forall n\}. \tag{A.46}$$

We notice that there are two subsequences $\{n_k\}$ and $\{n_j\}$ of $\{n\}$ such that

$$\bar{a} = \overline{\lim}\, a_{n_k}, \underline{a} = \lim\, a_{n_j}.$$

Furthermore when $\bar{a} = \underline{a}$ this is also the value of the unique limit of $\{a_n\}$ for $n \to \infty$.

A.3 Lebesgue Measure

There are many possible approaches to the definition of Lebesgue measure, measurable sets, integral, etc. Here we will follow the most classical, originating directly from Lebesgue's work. A classical textbook on the subject is [62].

To keep things easier we can think that all the sets described hereafter are subsets of a bounded set B_0 (e.g., a large open cube) in $\mathbb{R}^1$, $\mathbb{R}^2$ or $\mathbb{R}^3$, although the concepts hold generally in $\mathbb{R}^n$.

So we will call *open intervals* any of the following sets

$$\text{in } \mathbb{R}^1, I = \{a < x < b\};$$

$$\text{in } \mathbb{R}^2, I = \{a < x < b; c < y < d\};$$

$$\text{in } \mathbb{R}^3, I = \{a < x < b; c < y < d\ e < z < f\}.$$

Note that, so defined, the intervals are open, but it would be the same if we had chosen to take them closed since their elementary measures,

$$m(I) = b - a, \quad m(I) = (b - a)(c - d), \quad m(I) = (b - a)(c - d)(f - e),$$

respectively, do not change if we add the extremes to I. Such relations are the building blocks of Lebesgue measure theory.

Definition A.15 We call a multinterval J, associated with a countable family of intervals $\{I_k, k = 1, 2, \ldots\}$, $I_k \subset B_0$, the set

$$J = \bigcup_k J_k.$$

Since I_k are open, J is open too and $J \subset B_0$.

Definition A.16 Given an arbitrary subset $D \subset B_0$ separated from the boundary of B_0, we say that $J = \bigcup_k J_k$ covers D if

$$D \subset J \subset B_0; \tag{A.47}$$

we let J_D denote the class of all J satisfying (A.47) and we put

$$L_e(D) = \inf_{j \in J_D} \sum_k m(J_k). \tag{A.48}$$

$L_e(D)$ is called the *exterior Lebesgue measure* of D.

Notice that, since $B_0 \in J_D$ for every subset D, and B_0 is bounded, $J_D$is never empty and $L_e(D)$ exists and is bounded for any $D \subset B_0$.

We notice as well that $L_e(D)$, as an exterior measure, has the characteristics described in the following proposition.

Proposition A.7 *$L_e(D)$ satisfies for every $D \subset B_0$ the following properties*

a)

$$L_e(D) \geq 0 \qquad (L_e(\emptyset) = 0); \tag{A.49}$$

b)

$$\begin{aligned} &\forall \{D_k\},\ D_k \subset B_0 \quad k = 1, 2, \ldots \\ &L_e(\bigcup_k D_k) \leq \sum_k L_e(D_k), \end{aligned} \tag{A.50}$$

typical of an exterior measure.

The fact that, given a $D \subset B_0$, there is always an open set $A \supset D$ such that, for a fixed $\varepsilon > 0$

$$|L_e(A)| < |L_e(D)| + \varepsilon \tag{A.51}$$

is clear, in fact it is enough to take $A = J$, suitably adding the overlapping boundaries of I_k, which by the way have zero measure, in order to keep J as an open set. However this does not mean that $L_e(A \backslash D)$ can be made as small as we like. This suggest the definition of a more restricted class of sets D.

Definition A.17 Let us call A_D the class of open sets $A \subset B_0$ that contain D (i.e., $A \in A_D \Longrightarrow A \supset D$); then we say that D is Lebesgue measurable if

$$\inf_{A \subset A_D} L_e(A \setminus D) = 0; \tag{A.52}$$

in this case we define the Lebesgue measure of D as

$$L(D) = L_e(D). \tag{A.53}$$

Proposition A.8 *The family $\mathscr{M}$ of all measurable sets is a σ-algebra, namely*

a) $\emptyset, B_0 \in \mathscr{M}$
b) $\forall \{D_k\},\ k = 1, 2, \ldots \qquad D_k \in \mathscr{M} \Longrightarrow \bigcup_k D_k \in \mathscr{M}$..
In particular, if $\{D_k\}$ are disjoint sets, $D_j \cap D_k = \emptyset,\ j \neq k$, we have

$$L(\bigcup_k D_k) = \sum_k L(D_k)(< L(B_0)).$$

c) $D \in \mathscr{M} \Longrightarrow D^c \in \mathscr{M}$ and $L(D^c) = L(B_0) - l_e(D)$;
d) $\forall E, L_e(E) = 0 \Longrightarrow E \in \mathscr{M}$ and $L(E) = 0$;
e) $\forall D \in \mathscr{M} \Longleftrightarrow \exists C$ *(closed)* $C \subset D$ *such that* $\forall \varepsilon$ *fixed*

$$L_e(E \setminus C) < \varepsilon; \tag{A.54}$$

furthermore, all open sets and all closed sets are measurable and so are countable unions and intersections of the above, mixed in any order, namely the so-called Borel sets.

In particular, by exploiting the property b) it is not difficult to see that, if we have an increasing family $\{D_k\}$, then $\lim_{k\to\infty} D_k = \bigcup_k D_k = D$ and

$$\lim_{k\to\infty} L(D_k) = L_e(D). \tag{A.55}$$

The same holds for a decreasing family, namely $D = \bigcap_k D_k$ implies

$$L(D) = \lim L(D_k). \tag{A.56}$$

We note that the above concepts can be extended to unbounded sets, by letting the sides of B_0 to tend to infinity.

Definition A.18 A real function $f(\boldsymbol{x})$; $\boldsymbol{x} \in B_0$, is said to be *Lebesgue measurable*, or simply *measurable*, if the sets

$$S_f(a) = \{\boldsymbol{x};\ f(\boldsymbol{x}) < a\} \in \mathscr{M} \tag{A.57}$$

are measurable for all $a \in \mathbb{R}$. Using the fact that $S_f(a)$ is increasing with a and (A.55), (A.56), we can easily prove that the following sets are also measurable:

$$\{f(\boldsymbol{x}) \le a\},\ \{f(\boldsymbol{x}) > b\},\ \{f(\boldsymbol{x}) \ge b\},\ \{a \le f(\boldsymbol{x}) < b\}. \tag{A.58}$$

A.4 Lebesgue Integral

We define the *Lebesgue integral* in three steps:

a) Let $f(\boldsymbol{x}) \in \mathscr{M}(\boldsymbol{x} \in B_0)$ be such that

$$0 \le f(\boldsymbol{x}) \le M < +\infty. \tag{A.59}$$

Let us partition the interval

$$\begin{aligned}\overline{OM} &= [m_0 m_1) \cup [m_1 m_2) \cup \cdots \cup [m_{n-1} m_n] \\ &= \bigcup_k \Delta_k\end{aligned}$$

and put

$$\delta = \max(m_k - m_{k+1}). \tag{A.60}$$

Moreover, let us define

$$B_k = \{\boldsymbol{x}\,;\, f(\boldsymbol{x}) \in \Delta_k\} \in \mathscr{M}; \tag{A.61}$$

note that by the definition of the intervals Δ_k, the sets B_k are measurable and disjoint and

$$\sum_k L(B_k) = L(B_0) < +\infty. \tag{A.62}$$

Now take arbitrary points $\xi_k \in B_k$ and define

$$\lim_{\delta \to 0} \sum_{k=1}^{n} \xi_k L(B_k) = \int_{B_0} f(\boldsymbol{x}) d_3 x, \tag{A.63}$$

where $d_3 x$ is the volume element in $\mathbb{R}^3$ or $\mathbb{R}^1, \mathbb{R}^2$, depending on where B_0 is. One can prove that (A.63) is well defined, i.e., the limit exists for every measurable f satisfying (A.59) and is independent of the choice of the partition $\{\Delta_k\}$ and of the points $\xi_k \in \Delta_k$,

b) Let $f(\boldsymbol{x})$ be of any sign but bounded, and put

$$f(\boldsymbol{x}) = f_+(\boldsymbol{x}) - f_-(\boldsymbol{x}), \tag{A.64}$$

with

$$f_+(\boldsymbol{x}) = \begin{cases} f(\boldsymbol{x}), & \text{if } f(\boldsymbol{x}) \geq 0, \\ 0, & \text{if } f(\boldsymbol{x}) < 0, \end{cases}$$

$$f_-(\boldsymbol{x}) = \begin{cases} -f(\boldsymbol{x}), & \text{if } f(\boldsymbol{x}) < 0, \\ 0, & \text{if } f(\boldsymbol{x}) \geq 0, \end{cases}$$

so that

$$0 \leq |f(\boldsymbol{x})| = f_+(\boldsymbol{x}) + f_-(\boldsymbol{x}) \leq M. \tag{A.65}$$

Since, then, f_+ and f_- are both non-negative and bounded, we can define their integrals by point a), and subsequently put

$$\int_{B_0} f(\boldsymbol{x}) d_3x = \int_{B_0} f_+(\boldsymbol{x}) d_3x - \int_{B_0} f_-(\boldsymbol{x}) d_3x. \tag{A.66}$$

We note that the integrability of $f(\boldsymbol{x})$ is equivalent to the integrability of $|f(\boldsymbol{x})|$.

c) Assume now that $f(\boldsymbol{x})$ is not bounded and define the truncated functions

$$f_{\pm M} = \begin{cases} f_+(\boldsymbol{x}), & \text{if } f_\pm(\boldsymbol{x}) \leq M, \\ M, & \text{if } f_\pm(\boldsymbol{x}) \geq M. \end{cases}$$

Then we say that $f(\boldsymbol{x})$ is Lebesgue integrable or summable, if there exist the limits

$$\lim_{M\to\infty} \int_{B_0} f_{\pm M}(\boldsymbol{x}) d_3x = \int_{B_0} f_\pm(\boldsymbol{x}) d_3x < +\infty, \tag{A.67}$$

and then we put again

$$\int_{B_0} f(\boldsymbol{x}) d_3x = \int_{B_0} f_+(\boldsymbol{x}) d_3x - \int_{B_0} f_-(\boldsymbol{x}) d_3x. \tag{A.68}$$

We note once more that $f(\boldsymbol{x})$ is integrable on B_0 if $|f(\boldsymbol{x})|$ is integrable over B_0.

Finally, letting the sides of B_0 tend to infinity if the integrals admit limits too, one arrives at the definition of the Lebesgue integral on $\mathbb{R}^1$, $\mathbb{R}^2$, $\mathbb{R}^3$, respectively. The class of functions integrable on B_0, which coincides with the class of absolutely integrable functions on B_0, will be denoted by $L^1(B_0)$, or simply L^1.

The following propositions constitute a collection of important properties of the Lebesgue integral that will be used in the text; all functions $f(\boldsymbol{x})$ in the following are supposed to be measurable. A property holding for all $\boldsymbol{x} \in B_0$, except for a set E of zero measure, is said to hold almost everywhere (a.e.).

Proposition A.9 (Beppo Levi) *Let $\{f_n(x)\}$ be a non-decreasing sequence in L^1 and assume that, for some C*

$$\int_{B_0} f_n(\boldsymbol{x}) d_3x \leq C. \tag{A.69}$$

Then there is a function $f(\boldsymbol{x}) \in L^1$ such that

$$\lim_{n\to\infty} f_n(\boldsymbol{x}) = f(\boldsymbol{x}) \quad a.e.\ in\ B_0 \tag{A.70}$$

and

$$\int_{B_0} f(\boldsymbol{x}) d_3x \leq C. \tag{A.71}$$

Proposition A.10 (Lebesgue) *Let $\{f_n(\boldsymbol{x})\}$ be a sequence in L^1 and assume that*

$$|f_n(\boldsymbol{x})| \leq g(\boldsymbol{x}) \in L^1; \tag{A.72}$$

assume further that

$$\lim_{n\to\infty} f_n(\boldsymbol{x}) = f(\boldsymbol{x})\, a.e. \tag{A.73}$$

Then $f(\boldsymbol{x}) \in L^1$ and

$$\lim_{n\to\infty} \int_{B_0} f_n(x) d_3x = \int_{B_0} f(\boldsymbol{x}) d_3x, \tag{A.74}$$

$$\int_{B_0} |f(\boldsymbol{x})|\, d_3x \leq \int_{B_0} g(\boldsymbol{x}) d_3x. \tag{A.75}$$

Proposition A.11 (Fatou) *Let $\{f_n(\boldsymbol{x})\}$ be a sequence of positive L^1 functions such that*

$$0 \leq \int_{B_0} f_n(\boldsymbol{x}) d_3x \leq C. \tag{A.76}$$

Assume that

$$\lim_{n\to\infty} f_n(\boldsymbol{x}) = f(\boldsymbol{x}) \quad a.e. \tag{A.77}$$

Then $f(\boldsymbol{x}) \in L^1$ *and*

$$\lim_{n\to\infty} \int_{B_0} f_n(x) d_3x = \int_{B_0.} f(\boldsymbol{x}) d_3x, \tag{A.78}$$

Proposition A.12 (Fubini) *Let*

$$B_0 = B_1 \times B_2,$$

i.e., $(\boldsymbol{x}, \boldsymbol{y}) \in B_0 \Longleftrightarrow \boldsymbol{x} \in B_1, \boldsymbol{y} \in B_2$. *Let* $f(\boldsymbol{x}, \boldsymbol{y})$ *be in* $L^1(B_0)$. *Then*

a) *For almost all fixed* $x \in B_1$ *the function* $\int_{B_2} f(\boldsymbol{x}, \boldsymbol{y}) d_3y$ *exists and is in* $L_1(B_1)$ *and viceversa, exchanging* $\boldsymbol{x}$ *and* $\boldsymbol{y}$;
b) *The integral over* B_0 *is equal to the two iterated integrals:*

$$\begin{aligned} \int_{B_0} f(\boldsymbol{x}, \boldsymbol{y}) dx dy &= \int_{B_1} \int_{B_2} f(\boldsymbol{x}, \boldsymbol{y}) d_3x d_3y = \\ &= \int_{B_1} \left[\int_{B_2} f(\boldsymbol{x}, \boldsymbol{y}) d_3y \right] d_3y = \\ &= \int_{B_2} \left[\int_{B_1} f(\boldsymbol{x}, \boldsymbol{y}) d_3x \right] d_3y. \end{aligned} \tag{A.79}$$

c) *If in addition* $f(\boldsymbol{x}, \boldsymbol{y}) \geq 0$, *then the existence of any integral in* (A.79) *implies the existence of all the others.*

Proposition A.13 *Let* $\{f_n(\boldsymbol{x})\}$ *be a sequence in* L^1, *with* $f_n(\boldsymbol{x}) \geq 0$, *and assume that*

$$\lim_{n\to\infty} \int_{B_0} f_n(\boldsymbol{x}) d_3x = 0. \tag{A.80}$$

Then

a) $f_n(\boldsymbol{x}) \to 0$ *in measure, i.e., for every* $\varepsilon > 0$,

$$\lim_{n\to\infty} L\{\boldsymbol{x};\ f_n(\boldsymbol{x}) > \varepsilon\} = 0. \tag{A.81}$$

b) *There is a subsequence* $\{n_k\}$ *of* $\{n\}$ *such that*

$$\lim_{k\to\infty} f_{n_k}(\boldsymbol{x}) = 0 \quad a.e.. \tag{A.82}$$

Remark A.5 Notice that as a consequence of Proposition A.13 we can claim that if $f(\boldsymbol{x}) \in L^1$, $f(\boldsymbol{x}) \geq 0$ and $\int_{B_0} f(\boldsymbol{x}) d\boldsymbol{x} = 0$, then $f(\boldsymbol{x}) = 0$ a.e.; in fact it is enough to take $f_n(\boldsymbol{x}) \equiv f(\boldsymbol{x})$, $\forall n$, in the above proposition.

Remark A.6 When we talk about L^1, it is convenient to consider as its elements, not individual integrable functions $f(\boldsymbol{x})$, but equivalence classes of functions, defined as

$$f \in L^1, \mathcal{G}_f \equiv \{g(\boldsymbol{x}) \in L^1, \int_{B_0} |f(\boldsymbol{x}) - g(\boldsymbol{x})| \, d_3x = 0\}; \tag{A.83}$$

namely, also following Remark A.4, every two functions that differ only on a set of measure zero, are considered as equivalent in L^1.

A.5 Infinite-Dimensional Linear Spaces

The Lebesgue measure and integral are used to build some functional spaces that we will present as generalizations of finite-dimensional linear spaces.

Definition A.19 A linear space X is said to be infinite dimensional if, for any N there is a collection $\{x_1, x_2, \ldots, x_N\}$ of linear independent vectors in X.

Definition A.20 The real linear space X is said to be normed if $\forall x \in X$ we can it is endowed with a norm, i.e., a function $\|\cdot\| : X \to \mathbb{R}$ such that

a) $\|x\| \geq 0$ for all $x \in X$, and $\|x\| = 0 \iff x = 0$;
b) $\|\lambda x\| = |\lambda| \, \|x\|$ for all $x \in X$ and all $\lambda \in \mathbb{R}$;
c) $\|x + y\| \leq \|x\| + \|y\|$ for all $x, y \in X$.

Definition A.21 A sequence $\{x_n\}$ in the normed space X is a Cauchy sequence if

$$\lim_{n,m \to \infty} \|x_n - x_m\| = 0. \tag{A.84}$$

Definition A.22 A sequence $\{x_n\} \subset X$ is said converge to the element $x \in X$ if

$$\lim_{n \to \infty} \|x - x_n\| = 0; \tag{A.85}$$

note that, by Definition A.20 c), every convergent sequence is also Cauchy.

Definition A.23 A normed space X is said to be complete, or a Banach space (B-space), if every Cauchy sequence in X converges to an element of X.

Proposition A.14 *The following normed spaces are B-spaces:*

a) the space ℓ^1 *of real sequences* $\{a_n, n = 1, 2, \ldots\} = \boldsymbol{a}$*, with the norm*

$$\|\boldsymbol{a}\|_{\ell^1} = \sum_{n=1}^{+\infty} |a_n|; \tag{A.86}$$

b) *the space* $L^1(B_0)$ *of measurable and summable functions on* B_0, *with the norm*

$$\|f\|_{L^1} = \int_{B_0} |f(\boldsymbol{x})|\, d\boldsymbol{x}\,; \tag{A.87}$$

c) *the space* $C(B_0)$ *of continuous functions on a closed set* $B_0 = \bar{B}_0$, *with the norm*

$$\|f(\boldsymbol{x})\|_C = \sup_{x \in B_0} |f(\boldsymbol{x})|\,; \tag{A.88}$$

if B_0 *is also bounded,* sup *is replaced by max.*

Definition A.24 The linear space X is said to be endowed with a real scalar product if $\forall x, y \in X$ we can define a real number, $\langle x, y\rangle$, such that

a) $\langle x, x, \rangle \geq 0$ and $\langle x, x\rangle = 0 \Longleftrightarrow x = 0$;
b) $\langle x, y\rangle = \langle y, x\rangle$;
c) $\langle x, \lambda y_1 + \mu y_2\rangle = \lambda \langle x, y_1\rangle + \mu \langle x, y_2\rangle$.

We observe that, given the symmetry b), $\langle x, y\rangle$ has to be linear in x, too.

We will need only occasionally spaces of complex elements; in this case b) has to be replaced by

$$\text{b}') \qquad \langle x, y\rangle^* = \langle y, x\rangle\,.$$

Then, if c) continues to hold, the scalar product is conjugate linear in x, namely

$$\text{c}') \qquad \langle \lambda x_1 + \mu x_2, y\rangle = \lambda^* \langle x, y\rangle + \mu^* \langle x_2, y\rangle\,.$$

Proposition A.15 *The following form of Cauchy–Schwarz inequality holds*

$$|\langle x, y\rangle|^2 \leq \langle x, x\rangle \langle y, y\rangle\,; \tag{A.89}$$

(A.89) *can be proved exactly as in Proposition* A.1.

Remark A.7 Inequality (A.89) implies immediately that a scalar product in X induces a norm by putting

$$\|x\| = \langle \boldsymbol{x}, \boldsymbol{x}\rangle^{1/2}; \tag{A.90}$$

indeed $\|x\|$ so defined enjoys the properties a), b), c) of Definition A.20.

Definition A.25 A linear space endowed, with scalar product $\langle , \rangle$, is called a Hilbert space (H-space), if it is complete with respect to the norm (A.90).

Remark A.8 Any closed linear subspace S of an H-space, is also an H-space.

Definition A.26 Two elements x, y of an H-space X are said to be orthogonal if

$$\langle x, y\rangle = 0\,; \tag{A.91}$$

the element x is said to be orthogonal to a subspace S if

$$\langle x, y\rangle = 0 \quad \forall y \in S\,; \tag{A.92}$$

given S (closed or not), the elements x that are orthogonal to S form a closed subspace of X, denoted $S^{\perp}$.

When S is closed, the subspace $S^{\perp}$ is called the *orthogonal complement* of S in X.

Proposition A.16 *Given a closed subspace S of the H-space X and an element $x \in X$, there is one and only one element $\hat{x} \in S$, such that [$\|x - \hat{x}\|$ is minimal. It turns out that $x - \hat{x} \in S^{\perp}$, so $\hat{x}$ is called the orthogonal projection of x on S. It is obvious that when $x \in S$ we have $\hat{x} = x$.*

Definition A.27 A Hilbert space X is said to be separable if there is a sequence $\{\xi_n\} \subset X$ which is everywhere dense, i.e.,

$$\forall x \in X, \exists \{n_k\} \subset \{n\} \text{ such that } \left\|\xi_{n_k} - x\right\| \to 0\,. \tag{A.93}$$

All H-spaces in this book are separable.

Proposition A.17 *The following spaces are separable* H*-spaces:*

a) the space ℓ^2 of real sequences $\boldsymbol{a} \equiv \{a_1, a_2, \ldots\}$ with the scalar product

$$\langle \boldsymbol{a}, \boldsymbol{b}\rangle = \sum_{n=1}^{+\infty} a_n b_n\,; \tag{A.94}$$

b) the space $L^2(B_0)$ of measurable functions on B_0 such that

$$\int_{B_0} f^2(\boldsymbol{x})d\boldsymbol{x} < +\infty\,; \tag{A.95}$$

the scalar product of L^2 is defined as

$$\langle f, g\rangle_{L^2} = \int_{B_0} f(\boldsymbol{x})g(\boldsymbol{x})d\boldsymbol{x}\,; \tag{A.96}$$

this is modified by taking f^* in (A.96) *when treating complex functions;*

c) the Sobolev space $H^{1,2}$ of functions such that

$$\int_{B_0} (f^2(\boldsymbol{x}) + |\nabla f(\boldsymbol{x})|^2) d\boldsymbol{x} < +\infty\,, \tag{A.97}$$

with the scalar product

$$\langle f, g \rangle_{H^{1,2}} = \int_{B_0} (f(x) g(\boldsymbol{x}) + \nabla f(\boldsymbol{x}) \cdot \nabla g(\boldsymbol{x})) d\boldsymbol{x}\,. \tag{A.98}$$

Obviously, $H^{1,2}(B_0) \subset L^2(B_0)$; moreover, the embedding is dense.

Remark A.9 If B_0 has finite Lebesgue measure, denoted here as $|B_0|$, then $L^1 \subset L^2$; in fact, by the Caudy–Schwarz inequality,

$$\left(\int_{B_0} |f(\boldsymbol{x})|\, d\boldsymbol{x} \right)^2 \leq |B_0| \int_{B_0} f(x)^2 d\boldsymbol{x}. \tag{A.99}$$

Proposition A.18 *Every separable H-space admits an orthonormal basis $\{e_n\}$, $\langle e_n, e_m \rangle = \delta_{nm}$, that can be obtained for instance by applying the Gram–Schmidt algorithm (see Proposition A.2) indefinitely.*

Proposition A.19 *Every element x of the H-space X can be expressed with respect to an orthonormal basis $\{e_n\}$ as the convergent (Fourier) series*

$$x = \sum_{n=1}^{+\infty} a_n e_n, \tag{A.100}$$

where

$$a_n = \langle x, e_n \rangle\,. \tag{A.101}$$

Moreover, if $y = \sum_{n=1}^{+\infty} b_n e_n$, then

$$\langle x, y \rangle = \sum_{n=1}^{+\infty} a_n b_n; \tag{A.102}$$

in particular

$$\|x\|^2 = \sum_{n=1}^{+\infty} a_n^2. \tag{A.103}$$

This implies that every separable H-space is isometrically isomorphic to ℓ^2.

A.6 Linear Operators

In this section X, Y denote two H-spaces.

Definition A.28 A linear operator $A : X \to Y$ is a map

$$y = Ax, \quad x \in X, \quad y \in Y \tag{A.104}$$

that is linear and homogeneous, namely

$$\forall \lambda, \mu \in \mathbb{R}, \quad \forall x_1, x_2 \in X, \quad A(\lambda x_1 + \mu x_2) = \lambda A x_1 + \mu A x_2. \tag{A.105}$$

The set of elements of X on which A is defined is called the *domain* of A, denoted $\mathscr{D}(A)$; the set of the elements y that correspond to some $x \in \mathscr{D}(A)$ is called the *range* or *image* of A, denoted $\mathscr{R}(A)$.

The linear subspace of X consisting of all elements x, $Ax = 0$, is called the *null space or kernel* of A, denoted $\mathscr{N}(A)$; when $\mathscr{N}(A) = \emptyset$, A is said to be into. If $\mathscr{R}(A) = Y$, A is said to be onto.

Definition A.29 Let A be a linear operator defined on X, with range in Y. Then A is said to be continuous on X if

$$\xi, x \in X, \|\xi - x\|_X \to 0 \Longrightarrow \|A\xi - Ax\|_Y \to 0. \tag{A.106}$$

Moreover, A is said to be bounded if there is a $c > 0$ such that

$$\|Ax\|_Y \leq c\|x\|_X \quad \text{for all} \quad x \in X. \tag{A.107}$$

When A is bounded, we call $\|A\|$ the number

$$\|A\| = \inf c = \sup_{\|x\|_X = 1} \|Ax\|_Y. \tag{A.108}$$

Proposition A.20 *A linear operator $A : X \to Y$ is bounded if and only if it is continuous at some point x_0, and therefore at any point $x \in X$.*

Proposition A.21 *The quantity $\|A\|$ defined by* (A.108) *satisfies the properties* a), b), c) *of Definition* A.20. *The set of all bounded linear operators between two B-spaces, X and Y, $\mathscr{L}(X, Y)$ is itself a B-space.*

Proposition A.22 (Banach) *If X, Y are two* B*-spaces and $A \in \mathscr{L}(X, Y)$, and if in addition A is into and onto Y, then there exists the inverse operator A^{-1} and $A^{-1} \in \mathscr{L}(Y, X)$.*

Definition A.30 Let $C \in \mathscr{L}(X, X)$ and assume that

$$\|C\| < 1; \tag{A.109}$$

then C is called a *contraction*.

Proposition A.23 *Let C be a contraction of X. Then:*

a) the equation

$$(I - C)x = x - Cx = y \tag{A.110}$$

has one and only one solution for every $y \in X$;

b) the solution of equation (A.110) *can be found as the limit of the sequence of successive approximations*

$$x_{n+1} = Cx_n + y \tag{A.111}$$

starting with any $x_1 \in X$;

c) the operator $(1 - C)^{-1}$, which exists and is bounded by Proposition A.22, *is given by the convergent (Neumann) series*

$$(I - C)^{-1} = \sum_{n=0}^{+\infty} C^n, \tag{A.112}$$

and

$$\left\|(I - C)^{-1}\right\| \leq \frac{1}{1 - \|C\|}. \tag{A.113}$$

Definition A.31 Let X, Y be two H-spaces; the transpose of an operator $A \in \mathscr{L}(X, Y)$ is the operator $A^T \in \mathscr{L}(Y, X)$ such that

$$\langle y, Ax\rangle_Y = \left\langle A^T y, x\right\rangle_X \forall x \in X, y \in Y; \tag{A.114}$$

if $A \in \mathscr{L}(X, X)$ and $A = A^T$, the operator is called *symmetric*.

The operator $A^T \in \mathscr{L}(Y, X)$ and its norm is equal to that of A, $\left\|A^T\right\| = \|A\|$.

Definition A.32 Let X be an H-space and $A \in \mathscr{L}(X, X)$, A is said to be positive if

$$\langle x, Ax\rangle_X > 0, \ \forall x \neq 0. \tag{A.115}$$

In this case A is into.

Proposition A.24 *If A is symmetric and positive, then there exists $A^{-1} \in \mathscr{L}(X, X)$ if and only if*

$$\langle x, Ax\rangle_X \geq a \langle x, x\rangle_X, \quad a > 0, \tag{A.116}$$

and, in this case

$$\left\| A^{-1} \right\| \leq \frac{1}{a}. \tag{A.117}$$

Definition A.33 Let K be a subset of a H-space X; we say that K is pre-compact if any sequence $\{x_n\} \subset K$ admits a subsequence $\{x_{n_k}\}$ which is Cauchy (and then convergent in X). If K is also closed, then it is said to be compact, and then $x_{n_k} \to x \in K$.

It is clear that all compact sets K are bounded.

Remark A.10 Any bounded closed set in $\mathbb{R}^n$ is compact by Bolzano–Weierstrass Theorem.

Definition A.34 An operator $C : X \to Y$ is said to be compact if it transforms any bounded set B in X into a pre-compact set K in Y. It is clear that any compact operator C belongs to $\mathscr{L}(X, Y)$.

Proposition A.25 *Let $C \in \mathscr{L}(X, X)$ be a compact operator and assume that $\mathscr{N}(I + C) = \emptyset$. Then $(I + C)^{-1} \in \mathscr{L}(X, X)$. Said in another way, if the equation*

$$(I + C)x = 0 \tag{A.118}$$

has only the solution $x = 0$ (uniqueness), then the equation

$$(I + C)x = y \tag{A.119}$$

has a solution, for any $y \in X$ (existence); this is part of a more general theorem called the Fredholm alternative.

Proposition A.26 *An integral operator C, mapping $f \in L^2(B_0)$ into $g \in L^2(B_1)$, $g = Cf$, has the form*

$$\boldsymbol{x} \in B_1, \quad g(\boldsymbol{x}) = \int_{B_0} c(\boldsymbol{x}, \boldsymbol{y}) f(\boldsymbol{y}) d\boldsymbol{y} . \tag{A.120}$$

If $c(\boldsymbol{x}, \boldsymbol{y}) \in L^2(B) \times L^2(B)$, i.e. $L^2(B_0 \times B_1)$, or

$$\int_{B_1} \int_{B_0} c^2(\boldsymbol{x}, \boldsymbol{y}) d\boldsymbol{x} d\boldsymbol{y} < +\infty, \tag{A.121}$$

then $C \in \mathscr{L}[L^2(B_0), L^2(B_1)]$ and is even compact.

Proposition A.27 *Let $C \in \mathscr{L}(X, X)$ be symmetric and compact; then there is a sequence of eigenvalue and eigenvector pairs, (λ_n, ξ_n), i.e.,*

a)

$$C\xi_n = \lambda_n \xi_n ; \tag{A.122}$$

b) Moreover, $\lambda_n \in \mathbb{R}$ and

$$|\lambda_n| \leq \|C\|, \qquad \forall n, \tag{A.123}$$

and

$$\lambda_n \to 0 \quad \text{when} \quad n \to \infty; \tag{A.124}$$

c) *$\{\xi_n\}$ is a complete orthonormal sequence in X*
d) *one has*

$$C = \sum_{n=1}^{+\infty} \lambda_n \xi_n \otimes \xi_n \tag{A.125}$$

where $\xi_n \otimes \xi_n$ denotes the projection operator on X defined by

$$(\xi_n \otimes \xi_n)(x) = \xi_n \langle \xi_n, x \rangle_X; \tag{A.126}$$

e) *if $\lambda_n \neq 0$ for all n, then C is invertible and*

$$C^{-1} = \sum_{n=1}^{+\infty} \lambda_n^{-1} \xi_n \otimes \xi_n; \tag{A.127}$$

clearly, C^{-1} is an unbounded operator in X.

The following proposition is the infinite-dimensional generalization of Proposition A.5.

Proposition A.28 *Let X, Y be H-spaces and $A \in \mathscr{L}(X, Y)$ be compact and assume that $\mathscr{N}(A) = \emptyset$. Then there is a sequence of singular elements $\{\lambda_n, \xi_n, \eta_n\} \in (R, X, Y)$, i.e.,*

a)

$$\begin{cases} A\xi_n = \lambda_n \eta_n \\ A^T \eta_n = \lambda_n \xi_n; \end{cases} \tag{A.128}$$

b)

$$\lambda_n > 0 \quad \text{and} \quad \lambda_n \to 0, n \to \infty;$$

c) *the sequence $\{\xi_n\}$ is orthonormal and complete in X,*
the sequence $\{\eta_n\}$ is orthonormal and complete in Y;

d)

$$A = \sum_{n=1}^{+\infty} \lambda_n \eta_n \otimes \xi_n \tag{A.129}$$

and

$$A^T = \sum_{n=1}^{+\infty} \lambda_n \xi_n \otimes \eta_n \,. \tag{A.130}$$

This proposition is the infinite-dimensional generalization of Remark A.4.

Definition A.35 Let X be an H-space; a sequence $\{x_n\} \leq X$ is said to be weakly convergent in X to the element x, and one writes $x_n \underset{X}{\overset{*}{\rightarrow}} x$ if

$$\langle x_n, \xi \rangle_X \longrightarrow \langle x, \xi \rangle_X, \quad \forall \xi \in X. \tag{A.131}$$

Proposition A.29 *The following holds:*

a) if $x_n \underset{n}{\overset{*}{\rightarrow}} x$ *then* $\{x_n\}$ *is bounded;*
b) if $\{x_n\}$ *is bounded, then there is a subsequence* $\{x_{n_k}\}$ *that is weakly convergent;*
c) if $x_n \overset{*}{\rightarrow} x$*, then*

$$\|x\|_X \leq \underline{\lim}_{n\to\infty} \|x_n\|_X;$$

d) if $x_n \overset{*}{\rightarrow} x$ *and at the same time* $\|x_n\|_X \rightarrow \|x\|_X$*, then* $x_n \rightarrow x$ *in the norm of* X.

Definition A.36 Let X be an H-space and $A \in \mathscr{L}(X, \mathbb{R})$, then A is called a bounded functional on X. Denoting $Ax = \ell(x)$, we put

$$\|\ell\|_* = \sup_{\|x\|_X = 1} |\ell(x)| \,. \tag{A.132}$$

The space of bounded functionals ℓ on X is called the *dual* of X, denoted X^*; one can prove that X^* is also an H-space, with the scalar product

$$\langle \ell, h \rangle_* = \frac{1}{4} \left(\|\ell + h\|_*^2 - \|\ell - h\|_*^2 \right), \tag{A.133}$$

where $\|\cdot\|_*$ is given by (A.132).

Proposition A.30 (Fisher–Riesz) *The space* X^* *can be represented by the space* X *itself, in the sense that for every* $\ell \in X^*$ *there is an* $x_\ell \in X$ *such that*

$$\forall x \in X \quad \ell(x) = \langle x_\ell, x \rangle_X; \tag{A.134}$$

furthermore, the correspondence $\ell \Longleftrightarrow x_\ell$ *is bijective and isometric, i.e.*

$$\|\ell\|_* = \|x_\ell\|_X, \tag{A.135}$$

implying that for every $\ell, h \in X^*$,

$$\langle \ell, h \rangle_* = \langle x_\ell, x_h \rangle_X, \tag{A.136}$$

A.7 Stochastics and Statistics

As anticipated, a few concepts of Probability Theory and Statistics are necessary for a plain reading of the text. Only essential definitions will be recalled here, without dwelling upon the very large subject of applications and examples.

Definition A.37 A probability space is a triple $(\Omega, \mathscr{A}, P)$, where

a) Ω is the set of the outcomes of a random event, also called *sample space*;
b) $\mathscr{A}$ is a σ-algebra of subsets of Ω, namely a family satisfying the following properties

   b1) $\Omega, \emptyset \in \mathscr{A}$,
   b2) $A \in \mathscr{A} \Longleftrightarrow A^c \in \mathscr{A}$,
   b3) for any sequence $\{A_n\} \le \mathscr{A}$,

$$\bigcup_n A_n \in \mathscr{A}.$$

   Note the similarity of the above with the properties that characterize Lebesgue measurable sets (Proposition A.8).
c) P is a measure, defined for all sets in $\mathscr{A}$, enjoying the following properties

   c1)

$$P(A) \ge 0 \quad \forall A \in \mathscr{A},$$
$$P(\Omega) = 1, \quad P(\emptyset) = 0;$$

   c2)

$$P(A^c) = 1 - P(A);$$

   c3) if $\{A_n\} \subset \mathscr{A}$ are disjoint, namely $A_i \cap A_j = \emptyset$, $i \ne j$, then

$$P\left(\bigcup_{n=1}^{+\infty} A_n\right) = \sum_{n=1}^{+\infty} P(A_n).$$

For technical reasons we will also assume that $\mathscr{A}$ is Lebesgue-closed, namely, that every set $E \subset \Omega$ of probability zero, i.e., such that

$$\forall \varepsilon > 0, \quad \exists A \in \mathscr{A} \Longrightarrow A \supset E, \quad P(A) < \varepsilon,$$

is also included in $\mathscr{A}$, so that any B that differs from $A \in \mathscr{A}$ by a set of probability zero, is also included in $\mathscr{A}$.

Definition A.38 A random variable (RV) on $(\Omega, \mathscr{A}, P)$ is a function $X(\omega)$, with values in $\mathbb{R}$, which is measurable with respect to P, i.e.,

$$\{\omega\,;\, X(\omega) \leq a\} \in \mathscr{A} \qquad \forall a \in \mathbb{R}$$

so that the following definition makes sense.

Definition A.39 If $X(\omega)$ is an RV $X(\omega)$, the distribution function of X or probability law is defined by the following formula

$$F_X(a) = P\{X(\omega) \leq a\}, \quad a \in \mathbb{R}. \tag{A.137}$$

Clearly, $F_X(a)$ is a non-decreasing, positive function, hence, by a famous theorem of Lebesgue, it is continuous a.e. on $\mathbb{R}$ and it has also a.e. a derivative $F'_X(u)$. The only possible discontinuities of $F(X)$ are positive jumps [118].

Definition A.40 The RV $X(\omega)$ is called proper if

$$\lim_{a\to-\infty} F_X(a) = 0, \qquad \lim_{a\to-\infty} F_X(a) = 1\,; \tag{A.138}$$

this implies that $X(\omega)$ cannot attain infinite values with a positive probability, i.e.,

$$P(X(\omega) = +\infty) = 0. \tag{A.139}$$

Definition A.41 The RV $X(\omega)$ is called regular if, after setting

$$f_X(x) = \frac{d}{dx} F_X(x) \quad \text{a.e.}, \tag{A.140}$$

we have

$$F_X(a) = \int_{+\infty}^{a} f_X(x)dx, \quad \forall a \in \mathbb{R}\,. \tag{A.141}$$

In this case $f_X(x)$ is called the *probability density function* of X (pdf). Notice that (A.141) implies that, for every subset B of the real line of zero Lebesgue measure we have

$$P(X(\omega) \in B) = 0. \tag{A.142}$$

Definition A.42 The moment of order n of $X(\omega)$ is given by the following integrals, provided they are finite:

$$\begin{aligned} E\{X^n\} &= \int_\Omega X(\omega)^n dP(\omega) = \\ &= \int_{\mathbb{R}} x^n dF_X(x) = \int_{\mathbb{R}} x^n f_X(x)dx\,; \end{aligned} \tag{A.143}$$

the last integral is indeed meaningful only if $X(\omega)$ is regular.

In particular we call mean of X, $\mu(x)$, the moment of order one,

$$\mu(X) = E\{X\} \tag{A.144}$$

and the variance of X, $\sigma^2(X)$, the centralized moment of order two,

$$\sigma^2(X) = E\{[X - E(X)]^2\}. \tag{A.145}$$

Definition A.43 We let $\mathscr{L}^2(\Omega)$ denote the family of RV's $\{X(\omega)\}$ that have a finite moment of order two:

$$X \in \mathscr{L}^2(\Omega) \Longleftrightarrow E\{X(\omega)^2\} < +\infty. \tag{A.146}$$

By virtue of the Cauchy–Schwarz inequality

$$E\{|X(\omega)|\}^2 \le E\{1\}E\{X^2(\omega)\} = E\{X(\omega)^2\}$$

we see that the RVs $X \in \mathscr{L}^2(\Omega)$ are always absolutely integrable and then integrable, i.e., they possess a mean.

The identity

$$E\{X^2(\omega)\} = \mu(X)^2 + \sigma^2(X) \tag{A.147}$$

implies that RVs $X \in \mathscr{L}^2(\Omega)$ have a finite variance.

The family $\mathscr{L}^2(\Omega)$ is an H-space, exactly as in the case of Lebesgue spaces, with scalar product

$$\begin{aligned} \langle X(\omega), Y(\omega)\rangle_{\mathscr{L}^2(\Omega)} &= E\{X(\omega)Y(\omega)\} = \\ &= \int X(\omega)Y(\omega)dP(\omega). \end{aligned} \tag{A.148}$$

The one-dimensional concept of RV is easily generalized to vector RVs taking values in $\mathbb{R}^n$.

Definition A.44 The law of $\boldsymbol{X}(\omega)$ is by definition

$$F_{\boldsymbol{X}}(\boldsymbol{x}) = F_{\boldsymbol{X}}(x_1, x_2, \ldots, x_n) = P(X_1 \leq x_1, x_2 \leq x_2, \ldots, X_n \leq x_n)\,; \tag{A.149}$$

the marginal distributions of the components X_i are given by

$$\begin{aligned} F_{X_i}(x_i) &= F_{\boldsymbol{X}}(+\infty, \ldots, x_i, \ldots, +\infty) \\ &= P(X_i < +\infty, \ldots, X_i \leq x, \ldots, x_n < +\infty) \end{aligned} \tag{A.150}$$

Definition A.45 The variable $\boldsymbol{X}(\omega)$ is called regular if it admits a pdf $f_{\boldsymbol{X}}(\boldsymbol{x})$, i.e.,

$$F_{\boldsymbol{X}}(\boldsymbol{x}) = \int_{-\infty}^{x_1} dt_1 \int_{-\infty}^{x_2} dt_2 \cdots \int_{-\infty}^{x_n} dt_n\, f_{\boldsymbol{X}}\,(t_1, t_2, \ldots, t_n)\,. \tag{A.151}$$

When $\boldsymbol{X}$ is regular, the probability of any (measurable) event $\boldsymbol{X}(\omega) \in A \subset \mathbb{R}^n$, can be written as

$$P(\boldsymbol{X}(\omega) \in A) = \int_A f_{\boldsymbol{X}}(\boldsymbol{x}) d_n x\,. \tag{A.152}$$

Definition A.46 Consider a regular two-dimensional random variable $(X, Y) \in \mathbb{R}^2$.
Recall that, by definition, the conditional probability of the event A given B is

$$P(A|B) = \frac{P(A \cap B)}{P(B)} \quad (P(B) \neq 0), \tag{A.153}$$

and that A and B are said to be stochastically independent if

$$P(A|B) = P(A) \Longleftrightarrow P(A \cap B) = P(A) \cdot P(B). \tag{A.154}$$

With this rule one can define the conditional probability

$$P(Y \leq y | X \in I) = \frac{\int_I \int_{-\infty}^{y} f_{XY}(t_1, t_2) dt_2 dt_1}{\int_I f_X(t_1) dt_1} \tag{A.155}$$

and with a limit for the interval I shrinking to x, if $f_X(x) \neq 0$,

$$P(Y \leq y | X = x) = \frac{\int_{-\infty}^{y} f_{XY}(x, t_2) dt_2}{f_X(x)}. \tag{A.156}$$

This is a distribution in y that admits of a pdf, called the *conditional* pdf of Y given $X = x$,

$$f_{Y|X}(Y|X) = \frac{f_{XY}(x, y)}{f_x(x)}. \tag{A.157}$$

We say that Y is stochastically independent of X if

$$f_{Y|X}(Y|X) = f_Y(y) \Longleftrightarrow f_{XY}(x, y) = f_X(x) f_Y(y). \tag{A.158}$$

The generalization of the above definitions to vector valued RV, $\boldsymbol{X}(\omega) \in \mathbb{R}^n$, is straightforward.

Definition A.47 We say that $\boldsymbol{X}(\omega) \in [\mathscr{L}^2(\Omega)]^n$ if

$$\int_\Omega |\boldsymbol{X}(\omega)|^2 dP(\omega) < +\infty. \tag{A.159}$$

Definition A.48 We define the mean $\mu(\boldsymbol{X})$ and the covariance matrix C_X of the RV $\boldsymbol{X}(\omega)$, when $\boldsymbol{X}(\omega) \in \left[\mathscr{L}^2(\Omega)\right]^n$ by the formulas

$$\mu(\boldsymbol{X}) = E\{\boldsymbol{X}(\omega)\} = \int_\Omega \boldsymbol{X}(\omega) dP(\omega) \tag{A.160}$$

and

$$C_X = E\{[\boldsymbol{X}(\omega) - \mu(\boldsymbol{X})][\boldsymbol{X}(\omega) - \mu(\boldsymbol{X})]^T\}. \tag{A.161}$$

In particular, the diagonal elements of C_X are the variances of the components.

Proposition A.31 *The covariance matrix C_X has the following properties:*

a) C_X is symmetric, it has a positive main diagonal

$$C_X^T = C_X. \tag{A.162}$$

b) C_X is positive definite:

$$\boldsymbol{\lambda}^T C_X \boldsymbol{\lambda} \geq 0, \quad \forall \boldsymbol{\lambda} \in \mathbb{R}^n. \tag{A.163}$$

c) If C_X is strictly positive, i.e.

$$\boldsymbol{\lambda}^T C_X \boldsymbol{\lambda} > 0, \quad \forall \boldsymbol{\lambda} \neq 0. \tag{A.164}$$

then $\boldsymbol{X}$ is a regular variable, namely the probability $P(\boldsymbol{X} \in B) = 0$ whenever the Lebesgue measure of B is zero in $\mathbb{R}^n$.

Notice that, if the components of $\boldsymbol{X}(\omega)$ are stochastically independent, one has $E\{(X_i - \mu_i)(X_k - \mu_k)\} = 0$, i.e. the off-diagonal elements of C_X are zero.

Proposition A.32 *Under an arbitrary linear transformation of the* RV $\boldsymbol{X}$, *namely*

$$\boldsymbol{Y} = A\boldsymbol{X} + \boldsymbol{a}, \tag{A.165}$$

the mean and the covariance of X *are transformed according to the rules*

$$\mu(Y) = A\mu(X) + a \tag{A.166}$$

and

$$C_Y = AC_X A^T. \tag{A.167}$$

Definition A.49 There is one particular important family of RV's in $\mathbb{R}^n$ that is defined exactly by their mean μ and covariance C, i.e., the family of normal or Gaussian variates. We say that X is regular normal with mean μ and covariance C if the pdf of X has the form

$$f_X(x) = \frac{1}{(2\pi)^{n/2}\sqrt{\det C}}\, e^{-\frac{1}{2}(x-\mu)^T C^{-1}(x-\mu)}, \tag{A.168}$$

with C a strictly positive matrix.

In particular, when $\mu = 0$ and $C = I$ we say that we have a standard normal random variable Z, with pdf

$$f_Z(z) = \frac{1}{(2\pi^{n/2})}\, e^{-\frac{1}{2}z^T z}. \tag{A.169}$$

Proposition A.33 (Best Linear Unbiased Prediction) *Let* $X \in \mathbb{R}^n$, $Y \in \mathbb{R}^m$ *be two RV's and consider a composite* RV

$$Z = \begin{bmatrix} X \\ Y \end{bmatrix} \in \mathbb{R}^{n+m}.$$

Assume that

$$\mu(Z) = \begin{bmatrix} \mu(X) \\ \mu(Y) \end{bmatrix} = 0 \tag{A.170}$$

and that

$$C_Z = \begin{bmatrix} C_X & C_{XY} \\ C_{YX} & C_Y \end{bmatrix} \tag{A.171}$$

is known and regular. Notice that (A.159) *defines implicitly* C_{YX} *given by* $C_{YX} = E\{[Y - \mu(Y)][X - \mu(X)]^T\}$ *and that* $C_{XY} = C_{YX}^T$.

The Best Linear Unbiased Predictor (BLUP) *of* Y *is a linear function of* X, $\hat{Y} = AX$, *when* A *is chosen according to rule*

$$E\{|Y - \hat{Y}|^2\} = \min. \tag{A.172}$$

It turns out that the A minimizing (A.172) *is*

$$A = C_{YX} C_X^{-1}, \tag{A.173}$$

i.e.,

$$\hat{Y} = C_{YX} C_X^{-1} X. \tag{A.174}$$

Notice that, due to the hypothesis (A.170), $E\{\hat{Y}\} = 0 = E\{Y\}$ *i.e.,* $\hat{Y}$ *is an unbiased predictor of* Y. *In particular, the prediction error*

$$e = Y - \hat{Y} \tag{A.175}$$

has covariance matrix

$$C_e = C_Y - C_{YX} C_X^{-1} C_{XY}. \tag{A.176}$$

Proposition A.34 (Least Squares Theory) *Let* $Y \in \mathbb{R}^m$ *be an* $\mathcal{L}^2(\Omega)$ RV *and assume that*

$$\mu(Y) = \{y\} = Ax + a, \quad x \in \mathbb{R}^n \tag{A.177}$$

Notice that x *is a vector of deterministic parameters and not a* RV.
Assume further that C_Y *is a known regular matrix.*
The best unbiased estimator $\hat{x}$ *of* x, (BLUE), *is the linear function of* Y

$$\hat{x} = BY + b, \tag{A.178}$$

that satisfies the unbiasedness condition

$$E\{\hat{x}\} = B(Ax + a) + b = 0, \quad \forall x \tag{A.179}$$

and minimizes the estimation error:

$$E\{|\hat{x} - x|^2\} = \min. \tag{A.180}$$

The solution is

$$\hat{x} = (A^T C_Y^{-1} A)^{-1} A^T C_Y^{-1} (Y - a) \tag{A.181}$$

and the covariance of the estimation error

$$\varepsilon = \hat{x} - x$$

is given by

$$C_{\boldsymbol{\varepsilon}} = (A^T C_{\boldsymbol{Y}}^{-1} A)^{-1}. \tag{A.182}$$

Many more concepts could be usefully reviewed here, however the above level of the mathematical preliminaries seems to us to be sufficient, considering also that some specific and more advanced mathematical laws are extensively explored in Appendices B and C.

Appendix B
The Theory of Random Fields and the Wiener–Kolmogorov Prediction Method

B.1 Introduction

A real function on a set T is a rule associating to each element $t \in T$ a real value $f(t) \in \mathbb{R}$.

T can be any set, for instance an interval on the real line $\mathbb{R}$ or the whole $\mathbb{R}$, a square in $\mathbb{R}^2$, the surface of a sphere, a subset of $\mathbb{R}^3$, or even a set in an infinite-dimensional space, though in this case $f(t)$ is usually called a functional. Leaving aside this last possibility, we consider T as some set in $\mathbb{R}^2$ or $\mathbb{R}^3$, on which it is convenient to assume that a notion of distance is introduced, for instance the one associated with the ordinary Euclidean topology.

The set $\mathscr{F}$ of all functions defined on T is a very large linear space, too undetermined to be useful for the search of a function on which we have only a partial information. Even if we restrict it to the subspace of measurable functions, $\mathscr{M}$, we gain little insight into problems of finding a "reasonable" function by watching the incomplete (and maybe inaccurate) information we have on it.

For instance, let T be the interval $I = [0, 1]$ and assume you know that at certain points $(t_1, t_2, \ldots, t_N)$ in I a function takes given values $f(t_1) = a_1, \ldots, (t_N) = a_N$. We would like to know $f(t)$ at some (or all) other points $\bar{t} \in I$. This is a classical "interpolation" problem.

If we only say that f is measurable on [0;1], than there is no hope to give a reasonable answer to this question. In fact, if $f_1(t)$ and $f_2(t)$ take at $\bar{t}$ " two completely" different values, there is no way to say whether one is a better guess than the other. In order to make one step further, we must be able to introduce some information telling us how the function f tends to associate one of its value $f(t)$ to the other values $f(t')$.

This could be done in functional terms, i.e., by describing the a priori degree of regularity of f (e.g., f is continuous or square integrable, with derivatives up to some order continuous or square integrable, etc.) and then deciding that we want

F. Sansò, D. Sampietro, *Analysis of the Gravity Field*, Lecture Notes in Geosystems Mathematics and Computing, https://doi.org/10.1007/978-3-030-74353-6

the "most regular" function, e.g., belonging to some Hilbert space, that matches the given information. This will be the object of Appendix C.

Another approach is that since $f(t)$ is unknown, maybe it should be considered as a random variable and, rather than assigning to it a specific value, we should describe our guess by means of a probability distribution, for which all values are considered as possible, certainly paying a particular attention to the mean value or to the most probable value.

In this approach therefore we consider a law associating to any t a random variable X_t, defined according to Kolmogorov's scheme (see [52, 100]), on some probability space $(\Omega, \mathscr{A}, P)$. To be specific, we shall assume that any such variable $X_t = x(t, \omega)$ $(t \in T, \omega \in \Omega)$ should also have finite second moment, namely

$$X_t \in \mathscr{L}^2(\Omega) \Longleftrightarrow E\left\{X_t^2\right\} = \int_\Omega x^2(t, \omega)\, dP(\omega) < +\infty. \tag{B.1}$$

Although not necessary, it will be easier for us to assume that each X_t has also zero mean, namely

$$E\{X_t\} = \int_\Omega x(t, \omega)\, dP(\omega) = 0, \tag{B.2}$$

so that (B.1) represents the variance of X_t too.

The set of random variables $\{X_t, t \in T\} \equiv \{x(t, \omega), t \in T\}$ is what is called a *random field* (RF), or a *random function*; when $T \subset \mathbb{R}$ and t is interpreted as a time, t is also called a *stochastic process*.

A standard approach to describe the probability distribution of a random field $\{X_t;\ t \in T\}$ is the one given by Kolmogorov, e.g., by means of the so-called *family of finite-dimensional marginal distributions*

$$F(t_1, \ldots, t_N; a_1, \ldots, a_N) = P\left(X_{t_1} \le a_1, \ldots, X_{t_N} \le a_N\right); \tag{B.3}$$

Kolmogorov's theorem states that, once the family $F(t_1, \ldots, t_N; a_1, \ldots, a_N)$ satisfies two certain compatibility conditions, it determines uniquely a probability distribution on $\mathscr{F}$. The difficulty with this approach is that, although we can find reasonable conditions guaranteeing that in reality the probability distributions of X_t is concentrated on suitable and more manageable subspaces than $\mathscr{F}$, e.g., on $L^2(T)$ or some Sobolev space $H^{n,p}(T)$, the approach cannot account for one fundamental process, namely, the so-called *continuous white noise* on T. In this Appendix we want to treat two types of models relating observations to our unknown function $x(t)$, by some observation equations: A) data are given at a finite number of points with a discrete noise, and B) data are given over the whole T with some noise.

We will follow an approach that allows the same treatment for both, A) and B), where the unknown function is considered as a "sample", $x(t) = x(t, \bar{\omega})$, drawn from a *Generalized Random Field* (GRF) (see [120]).

As we will see in Sect. B.2, this object does not live alone, and not always it has values as random variables for each $t \in T$, but rather it is defined through the coupling with some function $h(t)$ belonging to an underlying Hilbert space H. One can then explore the conditions under which a GRF is also an ordinary RF and find the relationships with such well-known concepts as first or second order moments of the RF.

The general problem we will attack and solve in this Appendix can be formulated as follows. Assume y is a vector of observable quantities that can be either finite- (model A) or infinite-dimensional (model B).

In either cases, we will assume that the "true" y belongs to some other Hilbert space H_y. Of y we know that a certain observation experiment has provided a vector y_0, which corresponds to y but with the addition of a vector of errors ν,

$$y_0 = y + \nu; \tag{B.4}$$

if H_y is just $\mathbb{R}^m$, for some m, the relation (B.4) is elementary, namely y_0, y, ν are vectors of real numbers, but when H_y is infinite-dimensional, since ν can be very wild and not belonging to H_y, the relation (B.4) needs to be interpreted and we take it for the moment just in a symbolic sense. Moreover, assume that the physics of the observation process tells us that $y \in H_y$ is in fact a linear function of another unknown object x, of which we know that it belongs to some Hilbert space H_x. With our problems this is always infinite-dimensional, i.e., x is a function of $t \in T$ and T has an infinite number of points, which sometimes could be a discrete sequence, but for the case we are interested in, they constitute a continuum.

So (B.4) becomes

$$y_0 = Ax + \nu. \tag{B.5}$$

We embed (B.5) into a stochastic model, by considering y_0, x and ν (all of them!) as samples from some random variables, that we will learn to describe as GRF's, Y, X, N.

Then (B.5) becomes

$$Y = AX + N; \tag{B.6}$$

moreover, we shall assume to have some knowledge on the stochastic structure of Y, X, N, namely to know their means (that we shall always assume to be zero) and their second-order moments (covariances).

The problem we want to tackle is: assume we want to know some linear functional of X_t, $L(X) = L_t(X_t)$, with L_t acting on the variable t; since X is random, $L(X)$ will be random, too.

More precisely, if $X_t = x(t, \omega)$ and ω is the variable describing the stochastic character of X_t, we have to understand that $L_t[x(t, \omega)] = L(X)(\omega)$.

Note that, due to the linearity of L_t, if one further assumes some kind of continuity, one has that

$$E\{L(X)\} = \int L_t[x(t,\omega)]\, dP(\omega) = L_t\left[\int x(t,\omega)\, dP(\omega)\right] = 0, \tag{B.7}$$

where $dP(\omega)$ is the probability law of ω on Ω.

Assume that to predict $L(X)$ we want to use some "clever" linear functional $\mathscr{M}$ of Y_τ such that $\mathscr{M}_\tau(Y_\tau) = \mathscr{M}(Y)$, i.e., we put

$$\hat{L}(X) = \mathscr{M}(Y). \tag{B.8}$$

For the functional $\mathscr{M}(Y)$ we can make the same remarks we did for $L(X)$ and, in particular, we have

$$E\{\mathscr{M}(Y)\} = 0. \tag{B.9}$$

Assume one decides to use some criterion to judge whether $\mathscr{M}$ is clever or not, for instance one uses the variance of the prediction error

$$\mathscr{E}^2 = E\left\{\left[L(X) - \hat{L}(X)\right]^2\right\}. \tag{B.10}$$

Then we want to find the answer to the following two questions: is there an $\mathscr{M}$ that minimizes the prediction error? What is the error $\mathscr{E}^2$, when $\mathscr{M}$ is the answer to the previous question?

Having the answer the to two questions we can use what we know, namely Y_0, to build an estimate

$$\hat{L}(X) = \mathscr{M}(y_0), \tag{B.11}$$

of which we can claim that it is a sample from a random variable that has a given variance $\mathscr{E}^2$ of the error.

We will do that in this Appendix, after having recalled the basics of GRF theory.

We close this Introduction by observing that on a historical ground, practically the same results have been obtained by Kolmogorov and by Wiener, who was reasoning along a path of signal analysis, exploiting the concept of stationarity, related to random processes through ergodic theory. We will see that Wiener's point of view will be essential, for instance, for the estimation of second-order moments, without which the whole theory becomes just an abstract construction. It is for this reason that we like to present the actual theory under the two names, although this is not customary in the mathematical literature on the subject.

We warn the reader that the literature on random fields is immense, from both the theoretical and the applied side. Here we refer basically to [52] and [120] for the mathematical side and to [147] for the applications to Geophysics, and to [133] for the applications to Geodesy.

B.2 Generalized Random Fields

Let a certain probability space $(\Omega, \mathscr{A}, P)$ be given, with P a probability distribution defined on the σ-algebra $\mathscr{A}$ of measurable subsets of Ω. Recall that a random variable (RV) X is a measurable function of ω; in our case we will consider mostly real RV's, i.e., only real functions $x(\omega)$.

The space $\mathscr{L}^2(\Omega)$ is the Hilbert space of RV's X such that

$$E\left\{X^2\right\} = \int_{\Omega} x^2(\omega)\, dP(\omega) < +\infty. \tag{B.12}$$

The scalar product in $\mathscr{L}^2(\Omega)$ is given for $X, Z \in \mathscr{L}^2(\Omega)$ by

$$\langle X, Z\rangle_{\mathscr{L}^2(\Omega)} = E\{XZ\} = \int_{\Omega} x(\omega)\, z(\omega)\, dP(\omega). \tag{B.13}$$

In case the RV's considered are complex-valued, (B.13) should be modified as

$$\langle X, Z\rangle_{\mathscr{L}^2(\Omega)} = E\left\{X^* Z\right\}, \tag{B.14}$$

with X^* the complex conjugate of X.

Since, by the Schwartz inequality and the fact that $E\left\{1^2\right\} = E\{1\} = 1$, one has

$$|E\{X\}| \le \left[E\left\{X^2\right\}\right]^{1/2},$$

we see that

$$\mu_X = E\{X\} = \langle 1, X\rangle_{\mathscr{L}^2(\Omega)}$$

is a continuous linear functional on $\mathscr{L}^2(\Omega)$.

In fact, in what follows, we will consider only a closed subspace of $\mathscr{L}^2(\Omega)$, namely that of RV's having zero mean:

$$\mathscr{L}_0^2(\Omega) \equiv \left\{X \in \mathscr{L}^2(\Omega);\, E\{X\} = 0\right\}. \tag{B.15}$$

Indeed, $\mathscr{L}_0^2(\Omega)$ is in its full right a Hilbert space and, with a little abuse of notation, we will call it again $\mathscr{L}^2(\Omega)$ without indices, as long as this will not cause any confusion.

Next let H be a separable Hilbert space.

Definition B.1 A generalized random field $X = X(h)$ on the space H is a bounded linear operator from H into $\mathscr{L}^2(\Omega)$,

$$\begin{cases} X : H \to \mathscr{L}^2(\Omega), \\ E\left\{X^2(h)\right\} = \|X(h)\|^2_{\mathscr{L}^2(\Omega)} \le c^2 \|h\|^2_H, \quad \forall h \in H. \end{cases} \tag{B.16}$$

Note that we include in the definition the injectivity of $X(h)$, namely

$$X(h) = 0 \Longrightarrow h = 0, \tag{B.17}$$

which, strictly speaking, is not necessary, although it simplifies the description of the nature of X.

Now consider two elements $h, f \in H$ and the corresponding RV's, $X(h)$, $X(f)$; one can define the bilinear form

$$B(h, f) = E\{X(h)X(f)\} = \langle X(h), X(f)\rangle_{\mathscr{L}^2(\Omega)}. \tag{B.18}$$

It is clear that $B(h, f)$ is bilinear on H and symmetric:

$$B(h, f) = B(f, h). \tag{B.19}$$

Moreover, B is continuous in both h and f, and jointly in $(h, f) \in H \times H$; in fact

$$\begin{aligned} |B(h, f)| &= \left|\langle X(h), X(f)\rangle_{\mathscr{L}^2(\Omega)}\right| \le \\ &\le c^2 \|h\|_H \cdot \|f\|_H \le \frac{1}{2}c^2\left(\|h\|^2_H + \|f\|^2_H\right). \end{aligned} \tag{B.20}$$

(B.20) implies that $B(h, f) \to 0$ for $h \to 0$, uniformly on any bounded set $\left\{f;\ \|f\|_H \le k\right\}$ and that the same is true if we switch the role of h and f; moreover, it is $B(h, f) \to 0$ too, if $(h, f) \to 0$.

Another property of $B(h, f)$ is that it is positive definite on the diagonal $\{f = h\}$:

$$B(h, h) \ge 0 \tag{B.21}$$

and in fact even strictly positive definite because, thanks to our injectivity hypothesis,

$$B(h, h) = 0 \Longrightarrow \|X(h)\|^2_{\mathscr{L}^2} = 0 \Longrightarrow X(h) = 0 \Longrightarrow h = 0. \tag{B.22}$$

With the relations (B.19) through (B.22) we can easily check that all the properties of a scalar product are verified. In fact, the Fisher-Riesz theorem on the representation of bounded linear functionals on H by means of elements of H (see [133], Theorem 2; 12.3) readily shows that there exists a linear operator C such that

$$B(h, f) = E\{X(h)X(f)\} = \langle h, Cf\rangle_H : \tag{B.23}$$

(B.19) says that C is selfadjoint,

$$C^T = C, \tag{B.24}$$

(B.20) says that C is continuous, i.e.,

$$\|C\| \le c^2, \tag{B.25}$$

and (B.21) and (B.22) together say that C is a strictly positive definite operator.

Definition B.2 The operator C satisfying (B.23) is called the *covariance operator* of $X(h)$ in H.

Definition B.3 We define on H a new scalar product, setting

$$\langle h, f \rangle_{H_C} = \langle h, C f \rangle_H = E\{X(h)\, X(f)\}, h, f \in H. \tag{B.26}$$

It is clear that the product (B.26) is well defined for every couple $h, f \in H$, but the question is: does (B.26) determine a different topology on H, namely convergent (Cauchy) sequences are the same, when using the norms induced on H by $\langle , \rangle_H$ and $\langle , \rangle_{H_C}$?

The answer is:

a) either C is strictly bounded below, i.e., there is a $c_0 > 0$ such that

$$\langle h, Ch \rangle_H = \|h\|^2_{H_C} \ge c_0^2 \|h\|^2_H, \qquad \forall h \in H \tag{B.27}$$

in other words,

$$C \ge c_0^2 I, \tag{B.28}$$

in which case the two topologies are equivalent because, thanks to (B.25) one has too

$$c_0^2 \|h\|^2_H \le \|h\|^2_{H_C} \le c^2 \|h\|^2_H, \tag{B.29}$$

and H is closed in both (i.e., when $\{h_n\}$ sequence has a limit $\bar{h} \in H$ in one topology it has the same limit in the other);

b) or C is not bounded below, by a positive constant, in which case there are H_C convergent sequences that are not H-convergent. Therefore, if we call H_C the closure of H in the H_C norm, we obtain a new larger Hilbert space

$$H_C \supset H, \tag{B.30}$$

where the embedding of H in H_C is dense.

Definition B.4 Let X be a GRF on H. If

$$C = c_0^2 I \Longleftrightarrow E\left\{X^2(h)\right\} = c_0^2 \|h\|_H^2, \tag{B.31}$$

we say that X is a white noise relative to H, with variance density c_0^2.

When $H = L^2(T)$ and (B.31) is verified, we say that X is white noise (in the sense of Wiener). In this case we have

$$E\left\{X(h)^2\right\} = c_0^2 \int_T h^2(t)\, dt. \tag{B.32}$$

We note that if we have a sequence $\{h_n\} \subset H$ such that $h_n \underset{H_C}{\rightarrow} \bar{h}$, i.e., $\{h_n\}$ is Cauchy in H_C, then, using (B.26),

$$E\left\{\left[X(h_{n+p}) - X(h_n)\right]^2\right\} = E\left\{X\left(h_{n+p} - h_n\right)^2\right\} = \left\|h_{n+p} - h_n\right\|_{H_C}^2 \rightarrow 0,$$

i.e. $\{X(h_n)\}$ is a Cauchy sequence in $\mathscr{L}^2(\Omega)$, so that the following extension by continuity of X to H_C is legitimate: for any $\bar{h} = \lim h_m$ we put

$$X(\bar{h}) \overset{\text{Def}}{=} \lim_{n\rightarrow\infty} X(h_n). \tag{B.33}$$

Since any $\bar{h} \in H_C$ is the limit of a sequence $\{h_n\} \in H$, the definition (B.33) can be extended to the whole H_C and we further on have

$$\|X(\bar{h})\|_{\mathscr{L}^2(\Omega)}^2 = E\left\{X^2(\bar{h})\right\} = \lim_{n\rightarrow\infty} E\left\{X(h_n)^2\right\} = \|\bar{h}\|_{H_C}^2. \tag{B.34}$$

Thus, we have proved.

Proposition B.1 *Every* GRF *originally defined on some Hilbert space H, admits of a unique extension to a Hilbert space H_C, in which H is densely embedded and relative to which it is a white noise.*

We emphasize that H_C is a very important Hilbert space, intimately related to the nature of the GRF, X. The relation (B.34) shows that there is an isometry between the space H_C itself and the collection of RV's $\{X(h), h \in H\} \subset \mathscr{L}^2(\Omega)$. In particular it is clear that

$$\mathscr{L}_X^2 \equiv \{X(h), h \in H\}, \tag{B.35}$$

is a linear space, because

$$aX(h_1) + bX(h_2) = X(a_1 h_1 + b h_2);$$

hence, $\mathscr{L}_X^2$ is just a subspace of $\mathscr{L}^2(\Omega)$.

But since $\mathscr{L}_X^2$ is isometrically isomorphic to H_C it has to be a Hilbert space too, namely a closed subspace of $\mathscr{L}^2(\Omega)$.

Definition B.5 A continuous linear stochastic functional L of the GRF X, defined on H is any RV, $Y \in \mathscr{L}^2(\Omega)$ obtained as an $\mathscr{L}^2(\Omega)$ limit of linear combinations of $\{X(h_i), h_i \in H\}$.

Such an L is also called an *admissible functional*. Due to our comments above on the isometry between $\mathscr{L}_X^2$ and H_C, we arrive at the following very important lemma.

Lemma B.1 *H_C represents the space of all admissible functionals on X, namely for every admissible functional L there is an element $\ell \in H_C$, such that*

$$\begin{cases} L(X) \equiv X(\ell), \\ \langle \ell, C\ell \rangle_H \equiv \|\ell\|_{HC}^2 < +\infty. \end{cases} \tag{B.36}$$

This lemma is important because if we want to write observation equations where an observable Y depends on the GRF X via some linear (or linearized) functional relation $Y = L(X)$, one has to know that $L(X)$ has at least almost surely finite values, as otherwise Y could not be observed.

In order to better understand the structure of X, we need to set up a certain formal spectral representation. Let us observe that when H is assumed to be separable (i.e., it contains a sequence that is everywhere dense) the space H_C, which contains H as a dense subset, has to be separable too.

Therefore, H_C contains sequences that are orthonormal and complete (*Complete OrthoNormal Systems*, CONS), i.e., orthonormal bases, [117, 161].

Let $\{\bar{\varphi}_n\}$ be a CONS in H_C and define the sequence

$$\bar{X}_n = X(\bar{\varphi}_n). \tag{B.37}$$

It is clear that $\bar{X}_n \in \mathscr{L}_X^2$ and, furthermore, recalling the isometry $\mathscr{L}_X^2 \Longleftrightarrow H_C$,

$$E\left\{\bar{X}_n \bar{X}_m\right\} = \langle \bar{\varphi}_n, \bar{\varphi}_m \rangle_{HC} = \delta_{nm}, \tag{B.38}$$

i.e., $\left\{\bar{X}_n\right\}$ is ON in $\mathscr{L}_X^2$. In addition, it is complete, because

$$\forall Y \in \mathscr{L}_X^2,\ \exists \bar{h} \in H_C \text{ such that } Y = X\left(\bar{h}\right)$$

and

$$\left\langle Y, \bar{X}_n \right\rangle_{\mathscr{L}_X^2} = E\left\{Y \bar{X}_n\right\} = \left\langle \bar{h}, \bar{\varphi}_n \right\rangle_{HC} = 0 \quad \forall n$$

implies $\bar{h} = 0$ and therefore $Y = 0$, too.

So $\left\{\bar{X}_n\right\}$ is a CONS in $\mathscr{L}_X^2$.

Consequently, one can always write

$$\begin{cases} \forall \bar{h} \in H_C, \bar{h} = \sum_{n=1}^{+\infty} \bar{h}_n \bar{\varphi}_n (t), \\ X(\bar{h}) = \sum_{n=1}^{+\infty} \bar{h}_n X(\bar{\varphi}_n) = \sum_{n=0}^{+\infty} \bar{h}_n \bar{X}_n. \end{cases} \tag{B.39}$$

Let us notice that, since $\{\bar{\varphi}_n\}$ is CONS in H_C one has $\sum_{n=1}^{+\infty} \bar{h}_n^2 < +\infty$, implying that the second line of (B.39) represents the Fourier series expansion of the $\mathscr{L}_X^2$ RV $X(\bar{h})$, with respect to the ON basis $\{\bar{X}_n\}$; this series is indeed convergent in $\mathscr{L}_X^2$.

At this point one could be tempted to represent the GRF X by the "function"

$$X(\omega, t) = \sum_{n=1}^{+\infty} \bar{X}_n(\omega) \bar{\varphi}_n(t). \tag{B.40}$$

In fact, interpreting $X(h)$ as a scalar product in H_C one could formally write

$$\begin{aligned} X(\bar{h}) &= \langle X(\omega, t), \bar{h}(t) \rangle_{H_C} = \sum_{n=1}^{+\infty} \bar{h}_n \langle X(\omega, t), \bar{\varphi}_n \rangle_{H_C} = \\ &= \sum_{n=1}^{+\infty} \bar{h}_n \bar{X}_n(\omega). \end{aligned} \tag{B.41}$$

In reality the expression $\langle X(\omega, t), \bar{h}(t) \rangle_{H_C}$ is only formal, in that, as a counterexample will demonstrate, $X(\omega, t)$ generally does not belong to H_C on a set of $\omega \in \Omega$ with probability 1. Therefore, we cannot take the expression of $\langle X(\omega, t), \bar{h} \rangle_{H_C}$ as a true scalar product, but we rather redefine such an object.

Definition B.6 For every $\bar{h} \in H_C$ we set

$$\langle X(\omega, t), \bar{h}(t) \rangle_{H_C} = X(\bar{h}), \tag{B.42}$$

even if $P\{X(\omega, t) \in H_C\} = 0$; moreover, we define the formal series (B.40) as the object that can only be coupled with $\bar{h} \in H_C$ according to

$$\left\langle \sum_{n=1}^{+\infty} \bar{X}_n(\omega) \bar{\varphi}_n(t), \bar{h} \right\rangle_{H_C} = \sum_{n=1}^{+\infty} \bar{X}_n(\omega) \bar{h}_n. \tag{B.43}$$

Before proceeding we need to justify our claim that in general the sample elements $\omega \to X(\omega, t)$ are not functions in H_C.

We show it by a counterexample.

Example B.1 Assume $\{X(\bar{h}), \bar{h} \in H_C\}$ is a normal family, i.e., the probability law of $X(\bar{h})$ is $\mathscr{N}\left(0, \|h\|_{H_C}^2\right)$.

Note that since $\lambda X(\bar{h}_1) + \mu X(\bar{h}_2) = X(\lambda \bar{h}_1 + \mu \bar{h}_2)$, and the last variable is normal, the above statement shows that any linear combination of elements of $\mathscr{L}_X^2$

is a normal variate and therefore even each vector $X\left(\bar{\boldsymbol{h}}\right) = \begin{bmatrix} X\left(\bar{h}_1\right) \\ X\left(\bar{h}_2\right) \\ \vdots \\ X\left(\bar{h}_n\right) \end{bmatrix}$ is normally distributed with zero mean and covariance matrix

$$E\left\{X\left(\bar{\boldsymbol{h}}\right) X\left(\bar{\boldsymbol{h}}^T\right)\right\}_{ij} = \left\{\langle \bar{h}_i, \bar{h}_j \rangle_{H_C}\right\}. \tag{B.44}$$

In particular, this implies that, if we take the full sequence $\bar{\boldsymbol{\varphi}} = \begin{bmatrix} \bar{\varphi}_1 \\ \bar{\varphi}_2 \\ \cdots \\ \bar{\varphi}_n \\ \vdots \end{bmatrix}$, then the infinite vector

$$\boldsymbol{X} = \begin{bmatrix} X\left(\varphi_1\right) \\ X\left(\varphi_2\right) \\ \vdots \\ X\left(\varphi_n\right) \\ \vdots \end{bmatrix} = \begin{bmatrix} \bar{X}_1 \\ \bar{X}_2 \\ \vdots \\ \bar{X}_n \\ \vdots \end{bmatrix}$$

is a normal variate on $\mathbb{R}^\infty$, with zero mean and covariance matrix equal to the identity (see (B.38)). Moreover the "components" $\bar{X}_n$, that are uncorrelated and jointly normal, are also stochastically independent.

Now we prove that for such a family

$$P\left\{X\left(\omega, t\right) \in H_C, t \in T\right\} = 0, \tag{B.45}$$

which can be rewritten as

$$P\left\{\|X\left(\omega, t\right)\|^2_{H_C} < +\infty\right\} = 0. \tag{B.46}$$

Considering that, if $X\left(\omega, t\right)$ would really belong to H_C, one should have

$$\|X\left(\omega, t\right)\|^2_{H_C} = \sum_{n=1}^{+\infty} \bar{X}_n^2\left(\omega\right),$$

we can further rewrite (B.46) as

$$P\left(\sum_{n=1}^{+\infty} \bar{X}_n^2 < +\infty\right) = 0. \tag{B.47}$$

On the other hand, for any N,

$$\sum_{n=1}^{+\infty} \bar{X}_n^2(\omega) > \sum_{n=1}^{N} \bar{X}_n^2(\omega).$$

Now fix any real $\Lambda > 0$; we have

$$P\left\{\sum_{n=1}^{+\infty} \bar{X}_n^2(\omega) < \Lambda\right\} \leq P\left\{\sum_{n=1}^{N} \bar{X}_n^2(\omega) < \Lambda\right\}. \tag{B.48}$$

Since $\bar{X}_n(\omega)$ are stochastically independent, $\bar{X}_n^2(\omega)$ are stochastically independent (and identically distributed) too, and one can write

$$P\left\{\sum_{n=1}^{N} \bar{X}_n^2(\omega) < \Lambda\right\} < P\left\{\bar{X}_1^2(\omega) < \Lambda, \bar{X}_2^2(\omega) < \Lambda, \ldots, \bar{X}_N^2(\omega) < \Lambda\right\} =$$
$$= P\left\{\bar{X}_1^2(\omega) < \Lambda\right\}^N . \tag{B.49}$$

Since $\bar{X}_1(\omega)$ is a standard normal variate,

$$P\left(\bar{X}_1^2(\omega) < \Lambda\right) = \rho < 1$$

so (B.49) says that $\sum_{n=1}^{N} X_n^2(\omega) \to 0$ in probability as $N \to \infty$ and (B.48) says that

$$\forall \Lambda > 0, P\left\{\sum_{n=1}^{+\infty} \bar{X}_n^2(\omega) < \Lambda\right\} = 0.$$

Since we can define a sequence of intervals $I_n = (-n, n]$ covering the real line, and such that for each n the probability that $\|X(\omega, t)\|_{H_C}^2$ belongs to I_n is zero, we deduce that (B.46) is true, i.e., $X(\omega, t)$ does not belong to H_C with probability 1.

We will come back later to (B.40) and look at this series as a function of ω, for fixed $t \in T$.

We close this section discussing some matters that will become useful in the sequel.

In particular, we will explore the consequences of some restrictive hypotheses one can make on the operator C.

As we have seen, C is by definition bounded, selfadjoint and positive on H; now we will assume C to be compact and to have finite trace, i.e., to be a nuclear operator on H.

As a consequence of compactness we know that C admits a system of eigenfunctions $\{\varphi_n\}$:

$$C\varphi_n = c_n^2 \varphi_n \tag{B.50}$$

($\{c_n^2\}$ is decreasing, $c_n^2 \to 0$ for $n \to \infty$). That $\{\varphi_n\}$ is ON in H is well known and that it is complete in H comes from

$$h_n = \langle h, \varphi_n \rangle_H = 0, \forall n \Longrightarrow \langle h, Ch \rangle_H = \sum_{n=1}^{+\infty} c_n^2 h_n^2 = 0 \Longrightarrow h = 0.$$

That C is nuclear means that

$$\mathrm{Tr}\, C = \sum_{n=1}^{+\infty} \langle \varphi_n, C\varphi_n \rangle = \sum_{n=1}^{+\infty} c_n^2 < +\infty. \tag{B.51}$$

Recall that C is then represented, in terms of $\{\varphi_n(t)\}$, by the kernel

$$C\left(t, t'\right) = \sum_{n=1}^{+\infty} c_n^2 \varphi_n(t)\, \varphi_n\left(t'\right) \tag{B.52}$$

in the sense that

$$Ch = \left\langle C\left(t, t'\right), h\left(t'\right)\right\rangle_H = \sum_{n=1}^{+\infty} c_n^2 h_n \varphi_n(t); \tag{B.53}$$

(B.53) implies that

$$\langle h, Ch \rangle_H = \sum_{n=1}^{+\infty} c_n^2 h_n^2. \tag{B.54}$$

We make now another hypothesis concerning the system $\{\varphi_n(t)\}$, namely that each $\varphi_n(t)$ is continuous and bounded on T:

$$t \in T, \quad |\varphi_n(t)| \le A. \tag{B.55}$$

By using this hypothesis together with (B.51), we immediately see that the series (B.52) is uniformly convergent, so that $C\left(t, t'\right)$ is a continuous function and

$$\left|C\left(t,t'\right)\right| \leq A^2 \sum_{n=1}^{+\infty} c_n^2 = \bar{c}^2 < +\infty. \tag{B.56}$$

The following properties of the kernel $C\left(t,t'\right)$ are very easy to prove:

$$1)\ C\left(t,t'\right) = C\left(t',t\right), \tag{B.57}$$

$$2)\ \left|C\left(t,t'\right)\right|^2 \leq C\left(t,t\right) C\left(t',t'\right); \tag{B.58}$$

$$3)\ \forall N, \qquad \forall \{\lambda_k; k = 1, 2 \ldots N\}, \quad \forall \{t_k \in T; k = 1, \ldots, N\},$$

$$\sum_{k,j=1}^{N} \lambda_k \lambda_j C\left(t_k, t_j\right) \geq 0. \tag{B.59}$$

Property 3) is called *definite positiveness of the function* $C\left(t,t'\right)$.

As we see the properties 1), 2), 3) above are strongly reminiscent of other two kernels frequently met in the mathematical geodesy and geophysics literature, as well as in the theory of random functions; the first appears in the theory of Reproducing Kernel Hilbert Spaces (RKHS) (see e.g. [74, 133]), of which we shall give here a brief summary, and the other is the concept of covariance function (see e.g. [52, 161]), that will be our concern from Sect. B.4 on.

Definition B.7 A Hilbert space $H(K)$ of functions on T is said to be endowed with a reproducing kernel, $K\left(t,t'\right)$, if

- $K\left(t,t'\right)$is symmetric, continuous and bounded on $T \times T$, i.e.,

$$\left|K\left(t,t'\right)\right| \leq c^2 \tag{B.60}$$

  for some positive constant c^2;
- for each fixed $t \in T$, $K\left(t,t'\right) \in H\left(K\right)$ as a function of t', and

$$\left\langle K\left(t,t'\right), h\left(t'\right)\right\rangle_{H(K)} \equiv h\left(t\right), \quad \forall t \in T, \quad \forall h \in H\left(K\right). \tag{B.61}$$

Proposition B.2 *Let* $H(K)$ *be a* RKHS *with reproducing kernel* $K\left(t,t'\right)$ *then every* $h \in H\left(K\right)$ *is a continuous function on* T*; conversely, assume that the evaluation functional*

$$e_t\left[h\right] \equiv h\left(t\right) \tag{B.62}$$

is continuous on $H(K)$*, i.e.,*

$$\left|e_t\left[h\right]\right| = \left|h\left(t\right)\right| \leq c \left\|h\right\|_{H(K)}, \tag{B.63}$$

then $H(K)$ *is endowed with a reproducing kernel* $K\left(t,t'\right)$.

Proposition B.3 *Assume $H(K)$ to be separable, so that $H(K)$ admits a* CONS $\{\psi_n(t)\}$*; then a necessary and sufficient condition for $H(K)$ to have a reproducing kernel $K(t, t')$ is that the series*

$$K(t, t') = \sum_{n=1}^{+\infty} \psi_n(t) \psi_n(t') \tag{B.64}$$

be convergent in $H(K)$, i.e.,

$$\sum_{n=1}^{+\infty} \psi_n(t)^2 \leq c^2, \quad \forall t \in T, \tag{B.65}$$

and the resulting function be continuous. In this case the symmetry of $K(t, t')$ is a consequence of (B.64) *and the continuity of $\psi_n(t)$ becomes a consequence of* (B.61), (B.65).

The representation (B.64) *holds for any* CONS *in $H(K)$; the corresponding reproducing kernel is unique.*

Proposition B.4 *A reproducing kernel $K(t, t')$ enjoys the three properties (B.57), (B.58), (B.59); viceversa, given a kernel $K(t, t')$ enjoying the three properties above, there is a Hilbert space $H(K)$ of which $K(t, t')$ is the reproducing kernel.*

As we are interested in understanding what is the Hilbert space $H(C)$, with C the kernel (B.52), we need to introduce still another Hilbert space, densely contained in H.

Definition B.8 We define the Hilbert space $H_{C^{-1}}$ as the subspace of H, closed in the norm

$$\|h\|^2_{C^{-1}} = \sum_{n=1}^{+\infty} c_n^{-2} h_n^2 = \left\langle h, C^{-1} h \right\rangle_H, \tag{B.66}$$
$$(h_n = \langle h, \varphi_n \rangle_H).$$

It is obvious that, since $c_n^2 \leq c_1^2$,

$$\|h\|^2_H = \sum_{n=1}^{+\infty} h_n^2 \leq c_1^2 \sum_{n=1}^{+\infty} c_n^{-2} h_n^2 = c_1^2 \|h\|^2_{H_{C^{-1}}},$$

so that

$$h \in H_{C^{-1}} \Longrightarrow h \in H,$$

i.e., $H_{C^{-1}} \subset H$.

Furthermore, since finite linear combinations of φ_n of the form $h^N = \sum_{n=1}^N a_n\varphi_n$, are in both $H_{C^{-1}}$ and H, and moreover they constitute a dense linear subspace of H, we conclude that $H_{C^{-1}}$ is also densely embedded in H. So in this way we arrive at the definition of a triplet of Hilbert spaces

$$H_{C^{-1}} \subset H \subset H_C, \tag{B.67}$$

densely embedded into one another, with norms related by

$$\frac{1}{c_1^2}\|H\|_{H_C}^2 = \frac{1}{c_1^2}\sum_{n=1}^{+\infty} c_n^2 h_n^2 \le \sum_{n=1}^{+\infty} h_n^2 = \|h\|_H^2 \le c_1^2 \sum_{n=1}^{+\infty} c_n^{-2} h_n^2 = c_1^2\|h\|_{H_{C^{-1}}}^2 . \tag{B.68}$$

Such a triplet is called a *standard triplet*.

Lemma B.2 *Let $H_{C^{-1}} \subset H \subset H_C$ be a standard triplet; then any continuous linear functional on $H_{C^{-1}}$ can be represented, after a suitable limit process, as the coupling via the H scalar product with an element of H_C.*

Proof First of all, let us notice that, as a consequence of the dense embedding, $\{\varphi_n\}$ is a CONS in H, but it is also a complete orthogonal system in the space, H_C and $H_{C^{-1}}$, where however it is not normalized, because

$$\|\varphi_n\|_{H_C}^2 = c_n^2, \quad \|\varphi_n\|_{H_{C^{-1}}}^2 = c_n^{-2}. \tag{B.69}$$

Therefore, we have CONS in H_C and $H_{C^{-1}}$ defined by

$$\{\bar{\varphi}_n\} = \left\{\frac{\varphi_n}{c_n}\right\} \text{ in } H_C \text{ and } \{\tilde{\varphi}_n\} = \{c_n\varphi_n\} \text{ in } H_{C^{-1}},$$

respectively.

Now take a generic $\tilde{g} \in H_{C^{-1}}$, i.e.,

$$\tilde{g} = \sum_{n=1}^{+\infty} \tilde{g}_n\tilde{\varphi}_n, \quad \text{with} \sum_{n=1}^{+\infty} \tilde{g}_n^2 < +\infty.$$

Since $\tilde{\varphi}_n = c_n\varphi_n$, we see that

$$\tilde{g} = \sum_{n=1}^{+\infty} \tilde{g}_n c_n \varphi_n = C^{1/2}\sum_{n=1}^{+\infty} \tilde{g}_n\varphi_n = C^{1/2} g, \tag{B.70}$$

where

$$g = \sum_{n=1}^{+\infty} \tilde{g}_n \varphi_n \in H, \quad \text{with } \|g\|_H^2 = \sum_{n=1}^{+\infty} \tilde{g}_n^2 = \|\tilde{g}\|_{H_{C^{-1}}}^2. \tag{B.71}$$

Thus, $C^{1/2}$ is an isometry $H \to H_{C^{-1}}$, and indeed $C^{-1/2}$ is an isometry $H_{C^{-1}} \to H$.

A similar reasoning holds for the relation between H_C and H, so that we have

$$H_{C^{-1}} = C^{1/2}(H) \quad \text{and} \quad H = C^{1/2}(H_C). \tag{B.72}$$

Now let $L(\tilde{g})$ be a continuous linear functional on $H_{C^{-1}}$; put

$$F(g) = L(\tilde{g}) = L\left(C^{1/2} g\right).$$

Since

$$|L(\tilde{g})|^2 \le a \|\tilde{g}\|_{H_{C^{-1}}}^2 = a \left\langle \tilde{g}, C^{-1}\tilde{g} \right\rangle_H$$

we have

$$|F(g)|^2 \le a \left\langle C^{1/2} g, C^{-1} C^{1/2} g \right\rangle_H = a \|g\|_H^2,$$

so that $F(g)$ is a continuous linear functional on H. By Riesz's theorem, there is an $h \in H$ such that

$$F(g) = L(\tilde{g}) = \langle h, g \rangle_H = \left\langle h, C^{-1/2}\tilde{g} \right\rangle_H = \left\langle C^{-1/2} h, \tilde{g} \right\rangle_H.$$

But

$$C^{-1/2} h = \bar{h} \in H_C,$$

so we have

$$L(\tilde{g}) = \left\langle \bar{h}, \tilde{g} \right\rangle_H. \tag{B.73}$$

To make this result rigorous, the proof can be repeated first for finite linear combinations of $\{\varphi_n\}$, and then taking the limits in the relevant spaces. □

As we have seen, we can therefore identify $H_{C^{-1}}^*$ with H_C; the reciprocal relation is also true and can be proved in the same manner, namely

$$H_{C^{-1}}^* = H_C \text{ and } H_C^* = H_{C^{-1}}. \tag{B.74}$$

This result will be useful later on. Having defined the space $H_{C^{-1}}$, we can now prove the following proposition.

Proposition B.5 *We have*

$$H(C) = H_{C^{-1}}, \tag{B.75}$$

i.e., $H_{C^{-1}}$ *is a* RKHS *with kernel* $C(t, t')$ *given by* (B.52).

Proof It is enough to recall that $\{\tilde{\varphi}_n\} = \{c_n \varphi_n\}$ is a CONS in $H_{C^{-1}}$ so that, by definition,

$$C(t, t') = \sum_{n=1}^{+\infty} c_n^2 \varphi_n(t)\, \varphi_n(t') = \sum_{n=1}^{+\infty} \tilde{\varphi}_n(t)\, \tilde{\varphi}_n(t'). \tag{B.76}$$

That the series is convergent depends on our hypothesis on the operator C; that $C(t, t')$ is the reproducing kernel of $H_{C^{-1}}$ is an consequence of Proposition B.3.

□

In order to avoid the false idea that a series like (B.64) is always convergent in H, we present the important counterexample where H is an L^2 space.

Example B.2 In this example we take $H = L^2(0, 1)$, to show that this space does not admit a reproducing kernel. That this is true is already obvious if one observes that the evaluation functional e_t, $(t \in [0, 1])$ cannot be continuous on $L^2(0, 1)$, because otherwise the space of continuous functions on [0, 1] should contain $L^2(0, 1)$, which is obviously false. Nevertheless, we want to use the condition (B.71), to show that this relation is violated for CONS in $L^2(0, 1)$, so that (B.64) cannot be a reproducing kernel. In fact it is well-known that the sequence

$$\psi_n(t) = \begin{cases} \cos 2\pi n t, & \text{if} \quad n \geq 0, \\ \sin 2\pi |n| t, & \text{if} \quad n < 0, \end{cases} \qquad t \in [0, 1],$$

is a CONS in $L^2(0, 1)$. But then

$$\sum_{n=-\infty}^{+\infty} \psi_n^2(t) = \sum_{n=0}^{+\infty} \left\{ (\sin 2\pi n t)^2 + (\cos 2\pi n t)^2 \right\} = +\infty$$

and (B.65) is not satisfied.

B.3 The Wiener–Kolmogorov Minimum Mean Square Prediction Error Principle

The beauty of the GRF approach lies in the fact that we can treat the prediction problem, already presented in Sect. B.1, irrespectively of the irregularity of the random fields involved, provided that they possess a covariance operator defined on some Hilbert space. So in this section we will impose on C no restriction other than it is a positive, selfadjoint bounded operator on some Hilbert space H; as we have seen, this includes for instance GRF's of WN of Wiener type.

Definition B.9 (The Prediction Problem) Let two Hilbert spaces H_x, H_y be given; let two GRF's X, Y be defined on the above spaces; let N be a further GRF on H_y, representing a disturbance in the process of observation of Y. The three GRF's are related by the linear equation

$$Y = AX + N, \tag{B.77}$$

where A is a bounded operator A: $H_x \to H_y$. For the three GRF's X, Y, N we know the corresponding covariance operators, namely

$$\begin{aligned} \forall h \in H_x, E\{X(h)\}^2 &= \langle h, C_X h\rangle_{H_x}; \\ \forall k \in H_y, E\left\{Y(k)^2\right\} &= \langle k, C_Y k\rangle_{H_y}; \\ \forall k \in H_y, E\{N(k)\} &= \langle k, C_N k\rangle_{H_y}; \end{aligned} \tag{B.78}$$

Since Y is "observable", we can determine, given Y, every $\mathscr{L}^2(\Omega)$ variable of the type

$$Y(k) = \langle Y, k\rangle_{H_y}, \quad k \in H_y; \tag{B.79}$$

given an $h \in H_x$ we want to find a $k(h) \in H_y$ such that $Y(k)$ can be considered a "predictor" of $X(h)$, namely, it is close to $X(h)$ in some stochastic sense. To be completed, the statement of the prediction problem requires first to define the quantity

$$U = AX; \tag{B.80}$$

furthermore, we need to specify the stochastic relation between U and N, which we will assume to be independent of one another. Finally, we need to introduce a quantity that measures the "closeness" of $Y(k)$ to $X(h)$.

The first question is solved by the formula

$$U(k) = AX(k) = X\left(A^T k\right). \tag{B.81}$$

(B.81) is a consistent definition: in fact, if A is continuous between two Hilbert spaces $H_x \to H_y$ and we identify the duals H_x^*, H_y^* with H_x, H_y themselves via Riesz's theorem, then A^T is indeed continuous $H_y \to H_x$, so that

$$E\left\{X\left(A^T k\right)^2\right\} = \left\langle A^T k, C_X A^T k\right\rangle_{H_x} = \\ = \left\langle k, AC_X A^T k\right\rangle_{H_y} \leq c \, \|k\|_{H_y}^2 , \tag{B.82}$$

i.e., $U = AX$ is a GRF on H_y, with covariance operator

$$C_U = AC_X A^T . \tag{B.83}$$

The last question is solved by the Wiener–Kolmogorov principle.

Remark B.1 Let us notice that the three covariance operators C_X, C_Y, C_N are not independent. In fact, since $Y = U + N = AX + N$ and U and N are stochastically independent, one has

$$E\left\{Y\,(k)^2\right\} = E\left\{U\,(k)^2\right\} + E\left\{N\,(k)^2\right\} = \\ = \langle k, C_U k\rangle_{H_y} + \langle k, C_N k\rangle_{H_y} , \quad \forall k \in H_y, \tag{B.84}$$

with C_U given by (B.83). Therefore we get,

$$C_Y = C_U + C_N = AC_X A^\top + C_N, \tag{B.85}$$

namely, given C_X, C_N and A the covariance C_Y is fixed.

For later use, let us compute here the cross-covariance operator C_{YX} defined as

$$E\,\{Y\,(k)\,X\,(h)\} = \langle k, C_{YX} h\rangle_{H_y} . \tag{B.86}$$

In view of the independence of X and N, and recalling (B.77), for any $k \in H_y$, and $h \in H_x$ we have

$$E\,\{Y\,(k)\,X\,(h)\} = E\,\{U\,(k)\,X\,(h)\} = E\,\{AX\,(k)\,X\,(h)\} = \\ E\left\{X\left(A^\top k\right) Y\,(h)\right\} = \left\langle A^\top k, C_X h\right\rangle_{H_x} = \langle k, AC_X h\rangle_{H_y} ;$$

therefore

$$C_{YX} = AC_X, C_{XY} = C_X A^\top . \tag{B.87}$$

Definition B.10 (The Wiener–Kolmogorov Principle) If $Y(k)$ has to be a "predictor" of $X(h)$, then it is logical to call the difference

$$\varepsilon = Y(k) - X(h), \tag{B.88}$$

the prediction error; since $Y(k), X(h) \in \mathcal{L}^2(\Omega)$, so is $\varepsilon = \varepsilon(\omega)$.

A natural index of smallness of the error is therefore the mean square value of ε,

$$\mathcal{E}^2 = E\left\{\varepsilon^2\right\}. \tag{B.89}$$

Since clearly $E\{\varepsilon\} = 0$, $\mathcal{E}^2$ is also the variance of the prediction error.

It is also clear that $\mathcal{E}^2$ is a functional of k; the Wiener–Kolmogorov principle therefore defines an optimal predictor $Y(k) = \widehat{X(h)}$ (the symbol $\widehat{\ }$ means predictor of) as the one that corresponds to a minimum of $\mathcal{E}^2(k)$.

We are ready now to state the main result of this section: we do that under a mild restriction on the nature of C_N, which strongly simplifies the treatment of the minimization of $\mathcal{E}^2$.

Theorem B.1 *Assume that C_N is strictly positively bounded below, i.e.,*

$$C_N \geq c_0^2 I_{H_y}, \tag{B.90}$$

meaning that N has a white noise component (on H_y) in it; then $\mathcal{E}^2(k)$ has a minimum at

$$k = C_Y^{-1} C_{YX} h = \left(A C_X A^\top + C_N\right)^{-1} A C_X h \tag{B.91}$$

corresponding to the predictor

$$\widehat{X(h)} = Y(k) = \left\langle \mathbf{Y}, C_Y^{-1} C_{YX} h\right\rangle_{H_y}; \tag{B.92}$$

the mean square error of such a predictor is

$$\mathcal{E}^2(k) = \left\langle h, \left(C_X - C_{XY} C_Y^{-1} C_{YX}\right) h\right\rangle_{H_x}. \tag{B.93}$$

We further add that (B.93) *entails that $\varepsilon(h) = \widehat{X(h)} - X(h)$ is a* GRF *on H_x with covariance operator*

$$C_\varepsilon = C_X - C_{XY} C_Y^{-1} C_{YX}. \tag{B.94}$$

Proof We start by observing that, (B.85) and (B.90) imply that

$$C_Y \geq c_0^2 I_{H_y},$$

namely C_Y is invertible and its inverse, $C_Y^{-1}\colon H_y \to H_y$, is bounded above by

$$C_Y^{-1} \le \frac{1}{c_0^2} I_{H_y}.$$

We recall that under this condition

$$H_{C_Y} \equiv \bar{H}_y \equiv H_y, \tag{B.95}$$

namely, the admissible functionals on Y are just the elements of H_y.

We first of all compute explicitly $\mathscr{E}^2(k)$.

One has, also recalling (8.47) and (B.85),

$$\begin{aligned}
\mathscr{E}^2(k) &= E\left\{[AX(k) + N(k) - X(h)]^2\right\} \\
&= E\left\{\left[X\left(A^\top k - h\right) + N(k)\right]^2\right\} = \\
&= \left\langle A^\top k - h, C_X\left(A^\top k - h\right)\right\rangle_{H_x} + \langle k, C_N k\rangle_{H_y} = \\
&= \left\langle A^\top k, C_X A^\top k\right\rangle_{H_x} - 2\left\langle A^\top k, C_X h\right\rangle_{H_x} + \langle h, C_X h\rangle_{H_x} + \langle k, C_N k\rangle_{H_y} = \\
&= \langle k, C_Y k\rangle_{H_y} - 2\langle k, C_{YX} h\rangle_{H_y} + \langle h, C_X h\rangle_{H_x}.
\end{aligned} \tag{B.96}$$

Hence, $\mathscr{E}^2(k)$ is just a positive quadratic functional of k and its minimum, when it exists, can be obtained by setting to zero the first variation, i.e.,

$$\left.\frac{d}{dt}\mathscr{E}^2(k + t\delta k)\right|_{t=0} = 0, \quad \forall \delta k \in H_y.$$

The computation is standard and gives the linear equation

$$C_Y k = C_{YX} h, \tag{B.97}$$

the solution of which, if it exists in H_y, is given by (B.91).

On the other hand, under our hypotheses,

$$k = C_Y^{-1} A C_X h \in H_y. \tag{B.98}$$

In fact, since $h \in H_x$ and C_X is bounded $C_X h \in H_x$; since A is bounded $H_x \to H_y$, we see that $AC_X h \in H_y$. Finally, since C_Y^{-1} is bounded $H_y \to H_y$, we see that (B.98) holds true. So $Y(k)$ is well-defined and we can write

$$\widehat{X(h)} = Y(k) = \left\langle Y, C_Y^{-1} C_{YX} h\right\rangle_{H_y}.$$

Thus, (B.91) and (B.92) are proved.

By inserting $k = C_Y^{-1} C_{YX} h$ into $\mathscr{E}^2(k)$, given by (B.96), one proves (B.93). □

We underline once more that the presence of some white noise in N has a regularizing effect on the solution, in that C_Y^{-1} is then bounded and (B.95) holds.

Remark B.2 Observe that since $\varepsilon \in \mathscr{L}^2(\Omega)$, $\mathscr{E}^2$ as given by (B.89) is just the square of the distance of $Y(k)$ from $X(h)$:

$$\mathscr{E}^2 = E\left\{\varepsilon^2\right\} = \|\varepsilon\|^2_{\mathscr{L}^2(\Omega)} = \|Y(k) - X(h)\|^2_{\mathscr{L}^2(\Omega)}. \tag{B.99}$$

So the search of the minimum of $\mathscr{E}^2$ corresponds to the search of the minimum distance (least squares) between the given element $X(h) \in \mathscr{L}^2(\Omega)$ and the element $Y(k) \in \mathscr{L}^2(\Omega)$, when k varies in H_y.

Since $Y(k) \in \mathscr{L}^2_Y$, which is a closed subspace of $\mathscr{L}^2(\Omega)$, we are searching the element in $\mathscr{L}^2_Y$ of minimum distance from $X(h)$.

Since $\mathscr{L}^2_Y$ is closed in $\mathscr{L}^2(\Omega)$, such an element exists and it is the orthogonal projection of $X(h)$ on $\mathscr{L}^2_Y$ (see [133], Theorem 1; 12.3).

Furthermore, since $\mathscr{L}^2_Y$ is isometric to $\bar{H}_y$, which in this case is H_y (see (B.95)), we find a $k \in H_y$, that minimizes $\mathscr{E}^2$. This is essentially the geometric interpretation of Theorem B.1.

Example B.3 We wish to apply the Wiener–Kolmogorov solution to a very degenerate case, in which $H_x \equiv \mathbb{R}^n$ an $H_y \equiv \mathbb{R}^m$.

Namely in this case the GRF's X, Y are just vector random variables

$$\boldsymbol{X} = \begin{bmatrix} X_1 \\ \vdots \\ X_n \end{bmatrix} \; \boldsymbol{Y} = \begin{bmatrix} Y_1 \\ \vdots \\ Y_m \end{bmatrix}; \text{ in this case also } N \text{ is a vector, } \; N = \begin{bmatrix} \nu_1 \\ \vdots \\ \nu_m \end{bmatrix}.$$

Moreover, the operator A: $H_x \to H_y$ becomes simply a matrix with m rows and n columns.

The elements of H_x, H_y are just deterministic vectors $\boldsymbol{h} \in H_x$, $\boldsymbol{k} \in H_y$ and the scalar products in H_x, H_y are the usual Euclidean products:

$$\langle h_1, h_2 \rangle_{H_x} = \boldsymbol{h}_1^\top \boldsymbol{h}_2, \langle k_1, k_2 \rangle_{H_y} = \boldsymbol{k}_1^\top \boldsymbol{k}_2.$$

Therefore, we have

$$X(h) = \boldsymbol{h}^\top \boldsymbol{X} = \sum_{i=1}^{n} h_i X_i$$

$$Y(k) = \boldsymbol{k}^\top \boldsymbol{Y} = \sum_{i=1}^{m} k_i Y_i.$$

Accordingly

$$E\{X(h)^2\} = \boldsymbol{h}^\top C_X \boldsymbol{h},$$
$$E\{Y(k)^2\} = \boldsymbol{k}^\top C_Y \boldsymbol{k},$$
$$E\{N(k)^2\} = \boldsymbol{k}^\top C_N \boldsymbol{k},$$
$$E\{Y(k)X(h)\} = \boldsymbol{k}^\top C_{YX} \boldsymbol{h},$$

where C_X, C_Y, C_N are covariance matrices and C_{YX} is a cross-covariance matrix in the usual sense.

An application of (B.92) then says that the best linear predictor of $\boldsymbol{h}^\mathrm{T}\boldsymbol{X}$ is

$$\boldsymbol{h}^\top \widehat{\boldsymbol{X}} = \boldsymbol{k}^\top \boldsymbol{Y}^\top = \boldsymbol{h}^\top C_{XY} \boldsymbol{C}_Y^{-1} \boldsymbol{Y}. \tag{B.100}$$

Since for $\boldsymbol{h}$ we can choose $\begin{bmatrix} 1 \\ 0 \\ \vdots \\ 0 \end{bmatrix}, \begin{bmatrix} 0 \\ 1 \\ \vdots \\ 0 \end{bmatrix}$ and so on, (B.100) implies that

$$\widehat{\boldsymbol{X}} = C_{XY} \boldsymbol{C}_Y^{-1} \boldsymbol{Y} \tag{B.101}$$

which is a perfectly known formula in linear regression theory (e.g., see [100]).

In particular, the known error covariance matrix

$$C_\varepsilon = C_X - C_{XY} C_Y^{-1} C_{YX}$$

corresponds perfectly to (B.94).

Based on Example B.3 one would be inclined to ask whether in formulas (B.92), (B.93) one could avoid the vector $h \in H_x$ and go directly to the prediction $\widehat{X}_t$ of X_t and its error covariance operator.

This is basically equivalent to asking whether one could predict X_t for every fixed $t \in T$. The answer is that in general this is not possible, but under some restrictive hypotheses on C_X, C_N the pointwise prediction of X_t becomes possible. This point will be more thoroughly analysed in the next section.

B.4 GRF's That Are RF's: Covariance Function

The first question we want to answer is: when does it happen that a GRF X is also an ordinary RF? This is the same as asking: when does it happen that for each fixed $t \in T$, $X(\omega, t) = X_t$, is an $\mathscr{L}^2(\Omega)$ RV? Or, otherwise, when is the evaluation functional of t, e_t, an admissible functional for X? Put in this way, we have already an answer from Lemma B.1: e_t is an admissible functional if we can represent it as an element of H_C. But, owing to Lemma B.2, the elements of $H_C$represent also the bounded functionals on $H_{C^{-1}}$, so $e_t \in H_C$ if $e_t(\bar{g})$, $\bar{g} \in H_{C^{-1}}$, is a bounded functional. But, due to Proposition B.2, the following equivalence holds: (e_t is bounded on $H_{C^{-1}}$) $\Longleftrightarrow$ ($H_{C^{-1}}$ has a reproducing kernel). However, we know, by Proposition B.5, that if the $C(t, t')$ defined by (B.52) exists and is continuous, then it is the reproducing kernel of $H_{C^{-1}}$, implying that $e_t$is an admissible functional. Since we have already studied sufficient conditions for the existence of $C(t, t')$, we arrive at the following proposition.

Proposition B.6 *e_t is an admissible functional for the GRF X, $\Longleftrightarrow$ there is a continuous function $C(t, t')$ given by (B.52).*

This happens, for instance, if $\Sigma_{n=1}^{+\infty} c_n^2 < +\infty$, i.e., (B.51) *is satisfied, and there is a* CONS $\{\varphi_n(t)\}$ *in H, with $\varphi_n(t)$ continuous and uniformly bounded with respect to n, i.e.,* (B.54) *is satisfied.*

Maybe one more direct, though heuristic, way to see that conditions (B.51), (B.54) are sufficient for the existence of $e_t(X) \in \mathscr{L}^2(\Omega)$, $\forall t \in T$, is just to recall that

$$X(h) = \sum_{n=1}^{+\infty} X_n h_n, \quad (h_n = \langle h, \varphi_n \rangle_H),$$

so that, when $\varphi_n(t)$ are continuous, we can put

$$X(h) = \left\langle \sum_{n=1}^{+\infty} X_n \varphi_n(t), h \right\rangle_H,$$

namely

$$X_t = \sum_{n=1}^{+\infty} X_n \varphi_n(t). \tag{B.102}$$

So again $X_t \in \mathscr{L}^2(\Omega)$ iff

$$E\{X_t^2\} = \sum_{n=1}^{+\infty} E\{X_n^2\}\varphi_n^2(t) = \sum_{n=1}^{+\infty} c_n^2 \varphi_n^2(t) < +\infty. \tag{B.103}$$

In any event, what interests us here is that, under the assumption that $C(t, t')$, exists the series (B.102) is $\mathscr{L}^2(\Omega)$ convergent and we can define $\{X_t, t \in T\}$ as an ordinary Random Field.

Definition B.11 (Covariance Function) The function $C(t, t')$, when it exists, is defined as the covariance function of the RF $\{X_t, t \in T\}$.

One has then

$$C(t, t') = E\{X_t, X'_t\} = E\{e_t(X)e'_t(X)\}. \tag{B.104}$$

In view of our discussion in (B.57), (B.58), and (B.59), we can state the following.

Proposition B.7 *Every covariance function is a symmetric, positive definite function and satisfies*

$$|C(t, t')|^2 \le C(t, t)C(t', t'), \tag{B.105}$$

In particular, (B.105) *expresses that the correlation coefficient of* X_t, $X_{t'}$ *is never larger in modulus than* 1.

The last remark clarifies that only if X_t is a linear function of $X_{t'}$, we may have $\rho_{t,t'} = \frac{E\{X_t, X_{t'}\}}{[E\{X_t^2\}E\{X_{t'}^2\}]^{1/2}} = \pm 1$, or the covariance matrix

$$C = \begin{bmatrix} C(t, t) & C(t, t') \\ C(t, t') & C(t', t') \end{bmatrix}$$

is singular. This generalizes to the case of any N-tuple of points $\{t_1, t_2, \ldots, t_N\}$ so that

$$C = \begin{bmatrix} C(t_1, t_1) & \cdots & C(t, t_N) \\ \vdots & \ddots & \\ C(t_1, t_N) & & C(t_N, t_N) \end{bmatrix}$$

can be singular only if there is a linear relation between $X_{t_1}, X_{t_2}, \ldots, X_{t_N}$. If we exclude such a case, the covariance function has to be strictly positive definite, i.e., (B.59) holds always with $>$ if the coefficients $\{\lambda_1, \ldots, \lambda_N\}$ are not all equal to zero.

Since we have claimed that a bounded covariance of $\{X_t\}$ exists if $e_t \in H_C$, we would like to see what is the representer of $e_t$in H_C.

Proposition B.8 *Under the given hypotheses on* C*,* $e_t \in H_C$ *and is represented by*

$$e_t(t') = \sum_{n=1}^{+\infty} \varphi_n(t)\varphi_n(t'). \tag{B.106}$$

Proof First of all, we note that in general the object in (B.106) is not a function; for instance, when $H = L^2(T)$ (see also Example B.2) it is known that $e_t(t') = \delta(t - t')$, i.e., it is a distribution. On the other hand, we know that $H_C$is a much larger space than H. So what we need to show is that the series in (B.106) is convergent in H_C. To this end it is enough to recall that $\{\bar{\varphi}_n\} = \{\frac{\varphi_n}{c_n}\}$ is a CONS in H_C, so that after rewriting (B.106) as

$$e_t(t') = \sum_{n=1}^{+\infty} \varphi_n(t) c_n \bar{\varphi}_n(t') \tag{B.107}$$

we see that (B.107) is the expansion of $e_t(t')$ in the basis $\{\bar{\varphi}_n\}$. Then $e_t(t') \in H_C$ if its Fourier coefficients are square summable, i.e.,

$$\sum_{n=1}^{+\infty} c_n^2 \varphi_n^2(t) < +\infty,$$

which is precisely the condition we imposed for the existence of $C(t, t')$. □

With the help of the covariance function one can generalize Proposition B.6 to the extent that we can characterize all admissible functionals $L \in H_C$. This is particularly useful, as we shall see in Example B.4, where the setting of the prediction problem corresponds to a finite number of observations, and not to the whole GRF Y, while we still want to predict the whole X.

To make things more readable we need first to introduce a particular notation (see [15]).

Definition B.12 (Krarup's Notation) Let L_t be a linear functional acting on functions of t; we denote

$$C(L, t') = L_t[C(t, t')] = \sum_{n=1}^{+\infty} c_n^2 L_t(\varphi_n(t))\varphi_n(t'). \tag{B.108}$$

Furthermore, if $M_{t'}$ is another linear functional, acting on functions of t', we put

$$C(L, M) = M_{t'}\{L_t[C(t, t')]\} = \sum_{n=1}^{+\infty} c_n^2 L_t[\varphi_n(t)] M_{t'}[\varphi_n(t')]. \tag{B.109}$$

Proposition B.9 *L is an admissible functional, i.e., it is represented by an element of H_C, iff*

$$C(L, L) = \sum_{n=1}^{+\infty} c_n^2 L_t[\varphi_n(t)]^2 < +\infty. \tag{B.110}$$

Moreover, if $L \in H_C$ and we consider the function $C(L, t')$, we have

$$C(L, t') \in H_{C^{-1}}; \tag{B.111}$$

in particular, $C(L, t')$ is a regular, continuous function.

Proof $L \in H_C$ means that it is a continuous functional on $H_{C^{-1}}$, namely, $|L(\tilde{g})| < +\infty$ for any $\tilde{g} \in H_{C^{-1}}$. Since $\{\bar{\varphi}_n\} = \{c_n\varphi_n\}$ is a CONS in $H_{C^{-1}}$, we can write

$$|L(\tilde{g})| = |\sum_{n=1}^{+\infty} \tilde{g}_n c_n L(\varphi_n)|. \tag{B.112}$$

We know that $\tilde{g} \in H_{C^{-1}}$ is equivalent to $\sum_{n=1}^{+\infty} \tilde{g}_n^2 < +\infty$, hence, (B.112) can be true for all $\tilde{g} \in H_{C^{-1}}$ if

$$\sum_{n=1}^{+\infty} c_n^2 L(\varphi_n)^2 = C(L, L) < +\infty.$$

Moreover, (B.108) shows that

$$C(L, t') = \sum_{n=1}^{+\infty} c_n L(\varphi_n)\tilde{\varphi}_n(t'); \tag{B.113}$$

since $\{\tilde{\varphi}_n(t)\}$ is a CONS in $H_{C^{-1}}$ and (B.110) is true when L is admissible, we see that (B.113) represents an $H_{C^{-1}}$-convergent series, i.e. $C(L, t') \in H_{C^{-1}}$. In particular, from Proposition B.6 and the relation (B.113) we deduce that the above series defining $C(L, t')$ is uniformly convergent and so $C(L, t')$ is continuous. □

We can now answer a question raised at the end of the previous section, namely, is $\widehat{X}$ a Random Field with point values?

Proposition B.10 *When X_t is an* RF, *with a bounded covariance function*

$$|C_X(t, t)| \le A \operatorname{Tr} C_X \le A\bar{c}^2, \tag{B.114}$$

for instance when Tr $C_X < \bar{c}^2$, *according to* (B.56), *then $\widehat{X}_t$ is also an* RF. *As a matter of fact, $\widehat{X}_t$ is in general a function of t smoother than X_t, namely $\widehat{X}_t \in H_{C^{-1}}$ almost surely.*

Proof Let us recall that the prediction error ε is a GRF defined as

$$\varepsilon = \widehat{X} - X; \tag{B.115}$$

in general, this GRF has a covariance operator C_ε that, according to Theorem B.1, is

$$C_\varepsilon = C_X - C_{XY}C_Y^{-1}C_{YX}. \tag{B.116}$$

Now referring to (B.92), and recalling the Definition B.6, we see that

$$\widehat{X} = C_{XY} C_Y^{-1} Y, \tag{B.117}$$

so that

$$C_{\widehat{X}} = C_{XY} C_Y^{-1} C_{YX}. \tag{B.118}$$

Therefore, (B.116) can be rewritten as

$$C_{\widehat{X}} + C_\varepsilon = C_X, \tag{B.119}$$

which is just the Pythagorean law for covariances, showing that $\widehat{X}$ and ε are not correlated. Since covariance operators are always positive, (B.119) implies that

$$C_{\widehat{X}} < C_X. \tag{B.120}$$

So if we assume that $\operatorname{Tr} C_X < +\infty$, we also have

$$\operatorname{Tr} C_{\widehat{X}} < +\infty$$

and $\widehat{X}$ admits point values $\widehat{X}_t$, i.e., $\widehat{X}_t$ is an ordinary RF.

We notice that (B.119) implies at the same time that

$$C_\varepsilon < C_X. \tag{B.121}$$

and ε is then also an ordinary RF.

As for the second statement, let us first recall that, under the hypothesis (B.114), $X_t \in H_X$ almost surely, because

$$E\{||X_t||^2_{H_x}\} = \sum_{n=1}^{+\infty} c_n^2.$$

On the other hand, $X_t \notin H_{C^{-1}}$; in fact,

$$||X_t||^2_{H_{C_X^{-1}}} = \sum_{n=1}^{+\infty} \frac{X_n^2}{c_n^2} \tag{B.122}$$

and since $E\{X_n^2\} = c_n^2$ this series is not likely to converge, as shown in Example B.1. But, on the contrary, we have $\widehat{X}_t \in H_{C_X^{-1}}$.

Indeed, compute

$$< \widehat{X}, \tilde{\varphi}_n >_{H_{C_X^{-1}}} =< C_X A^T C_Y^{-1} Y, \tilde{\varphi}_n >_{H_{C_X^{-1}}} =$$
$$=< C_X A^T C_Y^{-1} Y, C_X^{-1} \tilde{\varphi}_n >_{H_x} =< A^T C_Y^{-1} Y, \tilde{\varphi}_n >_{H_x} =$$
$$= c_n < Y, C_Y^{-1} A\varphi_n >_{H_y} = c_n Y(C_Y^{-1} A\varphi_n).$$

Now $C_Y^{-1}A$ is a bounded operator, $H_x \to H_y$ and furthermore $||\varphi_n||_{H_x} = 1$, so

$$E\{Y(C_Y^{-1} A\varphi_n)^2\} \leq a||C_Y^{-1} A\varphi_n||^2_{H_y} \leq b,$$

uniformly in n. Therefore,

$$E\{||\widehat{X}||^2_{H_{C_X^{-1}}}\} = \sum_{n=1}^{+\infty} c_n^2 E\{Y(C_Y^{-1} A\varphi_n)^2\} \leq b \sum_{n=1}^{+\infty} c_n^2 < +\infty \tag{B.123}$$

and $\widehat{X} \in H_{C_X^{-1}}$ with probability 1. □

In the next example we see a specification of the Wiener–Kolmogorov theory, presented in Sect. B.3, with the model of observation equations that is mostly used in Geophysics and Geodesy, where it is known as *general collocation theory* [133]. In fact, although we aim at predicting the RF X_t as a continuum in $t \in T$, usually the information we have on the observable GRF Y is limited to a finite number of observations $Y(k_i)$, $i = 1, 2, \ldots, N$. This means that we can write a discrete, finite set of observation equations

$$i = 1, 2, \ldots, N, \quad Y_i = L_i(X) + \nu_i, \tag{B.124}$$

so that now H_y can be taken as $\mathbb{R}^N$, $L_i$is a finite set of N admissible functionals of the RF X_t and ν_i is an ordinary random vector of noises.

For the sake of brevity we will prefer a vector version of (B.124), namely

$$\boldsymbol{Y} = \boldsymbol{L}(X) + \boldsymbol{\nu}. \tag{B.125}$$

Again, at the level of notation we will extend Krarup's rule (Definition B.11), to vectors, according to

$$C(\boldsymbol{L}, t) = \begin{bmatrix} \vdots \\ C(L_i, t) \\ \vdots \end{bmatrix} \tag{B.126}$$

$$C(\boldsymbol{L}, \boldsymbol{L}^T) = \begin{bmatrix} & \vdots & \\ \cdots & C(L_i, L_k) & \cdots \\ & \vdots & \end{bmatrix}. \tag{B.127}$$

If we assume that X_t is an RF with bounded covariance function $C_X(t, t')$ and it is independent of $\boldsymbol{\nu}$, like in Example B.3, we find that the covariance operator of $\boldsymbol{Y}$ is just a matrix and its elements are

$$E\{Y_i Y_k\} = L_i\{L_k C(t, t')\} + C_{\nu ik},$$

namely

$$C_Y = C_X(\boldsymbol{L}, \boldsymbol{L}^T) + C_\nu. \tag{B.128}$$

Similarly, $C_{YX_{t'}}$ is given by

$$C_{YX_{t'}}(t') = C_X(\boldsymbol{L}, t'). \tag{B.129}$$

Example B.4 The idea is to represent the Wiener–Kolmogorov result, namely the optimal predictor (B.101), when $\boldsymbol{Y} \in \mathbb{R}^N$. Most of the work has already been done deriving (B.101) through (B.129).

We have then

$$\begin{aligned} \widehat{X_t} &= C_X(t, \boldsymbol{L}^T)[C_X(\boldsymbol{L}, \boldsymbol{L}^T) + C_\upsilon]^{-1}\boldsymbol{Y} = \\ &= \sum_{i,k=1}^{N} C_X(t, t_i')\{C_X(\boldsymbol{L}, \boldsymbol{L}^T) + C_\upsilon]_{ik}^{-1} Y_k. \end{aligned} \tag{B.130}$$

The corresponding prediction error at a point $t \in T$ is given by (B.94) with the help of the evaluation functional e_t, namely

$$\begin{aligned} \mathcal{E}_t^2 &=< e_t, C_\varepsilon e_t >_{H_X} = C_X(t, t) - C_X(t, \boldsymbol{L}^T) C_Y^{-1} C_X(\boldsymbol{L}, t) = \\ &= C_X(t, t) - \sum_{i,k=1}^{N} C_X(t, L_i) C_X(t, L_k) C_{Y,ik}^{-1}. \end{aligned} \tag{B.131}$$

One particular case of this example is when $L_i = e_{t_i}$, namely the observation equations are just the values of X_t at t_i with an additional noise ν_i; indeed, in this case one has $C_X(t, L_i) = C_X(t, t_i)$ and $C_X(L_i, L_k) = C_X(t_i, t_k)$.

Therefore the interpolation problem, with noise filtering, has the solution

$$\widehat{X_t} = \sum_{i,k=1}^{N} C_X(t, t_i)[C_X(t_i, t_k) + C_{vik}]^{(-1)} Y_k$$

and

$$\mathscr{E}_t^2 = C_X(t, t) - \sum_{i,k=1}^{N} C_X(t, t_i)[C_X(t_i, t_k) + C_{vik}]^{(-1)} C(i_k, t)$$

Readers are advised to derive by themselves (B.130), (B.131) starting from

$$\begin{aligned}
&\boldsymbol{Y} = \boldsymbol{L}(X) + \boldsymbol{\nu}, \\
&\widehat{X}_t = \boldsymbol{\lambda}^T \boldsymbol{Y}, \\
&\varepsilon_t = \boldsymbol{\lambda}^t \boldsymbol{Y} - X_t = [\boldsymbol{\lambda}^T \boldsymbol{L}] - e_t(X) + \boldsymbol{\lambda}^T \boldsymbol{\nu}, \\
&e_t(X) = X_t, \\
&\mathscr{E}_t^2 = \boldsymbol{\lambda}^T C_X(\boldsymbol{L}, \boldsymbol{L}^T)\boldsymbol{\lambda} - 2\boldsymbol{\lambda}^T C_X(\boldsymbol{L}, t) + C(t, t) + \boldsymbol{\lambda}^T C_\nu \boldsymbol{\lambda}.
\end{aligned}$$

Thus minimization of $\mathscr{E}_t^2$ with respect to $\boldsymbol{\lambda}$ leads again to (B.130), (B.131). This approach, with no reference to GRF's, is as a matter of fact the classical one followed, e.g., by H. Moritz already in 1963 for the interpolation and prediction of gravity anomalies (in this respect see also [133]).

We close the section discussing an important matter regarding the Wiener–Kolmogorov theory. The computability of the optimal, Wiener–Kolmogorov predictor (B.130) and of its mean square error (B.131) depends on the availability of C_X and C_ν.

While C_ν is typically, at least approximately, known, or guessed, from the physics of the observation process, the covariance $C_X(t, t')$ is generally as unknown as X_t itself.

We will see in the next section, generalizing to RF the theory of stationary stochastic processes how C_X can be approximately estimated from a finite set of observed data $\{Y_{t_i}; i = 1, \ldots, N\}$, at least if the class of RF's to which X_t is assumed to belong is suitably restricted.

The procedure however clearly shows that $C_X(t, t')$ will be only roughly estimated, with an error δC_X that can be a significant percentage of C_X.

The question then arises, whether an imprecise knowledge of C_X will spoil the quality of the Wiener–Kolmogorov predictor.

Fortunately, the answer is in the negative, because there is a lemma proving a certain insensitivity of the WK predictor to errors in C_X (see [135]). We prove here this lemma under the simplifying hypothesis that $A = I$; in this case the observations Y_{t_i} are just

$$Y_i = X_{t_i} + \nu_i, \tag{B.132}$$

as discussed as well in the Example B.4 and our target is to predict $\widehat{X}_t$ at any t, namely, to filter the data (i.e., to reduce the effect of the noise) and interpolate the unknown field.

Lemma B.3 (Insensitivity of the WK Filter to C_X) *Denoting by $O(q)$ the order of magnitude of a quantity q, let $\alpha = O(\delta C_X C_X^{-1})$ be the order of magnitude of the relative error of C_X and let $\beta = O(C_\nu C_y^{-1})$, i.e., the ratio of the magnitude of noise ν with respect to the magnitude of the observations $\boldsymbol{Y}$. Then the error $\delta\widehat{X}_t$ of the* WK *predictor $\widehat{X}_t$, obeys the estimate*

$$O(\delta\widehat{X}_t) = \alpha \cdot \beta \cdot O(\widehat{X_t}). \tag{B.133}$$

Proof Let us call $\boldsymbol{X}$ the vector of the values of X_t at $t_1, t_2, \ldots, t_N$, so that our observation equation (B.132) becomes

$$\boldsymbol{Y} = \boldsymbol{X} + \boldsymbol{\nu}.$$

Let $C_X = [C_X(t_i, t_k)]$ and $C_Y = C_X + C_\nu$. Then $C_{XY} = C_X$ and the finite-dimensional filter of $\boldsymbol{Y}$ is

$$\widehat{\boldsymbol{X}} = C_X C_Y^{-1} \boldsymbol{Y}. \tag{B.134}$$

By differentiating (B.134) with respect to C_X we want to prove that $\delta\widehat{X}$ satisfies (B.133).

In fact, considering that $\delta C_Y = \delta C_X$,

$$\begin{aligned}\delta\widehat{\boldsymbol{X}} &= \delta C_X C_Y^{-1}\boldsymbol{Y} - C_X C_Y^{-1}\delta C_X C_Y^{-1}\boldsymbol{Y} = \\ &= (I - C_X C_Y^{-1})\delta C_X C_X^{-1} C_X C_Y^{-1}\boldsymbol{Y} = (I - C_X C_Y^{-1})(\delta C_X C_X^{-1})\widehat{\boldsymbol{X}}.\end{aligned} \tag{B.135}$$

On the other hand

$$I - C_X C_Y^{-1} = I - (C_X + C_\nu)C_Y^{-1} + C_\nu C_Y^{-1} = C_\nu C_Y^{-1},$$

so that (B.135) reads

$$\delta\widehat{\boldsymbol{X}} = (C_\nu C_Y^{-1})(\delta C_X C_X^{-1})\widehat{\boldsymbol{X}}. \tag{B.136}$$

Passing to the order of magnitude of the various factors of (B.136), we obtain (B.133). □

Example B.5 To see the significance of the "insensitivity lemma" let us put numbers in an example.

Assume that the standard deviation of the noise ν is 50% of that of the signal X. Assume further that you know C_X with a 20% error. Then

$$\alpha = O(\delta C_X C_X^{-1}) = 0.2$$

and

$$\beta = O(C_\nu C_Y^{-1}) \sim \frac{0.5^2 O(C_X)}{(0.5^2 + 1^2) O(C_X)} = 20\%.$$

Therefore, we have

$$O(\delta X) = 4\% O(\widehat{X}),$$

namely the error δX is much smaller than the signal.

B.5 Homogenous Isotropic Random Fields (HIRF)

The concept of HIRF is a direct generalization of that of stationary stochastic processes. The latter says that if $\{X_t, t \in \mathbb{R}\}$ is a stochastic process, then the statistical behaviour of $\{X_{t-t_0}, t \in \mathbb{R}\}$ is identical for and fixed t_0 to that of X_t, i.e., the statistics of X_t (e.g., mean, covariance, finite order distributions, etc.) are invariant under time translation. The homogeneity and isotropy concept says that, if T is $\mathbb{R}^n$ (the case $T = \mathbb{R}^2$ is particularly important in this text) and if

$$\tau = t_0 + \mathscr{R} t \quad (\mathscr{R}^T = \mathscr{R}^{-1}) \tag{B.137}$$

is a roto-translation of $\mathbb{R}^n$ (t_0 is the translation and the matrix $\mathscr{R}$ represent a proper rotation), then

$$X_\tau \sim X_t,$$

in the sense that they possess the same finite-dimensional distributions and, in particular, they have the same statistics, e.g., the same mean and the same covariance function.

Since these two statistics are those that primarily interest us for the application of linear prediction theory, we will study in more depth what is the effect that the homogeneity and isotropy hypotheses produce on these two statistics.

Since we will use these concepts for GRF's too, particularly for white or coloured noises, we will start our definitions in a form adequate to such fields, to come later to a specification for more manageable classes of fields.

Definition B.13 Let $T = \mathbb{R}^n$ and

$$\mathscr{T} = \mathscr{T}(t_0, \mathscr{R}) \tag{B.138}$$

be a roto-translation transformation (B.137), i.e.,

$$\tau = \mathscr{T}(t_0, \mathscr{R})t = t_0 + \mathscr{R}t; \tag{B.139}$$

let X be a GRF on a Hilbert space H of functions of $t \in \mathbb{R}^n$, such that $h(t) \in H \Longrightarrow h(\mathscr{T}t) \in H$; we say that X possesses the (HI) property in the mean if

$$\forall h \in H, \qquad E\{X[h(\mathscr{T}t)]\} = E\{X[h(t)]\}. \tag{B.140}$$

We don't need a formal proof to understand that if $X(h)$ has zero mean for all $h \in H$, then so does $X[h(\mathscr{T}t)]$, because $h(\mathscr{T}t) \in H$. So in the case we treat here, all GRF's are (*HI*) in the mean and more precisely they have zero mean when coupled with any $h \in H$.

Definition B.14 Given a transformation group $\{\mathscr{T}\}$, like roto-translations in $\mathbb{R}^n$, we define an operator $\mathscr{T}$, denoted by the same symbol as the transformation, through its action on $h(t) \in H$, by

$$\mathscr{T}^{-1}h(t) \equiv h(\mathscr{T}t). \tag{B.141}$$

Moreover, we say that (the topology of) H is invariant under $\{\mathscr{T}\}$ if

$$||\mathscr{T}^{-1}h(t)||_H^2 = ||h(\mathscr{T}t)||_H^2 = ||h(t)||_H^2. \tag{B.142}$$

Remark B.3 It is obvious from (B.142) that, when (the topology of) H is invariant under $\{\mathscr{T}\}$, the operator $\mathscr{T}^{-1}$ is unitary, i.e., $\mathscr{T}$ itself is unitary and

$$\mathscr{T}^T\mathscr{T} \equiv \mathscr{T}\mathscr{T}^T \equiv I_H. \tag{B.143}$$

In addition, it is clear that for any $h, k \in H$,

$$\langle h(\mathscr{T}t), k(\mathscr{T}t)\rangle_H \equiv \langle h(t), k(t)\rangle_H\,. \tag{B.144}$$

Example B.6 Take $T \equiv \mathbb{R}^2$ to fix the ideas, and $H \equiv L^2(\mathbb{R}^2)$; then, since the area element d_2t is invariant under roto-translations, i.e., the Jacobian of the transformation (B.137) is equal to 1, or equivalently

$$d_2\tau = d_2t,$$

one has

$$\begin{aligned}||h(t_0 + \mathscr{R}t)||_H^2 &= \int_{\mathbb{R}^2} h^2(t_0 + Rt)d_2t = \\ &= \int_{\mathbb{R}^2} h^2(\tau)d_2\tau = ||h(t)||_H^2.\end{aligned}$$

So $L^2(\mathbb{R}^2)$ is invariant under roto-translations.

The same is true for instance, for the Sobolev space $H = H^{1,2}(\mathbb{R}^2)$, the space of functions that are in L^2 together with their first derivatives, equipped with the norm

$$h||^2_{H^{1,2}} \equiv \int_{\mathbb{R}^2} [h^2(t) + |\nabla h(t)]^2 d_2 t.$$

Definition B.15 Let H be invariant under $\{\mathscr{T}\}$ and X be a GRF on H. We say that X is (HI) in the covariance operator C, if

$$\forall h \in H, \qquad E\{X^2[h(\mathscr{T}t)]\} \equiv E\{X^2[h(t)]\} = \langle h, Ch\rangle_H. \tag{B.145}$$

Proposition B.11 *The* GRF X *is* (HI) *in the covariance operator, iff*

$$\mathscr{T}\, C\, \mathscr{T}^T = C, \tag{B.146}$$

or C *commutes with* $\mathscr{T}$*:*

$$\mathscr{T}\, C = C\, \mathscr{T}. \tag{B.147}$$

Proof Since $\mathscr{T}$ is unitary,

$$h(\mathscr{T}t) = \mathscr{T}^T h(t),$$

so that

$$\begin{aligned} E\{X^2[h(\mathscr{T}t)]\} &\Longrightarrow \left\langle \mathscr{T}^T h, C\mathscr{T}^T h \right\rangle_H \equiv \\ &\equiv \left\langle h, \mathscr{T}\, C \mathscr{T}^T h \right\rangle_H \end{aligned} \tag{B.148}$$

Now recall that if A is a self-adjoint operator, i.e., $A^T = A$, then

$$\langle h, Ah\rangle = 0 \text{ for all } h \in H \Longrightarrow A = 0. \tag{B.149}$$

Indeed, if $\langle h, Ah\rangle = 0$, $\forall h \in H$, *then also,*

$$\langle k, Ah\rangle_H = \frac{1}{4}\{\langle h+k, A(h+k)\rangle_H - \langle h-k, A(h-k)\rangle_H\} = 0$$

$\forall k, h \in H$ and (B.149) follows. But then, since (B.148) has to be equal to $\langle h, Ch\rangle_H$, (B.146) is proved.

(B.147) comes from (B.146) upon multiplying by $\mathscr{T}$ on the right. □

Remark B.4 We note that if X is a WN with respect to the invariant H, then it is (HI) in the covariance. In fact, in this case $C_X = I_H$ and

$$\mathscr{T} C_X \mathscr{T}^T \equiv \mathscr{T}\mathscr{T}^T \equiv I_H \equiv C_X. \tag{B.150}$$

Since from now on we will treat mainly RF that have covariance operators of the form

$$C_X = c_0^2 I + C,$$

with C the operator with bounded kernel $C(t, t')$, and we already know that the WN component of C is (HI), we are interested in how the condition (B.111) translates into a condition on the covariance function.

Proposition B.12 *The* RF $\{X_t\}$ *is* (HI) *in the covariance, iff its covariance* $C(t, t')$ *is a function of* $|t - t'|$ *only, i.e.,*

$$C(t, t') = F(|t - t'|). \tag{B.151}$$

Proof Since

$$|\mathscr{T}t - \mathscr{T}t'| \equiv |t - t'|$$

the if part is trivial. For the only if part, we observe that necessarily

$$\begin{aligned} C(t, t') &\equiv E\{X_t X_{t'}\} = \\ &= E\{X_{t_0+\mathscr{R}t} X_{t_0+\mathscr{R}t'}\} \end{aligned} \tag{B.152}$$

for all $t_0, \mathscr{R}$. But then, taking $t_0 = -\mathscr{R}t'$ in (B.152) we get

$$\begin{aligned} C(t, t') &= E\{X_{\mathscr{R}(t-t')} X_0\} = \\ &= C(\mathscr{R}(t - t'), 0) = G[\mathscr{R}(t - t')]. \end{aligned} \tag{B.153}$$

Since $G[\mathscr{R}(t - t')]$ must have the same value whatever is $\mathscr{R}$, i.e., rotating $t - t'$ to all points on the sphere of radius $|t - t'|$, one has to have

$$G[\mathscr{R}(t - t')] = F(|t - t'|),$$

as we needed to prove. □

B.6 The Ergodic Theory for HIRF's and the Empirical Covariance Estimation

In this section we will assume, just to fix the ideas, that $T = R^2$, generalizations to R_n being straightforward.

Moreover, we shall assume that $\{X_t\}$ is a RF on L^2, namely $H = L^2$ (or a smaller space), so that

$$\begin{aligned} \forall f \in L^2 \qquad & E\{X_t(f)^2\} = \langle f, C_X f\rangle_{L^2} \equiv \\ & \equiv \int_{\mathbb{R}^2}\int_{\mathbb{R}^2} f(t)C_X(t,t')f(t')dt'dt. \end{aligned} \tag{B.154}$$

We already know that $L^2(\mathbb{R}^2)$ is invariant under roto-translations (see Example B.6); if we further assume that $\{X_t\}$ is (HI), we know as well, by Proposition B.12, that the covariance function of X has the form

$$C_X(t,t') = C_X(|t-t'|) \tag{B.155}$$

and (B.154) writes

$$E\{X_t(f)^2\} = \int_{\mathbb{R}^2}\int_{\mathbb{R}^2} f(t)C_X(|t-t'|)f(t')dtdt'. \tag{B.156}$$

Since the lemma we will prove below will use essentially the homogeneity property, i.e., the translation invariance, at the level of notation we will use the form $C_X(t-t')$ instead of (B.155), both being somewhat imprecise as they use the same symbol for different functions. Yet in this way the lemma will be valid for any RF enjoying the (H) property only.

The problem we tackle in this section is to estimate the covariance function from "data". We start by assuming that $\{X_t\}$ is known on a bounded neighborhood of the origin Q, that, to be specific, we take as a square of side $2L$:

$$Q \equiv \{t;\ |t_1| \leq L,\ |t_2| \leq L\}. \tag{B.157}$$

We will then derive an integral estimator of $C(\tau)$ that is unbiased and even asymptotically consistent, for $L \to \infty$. The discretization of the estimator, when only a finite number of data values is known, will then be derived at the end of the section; in this process the further isotropy hypothesis will prove to be useful.

Although we know that $E\{X_t\} = 0$, so that any continuous, homogeneous linear functional of $X_t$is an unbiased estimator of the mean of X_t, we start from this case, to use later the result of the following proposition.

Proposition B.13 *Let $C_X(t)$ be an absolutely integrable function on $\mathbb{R}^2$, i.e.,*

$$\int_{\mathbb{R}^2} |C_X(t)|d_2t < +\infty. \tag{B.158}$$

Then the functional of $\{X_t\}$ given by,

$$m_Q(X_t) = \frac{1}{|Q|}\int_Q X_t d_2t \tag{B.159}$$

($|Q| = 4L^2$), is an unbiased estimator of $E\{X_t\} = 0$, i.e.,

$$E\{m_Q(X_t)\} = 0, \tag{B.160}$$

and it is also asymptotically consistent, in the sense that

$$\lim_{L\to\infty} E\{m_Q(X_t)^2\} = 0. \tag{B.161}$$

Proof Let us note that the function

$$M_Q(\boldsymbol{t}) = \frac{1}{|Q|}\chi_Q(\boldsymbol{t}), \tag{B.162}$$

with $\chi_Q(\boldsymbol{t})$ the characteristic function of Q, is an L^2 function or, more precisely, that

$$\int_{\mathbb{R}^2} M_Q(\boldsymbol{t})d_2t = 1, \int_{\mathbb{R}^2} M_Q^2(\boldsymbol{t})d_2t = \frac{1}{|Q|}. \tag{B.163}$$

Then

$$m_Q(X_t) = X_t[M_Q(\boldsymbol{t})] \tag{B.164}$$

is an $\mathscr{L}^2(\Omega)$ RV and (B.160) is true.

On the other hand, exploiting (B.156), one has

$$\begin{aligned} E\{m_Q(X_t)^2\} &= \frac{1}{|Q|^2}\int_Q\int_Q C_X(\boldsymbol{t}-\boldsymbol{t}')d_2t \le \\ &\le \frac{1}{|Q|^2}\int_Q\left(\int_{\mathbb{R}^2}|C_X(\boldsymbol{t}-\boldsymbol{t}')|d_2t'\right)d_2t = \\ &= \frac{1}{|Q|}\int_{\mathbb{R}^2}|C_X(\boldsymbol{s})|d_2s. \end{aligned} \tag{B.165}$$

Given the hypothesis (B.158) and observing that $|Q| = 4L^2$, (B.161) follows. □

Now let us introduce the following RF, function of $\{X_t\}$:

$$Y_t = X_t X_{t+\tau} - C_X(\tau); \tag{B.166}$$

here τ is regarded as a fixed vector.

Proposition B.14 *Assume $\{X_t\}$ is homogeneous up to the fourth order and has finite and uniformly bounded fourth-order moment, namely*

$$E\{X_t^4\} \le \bar{c}^4 < +\infty. \tag{B.167}$$

Then $\{Y_t\}$ is a zero-mean RF *in $\mathscr{L}^2(\Omega)$, i.e.,*

$$E\{Y_t\} = 0, \quad E\{Y_t^2\} < +\infty \tag{B.168}$$

and $\{Y_t\}$ has a homogenous covariance function

$$E\{Y_t Y_t'\} = C_Y(t - t'). \tag{B.169}$$

Proof That $\{Y_t\}$ ha zero mean is straightforward, because

$$E\{X_t X_{t+\tau}\} - C_X(\tau) = 0.$$

That it has a finite variance comes from Schwarz' inequality, i.e.,

$$\begin{aligned} E\{Y_t^2\} &= E\{X_t^2 X_{t+\tau}^2\} - C_X^2(\tau) \le \\ &\le E\{X_t^4\}^{1/2} E\{X_{t+\tau}^4\}^{1/2} \le \bar{c}^4. \end{aligned}$$

That $\{Y_t\}$ is homogenous in the covariance comes from the analogous property of $\{X_t\}$ up to the fourth order, namely

$$\begin{aligned} E\{Y_t Y_t'\} &= E\{X_t X_{t+\tau} X_{t'} X_{t'+\tau}\} - C_X^2(\tau) \\ &= E\{X_0 X_\tau X_{t'-t} X_{t'+\tau-t}\} - C_X^2(\tau). \end{aligned} \tag{B.170}$$

which, when τ is fixed, is depends only on $t' - t$. □

We can now prove the main result of this section, which is basically equivalent to the weak ergodic property of stochastic processes (see [100]).

Lemma B.4 *Assume that $C_Y(t)$ is absolutely integrable on $\mathbb{R}^2$ and the hypotheses of Proposition* B.14 *are satisfied. Then*

$$m_Q(Y_t) = \frac{1}{|Q|} \int_Q Y_t \, d_2 t = \frac{1}{|Q|} \int_Q X_t X_{t+\tau} \, d_2 t - C_X(\tau) \tag{B.171}$$

is an unbiased and asymptotically consistent estimator of 0, *i.e.*

$$\widehat{C}_{Q,X}(\tau) = \frac{1}{|Q|} \int_Q X_t X_{t+\tau} \, d_2 t \tag{B.172}$$

is an unbiased and asymptotically consistent estimator of $C_X(\tau)$, for fixed τ, when $L \to \infty$

Proof The proof is obtained by combining Propositions B.13 and B.14 and the further observation that

$$E\{m_Q(Y_t)\} = 0 \Longrightarrow E\{\widehat{C}_{Q,X}(\tau)\} = C_X(\tau),$$
$$E\{m_Q^2(Y_t)\} \to 0 \Longrightarrow E\{\widehat{C}^2_{Q,X}(\tau)\} - C_X^2(\tau) \equiv \sigma^2\{\widehat{C}_{Q,X}(\tau)\} \to 0.$$

□

Remark B.5 One might argue that the hypothesis (B.170) is difficult to verify on the basis of the knowledge of $\{X_t\}$ alone. Nevertheless, we want to strengthen that (B.170) basically says that the functional relation of X_t and $X_{t'}$ up to the fourth order is in fact decaying fast enough when t and t' are separated by a large distance. In any way, let us elaborate this condition when $\{X_t\}$ is a Gaussian homogeneous process. In this case in fact the following remarkable formula holds:

$$\begin{aligned} E\{X_{t_1}X_{t_2}X_{t_3}X_{t_4})\} = C_X(t_2 - t_1)C_X(t_4 - t_3) + \\ + C_X(t_4 - t_1)C_X(t_3 - t_2) + C_X(t_3 - t_1)C_X(t_4 - t_2). \end{aligned} \tag{B.173}$$

This can be proved by considering the moment generating function of the normal vector variable $\boldsymbol{X} = [X_1, X_2, X_3, X_4]^T$, with distribution $\mathscr{N}[O, C]$, where $C_{ik} \equiv C_X(t_i - t_k)$. One has

$$E\{e^{\boldsymbol{\lambda}^T \boldsymbol{X}}\} = e^{\frac{1}{2}\boldsymbol{\lambda}^T C \boldsymbol{\lambda}} = g_X(\boldsymbol{\lambda}), \qquad \forall \boldsymbol{\lambda} \in \mathbb{R}^4, \tag{B.174}$$

so that

$$D_{\lambda_1} D_{\lambda_2} D_{\lambda_3} D_{\lambda_4} g_X(\boldsymbol{\lambda})|_{\lambda=0} = E\{X_1 X_2 X_3 X_4\}; \tag{B.175}$$

a lengthy computation gives (B.173).

If we apply (B.173) for $t_1 = t$, $t_2 = t + \tau$, $t_3 = t'$, $t_4 = t' + \tau$ we find

$$\begin{aligned} C_Y(t, t') &= E\{Y_t Y_{t+\tau}\} = E\{X_t Y_{t+\tau} X_{t'} X_{t'+\tau}\} - C_X^2(\tau) = \\ &= C_X^2(\tau) + C_X(t' + \tau - t)C_X(t' - t - \tau) + \\ &\quad + C_X(t' - t)C_X(t' - t) - C_X^2(\tau) = \\ &= C_X(t' - t)^2 + C_X(t' - t + \tau)C_X(t' - t - \tau) \end{aligned} \tag{B.176}$$

This proves once more that $C_Y(t, t') = C_Y(t - t')$, but even more, by applying the Schwarz inequality one gets

$$\int_{\mathbb{R}^2} |C_Y(\boldsymbol{t})| d_2 t \leq \int_{\mathbb{R}^2} C_X^2(\boldsymbol{t}) d_2 t + \int_{\mathbb{R}^2} |C_X(\boldsymbol{t}+\tau) C_X(\boldsymbol{t}-\tau)| d_2 t$$
$$\leq 2 \int_{\mathbb{R}^2} C_X^2(\boldsymbol{t}) d_2 t.$$

Hence $C_Y(\boldsymbol{t})$ is absolutely integrable if $C_X(\boldsymbol{t}) \in L^2(\mathbb{R}^2)$; moreover, since

$$|C_X(\boldsymbol{t})| \leq C(0) < +\infty,$$

it is enough to assume that $C_X(\boldsymbol{t})$ is absolutely integrable to guarantee the same property for $C_Y(\boldsymbol{t})$.

Remark B.6 Let us assume that the RF $\{X_t\}$ is both homogenous and isotropic in the covariance, i.e.,

$$C_X(\boldsymbol{t}, \boldsymbol{t}') = C_X(|\boldsymbol{t} - \boldsymbol{t}'|). \tag{B.177}$$

Let α denote the angle between any $\boldsymbol{\tau} \in \mathbb{R}^2$ with the first axis (t_1), so that with $\boldsymbol{\tau}_\alpha$ we can denote a vector in $\mathbb{R}^2$ with modulus $|\boldsymbol{\tau}|$ and angle α. Let us now return to (B.172), which we write

$$\widehat{C}_{Q,X}(\tau_\alpha) = \frac{1}{|Q|} \int_Q X_t X_{t+\tau_\alpha} d_2 t. \tag{B.178}$$

If (B.177) is true, then for any α,

$$E\{\widehat{C}_{Q,X}(\tau_\alpha)\} = C_X(|\tau|),$$

so that, averaging with respect to α,

$$\frac{1}{2\pi} \int_0^{2\pi} E\{\widehat{C}_{Q,X}(\tau_\alpha)\} d\alpha = E\{\frac{1}{2\pi} \int \widehat{C}_{Q,X}(\tau_\alpha)\} d\alpha = C_X(|\tau|).$$

This proves that

$$\widehat{C}_{Q,X}(|\tau|) = \frac{1}{2\pi} \int_0^{2\pi} \widehat{C}_{Q,X}|(\tau_\alpha)| \, d\alpha = \frac{1}{2\pi} \int_0^{2\pi} \frac{1}{|Q|} \int_Q X_t X_{t+\tau_\alpha} d_2 t \, d\alpha \tag{B.179}$$

is also an unbiased estimator of $C_X(|\tau|)$.

The asymptotic consistency of (B.179) is also not difficult to prove, considering that

$$\left(\frac{1}{2\pi} \int_0^{2\pi} \widehat{C}_{Q,X}(\tau_\alpha) \, d\alpha \right)^2 \leq \frac{1}{2\pi} \int_0^{2\pi} \widehat{C}_{Q,X}(\tau_\alpha)^2 d\alpha.$$

The advantage of (B.179) over (B.172) will be soon clarified discussing the discrete version of such a formula.

Here we just note one further way of writing (B.179), i.e.,

$$\widehat{C}_{Q,X}(|\tau|)\frac{1}{|Q|}\int_Q X_t\left(\frac{1}{2\pi}\int_0^{2\pi} X_{t+\tau_\alpha}d\alpha\right)d_2t. \tag{B.180}$$

Namely, the estimator (B.180) is obtained by the following operations: first pick up a point $t \in Q$ and consider the circle of centered $\boldsymbol{t}$ and radius $|\tau|$; then take the average of $X_{t'}$ on this circle; finally multiply this average by $X_t$and average the resulting function of t on Q.

Remark B.7 (Discretization) Driven by the description of the estimator (B.180) given above, we propose the following discretized version, to be applied when data are not given as a continuum on Q, but are rather a sample of N values of $X_t$at the points $(t_1, t_2, \ldots, t_N)$, densely distributed in a region Q.

"Densely" here means that the correlation between samples at nearest neighbouring points is high on the average, e.g. larger than 75%.

Pick up a point $\boldsymbol{t}_i$ where X_t has been sampled; fixing $|\tau|$ define the interval $I_\Delta(|\tau|) = [|\tau| - \frac{\Delta}{2}, |\tau| + \frac{\Delta}{2}]$ for a suitably small Δ; take all the points $\boldsymbol{t}_j$ in the annulus

$$\Gamma_\Delta(t_i) = \{\boldsymbol{t}_j;\ |\boldsymbol{t}_j - \boldsymbol{t}_i| \in I_\Delta(|\tau|)\} \tag{B.181}$$

and let N_i be the number of $\boldsymbol{t}_j \in \Gamma_\Delta(\boldsymbol{t}_i)$; form the average of X_t on $\Gamma_\Delta(\boldsymbol{t}_i)$, namely

$$\bar{X}_{t_i} = \frac{1}{N}\sum_{t_j\in\Gamma_\Delta(t_i)} X_{t_j}; \tag{B.182}$$

finally, take the product of X_{t_i} with $\bar{X}_{t_i}$ and average over all t_i.

The estimate thus obtained is

$$\widehat{C}_X(|\tau|)\frac{1}{N}\sum_{i=1}^N X_{t_i}\bar{X}_{t_i} = \frac{1}{N}\sum_{i=1}^N X_{t_i}\left(\frac{1}{N_i}\sum_{t_j\in\Gamma_\Delta(t_i)} X_{t_j}\right) \tag{B.183}$$

and it constitutes a direct discretization of (B.180).

A wise choice of Δ is essential in order to get a good estimator of $C_X(|\tau|)$; Δ should be large enough to guarantee a sufficiently high value of N_i, but at the same time it should be small enough to reduce the dithering problem implicit in (B.183). In fact, if we take the expectation of (B.183) we get

$$E\{\widehat{C}_X(|\tau|) = \frac{1}{N}\sum_{i=1}^N \frac{1}{N_i}\sum_{\boldsymbol{t}_j\in\Gamma_\Delta(\boldsymbol{t}_i)} C(|\boldsymbol{t}_i - \boldsymbol{t}_j|).$$

Now, since $|\boldsymbol{\tau}| - \frac{\Delta}{2} \leq |\boldsymbol{t}_i - \boldsymbol{t}_j| \leq |\boldsymbol{\tau}| + \frac{\Delta}{2}$ we have indeed $C(|\boldsymbol{t}_i - \boldsymbol{t}_j|) \cong C(|\boldsymbol{\tau}|)$, implying that $C_X(|\boldsymbol{\tau}|)$ is an almost unbiased estimator of $C_X(|\boldsymbol{\tau}|)$, but such a relation is more precise if Δ is small.

Let us further notice that $|\boldsymbol{\tau}|$ cannot be very large relative to the size of Q (the area in which the points $\{t_i\}$ are densely distributed). For instance, if Q is a square of side $2L$, when $|\boldsymbol{\tau}| \; > L$ in no point $\boldsymbol{t}_j \in Q$ one can find a full circle centered at $\boldsymbol{t}_i$ and of radius $|\boldsymbol{\tau}|$, all contained in Q.

A good rule is to avoid estimating $C_X(|\boldsymbol{\tau}|)$ for $|\boldsymbol{\tau}| > L/2$. A good practical rule is to choose Δ so that

$$\frac{L}{50} \leq \Delta \leq \frac{L}{20},$$

corresponding to the estimate of 10–25 empirical values of the covariance.

To conclude, let us observe that the number of couples $\boldsymbol{t}_i, \boldsymbol{t}_j$ such that $|t_i - t_j| \in I_\Delta(|\boldsymbol{\tau}|)$ is much larger than the number of couples such that $\boldsymbol{t}_i - \boldsymbol{t}_j \approx \boldsymbol{\tau}$; for this reason, averaging on $\Gamma_\Delta(\boldsymbol{t}_i)$ has the effect to improve the quality of the estimator $C_X(|\boldsymbol{\tau}|)$ and we have discretized (B.179) instead of (B.172).

More on the empirical estimation of the covariance function can be found the text, in Sect. 6.6.

Remark B.8 (Discrete Noisy Data) Assume that, instead of knowing directly the samples of X_t, one has the values

$$Y_{t_i} = X_{t_i} + \nu_i, \tag{B.184}$$

i.e., a noisy version of X_t. For the sake of simplicity, assume that ν_i are uncorrelated and with the same variance σ_ν^2. Then we notice that by applying (B.183) to Y_t, for $\boldsymbol{\tau} = 0$,

$$\widehat{C}_Y(|\boldsymbol{\tau}|) \sim C_Y(|\boldsymbol{\tau}|) = C_X(|\boldsymbol{\tau}|), \tag{B.185}$$

because for $\boldsymbol{t}_j \neq \boldsymbol{t}_i$ one has $E\{\nu_j \nu_i\} = 0$. If, on the contrary, we compute $\widehat{C}_Y(0)$, we have

$$\widehat{C}_Y(0) = \frac{1}{N} \sum_{i=1}^{N} Y_{t_i}^2 \sim C_X(0) + \sigma_\nu^2. \tag{B.186}$$

Therefore if $\boldsymbol{\tau} \; \neq 0$ the empirical covariance of Y_t provides an estimate of $C_X(|\boldsymbol{\tau}|)$; if this can be modelled and extrapolated to $\boldsymbol{\tau} = 0$, one can use (B.186) to estimate σ_ν^2.

One of the essential features characterizing a covariance function is its positive definiteness; in fact, this is the property that makes the mean square prediction error of the WK theory smaller than the variance of X_t (see (B.94)), a requirement that is essential for asserting that the prediction error of $\widehat{X}_t$ is better than the crude use of 0 as predictor.

What is a positive definite function has been defined by formula (B.59). The direct application of the definition though is quite difficult and we will find in the next section a much easier equivalent condition in terms of Fourier Transforms, at least for HIRF's.

On the other hand, the use of a generic interpolator to create a continuous model $\widehat{C}_X(|\boldsymbol{\tau}|)$ from the empirical estimate $\{\widehat{C}_X(|\boldsymbol{\tau}|_k)\}$ very easily produces a non-positive definite function. This is why a common practice is to introduce a family of functions $C_{X,M}(|\boldsymbol{\tau}|, \boldsymbol{\theta})$, depending on parameters $\boldsymbol{\theta}$, that are known a priori to be positive definite, and then to adapt $\boldsymbol{\theta}$ in such a way that $C_{X,M}(|\boldsymbol{\tau}|, \boldsymbol{\theta})$ interpolates the empirical values $\{\widehat{C}_X(|\boldsymbol{\tau}|_k)\}$. Typical features of the empirical covariance that we would like to see well approximated by the model covariance are:

- the value of $\widehat{C}_X(0)$ (if available), which is also called *power of the signal;*
- the correlation length, namely the value of $|\boldsymbol{\tau}|_c$ such that

$$\widehat{C}_X(|\tau|_c) = \frac{1}{2}\widehat{C}_X(0);$$

- the value of the first and subsequent zeros, when they are well visible in the empirical covariance;
- the location and values of further relative maxima and minima, when they are well visible in the covariance.

The reader is warned that often, when the empirical covariance function is longtailed, we find oscillating values at large lags $|\boldsymbol{\tau}|_k$, with a small magnitude compared to $\widehat{C}_X(0)$ (e.g., $|\widehat{C}_X(|\boldsymbol{\tau}|_k) < 0.1\widehat{C}_X(0)$).

This feature depends mostly on the fact that, when $|\boldsymbol{\tau}|$ increases beyond a certain value, the number of couples $(\boldsymbol{t}_i, \boldsymbol{t}_j)$ on which the estimator (B.183) is evaluated drops and consequently its estimation error inflates, so that average values close to zero can easily result in positive or negative estimates.

B.7 The Fourier Transform of a HIRF and the Spectrum

Let us consider a GRF X on $L^2(\mathbb{R}^2)$; by definition,

$$\forall f \in L^2\left(\mathbb{R}^2\right), \quad E\left\{X(f)^2\right\} = \langle f, C_X f\rangle_{L^2(\mathbb{R}^2)} = \|f\|^2_{L^2(\mathbb{R}^2)}\,. \tag{B.187}$$

On the other hand, by Perseval's identity, one has

$$\|f\|^2_{L^2(\mathbb{R}^2)} = \int_{\mathbb{R}^2} \left|\hat{f}(\boldsymbol{p})\right|^2 d_2 p \tag{B.188}$$

where as usual we have denoted by $\hat{f}(\boldsymbol{p})$ the Fourier transform of $f(t)$.

In particular in this section the hat symbol will not be used to denote "the estimator of". It follows that we can transfer as well $X(f)$ to a bounded operator acting on the space of Fourier transforms in $L^2(\mathbb{R}^2)$, i.e., we can define

$$\hat{X}\left(\hat{f}\right) = X(f), \quad f \in L^2\left(\mathbb{R}^2\right), \quad \hat{f} = \mathscr{F}(f). \tag{B.189}$$

Note that, since $\hat{f}$ is in general a complex function, in reality $\hat{f} \in \hat{L}^2(\mathbb{R}^2)$, a proper closed subspace of the complex functions in $L^2(\mathbb{R}^2)$; namely $\hat{L}^2(\mathbb{R}^2)$, as the space of transforms of real functions, is the subspace of the complex $L^2(\mathbb{R}^2)$ consisting of the functions satisfying

$$f^*(\boldsymbol{p}) = f(-\boldsymbol{p}); \tag{B.190}$$

we note that real even functions do satisfy (B.190) and so if they are in L^2, they belong to the subspace $\hat{L}^2$ too.

As a consequence, also $\hat{X}$ as a GRF is complex-valued and, when it has a realization in terms of a regular random function $\hat{X}$, satisfies the relation (B.190) too.

Definition B.16 By definition, recalling also (B.81), for any $\hat{f} \in \hat{L}^2(\mathbb{R}^2)$,

$$\hat{X}\left(\hat{f}\right) = \hat{X}\{\mathscr{F}(f)\} = \mathscr{F}^T\hat{X}(f) \equiv X(f).$$

That is, by definition, $\mathscr{F}^T\hat{X} \equiv X$, and so we call the GRF $\hat{X}$ the Fourier transform of X.

Proposition B.15 *Considering that*

$$E\left\{X(f)^2\right\} = \langle f, C_X f\rangle_{L^2(\mathbb{R}^2)} = E\left\{\left|\hat{X}\left(\hat{f}\right)\right|^2\right\} = \left\langle \hat{f}, C_{\hat{X}}\hat{f}\right\rangle_{\hat{L}^2(\mathbb{R}^2)} \tag{B.191}$$

we find

$$C_X = \mathscr{F}^T C_{\hat{X}}\mathscr{F}. \tag{B.192}$$

Proof Using $\hat{f} = \mathscr{F}(f)$ and $\langle f, g\rangle_{L^2} = \left\langle \hat{f}, \hat{g}\right\rangle_{\hat{L}^2}$, we obtain

$$\begin{aligned}\left\langle \hat{f}, C_{\hat{X}}\hat{f}\right\rangle_{\hat{L}^2} &= \langle \mathscr{F} f, C_{\hat{X}}\mathscr{F} f\rangle_{\hat{L}^2} = \\ &= \left\langle f, \mathscr{F}^T C_{\hat{X}}\mathscr{F} f\right\rangle_{\hat{L}^2} \equiv \langle f, C_X f\rangle_{L^2},\end{aligned}$$

i.e., (B.192) holds. □

Proposition B.16 *Assume that X has the* (HI) *property and its covariance function is a regular integrable function, so that it has a Fourier transform*

$$S_X(p) = \int_{\mathbb{R}^2} e^{i2\pi \boldsymbol{p}\cdot\boldsymbol{\tau}} C_X(\tau)\, d_2\tau, \tag{B.193}$$

with $S_X(p)$ a real function of $|\boldsymbol{p}| = p$ only. Then the covariance operator of $\hat{X}$, $C_{\hat{X}}$, corresponds to a distributional covariance function, namely

$$C_{\hat{X}}(\boldsymbol{p}-\boldsymbol{q}) = S_X(p)\,\delta(\boldsymbol{p}-\boldsymbol{q}); \tag{B.194}$$

we call $S_X(p)$ the spectrum of $\hat{X}$ and we say that $\hat{X}$ is a coloured noise on $\hat{L}^2(\mathbb{R}^2)$, with power density $S_X(p)$. Note that, since $S_X(p)$ is real, $C^*_{\hat{X}} \equiv C_{\hat{X}}$.

Proof We have to prove (B.194).

If X is (HI) one has

$$\langle f, C_X f\rangle_{L^2(\mathbb{R}^2)} = \int\int f(\boldsymbol{t}')\, C_X(\boldsymbol{t}'-\boldsymbol{t})\, f(\boldsymbol{t})\, d_2t\, d_2t'.$$

Since by the convolution theorem one has

$$\mathscr{F}\left\{\int C_X(\boldsymbol{t}'-\boldsymbol{t}) f(\boldsymbol{t})\, d_2t\right\} = S(p)\,\hat{f}(\boldsymbol{p}),$$

we deduce that

$$\begin{aligned}\langle f, C_X f\rangle_{L^2} &= \langle \mathscr{F}(f), \mathscr{F}\{C_X f\}\rangle_{\hat{L}^2} = \\ &= \int_{\mathbb{R}^2} \hat{f}(\boldsymbol{p})^* S_X(\boldsymbol{p})\, \hat{f}(p)\, d_2p \equiv \\ &\equiv \int\int \hat{f}^*(\boldsymbol{q})\, S_X(p)\, \delta(\boldsymbol{q}-\boldsymbol{p})\, \hat{f}(\boldsymbol{p})\, d_2p\, d_2q \\ &= \left\langle \hat{f}, C_{\hat{X}} \hat{f}\right\rangle_{\hat{L}^2},\end{aligned}$$

which yields (B.194). □

Let us comment, about Proposition B.16, that even if $\{X_t\}$ has regular functions as samples, i.e., $x(t,\omega)$ are regular functions of t for ω fixed, we cannot expect that these functions possess an ordinary Fourier transform, because they tend to be statistically equal on a bounded set and on all of its translates up to infinity.

This is why the covariance of an (HI) field can only have a distributional form.

Remark B.9 Recall that when X is isotropic in the covariance, namely $C_X(\boldsymbol{\tau}) = C_X(|\tau|)$, then also $S_X(\boldsymbol{p})$ depends on the $|\boldsymbol{p}|$ only. As always, we will use the same symbols $f(\boldsymbol{\tau})$ and $f(\tau)$ to mean $f(\boldsymbol{\tau}) = f(|\boldsymbol{\tau}|)$. The context will clarify which one of the two functions is currently in use.

$S_X(p)$ is an important statistics of the GRF X, but one of its properties is fundamental because it helps the characterization of its covariance function, when it exists.

Theorem B.2 (Wiener, Khinchin) *A necessary and sufficient condition for $S_X(p)$ to be the* FT *of a positive definite $C_X(\tau)$ is that $S_X(p)$ is non-negative, i.e.,*

$$S_X(p) \geq 0 \quad \forall p. \tag{B.195}$$

Proof We prove here only the sufficiency part, namely that (B.195) implies that $C_X(\tau)$ is positive definite. Assume that (B.195) is true and take N points $(\boldsymbol{t}_1, \boldsymbol{t}_2, \ldots, \boldsymbol{t}_N)$ and a vector $\boldsymbol{\lambda} = (\lambda_1, \lambda_2, \ldots, \lambda_N)^T$ of real constants. Then

$$C_X\left(\boldsymbol{t}_j - \boldsymbol{t}_k\right) = \int e^{-i2\pi \boldsymbol{p}\cdot(\boldsymbol{t}_j - \boldsymbol{t}_k)} S_X(p)\, d_2p, \tag{B.196}$$

so that

$$\begin{aligned}\sum_{j,k=1}^{N} \lambda_j \lambda_k C_X\left(\boldsymbol{t}_j - \boldsymbol{t}_k\right) &= \int \left(\sum_{j,k=1}^{N} \lambda_j \lambda_k e^{-i2\pi \boldsymbol{p}\cdot(\boldsymbol{t}_j - \boldsymbol{t}_k)}\right) S_X(p)\, d_2p \\ &= \int \left|\sum_{j=1}^{N} \lambda_j e^{-i2\pi \boldsymbol{p}\cdot\boldsymbol{t}_j}\right|^2 S(p)\, d_2p \geq 0,\end{aligned}$$

i.e., $C_X(\tau)$ is positive definite. □

Based on the above theorem, some properties of the covariance functions can be proved that are particularly useful to construct more complicated models from elementary covariances.

Proposition B.17 *Let $C_1(\tau)$, $C_2(\tau)$ be covariance functions, i.e., symmetric and positive definite functions. Then*

$$\forall \lambda,\ \mu > 0,\ \lambda C_1(\tau) + \mu C_2(\tau) = C(\tau), \tag{B.197}$$

$$C_1(\tau)\, C_2(\tau) = C(\tau), \tag{B.198}$$

and

$$\int C_1(\boldsymbol{\tau} - \boldsymbol{s})\, C_2(\boldsymbol{s})\, d_2s = C(\tau) \tag{B.199}$$

are also covariances.

Proof (B.197) is obvious. (B.198) and (B.199) come from the convolution theorem of Fourier transforms. In fact, let $S_1(p)$ and $S_2(p)$ denote the FT of C_1 and C_2, respectively, so that $S_1(p) \geq 0$, and $S_2(p) \geq 0$. Then

$$S(p) = \mathscr{F}(C_1(\tau) C_2(\tau)) = S_1 * S_2 = \int S_1(\boldsymbol{p}-\boldsymbol{q}) S_2(\boldsymbol{q}) d_2 q \geq 0$$

and

$$S(p) = \mathscr{F}(C_1 * C_2) = S_1(p) S_2(p) \geq 0.$$

□

Remark B.10 Let us consider a white noise N_t on $\mathbb{R}^2$ with variance density c_0^2, i.e., the GRF with covariance operator

$$C_N = c_0^2 I. \tag{B.200}$$

According to (B.192) we have for $\hat{N}$, the Fourier transform of N,

$$C_{\hat{N}} = c_0^2 \mathscr{F} \mathscr{F}^T = c_0^2 I \tag{B.201}$$

that is, the FT of a white noise N is again a white noise $\hat{N}$ with the same variance density c_0^2.

The same statement in terms of covariance and spectrum reads: if

$$C_N(\tau) = c_0^2 \delta(\boldsymbol{\tau}), \tag{B.202}$$

then

$$S_N(p) = c_0^2 \mathscr{F}\{\delta(\tau)\} = c_0^2, \tag{B.203}$$

i.e., the spectrum is constant and equal to the variance density.

Remark B.11 Assume now that, contrary to the previous Remark, $\{X_t\}$ is a RF, with bounded covariance function $C_X(\tau)$.

Assume further that $C_X(\tau)$ is integrable on $\mathbb{R}^2$; then as we know from FT theory the spectrum $S_X(p)$ is a bounded continuous function. But even more, $S_X(p)$ turns out to be also integrable; in fact

$$\int_{\mathbb{R}^2} S_X(p) d_2 p = C_X(0) < +\infty. \tag{B.204}$$

This means, for instance that if more generally $\{Y_t\}$ is a GRF with covariance operator

$$C_Y = c_0^2 I + C_X \tag{B.205}$$

with C_X corresponding to a bounded covariance function $C_X(\tau)$ that we assume to be integrable too, we can at once claim that the spectrum of $\{X_t\}$ is given by

$$S_X(p) = c_0^2 + \hat{C}_X(p) = c_0^2 + S_X(p) \tag{B.206}$$

where $S_X(p) = \hat{C}_X(p)$ is just the FT of $C_X(\tau)$.

Example B.7 Based on Remark B.11, let us revisit the filtering problem, but this time in the Fourier transform domain. So assume we have a GRF $\{Y_t\}$ and its FT $\left\{\hat{Y}_p\right\}$ related to $\{X_t\}$ by

$$Y_t = X_t + N_t, \quad \hat{Y}_p = \hat{X}_p + \hat{N}_p. \tag{B.207}$$

Assume further that the covariance structure of Y_t is that of (B.205), so that (B.206) holds. We want to find the optimal (WK) predictor of $\hat{X}_p$. Since in this section $\hat{X}_p$ represents the FT of X_t, only for this example we will denote by $\tilde{X}_p$ the WK predictor. Our general WK formula is

$$\tilde{X}_p = C_{\hat{X}\hat{Y}} C_{\hat{Y}}^{-1} \hat{Y}. \tag{B.208}$$

In this case

$$C_{\hat{Y}}(\boldsymbol{p}, \boldsymbol{q}) = S_Y(p)\delta(\boldsymbol{p} - \boldsymbol{q}) = \left[c_0^2 + S_X(p)\right]\delta(\boldsymbol{p} - \boldsymbol{q})\,; \tag{B.209}$$

note that the inverse operator $C_{\hat{Y}}^{-1}$ then has the kernel

$$C_Y^{(-1)}(\boldsymbol{p}, \boldsymbol{q}) = \left[c_0^2 + S_X(p)\right]^{-1} \delta(\boldsymbol{p} - \boldsymbol{q})\,. \tag{B.210}$$

In fact

$$\int_{\mathbb{R}^2} C_{\hat{Y}}^{(-1)}(\boldsymbol{p}, \boldsymbol{q})\, C_{\hat{Y}}(\boldsymbol{q}, \boldsymbol{s})\, d_2q =$$
$$= \int_{\mathbb{R}^2} [c_0 + S_X(p)]^{-1} \delta(\boldsymbol{p} - \boldsymbol{q}) \cdot \left[c_0^2 + S_X(p)\right] \delta(\boldsymbol{q} - \boldsymbol{s})\, d_2q = \delta(\boldsymbol{p} - \boldsymbol{s}), \tag{B.211}$$

which is the kernel of the identity.

Moreover, since $\hat{X}_p$ and $\hat{N}_p$ are uncorrelated,

$$C_{\hat{X}\hat{Y}} \sim S_X(p)\, \delta(\boldsymbol{p} - \boldsymbol{q})\,. \tag{B.212}$$

Therefore (B.208) becomes

$$\tilde{X}_p = \int_{\mathbb{R}^2} S_X(p)\,\delta(\boldsymbol{p}-\boldsymbol{q})\,d_2q \int_{\mathbb{R}^2} \left[c_0^2 + S_X(q)\right]^{-1} \delta(\boldsymbol{q}-\boldsymbol{s})\,\hat{Y}_s d_2s =$$
$$= \frac{S_X(p)}{c_0^2 + S_X(p)} \hat{Y}_p. \tag{B.213}$$

Thus formula is generally known in the literature as the *Wiener filter*, [100].

One very important issue in spectral theory for us is how to find (approximate) estimators of $S_X(p)$. This is in fact an essential step if we want to apply to gravity field prediction problems, an "automatic" covariance estimator like the one presented in Sect. 6.6.

We will present two results, one under the hypothesis that we know $\hat{X}_t$ over the whole $\mathbb{R}^2$ (case A); the other, assuming that $\{X_t\}$ is so regular as to have point values and we know a sample of them on the square, regular grid

$$G = \{\boldsymbol{t_k} = \boldsymbol{k}\Delta;\, \boldsymbol{k} = (k_1, k_2)\,;\, -N \le k_1 \le N, -N \le k_2 \le N\} \tag{B.214}$$

where Δ is the side of the cell in $\mathbb{R}^2$ and $2N+1$ is the number of nodes along each of the two axes. In such a case (we will call it case B), instead of the continuous Fourier Transform of $\{X_t\}$, we have a Discrete Fourier Transform of X_{t_k} which, we remind, can be defined as

$$\hat{X}_p = \frac{1}{(2N+1)^2} \sum_{\boldsymbol{t_k} \in G} e^{i\frac{2\pi}{T}\boldsymbol{p}\cdot\boldsymbol{t_k}} X_{t_k} \tag{B.215}$$

where

$$T = (2N+1)\,\Delta$$

is the size of the total side of the grid G.

If we call $f_0 = \frac{1}{T}$ the basic quantum of frequency, (B.215) is usually computed on the grid $\hat{G} \equiv \{f_0\boldsymbol{j};\ \ -N \le j_1 \le N,\ \ -N \le j_2 \le N\}$ in the frequency domain, so obtaining a one-to-one correspondence between the vectors X_{t_k} defined on G, and the vector X_{p_j} defined on G, i.e.,

$$\left(\boldsymbol{p}_j \in \hat{G}\right) \quad \hat{X}_{\boldsymbol{p}_j} = \frac{1}{(2N+1)^2} \sum_{t_k \in G} e^{i2\pi\,\boldsymbol{p}_j\cdot\boldsymbol{t}_k} X_{\boldsymbol{t}_k}$$
$$= \frac{1}{T^2} \sum_{t_k \in G} e^{i2\pi\,\boldsymbol{p}_j\cdot\boldsymbol{t}_k} X_{\boldsymbol{t_k}} \Delta^2 \tag{B.216}$$
$$\cong \frac{1}{T^2} \int_Q e^{i2\pi\,\boldsymbol{p}_j\cdot\boldsymbol{t}} X_t d_2t$$

where

$$Q \equiv \left\{ -\frac{T}{2} \le t_1 \le \frac{T}{2}; -\frac{T}{2} \le t_2 \le \frac{T}{2} \right\},$$

i.e., it is the region covering the grid G.

Let us note for later use that $|Q| = T^2$.

Case A)

We assume that $\hat{X}_p$ is known, so that $\hat{X}_p(\varphi)$ is a known $\mathscr{L}^2(\Omega)$ variable.

Consider the annulus $\mathscr{C}_\delta(\bar{p})$,

$$||\boldsymbol{p}| - \bar{p}| \le \frac{\delta}{2},$$

of width δ and central circle of radius $\bar{p}$.

The function

$$M_{\mathscr{C}_\delta(\bar{p})} = \frac{\chi_{\mathscr{C}_\delta(\bar{p})}(p)}{|\mathscr{C}_\delta(\bar{p})|},$$

with $\chi_{\mathscr{C}_\delta(\bar{p})}(p)$ the characteristic function of $\mathscr{C}_\delta(\bar{p})$, is indeed in $\hat{L}^2$, for any $\delta > 0$, so that we can write

$$\begin{aligned} E\left\{\left|\hat{X}\left(C_{M_\delta(\bar{p})}\right)\right|^2\right\} &= \frac{1}{|\mathscr{C}_\delta(\bar{p})|^2}\int_{\mathbb{R}^2}\int_{\mathbb{R}^2} S_X(p)\,\delta(\boldsymbol{p}-\boldsymbol{q})\,\chi_{\mathscr{C}_\delta(\bar{p})}(\boldsymbol{p}) \\ &\quad \times \chi_{\mathscr{C}_\delta(\bar{p})}(\boldsymbol{q})\,d_2q\,d_2p = \\ &= \frac{1}{|\mathscr{C}_\delta(\bar{p})|^2}\int_{\mathscr{C}_\delta(\bar{p})} S_X(p)\,d_2p \cong \\ &\cong \frac{S_X(\bar{p})}{|\mathscr{C}_\delta(\bar{p})|}. \end{aligned} \tag{B.217}$$

The approximation in (B.217) is good if $S_X(p)$ is smooth at $\bar{p}$ and δ is sufficiently small.

Considering that

$$\hat{X}\left(M_{\mathscr{C}_\delta(\bar{p})}\right) = \frac{1}{|\mathscr{C}_\delta(\bar{p})|}\int_{\mathscr{C}_\delta(\bar{p})} \hat{X}_p\,d_2p \tag{B.218}$$

we deduce from (B.217) that

$$\frac{1}{|\mathscr{C}_\delta(\bar{p})|}\left|\int_{\mathscr{C}_\delta(\bar{p})} \hat{X}_p\,d_2p\right|^2 = \hat{S}_X(\bar{p}) \tag{B.219}$$

is an approximate estimator of $S_X(\bar{p})$.

Case B)
Let us start from the approximate expression (B.216) and compute

$$E\left\{\left|\hat{X}_{p_j}\right|^2\right\} \cong \frac{1}{T^4}\int_Q\int_Q e^{i2\pi p_j\cdot(t-t')}E\left\{X_t X_{t'}\right\}d_2t d_2t' = \\ = \frac{1}{T^4}\int_Q\int_Q e^{i2\pi p_j\cdot(t-t')}C_X\left(t-t'\right)d_2t d_2t' \tag{B.220}$$

We note that the result of the integration of (B.220) has to be real and even positive, so we know that only the real part of the exponential, i.e., $\cos 2\pi \boldsymbol{p}_j \cdot \left(\boldsymbol{t}-\boldsymbol{t}'\right)$, has a role in such operation. Therefore, the integrand of the real part of (B.220), namely $C_X\left(\boldsymbol{t}-\boldsymbol{t}'\right)\cdot\cos 2\pi p_j\cdot\left(\boldsymbol{t}-\boldsymbol{t}'\right)$, is an even function of $\boldsymbol{t}-\boldsymbol{t}'$. In principle we have to compute the integral

$$I = \int_Q\int_Q F\left(\boldsymbol{t}-\boldsymbol{t}'\right)d_2\boldsymbol{t}d_2\boldsymbol{t}', \tag{B.221}$$

where $F\left(\boldsymbol{t}-\boldsymbol{t}'\right) = F\left(\boldsymbol{t}'-\boldsymbol{t}\right)$.

Since $d_2\boldsymbol{t} = dt_1dt_2$ and $d_2\boldsymbol{t}' = dt_1'dt_2'$, we can perform first the integration with respect to the couple dt_1dt_1' and then repeat the operation with the other couple dt_2dt_2', namely

$$I = \int_{-T/2}^{T/2}\int_{-T/2}^{T/2}\left(\int_{-T/2}^{T/2}\int_{-T/2}^{T/2}F\left(t_1-t_1', t_2-t_2'\right)dt_1dt_1'\right)dt_2dt_2'. \tag{B.222}$$

We report the result for the one-dimensional case, namely, given an even function $H(t-t')$, we get

$$J = \int_{-T/2}^{T/2}\int_{-T/2}^{T/2}H\left(t-t'\right)dtdt' = T\int_{-T}^{T}H(s)\left(1-\frac{|s|}{T}\right)ds. \tag{B.223}$$

This is obtained by dividing the square Q into two triangles $\mathscr{T}$ (see Fig. B.1); then one has

$$J = 2\int_{\mathscr{T}} H\left(t-t'\right)dtdt';$$

by using the transformation of variables $s = t - t'$; $\tau = t$, one gets

$$J = 2\int_0^T H(s)(T-s)\,ds; \tag{B.224}$$

using the corresponding integral from $-T$ to O yields (B.223).

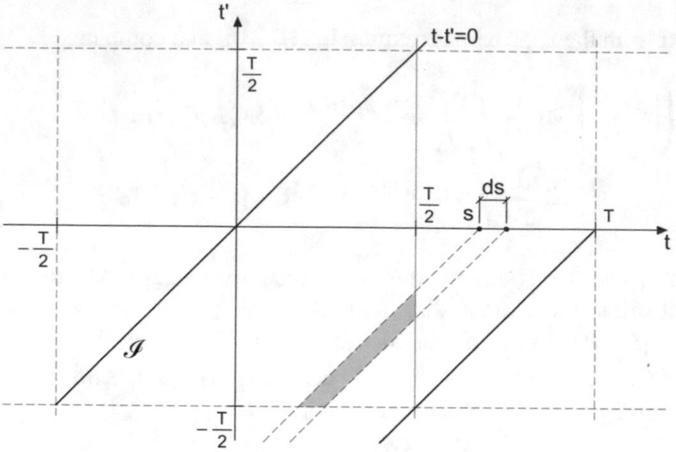

Fig. B.1 An elementary interpretation of the integral (B.224). Note that the integrand is a function of s only, i.e., it is constant along parallels to the main diagonal $t = t'$. Note also that the area of the dashed stripe is $dA = ds \cdot (T - s)$

By applying twice (B.223) to (B.222) we obtain

$$I = T^2 \int_{-T}^{T} \int_{-T}^{T} e^{i2\pi \boldsymbol{p}\cdot\boldsymbol{s}} F\left(|s_1|, |s_2|\right) \left(1 - \frac{|s_1|}{T}\right) \left(1 - \frac{|s_2|}{T}\right) ds_1 ds_2.$$

Summing up and returning to (B.220), we get

$$E\left\{\left|\tilde{X}_{p_j}\right|^2\right\} \cong \frac{1}{T^2} \int_{-T}^{T} \int_{-T}^{T} e^{i2\pi \boldsymbol{p}_j\cdot\boldsymbol{s}} C_X(s) \left(1 - \frac{|s_1|}{T}\right) \left(1 - \frac{|s_2|}{T}\right) ds_1 ds_2. \tag{B.225}$$

Since, on the integration domain,

$$\left| e^{i2\pi \boldsymbol{p}_j\cdot\boldsymbol{s}} C_X(s) \left(1 - \frac{|s_1|}{T}\right) \left(1 - \frac{|s_2|}{T}\right) \right| \leq |C_X(s)|,$$

which is supposed to be integrable on $\mathbb{R}^2$, we see that

$$\lim_{T\to\infty} \int_{-T}^{T} \int_{-T}^{T} e^{i2\pi \boldsymbol{p}_j\cdot\boldsymbol{s}} C_X(s) \left(1 - \frac{|s_1|}{T}\right) \left(1 - \frac{|s_2|}{T}\right) ds_1 ds_2 =$$

$$= \int_{\mathbb{R}^2} e^{i2\pi \boldsymbol{p}_j\cdot\boldsymbol{s}} C_X(s)\, d_2 s = S_X\left(p_j\right),$$

so that (B.225) can be further approximated, when T is large, by

$$E\left\{\left|\hat{X}_{p_j}\right|^2\right\} \cong \frac{S_X\left(\boldsymbol{p}_j\right)}{T^2}. \tag{B.226}$$

Now we can recall that, if $C_X\left(\boldsymbol{s}\right)$ is a function of $s = |\boldsymbol{s}|$ only, i.e., X is (HI), then $S_X\left(\boldsymbol{p}\right)$ is also function of $p = |\boldsymbol{p}|$ only.

Therefore, starting from (B.226), we can build an approximate estimator of $S_X(\bar{p})$ by using the same trick that we introduced for case A. Namely we consider an annulus $\mathscr{C}_\delta\left(\bar{p}\right)$ with central radius $\bar{p}$ and width δ; if δ is large enough, there will be many points, in fact $N_\delta\left(\bar{p}\right)$ points, $\boldsymbol{p}_j \in \mathscr{C}_\delta\left(\bar{p}\right)$, for each of which (B.226) holds.

Then as an approximate estimator of $S_X\left(\bar{p}\right)$ one can take

$$\hat{S}_X\left(\bar{p}\right) = \frac{T^2}{N_\delta\left(\bar{p}\right)} \sum_{p_j \in \mathscr{C}_\delta(\bar{p})} \left|\tilde{X}_{p_j}\right|^2; \tag{B.227}$$

this is certainly a much more robust estimator than that based on one point value of $\hat{X}_{\boldsymbol{p}}$ only.

B.8 Table of Covariances and Related Spectra

We conclude this Appendix with a short list of parametric families of functions that can be used as covariances of HIRF and the corresponding spectra. Here we will use the notation $\mathscr{H}\left(x\right)$ for the Heaviside function, namely

$$\mathscr{H}\left(x\right) = \begin{cases} 1, & \text{if } x > 0, \\ 1/2, & \text{if } x = 0, \\ 0, & \text{if } x < 0. \end{cases} \tag{B.228}$$

Furthermore, one of the spectra will be defined by the function $K_0(x)$, namely a so-called *modified Bessel function* of order 0, which can be represented as

$$K_0\left(x\right) = \int_0^{+\infty} \frac{\cos xt}{\sqrt{t^2+1}} dt. \tag{B.229}$$

This function is tabulated, see e.g., [2].

Salient characteristics of such a function are that it is symmetric, $K_0\left(x\right) \to +\infty$ when $x \to 0$ and that it decays exponentially too for $x \to +\infty$, through positive values.

$C(r)$ $\left(r=\sqrt{x_1^2+x_2^2}\right)$	$S(p)$ $\left(p=\sqrt{p_1^2+p_2^2}\right)$
$A J_0(2\pi\bar{p}r)$	$\frac{A}{2\pi}\frac{\delta(p-\bar{p})}{\bar{p}}$
$A\frac{J_1(2\pi\bar{p}r)}{r}$	$A\frac{\mathscr{H}(\bar{p}-p)}{\bar{p}}$
$A\frac{\sin 2\pi\bar{p}r}{r}$	$A\frac{\mathscr{H}(\bar{p}-p)}{\sqrt{\bar{p}^2-p^2}}$; $(p\le\bar{p})$
$Ae^{-\bar{p}r}$	$\frac{1}{2\pi}\frac{A\cdot\bar{p}}{(p^2+\bar{p}^2)^{3/2}}$
$Ae^{-\bar{p}^2r^2}$	$\frac{A\pi}{\bar{p}^2}e^{-\pi^2\frac{p^2}{\bar{p}^2}}$
$A\frac{1}{r^2+R^2}$	$AK_0(2\pi Rp)$
$A\frac{1}{(r^2+R^2)^{3/2}}$	$2\pi\frac{A}{R}e^{-2\pi Rp}$

More models of covariance functions can be derived by applying the rules of Proposition B.17.

One further observes that if $C(r)$ is a covariance function, then

$$-\Delta C(r)=-\left[C''(r)+\frac{1}{r}C'(r)\right]=K(r) \tag{B.230}$$

is also a covariance. In fact, taking the two-dimensional Laplacian of

$$C(r)=\int_{\mathbb{R}^2}e^{-i2\pi\,\boldsymbol{p}\cdot\boldsymbol{r}}S(p)\,d_2p \tag{B.231}$$

one gets

$$-\Delta C(r)=\int_{\mathbb{R}^2}e^{-i2\pi\,\boldsymbol{p}\cdot\boldsymbol{r}}4\pi^2p^2S(p)\,d_2p; \tag{B.232}$$

thus, Fourier Transform of $K(r)$ is $4\pi p^2S(p)\ge 0$, so that $K(r)$ is positive definite, too.

With this rule for instance one can prove that

$$-\Delta\left(Ae^{-\bar{p}^2r^2}\right)=A2p^2\left(1-2\bar{p}^2r^2\right)e^{-\bar{p}^2r^2} \tag{B.233}$$

is a covariance function.

With the model (B.233) we can shape a covariance that starts positively from $2\bar{p}^2A$, goes to zero at $r_0=\frac{1}{\sqrt{2}\bar{p}}$ and then decays at the infinity through negative values after reaching a minimum at $r=\frac{\sqrt{3}}{\bar{p}}$; such features are sometimes present in empirical covariance functions.

Appendix C
The Tikhonov Regularization and Morozov's Discrepancy Principle

C.1 Introduction

Our aim is to find approximate solutions to the equation

$$\begin{aligned} & y = A(x), \\ & x \in X, \ y \in Y, \end{aligned} \tag{C.1}$$

where X and Y are given functional spaces (in our case Hilbert spaces), when we know that A is a continuous injective (i.e., $A(x) = 0 \Longrightarrow x = 0$) operator from X to Y, but its inverse A^{-1} is unbounded $Y \to X$.

Often, particularly in our applications, due to some prior information, the approximate solution x will be required to satisfy some constraint

$$x \in K \subset X. \tag{C.2}$$

The reason why we talk about an approximate solution is that generally we do not have a perfect knowledge of the known term $y \in Y$, which would imply also $y \in A(K)$; rather, we have an element $y_0 \in Y$ which we know to have a discrepancy v with respect to y, so that instead of (C.1), we have the "observation equation"

$$\begin{cases} y_0 = y + v = A(x) + v, \\ x \in K \subset X, v \in Y, \ A \text{ continuous } X \to Y. \end{cases} \tag{C.3}$$

In this sense, if we ignore the functional properties of X and Y, the problem resembles a classical least squares problem, frequently treated in Geodesy.

Many deterministic regularization techniques are known in the scientific literature, among which we have chosen the Tikhonov regularization, due to its

F. Sansò, D. Sampietro, *Analysis of the Gravity Field*, Lecture Notes in Geosystems Mathematics and Computing, https://doi.org/10.1007/978-3-030-74353-6

widespread applications as well as to its relation to the stochastic techniques discussed in Appendix B.

The literature on the subject is very large, from the classical treatise of Tikhonov [150] to more recent books, like [61] and [147]. Also many different presentations can be done of Tikhonov's variational principle to find solutions under different hypotheses. Here we present two such approaches, one for a nonlinear operator $A(x)$, the other for a linear operator Ax, particularly suitable for applications to the inverse gravimetric problem in a two-layer model.

C.2 Tikhonov Principle for a Nonlinear Equation

We are looking for an approximate solution of (C.3) when $A(x)$ is a nonlinear operator. The basic idea of Tikhonov's approach is to look for the minimum of a functional $J_\lambda(y_0, x)$ which tries to balance the requirement that the discrepancy v should be small in some Hilbert space Y, the data space, with the requirement that x should be suitably smooth, namely its norm in some other Hilbert space X, the solution space, should not be too large.

Accordingly, we put

$$J_\lambda(y_0, x) = \|y_0 - A(x)\|_Y^2 + \lambda \|x\|_X^2 \tag{C.4}$$

for some positive constant λ, which for the moment remains arbitrary, and we look for the minimum of $J_\lambda(y_0, x)$, with respect to x varying in some prescribed set $K \subset X$. We notice that the use of the term $\|x\|_X^2$ in (C.4) as a regularizer in Tikhonov's functional $J_\lambda(y_0, x)$, is not the most general, but it covers all the cases of interest for us.

We aim at proving the following theorem.

Theorem C.1 *Assume X, Y are Hilbert spaces and $A(x)$ is a nonlinear operator defined on X, such that $A(0) = 0$, and injective, i.e., $A(x_1) = A(x_2) \Longrightarrow x_1 = x_2$; assume that K is a closed, convex set in X and that X is compactly embedded into another Hilbert space H,*

$$K \subset X \subset H. \tag{C.5}$$

Moreover, assume that $A(x)$ is also continuous from H to Y. Then there is an $x^* \in K$ such that

$$J_\lambda\left(y_0, x^*\right) = \bar{J} = \inf_{x \in K} J_\lambda(y_0, x); \tag{C.6}$$

moreover, every minimizing sequence $\{x_k\} \subset K$,

$$\lim_{n\to\infty} J_\lambda (y_0, x_n) = \bar{J} \tag{C.7}$$

contains a subsequence converging to some minimum point x^* in the topology of X.

Proof Let $\bar{J}$ be the non-negative infimum of $J_\lambda (y_0, x)$ on K, and take a sequence $\{x_n\} \subset K$ which is minimizing in the sense that (C.7) holds.

The sequence $\{x_n\}$ is bounded in X because, at least from a some n on,

$$\lambda \|x_n\|_X \leq J_\lambda (y_0, x_n) \leq J_\lambda (y_0, 0) = \|y_0\|_Y^2 .$$

The second inequality holds because $\bar{J} \leq J_\lambda (y_0, 0)$ by the definition of $\bar{J}$.

From $\{x_n\}$ we can extract a subsequence (denoted again by $\{x_n\}$) that is weakly convergent to some $x^* \in X$; since $\{x_n\} \in K$ and K is closed and convex in X, $x^* \in K$, too; this is the Krein–Milman's theorem [161]. Since X is compactly embedded in H from $\{x_n\}$ we can extract one further subsequence, again denoted by $\{x_n\}$, that is strongly convergent to x^* in H:

$$x_n \underset{H}{\longrightarrow} x^*. \tag{C.8}$$

Since $A(x)$ is continuous $H \to Y$, (C.8) implies

$$A(x_n) \underset{Y}{\longrightarrow} A\left(x^*\right). \tag{C.9}$$

On the other hand, since $x_n \overset{*}{\underset{X}{\rightarrow}} x^*$,

$$\left\|x^*\right\|_X^2 = \lim \left\langle x_n, x^*\right\rangle_X \leq \underline{\lim} \left\|x^*\right\|_X \|x_n\|_X$$

by the Schwarz inequality, so that

$$\left\|x^*\right\|_X \leq \underline{\lim} \|x_n\|_X . \tag{C.10}$$

Consequently,

$$\begin{aligned} \lim J_\lambda (y_0, x_n) &\equiv \left(\underline{\lim} \; \|y_0 - A(x_n)\|_Y^2 + \lambda \|x_n\|_X^2\right) \geq \\ &\geq \underline{\lim} \; \|y_0 - A(x_n)\|_Y^2 + \lambda \left\|x^*\right\|_Y^2 , \end{aligned} \tag{C.11}$$

but, by (C.9)

$$\lim \|y_0 - A(x_n)\|_Y^2 = \left\|y_0 - A\left(x^*\right)\right\|_Y^2 ,$$

so that

$$\lim J_\lambda(y_0, x_n) = \bar{J} \geq \left\| y_0 - A\left(x^*\right) \right\|_Y^2 + \lambda \left\| x^* \right\|^2 = J_\lambda\left(y_0, x^*\right). \tag{C.12}$$

Since, recalling that $x^* \in K$, the opposite inequality

$$J_\lambda\left(y_0, x^*\right) \geq \bar{J}$$

has to hold as well, we conclude that

$$J_\lambda\left(y_0, x^*\right) = \bar{J},$$

i.e., (C.6) holds.

Moreover, since

$$\lim J_\lambda(y_0, x_n) = J_\lambda\left(y_0, x^*\right)$$

and

$$\| y_0 - A(x_n) \|_Y^2 \longrightarrow \left\| y_0 - A\left(x^*\right) \right\|_Y^2,$$

also

$$\| x_n \|_X^2 \longrightarrow \left\| x^* \right\|_X^2. \tag{C.13}$$

But then

$$\left\| x_n - x^* \right\|_X^2 = \| x_n \|_X^2 - 2\left\langle x_n, x^* \right\rangle_X + \left\| x^* \right\|_X^2 \longrightarrow 0$$

because of (C.13) and because $x_n \underset{X}{\overset{*}{\rightarrow}} x^*$.

Thus, the strong convergence of x_n to x^* in X is proved, too. □

Remark C.1 Let us notice that by means of the embedding of X into H we are in fact requiring that $A(x)$ is compact from X to Y; this form however is particularly suited to our problem.

C.3 Tikhonov Principle for a Linear Problem

Here we assume that $A(x)$ is linear from X to Y, namely

$$A(x) = Ax; \tag{C.14}$$

in addition we assume that A is just continuous $X \to Y$. In this case a variant of the proof of Theorem C.1 is available that exploits the convexity of the functional $J_\lambda(y_0, x)$ as a function of x.

Let us recall that a functional $\Phi(x)$ is called (strictly) convex if (see [65])

$$\begin{cases} \Phi\left(tx + (1-t)x'\right) < t\Phi(x) + (1-t)\Phi\left(x'\right), \\ \forall x, x' \text{ fixed; } \forall t, \quad 0 < t < 1. \end{cases} \tag{C.15}$$

We remind as well that one classical way of proving (C.15) is to look at the function

$$F(t) = \Phi\left(tx + (1-t)x'\right), \tag{C.16}$$

and, when $F(t)$ is twice differentiable, show that

$$F''(t) > 0; \quad 0 < t < 1. \tag{C.17}$$

So we preliminarily establish the following proposition.

Proposition C.1 *Let us consider $J_\lambda(y_0, x)$ as a functional of x, defined on a convex closed set $K \subset X$; this functional is convex.*

Proof Let us set, for two fixed points $x \neq x' \in K$, and $0 < t < 1$,

$$\begin{aligned} F(t) &= \left\|y_0 - A\left[tx + (1-t)x'\right]\right\|_Y^2 + \lambda\left\|tx + (1-t)x'\right\|_X^2 = \\ &= \left\|\left(y_0 - Ax'\right) - tA\left(x - x'\right)\right\|_Y^2 + \lambda\left\|x' + t\left(x - x'\right)\right\|_X^2 = \\ &= \left\|y_0 - Ax'\right\|_Y^2 - 2t\left\langle y_0 - Ax', A\left(x - x'\right)\right\rangle_Y + t^2\left\|A\left(x - x'\right)\right\|_Y^2 + \\ &\quad + \lambda\left(\left\|x'\right\|_X^2 + 2t\left\langle x', \left(x - x'\right)\right\rangle_X + t^2\left\|x - x'\right\|_X^2\right). \end{aligned} \tag{C.18}$$

Since

$$F''(t) = 2\left\|A\left(x - x'\right)\right\|_Y^2 + 2\lambda\left\|x - x'\right\|_X^2 > 0, \quad \forall t \in (0, 1), \tag{C.19}$$

the proposition is proved. □

Note that the theorem holds even if A is not injective; in fact, if $x \neq x'$, then $\left\|x - x'\right\|_X > 0$ and (C.19) holds true.

We pass now to prove the main theorem.

Theorem C.2 *Let A be a continuous linear operator $X \to Y$. Then any minimizing sequence $\{x_n\}$ in the closed convex set K is a Cauchy sequence converging to an $x^* \in K$, such that*

$$J_\lambda\left(y_0, x^*\right) = \bar{J}. \tag{C.20}$$

In addition, such a minimum point x^* is unique.

Proof Let $\{x_n\}$ be a minimizing sequence in K; we first show that

$$\lim_{n,m\to\infty} J_\lambda\left(y_0, \frac{x_n + x_m}{2}\right) = \bar{J}. \tag{C.21}$$

In fact, since $x_n,\ x_m \in K$ and K is convex $\frac{x_n+x_m}{2} \in K$. So

$$J_\lambda\left(y_0, \frac{x_n + x_m}{2}\right) \geq \bar{J}, \tag{C.22}$$

by definition of $\bar{J}$. On the other hand, by Proposition C.1,

$$J_\lambda\left(y_0, \frac{x_n + x_m}{2}\right) < \frac{1}{2} J\left(x_n\right) + \frac{1}{2} J\left(x_m\right). \tag{C.23}$$

By taking the limit of (C.22), and (C.23) one gets

$$\bar{J} = \lim_{n,m\to\infty} \left[\frac{1}{2} J\left(x_n\right) + \frac{1}{2} J\left(x_n\right)\right] \geq \overline{\lim} J_\lambda\left(y_0, \frac{x_n + x_m}{2}\right) \geq$$
$$\geq \underline{\lim} J_\lambda\left(y_0, \frac{x_n + x_m}{2}\right) \geq \bar{J}$$

and (C.21) is proved.

Now we use the identity

$$\|\xi\|^2 + \|\eta\|^2 = \frac{1}{2}\left(\|\xi + \eta\|^2 + \|\xi - \eta\|^2\right), \tag{C.24}$$

which is characteristic of all Hilbert spaces.

Then we have

$$\begin{aligned} \|y_0 - Ax_n\|_Y^2 + \|y_0 - Ax_m\|_Y^2 &= \frac{1}{2}\Big(\|2y_0 - A\left(x_n - x_m\right)\|_Y^2 + \\ + \|A\left(x_n - x_m\right)\|_Y^2\Big) &= 2\left\|y_0 - A\frac{x_n + x_m}{2}\right\|_Y^2 + \frac{1}{2}\|A\left(x_n - x_m\right)\|_Y^2 \end{aligned} \tag{C.25}$$

and

$$\lambda\|x_n\|_X^2 + \lambda\|x_m\|_X^2 = 2\lambda\left\|\frac{x_n + x_m}{2}\right\|_X^2 + \frac{\lambda}{2}\|x_n - x_m\|_X^2. \tag{C.26}$$

Adding these two relations, one gets

$$J\left(x_n\right)+J\left(x_m\right)=2J\left(\frac{x_n+x_m}{2}\right)+\frac{1}{2}\left\|A\left(x_n-x_m\right)\right\|_Y^2+\frac{\lambda}{2}\left\|x_n-x_m\right\|_X^2, \tag{C.27}$$

whence

$$\frac{\lambda}{2}\left\|x_n-x_m\right\|_X^2\leq J\left(x_n\right)+J\left(x_m\right)-2J\left(\frac{x_n+x_m}{2}\right), \tag{C.28}$$

which tends to zero for $n, m \to \infty$, because of (C.20).

This proves that $\{x_n\}$ is Cauchy, and so it has a limit x^* in X. Since $K \subset X$ is closed, we have $x^* \in K$ too; since $J_\lambda\left(y_0, x\right)$ is obviously continuous in x (remember that A is continuous $X \to Y$) then

$$\bar{J}=\lim_{n\to\infty} J_\lambda\left(y_0, x_n\right)=J_\lambda\left(y_0, x^*\right), \tag{C.29}$$

i.e., (C.20) is proved.

Finally, we cannot have two minimum points, x^* and x'^*, of $J_\lambda\left(y_0, x\right)$ because otherwise, by the strict convexity of J_λ as a functional of x,

$$J_\lambda\left(y_0, \frac{x^*+x'^*}{2}\right)<\frac{1}{2}J_\lambda\left(x^*\right)+\frac{1}{2}J_\lambda\left(x'^*\right)=\bar{J}, \tag{C.30}$$

which is absurd, because $\frac{x^*+x'^*}{2} \in K$. □

Remark C.2 To the benefit of those more acquainted with Hilbert space geometry (and with least squares theory) we maintain that, put in the right perspective, Theorem C.2 is nothing but a very well-known theorem claiming that, given a closed subspace G of a Hilbert space H and an element $h \in H, h \notin G$, there is one and only one $h^* \in G$ that minimizes the distance to h, which is also the orthogonal projection of h on G.

In fact, let us define H as

$$H=X\times Y;\ \ h=\begin{bmatrix} x \\ y \end{bmatrix}\quad (x\in X, y\in Y). \tag{C.31}$$

H can be made a Hilbert space with scalar product

$$\left\langle h, h'\right\rangle_H=\left\langle\begin{bmatrix} x \\ y \end{bmatrix}, \begin{bmatrix} x' \\ y' \end{bmatrix}\right\rangle_H=\lambda\left\langle x, x'\right\rangle_X+\left\langle y, y'\right\rangle_Y. \tag{C.32}$$

Now consider the operator A defined on H, by the rule

$$k = \mathscr{A}\, h = \begin{bmatrix} I & 0 \\ A & 0 \end{bmatrix} \begin{bmatrix} x \\ y \end{bmatrix} \equiv \begin{bmatrix} x \\ Ax \end{bmatrix}; \tag{C.33}$$

since A is continuous and $\|x - x'\|_X < \|\mathscr{A}\, h - \mathscr{A}\, h'\|_H$, it is clear that (C.33) describes a closed subspace of H, called the graph G of the operator $\mathscr{A}$. Therefore given any $h_0 = \begin{bmatrix} 0 \\ y_0 \end{bmatrix}$ we can find one and only one $h^* = \begin{bmatrix} x^* \\ Ax^* \end{bmatrix} \in G$ of minimum distance from h_0, i.e., minimizing

$$J\,(h) \equiv \|h_0 - h\|_H^2 = \lambda\, \|x\|^2 + \|y_0 - Ax\|_J^2 = J_\lambda\,(y_0, x) \text{ over } h \in G. \tag{C.34}$$

In Fig. C.1 we display this result that, as we see, is the same of that of Theorem C.2.

The restriction of x to a convex subset $K \subset X$ which projects onto (K, AK) in G, does not change the conclusion, except for the fact that, depending on y_0, it might happen that Ax^* belongs to the boundary of AK (see Fig. C.1).

In fact the problem of the search of h^* such that

$$J\left(h^*\right) \equiv \min_{h \in (K, AK)} \|h_0 - h\|_H\,, \tag{C.35}$$

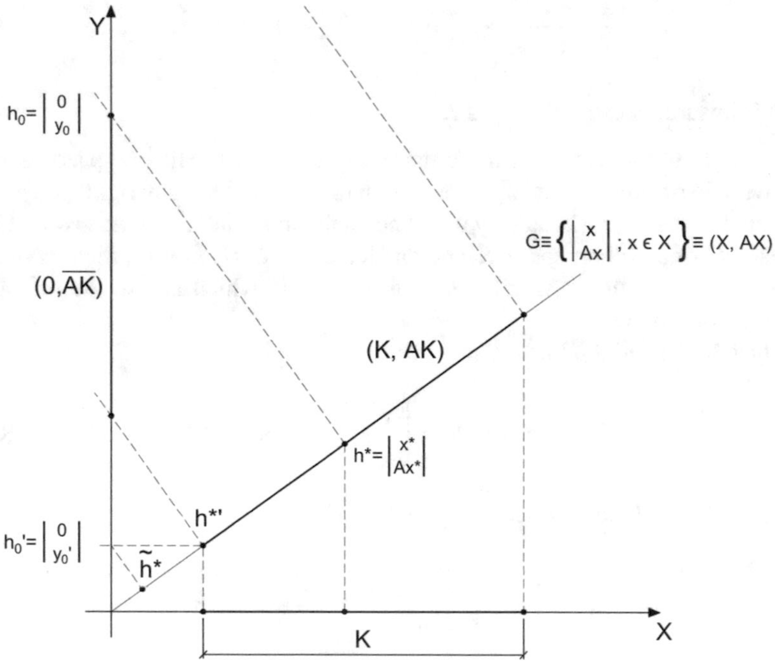

Fig. C.1 The least squares interpretation of Tikhonov's Theorem; when h_0' is outside $\left(0, \overline{AK}\right)$ the minimum distance point (h'^* in the figure) lies on the boundary of (K, AK), i.e., x'^* lies on the boundary of K

i.e., the constrained version of (C.34), has a solution, the existence (and uniqueness) of which is proved exactly as in Theorem C.2.

Succinctly, the proof goes through the following steps:

- a minimizing sequence $\{h_n\}$, $J\{h_n\} \searrow \bar{J}$, has to be bounded so that a subsequence converges weakly to some h^*;
- $h^* \in (K, AK)$,because this a set is closed and convex in H;
- $J\left(\frac{h_n+h_m}{2}\right) \searrow \bar{J}$,as in (C.23)

$$J(h_n) + J(h_m) = \frac{1}{2}\|2h_0 + (h_n + h_m)\|_H^2 + \frac{1}{2}\|h_n - h_m\|_H^2 =$$
$$= 2J\left(\frac{(h_n + h_m)}{2}\right) + \frac{1}{2}\|h_n - h_m\|_H^2$$

so that $\{h_n\}$ is a Cauchy sequence and then $h_n \overset{H}{\to} h^*$, and then $J(h^*) = \bar{J} = \lim J(h_n)$.

Since (K, AK) is closed in H, when $\bar{J} > 0$ it means that $h_0 \notin (K, AK)$. Moreover, let us notice that the set $\left(O, \overline{AK}\right)$ in Fig. C.1 is the set in Y where the points y_0 project orthogonally in $X \times Y$ on (K, AK). So if $y_0' \notin (0, \overline{AK})$ it means that it projects on G in a point $\tilde{h}^*$ outside (K, AK) so that $\forall h^* \in (K, AK)$,

$$\|h_0' - h^*\|^2 = \left\|h_0' - \tilde{h}^*\right\|^2 + \left\|\tilde{h}^* - h^*\right\|^2.$$

Therefore minimizing the LHS of the above relation with respect to $h^* \in (K, AK)$ is the same as minimizing $\left\|\tilde{h}^* - h^*\right\|^2$. Since $\tilde{h}^* \in G$, which is a closed subspace of $X \times Y$, we understand that the element h^* minimizing such a norm has to be on the boundary of (K, AK) in G which is closed and convex. Namely, the minimum distance of h_0' from (K, AK) is attained in $h^{*\prime}$ which is on the boundary of such a set relative to G (see Fig. C.1).

Remark C.3 Sometimes we have a prior information directly on the solution x of the general problem (C.3), whether it is nonlinear or linear. By this we mean that we know a certain $\tilde{x} \in K$ and by the Tikhonov principle we want to express not only that x has to be smooth, i.e., of bounded X norm, but also that x has to be close to $\tilde{x}$, in the space X.

This leads to a modification of the Tikhonov functional (C.4); namely, we put

$$\tilde{J}_\lambda(y_0, x) = \|y_0 - A(x)\|_Y^2 + \lambda\|x - \tilde{x}\|_X^2. \tag{C.36}$$

This a modification does not change the mathematical structure of the problem, because one can always change the unknown x according to

$$x = \tilde{x} + \delta x, \delta x \in \{K - \delta\tilde{x}\}, \tag{C.37}$$

and call

$$B(\delta x) = A(\tilde{x} + \delta x). \tag{C.38}$$

As a consequence, (C.36) becomes

$$J(y_0, \delta x) = \|y_0 - B(\delta x)\|_Y^2 + \lambda \|\delta x\|_X^2 \tag{C.39}$$

and the analysis of (C.39) is reconducted to what we have already seen. Let us notice that in this case we have generally $B(0) = A(\tilde{x}) \neq 0$, so that the proof of Theorem C.1 has to be changed accordingly. Despite the elementary character of this remark, in many practical situations it is quite useful, because we really have some approximate information $\tilde{x}$ on x and this has to be somehow included into the elaboration of our quasi-solution.

C.4 Tikhonov Solutions for Linear Problems

A common approach to the solution of a nonlinear problem, i.e., the minimization of (C.4), is to linearize the nonlinear operator $A(x)$ and then set up some iterative scheme. Therefore, apart from the problem of convergence of the iterations, even the solution of the nonlinear case goes through the solution of a linear one, namely the search of the minimum of

$$J_\lambda(y_0, x) = \|y_0 - Ax\|_Y^2 + \lambda \|x\|_X^2 . \tag{C.40}$$

We study the solution of the minimum of (C.40) in the unconstrained case. The variational argument is classical and basically the same as in a finite-dimensional space: one defines

$$f(t) = J_\lambda(y_0, x^* + t\xi) \tag{C.41}$$

and one requires that, for any $\xi \in X$,

$$\frac{1}{2} f'(0) = -\langle A\xi, y_0 - Ax\rangle_Y + \lambda \langle \xi, x\rangle_X = 0. \tag{C.42}$$

By using the definition of transpose operator A^T we derive the well-known normal equation

$$A^T Ax + \lambda x = A^T y_0. \tag{C.43}$$

Notice that sometimes we write observation equations in the Fourier domain, where functions are generally complex, so that the concept of adjoint operator A^* should be used instead of that of transpose operator.

Since A is continuous and injective, from X to Y, $A^T A$ is also continuous and injective, from X to X, and obviously strictly positive definite, since

$$x \notin 0, \quad \left\langle x, A^T Ax \right\rangle = \|Ax\|^2 > 0. \tag{C.44}$$

Since by assumption A has an unbounded inverse, so does $A^T A$ yet, the relation

$$A^T A + \lambda I > \lambda I \,, \tag{C.45}$$

which follows from (C.44), says that $A^T A + \lambda I$ has an inverse, the resolvent R_λ of $A^T A$, and this inverse is even a bounded operator $X \to X$, since

$$R_\lambda = (A^T A + \lambda I)^{-1} \leqslant \frac{1}{\lambda} I, \tag{C.46}$$

so that

$$\|R_\lambda\| \leqslant \frac{1}{\lambda}. \tag{C.47}$$

Therefore we can affirm that Tikhonov's solution is

$$x_\lambda = R_\lambda A^T y_0. \tag{C.48}$$

Incidentally, we note that, since X and Y are Hilbert spaces, the transpose operator $A^T : Y \to X$ is known to be bounded and in fact

$$\left\| A^T \right\| = \|A\| \,. \tag{C.49}$$

When dealing with complex Hilbert spaces, the same relation holds for the adjoint operator A^*. So we see that our solution x_λ, given by the Tikhonov solver

$$\mathscr{I}_\lambda = R_\lambda A^T, \quad x_\lambda = \mathscr{I}_\lambda y_0, \tag{C.50}$$

is defined, and bounded in X, $y_0 \in Y$; in fact

$$\|x_\lambda\|_X \leqslant \frac{1}{\lambda} \|A\| \, \|y_0\|_Y. \tag{C.51}$$

Moreover, upon multiplying to the identity

$$R_{\lambda'}^{-1} - R_\lambda^{-1} = (\lambda' - \lambda) I,$$

on the left by $R_{\lambda'}$ and on the right by R_λ, we obtain

$$R_\lambda - R_{\lambda'} = (\lambda' - \lambda) R_{\lambda'} R_\lambda, \tag{C.52}$$

which, combined with (C.47), gives

$$\|R_\lambda - R_{\lambda'}\| \leqslant \frac{\lambda' - \lambda}{\lambda' \lambda}. \tag{C.53}$$

The relation (C.53) says that as a function of λ the operator R_λ is continuous, in fact Lipschitz continuous, in the operator norm, for $\lambda > 0$.

Furthermore, since $x_\lambda - x_{\lambda'} = (\mathscr{I}_\lambda - \mathscr{I}_{\lambda'}) y_0 = (R_\lambda - R_{\lambda'}) A^T y_0$, we have

$$\|x_\lambda - x_{\lambda'}\|_X \leqslant \frac{|\lambda - \lambda'|}{\lambda' \lambda} \left\| A^T y_0 \right\|_X. \tag{C.54}$$

Moreover, since $\mathscr{I}_\lambda - \mathscr{I}_{\lambda'} = (R_\lambda - R_{\lambda'}) A^T$, also $\mathscr{I}_\lambda$ is continuous in λ, in the sense of the operator norm, $Y \to X$.

Now that we know that the solution x_λ of Tikhonov's principle exists, we have to understand whether and how x_λ has some relation to x, which is the sought unknown of the problem. Indeed, since y_0 contains an error term $\upsilon \in Y$, we cannot hope to approximate arbitrarily well x by x_λ unless we assume that at the same time

$$\|\upsilon\|_Y = \delta \to 0. \tag{C.55}$$

So we are led to study the function

$$F(\lambda, \delta) = \|x_\lambda(y_0) - x\|_X^2, \tag{C.56}$$

with the purpose, to be accomplished in the next section, of determining whether there is a "strategy", namely a function

$$\lambda = \lambda(\delta) \tag{C.57}$$

such that

$$\left\|x_{\lambda(\delta)} - x\right\|_X = F[\lambda(\delta), \delta] \to 0 \tag{C.58}$$

for $\delta \to 0$. When such a function exists we say that $\lambda(\delta)$ is an *admissible strategy*.

The existence of at least one admissible strategy, e.g., the one proposed by Morozov [87], will be studied in the next section; in preparation for that, here we start studying the so-called *error function*, namely

$$\mathscr{E}^2(\lambda) = \|y_0 - A x_\lambda\|^2, \tag{C.59}$$

which, as we see, is one of the two terms in Tikhonov's principle. Since $y_0 - Ax_\lambda$ is the residual of the observation equation when we use x_λ as an approximate solution, it is only natural that we are also interested in whether this function goes to zero for $\delta \to 0$. We show in the next proposition that this is always the case when $\lambda \to 0$.

Proposition C.2 *The following properties of $\mathscr{E}^2(\lambda)$, hold:*

$$\text{a)}\quad \mathscr{E}^2(\lambda)\ \textit{is continuous on}\ (0, \infty); \tag{C.60}$$

$$\text{b)}\quad \mathscr{E}^2(\lambda)\ \textit{is increasing on}\ (0,\ \infty); \tag{C.61}$$

$$\text{c)}\quad \mathscr{E}^2(\lambda) \to \|y_0\|_Y^2\ \textit{when}\ (\lambda \to \infty); \tag{C.62}$$

$$\text{d)}\quad \mathscr{E}^2(\lambda) \to 0\ \textit{when}\ (\lambda \to 0). \tag{C.63}$$

Proof

a) In view of (C.54), Ax_λ is a continuous function of λ in Y, so (C.60) is obvious because for whatever normed space

$$\|\xi\| \leqslant \|\xi - \xi'\| + \|\xi'\| \Longrightarrow \big|\, \|\xi\| - \|\xi'\| \,\big| \leqslant \|\xi - \xi'\|,$$

i.e., the function $\xi \to \|\xi\|$ is continuous.

b) Observe preliminarily that, as it is clear also from (C.52), one has, for $\lambda \neq 0$,

$$\dot{R}_\lambda = -R_\lambda^2, \tag{C.64}$$

where the dot stands for the derivative with respect to λ.

Therefore, by (C.48),

$$\dot{x}_\lambda = \dot{R}_\lambda A^T y_0 = -R_\lambda^2 A^T y_0 = -R_\lambda x_\lambda. \tag{C.65}$$

But then

$$\begin{aligned} D_\lambda \mathscr{E}^2(\lambda) &= 2\langle y_0 - Ax_\lambda, -A\dot{x}_\lambda\rangle_Y = \\ &= -2\Big\langle A^T y_0 - A^T A x_\lambda\,,\ \dot{x}_\lambda\Big\rangle_X = \\ &= -2\lambda\langle x_\lambda, \dot{x}_\lambda\rangle_X = \\ &= 2\lambda\langle x_\lambda, R_\lambda x_\lambda\rangle_X, \end{aligned} \tag{C.66}$$

where we used the fact that x_λ is a solution of (C.43).

Now notice that the identities

$$R_\lambda \equiv R_\lambda R_\lambda^{-1} R_\lambda = R_\lambda (A^T A + \lambda I) R_\lambda, \quad R_\lambda^T = R_\lambda$$

imply that R_λ is strictly positive definite, because so is $A^T A + \lambda I$; therefore, (C.66) says that $D_\lambda \mathscr{E}^2(\lambda) > 0$ and (C.61) is proved.

c) Observe that

$$J_\lambda(y_0, x_\lambda) = \mathscr{E}^2(\lambda) + \lambda \, \|x_\lambda\|_X^2 \leqslant J_\lambda(y_0, 0) = \|y_0\|_Y^2 \, . \tag{C.67}$$

But then, when $\lambda \to \infty$

$$\|x_\lambda\|_X^2 \to 0, \tag{C.68}$$

i.e., $x_\lambda \to 0$, so that

$$\mathscr{E}^2(\lambda) = \|y_0 - Ax_\lambda\|_Y^2 \to \|y_0\|_Y^2 \, . \tag{C.69}$$

d) Notice that the following identity holds

$$(A^T A + \lambda I)^{-1} A^T = A^T (AA^T + \lambda I)^{-1}, \tag{C.70}$$

which can be proved by multiplying by $A^T A + \lambda I$ on the left and by $AA^T + \lambda I$ on the right.

Denote $M_\lambda = (AA^T + \lambda I)^{-1}$. It is clear that

$$M_\lambda^{-1} > \lambda I \Longrightarrow \|M_\lambda\| < \frac{1}{\lambda}. \tag{C.71}$$

Moreover, let us underline that we can assume y_0 to belong to the closure of the range of A in Y,

$$y_0 \in Y_A = [\mathscr{R}(A)]_Y; \tag{C.72}$$

even more, we will show that we can take Y from the beginning to be $Y = \bar{Y}_A$.

In fact, if this is not the case, i.e., $Y_A \subset Y$ strictly, we can always introduce the element $\bar{y}_0$, which is the orthogonal projection of y_0 onto $\bar{Y}_A$. But then, by the Pythagorean theorem,

$$\forall x, \quad \|y_0 - Ax\|_Y^2 = \|y_0 - \bar{y}_0\|_Y^2 + \|\bar{y}_0 - Ax\|_Y^2$$

which implies that

$$\begin{aligned} J_\lambda(y_0, x) &= \|y_0 - \bar{y}_0\|_Y^2 + \|\bar{y}_0 - Ax\|_Y^2 + \lambda \, \|x\|_X^2 \\ &= \|y_0 - \bar{y}_0\|_Y^2 + \bar{J}_\lambda(\bar{y}_0, x). \end{aligned} \tag{C.73}$$

Therefore the minimization of $J_\lambda(y_0, x)$ is equivalent to the minimization of $\bar{J}_\lambda(\bar{y}_0, x)$ and all what we said up to here remains true, with the addition that $\bar{y}_0 \in \bar{Y}_A$.

So we can redefine Y, setting $Y = \bar{Y}_A$. Given that, we see that

$$\overline{A^T Y} \equiv X, \tag{C.74}$$

i.e., $A^T Y$ is dense in X. In fact, if

$$\left\langle \xi, A^T y \right\rangle_X = 0, \quad \forall y \in Y,$$

we also have

$$\langle A\xi, y \rangle_Y = 0 \Longrightarrow A\xi = 0 \Longrightarrow \xi = 0, \quad \forall y \in Y \tag{C.75}$$

because $A\xi \in Y$ and A is injective by hypothesis.
The relation (C.74) implies that

$$\overline{\mathscr{R}(AA^T)} = Y. \tag{C.76}$$

In fact, using the symbol $\overset{\text{d}}{\subset}$ to express that a set is densely included in another set, the definition of the range and the continuity of both A and A^T simply that,

$$\mathscr{R}(AA^T) = A(\mathscr{R}(A^T)) \overset{\text{d}}{\subset} A(X) \overset{\text{d}}{\subset} Y,$$

so (C.76) is proved. Therefore, given any $y_0 \in Y$ and any $\varepsilon > 0$, we can find $\bar{y} \in \mathscr{R}(AA^T)$ such that

$$\|y_0 - \bar{y}\|_Y < \varepsilon. \tag{C.77}$$

We note as well that, since $\bar{y} \in \mathscr{R}(AA^T)$, there exists a $w \in Y$ such that

$$\bar{y} = AA^T w. \tag{C.78}$$

Once ε is given, $\bar{y}$ can be fixed and so w.
Now we write

$$y_0 - Ax_\lambda = (I - AR_\lambda A^T) y_0\,; \tag{C.79}$$

but by (C.70) $R_\lambda A^T = A^T M_\lambda$, so that

$$y_0 - Ax_\lambda = (I - AA^T M_\lambda) y_0 = \lambda M_\lambda y_0, \tag{C.80}$$

because $AA^T M_\lambda + \lambda M_\lambda = I$ by definition. Therefore,

$$y_0 - Ax_\lambda = \lambda M_\lambda (y_0 - \bar{y}) + \lambda M_\lambda \bar{y} = \eta_1 + \eta_2 \tag{C.81}$$

and

$$\mathscr{E}(\lambda) = \|y_0 - Ax_\lambda\|_Y \leqslant \|\eta_1\|_Y + \|\eta_2\|_Y. \tag{C.82}$$

But, by (C.71),

$$\|\eta_1\|_Y = \|\lambda M_\lambda (y_0 - \bar{y})\| \leqslant \lambda \cdot \frac{1}{\lambda} \|y_0 - \bar{y}\| < \varepsilon. \tag{C.83}$$

Moreover, since by (C.71), (C.78) and the identity $M_\lambda AA^T = M_\lambda (M_\lambda^{-1} - \lambda I)$,

$$\begin{aligned} \|M_\lambda \bar{y}\|_Y = \left\| M_\lambda AA^T w \right\|_Y &= \|w - \lambda M_\lambda w\|_Y \leqslant \\ &\leq \|w\|_Y + \lambda \cdot \frac{1}{\lambda} \|w\|_Y = 2\|w\|_Y, \end{aligned}$$

we have too, for $\lambda \to 0$

$$\|\eta_2\|_Y = \|\lambda M_\lambda \bar{y}\|_Y \leqslant \lambda 2 \|w\|_Y \to 0. \tag{C.84}$$

Summarizing (C.82), (C.83), (C.84) we have proved that, when $\lambda \to 0$,

$$\overline{\lim} \|y_0 - Ax_\lambda\|_Y = \overline{\lim} \|\lambda M_\lambda y_0\|_Y \leqslant \varepsilon \tag{C.85}$$

and, since ε was arbitrary, (C.63) holds.

□

Proposition (C.2) shows that the typical behaviour of $\mathscr{E}^2(\lambda)$ is that plotted in Fig. C.2, namely to go smoothly and monotonically from 0 to $\|y_0\|_Y^2$. This result is fundamental for the choice of Morozov's strategy, which will be illustrated in the next section.

C.5 The Morozov Discrepancy Principle

As we have seen at the beginning of this Appendix, the data y_0 admit a discrepancy v with respect to the "true" value $y = Ax$; according to (C.55), we put $\|v\|_Y = \delta$. On the other hand, when we compute a Tikhonov solution x_λ of the problem, the observation equations $y = Ax$ are verified with $y = y_0$, $x = x_\lambda$ only apart from an estimated discrepancy

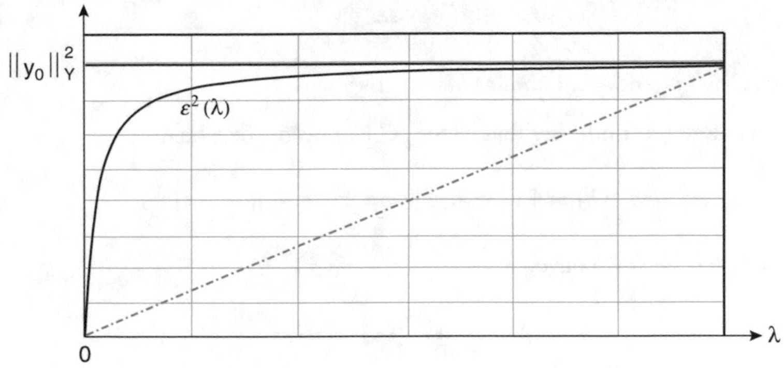

Fig. C.2 The plot of $\mathscr{E}^2(\lambda) = \|y_0 - Ax_\lambda\|_y^2$

$$v_\lambda = y_0 - Ax_\lambda; \tag{C.86}$$

according to (C.59), we set $\|v_\lambda\|_Y = \mathscr{E}(\lambda)$, the function that we have studied in the previous section.

Definition C.1 (Discrepancy Principle) We choose for λ the value for which

$$\mathscr{E}(\lambda) = \delta. \tag{C.87}$$

By the discussion Sect. C.4, Eq. (C.87) has one and only one solution provided that

$$\delta < \|y_0\|_Y \,. \tag{C.88}$$

This condition is actually not too restrictive, because if we had $\|v\|_Y \geq \|y_0\|_Y$, a simple reasonable solution of the observation equation

$$y_0 = Ax + v$$

would be $x = 0$, leaving an estimated discrepancy smaller, in norm, than the actual discrepancy. It has to be stressed that in practical situations we do not really know the value of δ, but we have at most a guess on its order of magnitude. Nevertheless, beyond the numerical problem of solving (C.87), Morozov's choice is of interest because it enables us to study the behaviour of the error of the solution, $F(\lambda, \delta)$ (see (C.56)) for $\delta \to 0$, as we will see in Theorem C.3. Notice that, widening the classical result of Tikhonov [150], we treat here the case where A is only continuous, because this is of interest for the matter of the book.

Theorem C.3 *Under the choice of the discrepancy principle* (C.87)*, if A is continuous and injective from X to Y, then any sequence x_{λ_n}, corresponding to $\delta_n \to 0$, contains a subsequence that converges strongly to x*

$$x_{\lambda_n} \to x, \tag{C.89}$$

namely (C.87) *provides an admissible strategy.*

Proof We note preliminarily that, since (C.87) holds, one has

$$\mathscr{E}^2(\lambda) = \|y_0 - Ax_\lambda\|_Y^2 = \delta^2 = \|y_0 - Ax\|_Y^2 , \tag{C.90}$$

so that when $\delta \to 0$ we have,

$$\lim_{\delta \to 0} \mathscr{E}^2(\lambda) = 0$$

and

$$\lim_{\delta \to 0} y_0 = y = Ax ;$$

therefore,

$$\|Ax - Ax_\lambda\|_Y \leq \|y_0 - Ax\|_Y + \|y_0 - Ax_\lambda\|_Y = 2\delta \to 0, \tag{C.91}$$

i.e.,

$$\lim_{\delta \to 0} Ax_\lambda = Ax \text{ (in } Y). \tag{C.92}$$

Now let us define a sequence

$$\{v_n\}, \quad \|v_n\|_Y = \delta_n \to 0, \tag{C.93}$$

with the corresponding values of λ_n, according to (C.87).

One has

$$J_{\lambda_n}\left(y_{0n}, x_{\lambda_n}\right) = \delta_n^2 + \lambda_n \left\|x_{\lambda_n}\right\|_X^2$$
$$\leq J_{\lambda_n}\left(y_{0n}, x\right) = \delta_n^2 + \lambda_n \|x\|_X^2 ,$$

implying that

$$\left\|x_{\lambda_n}\right\|_X \leq \|x\|_X .$$

This inequality has two consequences: the first is that there is a subsequence of $\{x_{\lambda_n}\}$ weakly converging to some limit x^* in X; we write, keeping the notation $\{x_{\lambda_n}\}$ for this subsequence,

$$x_{\lambda_n} \underset{X}{\overset{*}{\to}} x^*; \tag{C.94}$$

the second is that

$$\overline{\lim} \left\| x_{\lambda_n} \right\|_X \leq \|x\|_X . \tag{C.95}$$

We claim that

$$x^* = x . \tag{C.96}$$

Indeed, by (C.92),

$$\left\langle w, Ax_{\lambda_n} \right\rangle_Y \to \langle w, Ax \rangle_Y , \quad \forall w \in Y. \tag{C.97}$$

Moreover, from (C.94) we see that

$$\left\langle w, Ax_{\lambda_n} \right\rangle_Y = \left\langle A^T w, x_{\lambda_n} \right\rangle_X \to \left\langle A^T w, x^* \right\rangle_X = \left\langle w, Ax^* \right\rangle_Y \tag{C.98}$$

for all $w \in Y$. So we must have

$$\langle w, Ax \rangle_Y = \left\langle w, Ax^* \right\rangle_Y , \quad \forall w \in Y,$$

Whence

$$Ax = Ax^* \Longrightarrow x = x^*$$

because A is injective by hypothesis.

On the other hand, it is well known that (cf. (C.10))

$$\underline{\lim} \left\| x_{\lambda_n} \right\|_X \geq \left\| x^* \right\|_X \equiv \|x\|_X . \tag{C.99}$$

Comparing (C.95) with (C.99) one sees that

$$\lim_{n \to \infty} \left\| x_{\lambda_n} \right\|_X = \|x\|_X . \tag{C.100}$$

But then, arguing as before, using (C.94), (C.95) and (C.96), we have

$$\begin{aligned} \left\| x_{\lambda_n} - x \right\|_X^2 &= \left\| x_{\lambda_n} \right\|_X^2 - 2 \left\langle x_{\lambda_n}, x \right\rangle_X + \|x\|_X^2 \to \\ &\to \|x\|_X^2 - 2 \|x\|_X^2 + \|x\|_X^2 = 0; \end{aligned}$$

in other words x_{λ_n} converges to x strongly in X. □

Remark C.4 We know from the study of $\mathscr{E}^2(\lambda) = \delta^2(\lambda)$ performed in Sect. C.4 that when $\lambda \to 0$, $\delta^2(\lambda) \to 0$ too and the same is true indeed for the inverse function $\lambda = \lambda(\delta)$.

Yet it is possible to be more precise in this respect, if, for instance, we make the further assumption that A is compact. In this case we can take advantage of the spectral representation

$$A = \sum_{n=1}^{+\infty} \mu_n v_n \otimes u_n \qquad (\mu_n > 0\,;\, \mu_n \to 0\,, n \to \infty), \tag{C.101}$$

where, since A is injective and the range $\mathscr{R}(A)$ is dense in Y,

- $\{u_n\}$ is a complete orthonormal system in X,
- $\{v_n\}$ is a complete orthonormal system in Y,

and we have

$$A^T A u_n = \mu_n^2 u_n\,,\; AA^T v_n = \mu_n^2 v_n. \tag{C.102}$$

So the relation

$$y = Ax\,,$$

becomes simply

$$y_n = \mu_n x_n \quad \big(y_n = \langle v_n, y\rangle_Y\,, \quad x_n = \langle u_n, x\rangle_X\big)\,. \tag{C.103}$$

Recalling the relations

$$\begin{aligned} v_\lambda &= y_0 - Ax_\lambda = (I - AR_\lambda A^T)y_0 = \\ &= (I - AA^T M_\lambda)y_0 = \lambda M_\lambda y_0 = \lambda M_\lambda y + \lambda M_\lambda v = \\ &= \lambda M_\lambda Ax + \lambda\delta M_\lambda \bar{v}, \end{aligned} \tag{C.104}$$

where we have put

$$v = \delta\bar{v} \quad (\|v\|_Y = \delta,\, \|\bar{v}\|_Y = 1)\,, \tag{C.105}$$

we can write in spectral terms

$$v_{n\lambda} = \frac{\lambda\mu_n}{(\mu_n^2 + \lambda)} x_n + \delta \frac{\lambda}{(\mu_n^2 + \lambda)} \bar{v}_n. \tag{C.106}$$

By its definition, the discrepancy principle is translated into the equation

$$\delta^2 = \|v_\lambda\|_Y^2 = \sum v_{n\lambda}^2. \tag{C.107}$$

Using (C.106) we have

$$\delta^2 = \sum \frac{\lambda^2 \mu_n^2}{\left(\mu_n^2 + \lambda\right)^2} x_n^2 + 2\delta \sum \frac{\lambda^2 \mu_n}{\left(\mu_n^2 + \lambda\right)^2} x_n \bar{v}_n + \\ + \delta^2 \sum \frac{\lambda^2}{\left(\mu_n^2 + \lambda\right)^2} \bar{v}_n^2. \tag{C.108}$$

It is convenient to put

$$\begin{cases} a(\lambda) = \sum \frac{\lambda^2}{(\mu_n^2+\lambda)^2} \bar{v}_n^2, \\ b(\lambda) = \sum \frac{\lambda^{3/2} \mu_n}{\left(\mu_n^2+\lambda\right)^2} x_n \bar{v}_n, \\ c(\lambda) = \sum \frac{\lambda \mu_n^2}{\left(\mu_n^2+\lambda\right)^2} x_n^2, \end{cases} \tag{C.109}$$

so that the defining Eq. (C.108) can now be written as

$$[1 - a(\lambda)]\delta^2 - 2\delta\sqrt{\lambda} b(\lambda) - \lambda c(\lambda) = 0$$

or, better,

$$[1 - a(\lambda)] \left(\frac{\delta}{\sqrt{\lambda}} \right)^2 - 2 \left(\frac{\delta}{\sqrt{\lambda}} \right) b(\lambda) - c(\lambda) = 0. \tag{C.110}$$

Thanks to the fact that

$$\frac{\lambda \mu_n^2}{\left(\mu_n^2 + \lambda\right)^2} \leq \frac{1}{2}; \quad \frac{\lambda^{3/2} \mu_n}{\left(\mu_n^2 + \lambda\right)^2} \leq \frac{3^{1,5}}{16}; \quad \frac{\lambda^2}{\left(\mu_n^2 + \lambda\right)^2} \leq 1 \tag{C.111}$$

and that

$$\sum \bar{v}_n^2 = 1, \quad \sum |x_n \bar{v}_n| \leq \sqrt{\sum x_n^2} < +\infty, \tag{C.112}$$

we see that the series in (C.109) are absolutely and uniformly convergent in λ, so that we can take the limit for $\lambda \to 0$ under the summation symbol. Then we find

$$a(\lambda) \to 0, \quad b(\lambda) \to 0, \quad c(\lambda) \to 0, \tag{C.113}$$

for $\lambda \to 0$.

Thus, we can assert that Morozov's principle provides an admissible strategy such that, for any constant $c > 0$ and sufficiently small λ,

$$\delta(\lambda) < c\lambda^{\frac{1}{2}}, \tag{C.114}$$

or, if you prefer, for any constant $k > 0$ and sufficiently small δ,

$$\lambda(\delta) > k\delta^2. \tag{C.115}$$

Namely, Morozov's choice of $\lambda(\delta)$ has a behaviour close to zero that does not reach the exponent 2. This can be compared to the stochastic approach, which is known to be equivalent to the Tikhonov-like principle (see Appendix B.3) of minimizing the quadratic form

$$\Phi(y_0, x) = (y_0 - Ax)C_\nu^{-1}(y - Ax) + x^T C_X^{-1} x. \tag{C.116}$$

For a moment, we think of y, x as finite-dimensional vectors to avoid functional intricacies and assume that (C.116) could be written as

$$\Phi(y_0, x) = \sigma_\nu^2 \, |y_0 - Ax|^2 + x^T C_X^{-1} x \, ; \tag{C.117}$$

now, if we take

$$|y_0 - Ax|^2 = \|y_0 - Ax\|_Y^2 \quad \text{and} \quad x^T C_X^{-1} x = \|x\|_X^2 \, ,$$

we find that the target function (C.117) can be considered as a Tikhonov principle with $\lambda \div \sigma_\nu^2$. Since σ_ν^2 is proportional to $|\nu|^2$, which we called δ^2 in Tikhonov's language, we see that there is a relation between Tikhonov's deterministic approach and the Wiener–Kolmogorov theory; however, the subtle difference in the exponent is in reality a symptom that a truly functional interpretation of W-K theory would lead us to completely different spaces than those used here.

Remark C.5 It might seem that the numerical solution of the discrepancy principle equation (C.87) is an impossible task; however, the job can be performed by means of the Newton–Raphson method, although this is indeed a heavy procedure, since at each step a new Tikhonov solution has to be computed. On the other hand, by exploiting a smart trick the method certainly leads to the true solution. The idea is to change the parameter λ, introducing a new parameter according to

$$\lambda = \frac{1}{\gamma}; \tag{C.118}$$

one then verifies that the function

$$f(\gamma) = J_{1/\gamma}(y_0, x) \, , \tag{C.119}$$

is a convex decreasing function, as in Fig. C.3.

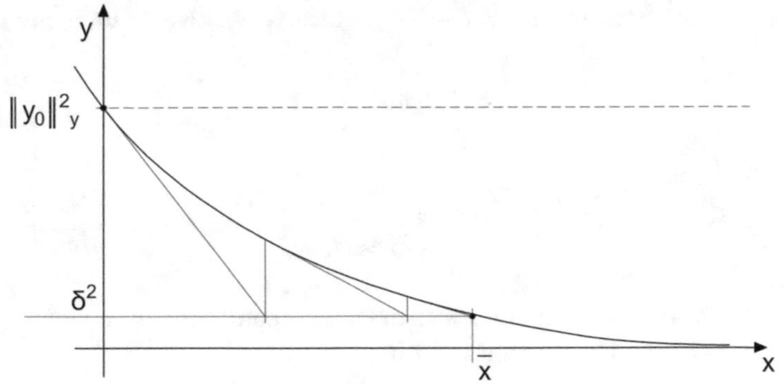

Fig. C.3 The plot of $f(\gamma)$; the function is decreasing and convex and the Newton approximation process starting from $\gamma = 0$ is convergent

In fact, if we denote by a prime the derivative with respect to γ and by a dot the derivative with respect to λ, we have, by the chain rule,

$$f'(\gamma) = \dot{J}_\lambda \left(-\frac{1}{\gamma^2}\right) \equiv -\lambda^2 \dot{J}_\lambda \tag{C.120}$$

and

$$f''(\gamma) = \ddot{J}_\lambda \left(\frac{1}{\gamma^4}\right) + \dot{J}_\lambda \frac{2}{\gamma^3} \equiv \lambda^3(\lambda \ddot{J}_\lambda + 2\dot{J}_\lambda). \tag{C.121}$$

So, recalling (C.65), (C.66) and noting that

$$\dot{J}_\lambda = D_\lambda \mathscr{E}^2(\lambda) + \|x_\lambda\|_X^2 + 2\lambda \langle x_\lambda, \dot{x}_\lambda \rangle_X = \|x_\lambda\|_X^2 \,,$$

we get

$$f'(\gamma) = -\lambda^2 \dot{J}_\lambda = -\frac{1}{\gamma^2} \|x_\lambda\|_X^2 \,, \tag{C.122}$$

i.e., $f(\gamma)$ is decreasing.

Moreover, recalling (C.65), and (C.122),

$$\ddot{J} = 2 \langle x_\lambda, \dot{x}_\lambda \rangle_X = -2 \langle x_\lambda, R_\lambda x_\lambda \rangle_X$$

so that, inserting the expressions for $\ddot{J}$ and $\dot{J}$ into (C.121), we obtain

$$f''(\gamma) = 2\lambda^3 \langle x_\lambda, (I - \lambda R_\lambda) x_\lambda \rangle_X \,. \tag{C.123}$$

Now it is enough to notice that $I - \lambda R_\lambda$ is clearly positive definite, because

$$\|R_\lambda\| < \frac{1}{\lambda},$$

(see (C.47)), so that

$$f''(\gamma) > 0,$$

namely $f(\gamma)$ is convex, continuous with first and second derivative, and the Newton approximation method is convergent for it.

References

1. Abramowitz, M., & Stegun, I. A. (1965). *Handbook of mathematical functions: With formulas, graphs, and mathematical tables* (Vol. 55). Courier Corporation.
2. Abramowitz, M., & Stegun, I. A., et al. (1972). *Handbook of mathematical functions: With formulas, graphs, and mathematical tables* (Vol. 55). New York: Dover.
3. Acar, R., & Vogel, C. R. (1994). Analysis of bounded variation penalty methods for ill-posed problems. *Inverse Problems, 10*(6), 1217.
4. Amante, C., & Eakins, B. W. (2009). *ETOPO1 1 arc-minute global relief model: Procedures, data sources and analysis*. US Department of Commerce, National Oceanic and Atmospheric Administration, National Environmental Satellite, Data, and Information Service, National Geophysical Data Center, Marine Geology and Geophysics Division Colorado.
5. Anderson, D. L. (1989). *Theory of the Earth*. Blackwell Scientific Publications.
6. Axler, S., Bourdon, P., & Ramey, W. (2001). *Harmonic function theory*. Graduate Texts in Mathematics (Vol. 137). Springer Science & Business Media.
7. Ballani, L., & Stromeyer, D. (1982). The inverse gravimetric problem: A Hilbert space approach. In *Proceedings of the International Symposium 'Figure of the Earth, the Moon, and other Planets'* (pp. 20–25).
8. Barthelmes, F., et al. (2012). International Centre for Global Earth Models (ICGEM).
9. Biagi, L., & Sansò, F. (2001). Tclight: A new technique for fast RTC computation. In *Gravity, geoid and geodynamics 2000* (pp. 61–66). Springer.
10. Bizouard, C., & Becker, O. (2008). Web tools and web service of the IERS Earth Orientation Centre. In *Measuring the Future, Proceedings of the Fifth IVS* (p. 199).
11. Blakely, R. J. (1996). *Potential theory in gravity and magnetic applications*. Cambridge University Press.
12. Blick, C., Freeden, W., & Nutz, H. (2017). Feature extraction of geological signatures by multiscale gravimetry. *GEM-International Journal on Geomathematics, 8*(1), 57–83.
13. Blick, C., Freeden, W., & Nutz, H. (2018). Gravimetry and exploration. In *Handbook of mathematical geodesy* (pp. 687–751). Springer.
14. Boedecker, G. (1993). The international absolute gravity basestation network (IAGBN): Status 1992. In *Geodesy and physics of the Earth* (pp. 61–62). Springer.
15. Borre, K. (2006). *Mathematical foundation of geodesy: Selected papers of Torben Krarup*. Springer Science & Business Media.
16. Bott, M. H. P., & Smith, R. A. (1958). The estimation of the limiting depth of gravitating bodies. *Geophysical Prospecting, 6*(1), 1–10.
17. Bouguer, P. (1749). *La figure de la terre*. CA Jombert, Paris.

F. Sansò, D. Sampietro, *Analysis of the Gravity Field*, Lecture Notes in Geosystems Mathematics and Computing, https://doi.org/10.1007/978-3-030-74353-6

18. Bracewell, R. N. (1986). *The Fourier transform and its applications* (Vol. 31999). New York: McGraw-Hill.
19. Braitenberg, C., Sampietro, D., Pivetta, T., Zuliani, D., Barbagallo, A., Fabris, P., Rossi, L., Fabbri, J., & Mansi, A. H. (2016). Gravity for detecting caves: Airborne and terrestrial simulations based on a comprehensive karstic cave benchmark. *Pure and Applied Geophysics, 173*(4), 1243–1264.
20. Brozena, J. M. (1992). The Greenland aerogeophysics project: Airborne gravity, topographic and magnetic mapping of an entire continent. In *From Mars to Greenland: Charting gravity with space and airborne instruments* (pp. 203–214). Springer.
21. Capponi, M. (2018). *Very Improved KINematic Gravimetry: A New Approach to Aerogravimetry*. PhD thesis, Università di Roma La Sapienza.
22. Capponi, M., Sampietro, D., & Sansò, F. (2021). Regularized solutions of the two layers inverse gravimetric problem in the space of bounded variation functions. In P. Novák, M. Crespi, N. Sneeuw, & F. Sansò (Eds.), *IX Hotine-Marussi Symposium on Mathematical Geodesy* (pp. 107–116). Cham: Springer International Publishing.
23. CarbonNet. (2012). Gippsland nearshore airborne gravity survey - data packages. Department of Primary Industries, Victoria.
24. Champeney, D. C. (1987). *A handbook of Fourier theorems*. Cambridge University Press.
25. Clairaut, A. C. (1743). *Théorie de la figure de la terre, tirée des principes de l'hydrostatique.* chez David fils, libraire, ruë Saint-Jacques à la Piume d'or.
26. Cressie, N., & Wikle, C. K. (2015). *Statistics for spatio-temporal data*. Wiley.
27. De Angelis, M., Bertoldi, A., Cacciapuoti, L., Giorgini, A., Lamporesi, G., Prevedelli, M., Saccorotti, G., Sorrentino, F., & Tino, G. M. (2008). Precision gravimetry with atomic sensors. *Measurement Science and Technology, 20*(2), 022001.
28. Dobrin, M. B., & Savit, C. H. (1988). *Introduction to geophysical prospecting* (4th ed.). New York: McGraw-Hill.
29. Donoho, D. L. (1995). Nonlinear solution of linear inverse problems by wavelet-vaguelette decomposition. *Applied and Computational Harmonic Analysis, 2*(2), 101–126.
30. Drinkwater, M. R., Haagmans, R., Muzi, D., Popescu, A., Floberghagen, R., Kern, M., & Fehringer, M. (2006). The GOCE gravity mission: Esa's first core Earth explorer. In *Proceedings of the 3rd International GOCE User Workshop* (pp. 6–8). Citeseer,
31. Dziewonski, A. M., & Anderson, D. L. (1981). Preliminary reference Earth model. *Physics of the Earth and Planetary Interiors, 25*(4), 297–356.
32. Engl, H. W., Hanke, M., & Neubauer, A. (1996). *Regularization of inverse problems* (Vol. 375). Springer Science & Business Media.
33. Ergtin, M., Okay, S., Sari, C., Oral, E. Z., Ash, M., Hall, J., & Miller, H. (2005). Gravity anomalies of the Cyprus arc and their tectonic implications. *Marine Geology, 221*(1–4), 349–358.
34. Eshagh, M. (2020). *Satellite gravimetry and the solid Earth: Mathematical foundations*. Elsevier.
35. Fairhead, J. D. (2016). *Advances in gravity and magnetic processing and interpretation*. EAGE Publications.
36. Fairweather, G., & Karageorghis, A. (1998). The method of fundamental solutions for elliptic boundary value problems. *Advances in Computational Mathematics, 9*(1–2), 69.
37. Farr, T. G., Rosen, P. A., Caro, E., Crippen, R., Duren, R., Hensley, S., Kobrick, M., Paller, M., Rodriguez, E., Roth, L., et al. (2007). The shuttle radar topography mission. *Reviews of Geophysics, 45*(2).
38. Feld, C., Mechie, J., Hübscher, C., Hall, J., Nicolaides, S., Gurbuz, C., Bauer, K., Louden, K., & Weber, M. (2017). Crustal structure of the Eratosthenes Seamount, Cyprus and S. Turkey from an amphibian wide-angle seismic profile. *Tectonophysics, 700*, 32–59.
39. Forsberg, R. (1984). A study of terrain reductions, density anomalies and geophysical inversion methods in gravity field modelling. Technical report, Ohio State Univ Columbus Dept. of Geodetic Science and Surveying.

40. Forsberg, R., & Kenyon, S. (1994). Evaluation and downward continuation of airborne gravity data-the Greenland example. In *Proc. of International Symposium on Kinematic Systems in Geodesy, Geomatics and Navigation (KIS94)* (pp. 531–538).
41. Förste, C., Bruinsma, S., Abrikosov, O., Flechtner, F., Marty, J.-C., Lemoine, J.-M., Dahle, C., Neumayer, H., Barthelmes, F., Konig, R., et al. (2014). EIGEN-6C4-the latest combined global gravity field model including GOCE data up to degree and order 1949 of GFZ potsdam and GRGS Toulouse. In *EGU General Assembly Conference Abstracts* (Vol. 16).
42. Fowler, C. M. (2004). *The solid Earth: An introduction to global geophysics*. Cambridge University Press.
43. Freeden, W., & Michel, V. (2012). *Multiscale potential theory: With applications to geoscience*. Springer Science & Business Media.
44. Freeden, W., & Nashed, M. Z. (2018). Inverse gravimetry as an ill-posed problem in mathematical geodesy. In *Handbook of mathematical geodesy* (pp. 641–685). Springer.
45. Freeden, W., & Nashed, M. Z. (2018). Operator-theoretic and regularization approaches to ill-posed problems. *GEM-International Journal on Geomathematics, 9*(1), 1–115.
46. Freeden, W., & Nashed, M. Z. (2020). Inverse gravimetry: Density signatures from gravitational potential data. In *Mathematische Geodasie/Mathematical Geodesy* (pp. 969–1052). Springer.
47. Freeden, W., & Sansò, F. (2019). Geodesy and mathematics: Interactions, acquisitions, and open problems. In *International Association of Geodesy Symposia: Proceedings of the IX Hotine-Marussi Symposium*. Springer.
48. Frigo, M., & Johnson, S. G. (1998). FFTW: An adaptive software architecture for the FFT. In *Proceedings of the 1998 IEEE International Conference on Acoustics, Speech and Signal Processing, ICASSP'98 (Cat. No. 98CH36181)* (Vol. 3, pp. 1381–1384). IEEE.
49. Gelman, A., Stern, H. S., Carlin, J. B., Dunson, D. B., Vehtari, A., & Rubin, D. B. (2013). *Bayesian data analysis*. Chapman and Hall/CRC.
50. Geman, S., & Geman, D. (1984). Stochastic relaxation, Gibbs distributions, and the Bayesian restoration of images. *IEEE Transactions on Pattern Analysis and Machine Intelligence, 6*, 721–741.
51. Geman, S., & Geman, D. (1987). Stochastic relaxation, Gibbs distributions, and the Bayesian restoration of images. In *Readings in computer vision* (pp. 564–584). Elsevier.
52. Gikhman, I. I., & Skorokhod, A. V. (2015). *The theory of stochastic processes I*. Springer.
53. Gilardoni, M., Reguzzoni, M., & Sampietro, D. (2016). GECO: A global gravity model by locally combining GOCE data and EGM2008. *Studia Geophysica et Geodaetica, 60*(2), 228–247.
54. Gilbarg, D., & Trudinger, N. S. (2015). *Elliptic partial differential equations of second order*. Springer.
55. Giusti, E., & Williams, G. H. (1984). *Minimal surfaces and functions of bounded variation*. Springer.
56. Götze, H.-J., & Lahmeyer, B. (1988). Application of three-dimensional interactive modeling in gravity and magnetics. *Geophysics, 53*(8), 1096–1108.
57. Goursat, E. (1925). *Cours d'analyse* (Vol. 2). Paris: Gauthier-Villars
58. Grad, M., Tiira, T., & ESC Working Group. (2009). The Moho depth map of the European plate. *Geophysical Journal International, 176*(1), 279–292.
59. Grant, F. S., & West, G. F. (1965). *Interpretation theory in applied geophysics*. McGraw-Hill.
60. Groetsch, C. W., & Groetsch, C. W. (1993). *Inverse problems in the mathematical sciences*. Springer.
61. Groetsch, C. W. (1984). *The theory of Tikhonov regularization for Fredholm equations*. Boston Pitman Publication.
62. Halmos, P. R. (2013). *Measure theory*. Springer.
63. Heiskanen, W. A., & Moritz, H. (1967). *Physical geodesy*. W.H. Freeman and Co. S. Francisco.
64. Hirt, C., Claessens, S., Fecher, T., Kuhn, M., Pail, R., & Rexer, M. (2013). New ultrahigh-resolution picture of Earth's gravity field. *Geophysical Research Letters, 40*(16), 4279–4283.

65. Hörmander, L. (1976). The boundary problems of physical geodesy. *Archive for Rational Mechanics and Analysis, 62*(1), 1–52.
66. Huang, Y., Chubakov, V., Mantovani, F., Rudnick, R. L., & McDonough, W. F. (2013). A reference Earth model for the heat-producing elements and associated geoneutrino flux. *Geochemistry, Geophysics, Geosystems, 14*(6), 2003–2029.
67. Isakov, V. (1990). *Inverse source problems*. Providence, RI: American Mathematical Soc.
68. Jacoby, W., & Smilde, P. L. (2009). *Gravity interpretation: Fundamentals and application of gravity inversion and geological interpretation*. Springer Science & Business Media.
69. Jarvis, A., Reuter, H. I., Nelson, A., & Guevara, E. (2008). Hole-filled SRTM for the globe version 4.
70. Jeffreys, H. (1998). *The theory of probability*. Oxford: OUP.
71. Jekeli, C. (2012). *Inertial navigation systems with geodetic applications*. Walter de Gruyter.
72. Kapoor, D. C. (1981). General bathymetric chart of the oceans (GEBCO). *Marine Geodesy, 5*(1), 73–80.
73. Kaula, W. M. (2013). *Theory of satellite geodesy: Applications of satellites to geodesy*. Courier Corporation.
74. Kirsch, A. (2011). *An introduction to the mathematical theory of inverse problems*. Springer Science & Business Media.
75. Kotevska, E. (2011). *Real Earth Oriented Gravitational Potential Determination*. PhD dissertation, Technische Universität Kaiserslautern.
76. Laske, G., Masters, G., Ma, Z., & Pasyanos, M. (2013). Update on crust1. 0—a 1-degree global model of Earth's crust. In *Geophys. Res. Abstr* (Vol. 15, p. 2658). EGU General Assembly Vienna, Austria.
77. Li, Y., & Oldenburg, D. W. (1998). 3-D inversion of gravity data. *Geophysics, 63*(1), 109–119.
78. MacMillan, W. D. (1958). *The theory of the potential*. Dover Books on Physics and Mathematical Physics. Dover Publications.
79. Mansi, A. H., Capponi, M., & Sampietro, D. (2018). Downward continuation of airborne gravity data by means of the change of boundary approach. *Pure and Applied Geophysics, 175*(3), 977–988.
80. Melchior, P. (2013). *The physics of the Earth's core: An introduction*. Elsevier.
81. Michel, V., & Fokas, A. S. (2008). A unified approach to various techniques for the non-uniqueness of the inverse gravimetric problem and wavelet-based methods. *Inverse Problems, 24*(4), 045019.
82. Michel, V. (2005). Regularized wavelet-based multiresolution recovery of the harmonic mass density distribution from data of the Earth's gravitational field at satellite height. *Inverse Problems, 21*(3), 997.
83. Mikhlin, S. G. (2014). *Integral equations: And their applications to certain problems in mechanics, mathematical physics and technology*. Elsevier.
84. Miranda, C. (2012). *Partial differential equations of elliptic type*. Springer Science & Business Media.
85. Mohorovičić, A. (1992). Earthquake of 8 October 1909. *Geofizika, 9*(1), 3–55.
86. Moritz, H. (1990). *The figure of the Earth: Theoretical geodesy and the Earth's interior*. Karlsruhe: Wichmann.
87. Morozov, V. A. (2012). *Methods for solving incorrectly posed problems*. Springer Science & Business Media.
88. Nagy, D. (1965). The evaluation of Heuman's lambda function and its application to calculate the gravitational effect of a right circular cylinder. *Pure and Applied Geophysics PAGEOPH, 62*(1), 5–12.
89. Nagy, D., Papp, G., & Benedek, J. (2000). The gravitational potential and its derivatives for the prism. *Journal of Geodesy, 74*(7–8), 552–560.
90. Nashed, M. Z. (2010). Inverse problems, moment problems, signal processing: Un menage a trois. In *Mathematics in Science and Technology (Proceedings of Satellite Conference of the Internat. Congress of Mathematicians), New Delhi, India* (pp. 2–19).

91. Nashed, M. Z., & Wahba, G. (1974). Regularization and approximation of linear operator equations in reproducing kernel spaces. *Bulletin of the American Mathematical Society, 80*(6), 1213–1218.
92. Nashed, M. Z. (1987). A new approach to classification and regularization of ill-posed operator equations. In *Inverse and ill-posed problems* (pp. 53–75). Elsevier.
93. Nassar, S., Schwarz, K.-P., & El-Sheimy, N. (2004). INS and INS/GPS accuracy improvement using autoregressive (AR) modeling of ins sensor errors. In *Proceedings of IONNTM* (pp. 936–944).
94. Nettleton, L. L. (1939). Determination of density for reduction of gravimeter observations. *Geophysics, 4*(3), 176–183.
95. Novikov, P. S. (1938). On the uniqueness of the inverse problem of potential theory. In *Dokl. Akad. NaukSSSR* (Vol. 18, pp. 165–168).
96. Oldenburg, D. W. (1974). The inversion and interpretation of gravity anomalies. *Geophysics, 39*(4), 526–536.
97. Oliver, M. A., & Webster, R. (1990). Kriging: A method of interpolation for geographical information systems. *International Journal of Geographical Information System, 4*(3), 313–332.
98. Pail, R., Bruinsma, S., Migliaccio, F., Förste, C., Goiginger, H., Schuh, W.-D., Höck, E., Reguzzoni, M., Brockmann, J. M., Abrikosov, O., et al. (2011). First GOCE gravity field models derived by three different approaches. *Journal of Geodesy, 85*(11), 819.
99. Papoulis, A. (1977). Generalized sampling expansion. *IEEE Transactions on Circuits and Systems, 24*(11), 652–654.
100. Papoulis, A. (1990). *Probability & statistics* (Vol. 2). Englewood Cliffs: Prentice-Hall.
101. Parker, R. L. (1974). Best bounds on density and depth from gravity data. *Geophysics, 39*(5), 644–649.
102. Parker, R. L. (1975). The theory of ideal bodies for gravity interpretation. *Geophysical Journal of the Royal Astronomical Society, 42*(2), 315–334.
103. Parker, R. L., & Parker, R. L. (1994). *Geophysical inverse theory* (Vol. 1). Princeton University Press.
104. Pavlis, N. K., Holmes, S. A., Kenyon, S. C., Factor, J. K. (2012). The development and evaluation of the Earth gravitational model 2008 (EGM2008). *Journal of Geophysical Research: Solid Earth, 117*(B4).
105. Pizzetti, P. (1894). *Sulla espressione della gravitá alla superficie del geoide supposto ellissoidico*. Tip. della R. Accademia dei Lincei.
106. Pizzetti, P. (1909). Corpi equivalenti rispetto alla attrazione newtoniana esterna. *Rendiconti dell'Accademia Nazionale dei Lincei, Roma, 18*, 211–215.
107. Pizzetti, P. (1910). Intorno alle possibili distribuzioni della massa nell'interno della terra. *Annali di Matematica Pura ed Applicata (1898-1922), 17*(1), 225–258.
108. Plouff, D. (1976). Gravity and magnetic fields of polygonal prisms and application to magnetic terrain corrections. *Geophysics, 41*(4), 727–741.
109. Plouff, D. (1977). Preliminary documentation for a FORTRAN program to compute gravity terrain corrections based on topography digitized on a geographic grid. Technical report, US Geological Survey.
110. Portniaguine, O., & Zhdanov, M. S. (1999). Focusing geophysical inversion images. *Geophysics, 64*(3), 874–887.
111. Press, W. H., Teukolsky, S. A., Vetterling, W. T., & Flannery, B. P. (2007). *Numerical recipes 3rd edition: The art of scientific computing*. Cambridge University Press.
112. Reguzzoni, M., Rossi, L., Baldoncini, M., Callegari, I., Poli, P., Sampietro, D., Strati, V., Mantovani, F., Andronico, G., Antonelli, V., et al. (2019). GIGJ: A crustal gravity model of the Guangdong province for predicting the geoneutrino signal at the Juno experiment. *Journal of Geophysical Research: Solid Earth, 124*(4), 4231–4249.
113. Reguzzoni, M., & Tselfes, N. (2009). Optimal multi-step collocation: Application to the space-wise approach for GOCE data analysis. *Journal of Geodesy, 83*(1), 13–29.

114. Reigber, Ch., Schwintzer, P., & Lühr, H. (1999). The champ geopotential mission. *Bolletino di Geofisica Teoricaed Applicata, 40*, 285–289.
115. Richter, M. (2016). Discretization of inverse problems. In *Inverse problems* (pp. 29–75). Springer.
116. Richter, M. (2016). *Inverse problems: Basics, theory and applications in geophysics*. Birkhäuser.
117. Riesz, F., & Sz Nagy, B. (1952). *Lecons d'analyse functionnelle*.
118. Riesz, F., & Sz Nagy, B. (1953). *Functional analysis* [1952] (Translated from the French L. F. Boron (Ed.), Ungar, 1990).
119. Rossi, L. (2017). *Bayesian Gravity Inversion by Monte Carlo Methods*. PhD thesis.
120. Rozanov, Yu. A. (1982). Markov random fields. In *Markov random fields* (pp. 55–102). Springer.
121. Rybakov, M., Voznesensky, V., Ben-Avraham, Z., & Lazar, M. (2008). The Niklas anomaly southwest of Cyprus: New insights from combined gravity and magnetic data. *Israel Journal of Earth Sciences, 57*(2), 125–138.
122. Sampietro, D., Capponi, M., Mansi, A. H., Gatti, A., Marchetti, P., & Sansò, F. (2017). Space-wise approach for airborne gravity data modelling. *Journal of Geodesy, 91*(5), 535–545.
123. Sampietro, D., Sona, G., & Venuti, G. (2007). Residual terrain correction on the sphere by an FFT algorithm. In *Proceedings of the 1st International Symposium on international gravity field service, Aug* (pp. 306–311).
124. Sampietro, D., Capponi, M., Triglione, D., Mansi, A. H., Marchetti, P., & Sansò, F. (2016). GTE: A new software for gravitational terrain effect computation: Theory and performances. *Pure and Applied Geophysics, 173*(7), 2435–2453.
125. Sampietro, D., Mansi, A., & Capponi, M. (2018). Moho depth and crustal architecture beneath the Levant Basin from global gravity field model. *Geosciences, 8*(6), 200.
126. Sansò, F. (1986). Statistical methods in physical geodesy. In *Mathematical and numerical techniques in physical geodesy* (pp. 49–155). Springer.
127. Sansò, F., Capponi, M., & Sampietro, D. (2019). Up and down through the gravity field (pp. 1–54). Berlin, Heidelberg: Springer.
128. Sansò, F., et al. (1980). The minimum mean square estimation error principle in physical geodesy (stochastic and non-stochastic interpretation).
129. Sansò, F. (2006). *Navigazione geodetica e rilevamento cinematico*. Polipress.
130. Sansò, F. (2014). On the regular decomposition of the inverse gravimetric problem in non-l^2spaces. *GEM-International Journal on Geomathematics, 5*(1), 33–61.
131. Sansò, F., & Migliaccio, F. (2020). The quantum measurement of gravity. In *Quantum measurement of gravity for geodesists and geophysicists* (pp. 105–133). Springer.
132. Sansò, F., Reguzzoni, M., Barzaghi, R., et al. (2019). *Geodetic heights*. Springer.
133. Sansò, F., & Sideris, M. G. (2013). *Geoid determination: Theory and methods*. Springer Science & Business Media.
134. Sansò, F., & Sideris, M. G. (2016). *Geodetic boundary value problem: The equivalence between Molodensky's and Helmert's solutions*. Springer.
135. Sansò, F., Venuti, G., & Tscherning, C. C. (2000). A theorem of insensitivity of the collocation solution to variations of the metric of the interpolation space. In *Geodesy beyond 2000* (pp. 233–240). Springer.
136. Schäffler, S. (2012). *Global optimization: A stochastic approach*. Springer Science & Business Media.
137. Schalkoff, R. J. (1989). *Digital image processing and computer vision*. New York: Wiley.
138. Schwarz, K. P., Colombo, O., Hein, G., & Knickmeyer, E. T. (1992). Requirements for airborne vector gravimetry. In *From Mars to Greenland: Charting gravity with space and airborne instruments* (pp. 273–283). Springer.
139. Sen, M. K., & Stoffa, P. L. (2013). *Global optimization methods in geophysical inversion*. Cambridge University Press.
140. Sjöberg, L. E., & Bagherbandi, M. (2017). *Gravity inversion and integration*. Springer.

141. Somigliana, C. (1948). Costanti terrestri e misure di gravità. *Il Nuovo Cimento (1943–1954), 5*(3), 111–120.
142. Stein, E. M. (1970). *Singular integrals and differentiability properties of functions* (Vol. 2). Princeton University Press.
143. Szwillus, W., Afonso, J. C. , Ebbing, J., & Mooney, W. D. (2019). Global crustal thickness and velocity structure from geostatistical analysis of seismic data. *Journal of Geophysical Research: Solid Earth, 124*(2), 1626–1652.
144. Talwani, M., & Ewing, M. (1960). Rapid computation of gravitational attraction of three-dimensional bodies of arbitrary shape. *Geophysics, 25*(1), 203–225.
145. Tanner, M. A. (1991). *Tools for statistical inference*. Springer.
146. Tapley, B. D., Bettadpur, S., Ries, J. C., Thompson, P. F., & Watkins, M. M. (2004). Grace measurements of mass variability in the Earth system. *Science, 305*(5683), 503–505.
147. Tarantola, A. (2005). *Inverse problem theory and methods for model parameter estimation*. SIAM.
148. Tchernychev, M. (1998). *GRAV3D-program for Calculation of Potential Fields Based on Grid Approach, User Guide*. Dep. of Earth and Ocean, Univ. of BC Vancouver, Canada,
149. Telford, W. M., Geldart, L. P., & Sheriff, R. E. (1990). *Applied geophysics* (Vol. 1). Cambridge University Press.
150. Tikhonov, A. N., & Arsenin, V. Y. (1977). *Solutions of ill-posed problems*. Winston: New York.
151. Torge, W., & Müller, J. (2012). *Geodesy*. Walter de Gruyter.
152. Tscherning, C. C. (1992). The gravsoft package for geoid determination. In *Proc 1st IAG Continental Workshop of the Geoid in Europe, Prague*.
153. Turcotte, D., & Schubert, G. (2014). *Geodynamics*. Cambridge University Press,
154. Uieda, L., Barbosa, V., & Braitenberg, C. (2016, July). Tesseroids: Forward-modeling gravitational fields in spherical coordinates. In *GEOPHYSICS* (pp. F41–F48).
155. Vanıcek, P., & Krakiwsky, E. J. (2015). *Geodesy: The concepts*. Elsevier.
156. Wackernagel, H. (2014). Geostatistics. *Wiley StatsRef: Statistics Reference Online* .
157. Wahba, G. (1990). *Spline models for observational data*. SIAM.
158. Watts, A. B. (2001). *Isostasy and flexure of the lithosphere*. Cambridge University Press.
159. Wei, M., & Schwarz, K. P. (1998). Flight test results from a strapdown airborne gravity system. *Journal of Geodesy, 72*(6), 323–332.
160. Whiteway, T. G. (2009). Australian bathymetry and topography grid. *Geoscience Australia, Canberra, 21*, 46.
161. Yosida, K. (1980). *Functional analysis*. Berlin, New York: Springer.
162. Zidarov, D. P. (1990). *Inverse gravimetric problem in geoprospecting and geodesy*. Amsterdam, New York: Elsevier

Index

F. Sansò, D. Sampietro, *Analysis of the Gravity Field*, Lecture Notes in Geosystems Mathematics and Computing, https://doi.org/10.1007/978-3-030-74353-6

Printed in the United States
by Baker & Taylor Publisher Services